MW01641424

Immunotherapy of Infections

INFECTIOUS DISEASE AND THERAPY

Series Editors

Brian E. Scully, M.B., B.Ch.
Harold C. Neu, M.D.

College of Physicians & Surgeons
Columbia University
New York, New York

1. Parasitic Infections in the Compromised Host, *edited by Peter D. Walzer and Robert M. Genta*
2. Nucleic Acid and Monoclonal Antibody Probes: Applications in Diagnostic Methodology, *edited by Bala Swaminathan and Gyan Prakash*
3. Opportunistic Infections in Patients with the Acquired Immunodeficiency Syndrome, *edited by Gifford Leoung and John Mills*
4. Acyclovir Therapy for Herpesvirus Infections, *edited by David A. Baker*
5. The New Generation of Quinolones, *edited by Clifford Siporin, Carl L. Heifetz, and John M. Domagala*
6. Methicillin-Resistant *Staphylococcus aureus*: Clinical Management and Laboratory Aspects, *edited by Mary T. Cafferkey*
7. Hepatitis B Vaccines in Clinical Practice, *edited by Ronald W. Ellis*
8. The New Macrolides, Azalides, and Streptogramins: Pharmacology and Clinical Applications, *edited by Harold C. Neu, Lowell S. Young, and Stephen H. Zinner*
9. Antimicrobial Therapy in the Elderly Patient, *edited by Thomas T. Yoshikawa and Dean C. Norman*
10. Viral Infections of the Gastrointestinal Tract: Second Edition, Revised and Expanded, *edited by Albert Z. Kapikian*
11. Development and Clinical Uses of Haemophilus b Conjugate Vaccines, *edited by Ronald W. Ellis and Dan M. Granoff*

12. *Pseudomonas aeruginosa* Infections and Treatment, *edited by Aldona L. Baltch and Raymond P. Smith*
13. Herpesvirus Infections, *edited by Ronald Glaser and James F. Jones*
14. Chronic Fatigue Syndrome, *edited by Stephen E. Straus*
15. Immunotherapy of Infections, *edited by K. Noel Masihi*

Additional Volumes in Production

Therapy of Bone Infections, *edited by Luis Jauregui*

Immunotherapy of Infections

edited by

K. Noel Masihi
Robert Koch Institute
Berlin, Germany

Marcel Dekker, Inc. New York • Basel • Hong Kong

Library of Congress Cataloging-in-Publication Data

Immunotherapy of infections / edited by K. Noel Masihi.
p. cm. — (Infectious disease and therapy ; v. 15)
Includes bibliographical references and index.
ISBN: 0-8247-9209-2
1. Communicable diseases—Immunotherapy—Congresses.
2. Infection—Immunotherapy—Congresses. I. Masihi, K. N. (K. Noel) II. Series.
RC112.I436 1994 94-16950
616.9′046—dc20 CIP

The publisher offers discounts on this book when ordered in bulk quantities. For more information, write to Special Sales/Professional Marketing at the address below.

This book is printed on acid-free paper.

Marcel Dekker, Inc.
270 Madison Avenue, New York, New York 10016

Current printing (last digit):
10 9 8 7 6 5 4 3 2 1

PRINTED IN THE UNITED STATES OF AMERICA

Series Introduction

Marcel Dekker, Inc., has for many years specialized in the publication of high-quality monographs in tightly focused areas in a variety of medical disciplines. These have been of great value to both the practicing physician and the research scientist as sources of detailed and up-to-date information presented in an attractive format. During the last decade, there has been a veritable explosion in knowledge in the various fields related to infectious diseases and clinical microbiology. Antimicrobial resistance, antibacterial and antiviral agents, AIDS, Lyme disease, infections in immunocompromised patients, and parasitic diseases are but a few of the areas in which an enormous amount of significant work has been published. The Infectious Disease and Therapy series covers carefully chosen topics that should be of interest and value to the practicing physician, the clinical microbiologist, and the research scientist.

Brian E. Scully, M.B., B.Ch.
Harold C. Neu, M.D.

Preface

The control of dangerous infectious diseases is an important concern for national health authorities and international agencies. Infections such as tuberculosis that ran rampant during the nineteenth century are resurging again with added vehemence. Public awareness of diseases long thought to be under control has been heightened by recent reports of outbreaks of diphtheria from areas in political turmoil, as well as the spread of cholera into virgin territories of South America. The World Health Organization estimates that infectious diseases account for almost half of all deaths that occur in developing countries. The global pandemic of AIDS infects over one million people a year and other sexually transmitted diseases continue to be among the most frequent infectious conditions. Antibiotic-resistant nosocomial infections remain significant worldwide and new human diseases (e.g., bacillary angiomatosis) continue to surface. Long-term intake of immunosuppressive and cytotoxic drugs manifest parasitic and bacterial infections as one of its most common complications. Multidrug resistance and infections with opportunistic pathogens in immunocompromised patients will predictably persist as problems into the twenty-first century. Knowledge of the intricate interaction of microbes with the immune system has engendered the realization that simple, straightforward strategies may not be adequate for effective control of existing and newly emerging infectious diseases. New concepts acting as adjuncts to established therapies are urgently needed.

Exploration of fundamental immunological processes at the molecular level has been germane to the current understanding of distinct immune responses in health and disease. Innovative regimens for effectively enhancing host resistance against infections are opening new perspectives. cDNA cloning of cytokines or appropriate cytokine receptor chains and their expression in bacteria, yeast, or mammalian cell lines have permitted production of an array of pure recombinant proteins. Development of novel substances by efficient synthetic methods has been pursued vigorously to add to the expanding repertoire of immunomodulators. There is an increasing interest in the elucidation of mechanisms of action of known immunomodulators and their efficacy in stimulating specific and nonspecific resistance. The burgeoning amount of new information offered in scientific publications and electronic media makes it imperative to assess the current trends at regular intervals and to promote an exchange of new ideas among investigators actively working in these fields.

The International Symposium on Immunotherapy of Infections was held in the spring of 1993 at the Robert Koch-Institute in Berlin, Germany. This meeting discussed novel approaches in the field of immunopharmacology of infectious diseases and assessed the evidence accrued over the last decade that indicates the emergence of cytokines and diverse immunomodulators as promising therapeutic agents.

The active participation of leading international authorities and research groups working in this field facilitated the discussion of recent developments and future strategies. Scientists from 20 countries presented their latest research results and reviewed the current state of the art under the sponsorship of the Federal Health Office and the International Society for Immunopharmacology.

While we must recognize the limitations and unique complexities of such exogenous intervention, targeted therapeutic exploitation of cytokines and judicious application of potent immunomodulatory drugs can lead to significant advances in the control of microbial diseases. The contributions presented in this book offer a critical appraisal of the current status of immunotherapy of infections.

K. Noel Masihi

Contents

Immunotherapy of Infections

I

CYTOKINES

1

Therapeutic Possibilities with Cytokines and Cytotoxic Lymphocytes in Infectious Diseases

Eva Lotzová
The University of Texas M.D. Anderson Cancer Center, Houston, Texas

I. RELEVANCE OF CYTOKINES IN THE TREATMENT OF INFECTIONS

Recent developments in biotechnology have led to production of an array of molecules—cytokines—that can induce differentiation and activation of hematopoietic and lymphoid compartments and subsequently lead to potentiation of host defense mechanisms against cancer and infections. Although cytokines are produced by mononuclear cells either constitutively or in response to activation signals, their induction has been shown to be impaired in patients with infectious diseases (1–5). Therefore, the patients with impaired cytokine production or those undergoing chemotherapy treatments (resulting in immunosuppression and high susceptibility to infections) may benefit from therapy with exogenously administered cytokines. Indeed, the preclinical models as well as some clinical experience indicate the relevance of cytokines in prophylaxis or therapy of various infections. In this chapter, the cytokines that may be relevant to therapy of infectious diseases are discussed.

A. Interleukin-1

Interleukin-1 (IL-1) is the cytokine produced by macrophages after stimulation with antigens or mitogens; this cytokine has been known for its broad range of activities, some of which may contribute to facilitation of host defense against infections. These activities include the role of IL-1 in the production of polymorphonuclear leukocytes (PMNs) in bone marrow, their accumulation at the site of infections, induction of release of IL-3 and granulocyte/macrophage growth factors, enhancement of the secretion of IL-2, and potentiation of monocyte phagocytosis and lymphocyte cytotoxic functions (6).

B. Interleukin-2

Interleukin-2 (IL-2) is the lymphokine that has been studied most extensively for its antitumor effects and has been shown to have a therapeutic potential in cancer (7–11). Limited information indicates that IL-2 may exert a protective effect against bacterial infections in a murine model (12). Considering the role of IL-2 in stimulation of natural killer (NK) cells (7,10,11,13) and participation of NK cells in antibacterial defense (14,15), the IL-2–induced protective activity against bacterial infection could be NK cell mediated. Because of the paucity of studies, it is difficult to suggest the therapeutic role of IL-2 in infections, although its effect on the growth and activation of lymphocytes and induction of interleukin gamma (IFN-γ) production indicates its possible benefit. Its therapeutic possibility also is in line with the observation that production of IL-2 is defective in some infections (1,2).

C. Interleukin-3

Interleukin-3 (IL-3) (multi-CSF [colony-stimulating factor]) is the glycoprotein produced by activated T cells and NK cells, which promotes differentiation and growth of bone marrow progenitors and enhances the function of monocytes and granulocytes (16–18). This molecule was suggested to play a role during parasitic infection (19).

D. Interleukins 5 to 8

Because of its role in eosinophil differentiation and function, IL-5 also is of potential interest in defense against infections (20,21); however, its clinical relevance remains to be defined.

Similarly, the role of IL-6, IL-7, and IL-8 in infectious diseases needs to be established; IL-6 may be of interest because of its ability to activate NK cells and T cells and to mediate acute inflammatory host response against infectious agents (22) and IL-7 and IL-8 because of their lymphocyte growth-promoting activity and chemotaxis of granulocytes and lymphocytes (23).

E. Interleukin-12

One of the cytokines discovered relatively recently is the NK cell stimulatory factor interleukin-12 (IL-12) (24,25). IL-12 is composed of two subunits, p35 and p40, and the coexpression of both subunits is required for production of this cytokine

(25). It is produced by B cells, B-cell lines, and monocytes. Monocytes produce high levels of this lymphokine after stimulation with *Staphylococcus aureus* and *Mycobacterium tuberculosis*. IL-12 also induces the production of IFN-γ by NK cells and T cells, proliferation of activated NK cells and T cells, and activates cytotoxic function of NK cells. All of these activities may play a role in antimicrobial defense.

The hematopoietic growth factors such as granulocyte-macrophage colony-stimulating factor (GM-CSF), granulocyte colony-stimulating factor (G-CSF), macrophage colony-stimulating factor (M-CSF) also are of relevance in infections for their abilities to stimulate proliferation and to enhance the functions of granulocytes and monocytes (26–30). The benefit of tumor necrosis factor alpha (TNF-α) in infectious diseases is quite controversial (31–34). The induction of this agent during the early phase of bacterial infections may actually accentuate the inflammation and enhance the severity of infection. However, TNF-α also has immunopotentiating properties and by activation of the functions of PMNs and lymphocytes may contribute to enhancement of antimicrobial defense. Interferons alpha, beta, and gamma may be quite important in some viral infections via their direct antimicrobial effects and activation of the host defense mechanisms, for example NK cells (35–37), as well as by enhancing the expression of the major histocompatibility complex (MHC) class I and class II antigens. IFN-γ has also been shown to have an effect in bacterial and fungal infections (38–40). Interferons are produced by a variety of leukocytes, including T cells, NK cells, B cells, and macrophages, especially after activation.

This brief overview indicates that cytokines or their combinations may have therapeutic value in the treatment of some infectious diseases. For instance, IL-1, IL-3, and hematopoietic growth factors, alone or in combination, may be beneficial in the treatment of infections associated with neutropenia or neonatal septicemia or in the prevention of infections accompanying aplastic anemia, chemotherapy, immunodeficiency, or burn injury. IL-2, IL-12, TNF-α, and IFNs may also be useful in potentiating host antimicrobial defense via stimulating the host effector cells. In addition, IFNs may have a direct inhibitory effect on the replication cycle of certain microbial agents.

However, it needs to be recognized that the therapy with exogenous cytokines also has limitations. The clinical experience indicates that in order to achieve the therapeutic benefit, the cytokines, as a rule, need to be administered in relatively high doses, which may result in severe adverse effects. Another obstacle to treatment with exogenous cytokines, especially those aimed at the potentiation of host defense mechanisms, is their possible inefficiency in highly immunocompromised hosts, lacking pertinent effector cells or containing disease-related factors preventing lymphocyte activation. Therefore, the development of alternative approaches to therapy of infectious disease as well as cancer is essential.

II. THERAPEUTIC POSSIBILITIES OF BONE MARROW–DERIVED LYMPHOCYTES IN INFECTIOUS DISEASES

The adoptive transfer of ex vivo stimulated lymphocytes may represent a plausible alternative to cytokine therapy in patients with infectious diseases. This approach would preclude the adverse effects associated with high cytokine doses and the interference of disease-associated factors with lymphocyte activation. Although the in vitro IL-2 activation and propagation of peripheral blood lymphocytes and their therapeutic use have been shown to be feasible, and the adoptive therapy trials demonstrated some responses in patients with cancer (10), the therapeutic benefit of this modality clearly needs to be facilitated.

We reasoned that the use of bone marrow compartment (as opposed to peripheral blood) as a source of effector cells may be beneficial in (1) inducing more effective cytotoxic lymphocytes from bone marrow precursors than from mature (more terminal) lymphocytes; (2) generating cytotoxic cells from bone marrow progenitors of patients whose mature lymphocytes are defective or not responsive to activation; and (3) generation of cytotoxic cells from bone marrow in instances where leukaphoresis is not possible.

Our initial studies showed that IL-2 was effective in activating cytotoxic lymphocytes from bone marrow compartment (41); more recently, we have shown that lymphocytes exhibiting more effective cytotoxicity against a broad range of targets, including virus-infected cells and resistant tumors, can be induced from bone marrow compartment by concurrent activation with IL-2 and a variety of tissue culture cell lines (42). Importantly, NK cells or T cells can be induced selectively depending on the cell line used for activation. Specifically, the Daudi/IL-2–stimulated cultures contained primarily $CD56^{+}/CD3^{-}$ NK cells and also $CD56^{+}/CD3^{+}$ T-cell subset. In contrast, those cultures only stimulated with BSM/IL-2 and AML/IL-2 or IL-2 were composed mainly of T-cell subsets, and contained only a small proportion of NK cells.

Since the phenotype is useful only in determining the lymphocyte subsets but not their function, we studied the efficacy of individual lymphocyte subsets, separated by fluorescence-activated cell sorter, to mediate the cytotoxicity. These studies showed that NK cells were the major oncolytic lymphocytes in Daudi/IL-2–stimulated cultures. Virtually no cytotoxicity was manifested by T cells. On the contrary, in AML/IL-2–activated cultures, both T cells and NK cells were cytotoxic, although T cells showed higher cytotoxic potential against acute myelogenous leukemia (AML) blasts. These studies indicate that various lymphocyte subsets with high cytotoxic activity can be generated from bone marrow compartment by simultaneous stimulation with cell lines or tumor cells. Importantly, these bone marrow–derived lymphocytes demonstrated high cytotoxic potential not only against the cell lines used for stimulation but also against a variety of other targets, which indicates their broad spectrum of cytotoxicity. It also is of note that the cytotoxicity of lymphocytes in bone marrow cultures activated with IL-2 only declined more rapidly than that in cultures stimulated with IL-2 and cell lines. The cultures stimulated with cell lines also displayed higher proliferative potential. Thus, the stimulator cells were effective not only in enhancing but also in maintaining the cytotoxic and proliferative activities of cytotoxic lymphocytes. These observations indicate that bone marrow culture system allows for propagation of highly activated cytotoxic lymphocytes for adoptive therapy purposes.

III. CONCLUSIONS

The therapeutic possibilities of cytokines has been investigated more actively in cancer than in infectious diseases; however, although limited, the information from animal models together with the properties of some cytokines (e.g., stimulation of proliferation, differentiation, and activation of hematopoietic and lymphoid cells) indicate their possible prophylactic and therapeutic role in infections. Despite the therapeutic potential of cytokines, it needs to be recognized that in order to be effective, the cytokines have to be administered in high doses resulting in many adverse effects. Additionally, some patients may not be responsive to in vivo cytokine therapy, because of their immunocompromised state or host inhibitory factors preventing lymphocyte proliferation, activation, and differentiation. In these instances, the cytokines may be used for in vitro lymphocyte activation and the induction of growth and differentiation in hematopoietic progenitors. These in vitro activated and propagated populations can then be adoptively transferred to patients to facilitate their antimicrobial defense. We have shown that highly cytotoxic lymphocytes with a broad range of killing activity, including virus-infected targets, can be induced from bone marrow compartment by simultaneous activation with IL-2 and tissue culture cell lines or tumor cells. Importantly, NK cells or T cells could be induced selectively depending on the cell line used for in vitro activation. Since NK cells and T cells have been shown to mediate activity against variety of viral, fungal, and bacterial infections (7), therapy with these lymphocytes alone or in combination with cytokines may facilitate antimicrobial defense. Alternatively or additionally, adoptive therapy with lymphocytes transfected with genes for cytokine production may also provide a new therapeutic modality.

ACKNOWLEDGMENT

This research was supported by grant CA55597 from the National Cancer Institute. The author is a recipient of Florence Maude Thomas Cancer Research Professorship.

REFERENCES

1. Lane HC, Depper JM, Greene WC, Whalen G, Waldmann TA, Fauci AS. Qualitative analysis of immune function in patients with the acquired immunodeficiency syn-

drome: Evidence for a selective defect in soluble antigen recognition. N Engl J Med 1986; 313:79.
2. Clerici M, Stocks NI, Zajac RA, Boswell RN, Lucey DR, Via CS, Shearer GM. Detection of three distinct patterns of T helper cell dysfunction in asymptomatic, human immunodeficiency virus-seropositive patients: Independence of $CD4^+$ cell numbers and clinical staging. J Clin Invest 1989; 84:1892.
3. Roilides E, Clerici M, DePalma L, Rubin M, Pizzo PA, Shearer GM. Helper T-cell responses in children infected with human immunodeficiency virus type 1. J Pediatr 1991; 118:724.
4. Depper JM, Leonard WJ, Kronke M, Waldmann TA, Green WC. Augmented T cell growth factor receptor expression in HTLV-1 infected human leukemic T cells. J Immunol 1984; 133:1691.
5. Linker-Israeli M, Bakke AC, Kitridou RC, Gendler S, Gillis S, Horwitz DA. Defective production of interleukin 1 and interleukin 2 in patients with systemic lupus erythematosus (SLE). J Immunol 1983; 130:2651.
6. Bagby GC, Jr. Interleukin-1 and hematopoiesis. Blood Rev 1989; 3:152.
7. Lotzová E, Herberman RB. Interleukin-2 and killer cells in cancer. Boca Raton, FL: CRC Press, 1990.
8. Lotzová E. Natural killer cells: Immunobiology and clinical prospects. *Cancer Invest* 1991; 9(2):173.
9. Lotzová E, Savary CA, Herberman RB. Induction of NK cell activity against fresh human leukemia in culture with interleukin 2. *J Immunol* 1987; 158:2718.
10. Rosenberg SA, Lotze MT, Muul LM, Chang AE, Avis FP, Leitman S, Linehan WM, Robertson CN, Lee RE, Rubin JT, Seipp CA, Simpson CG, White DE. A progress report on the treatment of 157 patients with advanced cancer using lymphokine-activated killer cells and interleukin-2 or high-dose interleukin-2 alone. N Engl J Med 1987; 316:889.
11. Lotzová E, Herberman RB. Natural Immunity, Cancer and Biological Response Modification. Basel: Karger, 1986.
12. Chong KT. Prophylactic administration of interleukin-2 protects mice from lethal challenge with gram-negative bacteria. Infect Immun 1989; 55:668.
13. Lotzová E, Savary CA. Growth kinetics, function and characterization of lymphocytes infiltrating ovarian tumors. In: Lotzová E, Herberman RB, eds. Interleukin-2 and killer cells in cancer. Boca Raton, FL, CRC Press, 1990:153.
14. Smith RA, Brzezicki MJ, Griggs N, Mahrer S. The role of natural killer cells in experimental murine salmonellosis. Nat Immun Cell Growth Regul 1989; 8:331.
15. Tarkkanen J, Saksela E, Lanier LL. Bacterial activation of human natural killer cells. Characteristics of the activation process and identification of the effector cells. J Immunol 1986; 137:2428.
16. Otsuka T, Miyajima A, Brown N, Otsu K, Abrams J, Saeland S, Caux C, De Waal Malefijt R, De Vries J, Meyerson P, Yokota K, Gemmel L, Rennick D, Lee F, Arai N, Arai KI, Yokota T. Isolation and characterization of an expressible cDNA encoding human IL-3: Induction of IL-3 mRNA in human T cell clones. J Immunol 1988; 140:2288.
17. Lopez AF, To LB, Yang YC, Gamble JR, Shannon MF, Burns GF, Dyson PG, Juttner CA, Clark S, Vadas MA. Stimulation of proliferation, differentiation, and

function of human cells by primate interleukin 3. Proc Natl Acad Sci USA 1987; 84:2761.

18. Sonada Y, Yang YC, Yong GG, Clark SC, Ogawa M. Analysis in serum-free culture of the targets of recombinant human hematopoietic growth factors: Interleukin-3 and granulocyte-macrophage colony-stimulating factor are specific for early developmental stages. Proc Natl Acad Sci USA 1988; 85:4360.
19. Guy-Grand D, Dy M, Luffau G, Vasalli P. Gut mucosal mast cells: Origin, traffic, and differentiation. J Exp Med 1984; 160:12.
20. Takatsu K. Interleukin 5 (IL-5) and its receptor. Microbial Immunol 1991; 35:593.
21. Weller PF. Cytokine regulation of eosinophil function. Clin Immunol Immunopathol 1992; 62:S55.
22. Bauer J, Herrmann F. Interleukin-6 in clinical medicine. Ann Hematol 1991; 62:203.
23. Balkwill FR, Burke F. The cytokine network. Immunol Today 1989; 10:299.
24. Chehimi J, Starr SE, Frank I, Rengaraju M, Jackson SJ, Llanes C, Kobayashi M, Perussia B, Young D, Nickbarg E, Wolf SF, Trinchieri G. Natural killer (NK) cell stimulatory factor increases the cytotoxic activity of NK cells from both healthy donors and human immunodeficiency virus-infected patients. J Exp Med 1992; 175:789.
25. Wolf N, Sieburth D, Perussia B, Yetz-Aldape J, D'Andrea A, Trinchieri G. Cell sources and inducers of natural killer cell stimulatory factor (NKSF/IL12) transcripts, subunits, and biological activity. Nat Immun 1992; 11(5):296.
26. Clark SC, Kamen R. The human hematopoietic colony-stimulating factors. Science 1987; 236:1229.
27. Metcalf D, Begley CG, Johnson GR, Nicola NA, Vadas MA, Lopez AF, Williamson DJ, Wong GG, Clark SC, Wang EA. Biologic properties in vitro of a recombinant human granulocyte-macrophage colony-stimulating factor. Blood 1986; 67:37.
28. Weisbart RH, Golde DW, Clark SC, Wong GG, Gasson JC. Human granulocyte-macrophage colony-stimulating factor is a neutrophil activator. Nature 1985; 314:361.
29. Roilides E, Walsh TJ, Pizzo PA, Rubin M. Granulocyte colony-stimulating factor enhances the phagocytic and bactericidal activity of normal and defective human neutrophils. J Infect Dis 1991; 163:579.
30. Karbassi A, Becker JM, Foster JS, Moore RN. Enhanced killing of *Candida albicans* by murine macrophages treated with macrophage colony-stimulating factor: Evidence for augmented expression of mannose receptors. J Immunol 1987; 139:417.
31. Pennica D, Nedwin GE, Hayflick JS, Seeburg PH, Derynck R, Palladino MA, Kohr WJ, Aggarwal BB, Goeddel DV. Human tumour necrosis factor: precursor structure, expression and homology to lymphotoxin. Nature 1984; 312:724.
32. Beutler BA, Milsark IW, Cerami A. Cachectin/tumor necrosis factor: Production, distribution, and metabolic fate in vivo. J Immunol 1985; 135:3972.
33. Silva AT, Bayston KF, Cohen J. Prophylactic and therapeutic effects of a monoclonal antibody to tumor necrosis factor-α in experimental gram-negative shock. J Infect Dis 1990; 162:421.
34. Degliantoni G, Murphy M, Kobayashi M, Francis MK, Perussia B, Trinchieri G. Natural killer (NK) cell-derived hematopoietic colony-inhibiting activity and NK cytotoxic factor. Relationship with tumor necrosis factor and synergism with immune interferon. J Exp Med 1985; 162:1512.

35. Lotzová E, Savary CA, Gutterman JU, Quesada JR, Hersh EM. Regulation of human natural killer cell cytotoxicity by recombinant leukocyte interferon clone A. J Biol Response Mod 1983; 2:482.
36. Lotzová E, Savary CA. Stimulation of NK cell cytotoxic potential of normal donors by two species of recombinant alpha interferon. J Interferon Res 1984; 4:201.
37. Perussia B, Kobayashi M, Rossi ME, Anegon I, Trinchieri G. Immune interferon enhances functional properties of human granulocytes: Role of FC receptors and effect of lymphotoxin, tumor necrosis factor, and granulocyte-macrophage colony-stimulating factor. J Immunol 1987; 138:765.
38. Murray HW. Interferon-γ, the activated macrophage, and host defense against microbial challenge. Ann Intern Med 1988; 108:595.
39. Morrison CJ, Brummer E, Stevens DA. In vivo activation of peripheral blood polymorphonuclear neutrophils by gamma interferon results in enhanced fungal killing. Infect Immun 1989; 57:2953.
40. Flesch IEA, Schwamberger G, Kaufmann SHE. Fungicidal activity of IFN-γ-activated macrophages. Extracellular killing of *Cryptococcus neoformans*. J Immunol 1989; 142:3219.
41. Lotzová E, Savary CA. Generation of NK cell activity from human bone marrow. J Immunol 1987; 139:279.
42. Fuchshuber PR, Lotzová E, Savary CA. Generation of MHC-nonrestricted and restricted oncolytic subsets from human bone marrow. Cell Immunol 1992; 139:30.

2

Impact of Direct Microorganism-Cytokine Interaction on Cytokine Immunotherapy of Infections

Michel Denis
University of Sherbrooke, Sherbrooke, Quebec, Canada

I. CYTOKINES AS IMMUNOTHERAPEUTIC TOOLS

Cytokines represent a group of small molecular weight proteins that regulate a wide variety of cellular functions and are active on numerous cell types (1). Many of these metabolites have growth factor activities, such as the lymphocyte growth factors interleukin-2 (IL-2) and interleukin-4 (IL-4), as well as metabolites that promote the growth and differentiation of uncommitted progenitor cells such as granulocyte-macrophage colony-stimulating factor (GM-CSF), granulocyte colony-stimulating factor (G-CSF), and interleukin-3 (IL-3) (2,3).

Another group of important cytokines is composed of tumor necrosis factor alpha (TNF-α), interleukin-6 (IL-6), and interleukin-1 (IL-1) (4); these proinflammatory cytokines are mediators of inflammatory responses that occur after an inflammatory or infectious stimulus. IL-1, IL-6, and TNF are believed to function as stress hormones of the immune system. They are produced mainly by macrophages and monocytes as a first-line response to infections (5). These three cytokines function individually or in concert as (1) activators of the hepatic acute phase response involved in the production of acute phase proteins, such as serum amyloid A, C-reactive protein, fibrinogen, and complement components; (2) endogenous pyrogens responsible for the fever response seen in infections; (3) stimulators of the production of important chemotactic agents such as interleukin-8 (IL-8), which is involved in excessive neutrophilia; (4) leukocyte activators; indeed, these molecules are involved in increased phagocytic and cytotoxic activity of neutrophils and macrophages, including the degranulation of neutrophils and their release of reactive oxygen intermediates (6,7). In addition, these cytokines are catabolic agents that are responsible for redirecting peripheral energy reserves to the acute metabolic demands of the inflammatory response. Obviously, an excess of this response is responsible for the cachexia that is associated with a number of chronic infectious diseases such as tuberculosis (8). One final role for these proinflammatory cytokines is as enhancers of the tissue remodeling that occurs after an infectious episode by virtue of their role in fibroblast proliferation, cartilage and bone resorption, and in the production of collagenase and protease (8). An obvious side effect of this response is tissue fibrosis, as well as excess tissue destruction that is observed in chronic infectious diseases (8).

An interesting feature of all cytokines studied thus far is their pleiotropic effects on various cell types. For instance, IL-2, which was originally described as a T-lymphocyte growth factor, has now been shown to possess a wide variety of effects on a large spectrum of cells. IL-2 may contribute to macrophage activation by binding to the P55 receptors present on these cells; it also enhances B-cell growth, causes a polyclonal immunoglobulin M (IgM) response, and it appears to have unexpected effects on vascular cells as well as nerve cells (3,9). These varied effects of any one single cytokine explain the difficulty of dissecting the impact of cytokine infusion on host resistance to infections or other endpoints.

Given their central role in protection versus a variety of infectious agents, cytokines have recently been used as protective agents or adjuvant material in protection versus infection. In vivo infusion of high doses of recombinant IL-2 has been shown to protect mice against gram-negative bacterial sepsis (10). Relatively high doses of IL-2 had to be used and the effect was not clarified but appeared to be related to a polyclonal IgM response and an enhanced clearance of bacteria from the infected tissues (11). IL-2 has also been used to enhance resistance against a number of more chronic infectious agents such as mycobacteria (12,13). In a mouse model of the disease, IL-2 has been shown to enhance resistance against *Mycobacterium avium* (14). Similar to findings in infections with *Escherichia coli,* the exact mechanisms were not clarified. In view of the previously described pleiotropic effects of IL-2, it may be that this cytokine enhances resistance in a variety of ways.

Findings in animal models have culminated in the use of IL-2 as an adjuvant in the reconstitution of a cutaneous cellular immunity in the setting of lepromatous leprosy (15). In this work, it was shown that IL-2 recapitulated a delayed-type hypersensitivity response and suggested the use of IL-2 as a therapeutic agent in immunosuppressed patients. Indeed, a recent study has shown beneficial effects in the use of IL-2 in severely immunosuppressed subjects (16). Low doses of IL-1 are beneficial in host resistance against lethal doses of *Klebsiella;* the mechanism did not appear to involve neutrophils nor a corticosteroid response (17). The macrophage growth factor (colony-stimulating factor [CSF-1]) has been shown to enhance the tissue clearance of *Candida albicans* and *Listeria* (18,19), but it paradoxically increased the growth of *Brucella* or *Mycobacterium avium* in the tissues of mice presumably by increasing the number of potential intracellular niches for these organisms (20,21).

The T-cell–derived lymphokine interferon gamma (IFN-γ) is an important lymphokine that is secreted by the TH_1 subtype of T cells, and, as with other cytokines, has a variety of effects on numerous cell types. One very well-described effect of IFN-γ is its ability to enhance macrophage microbiostatic or microbicidal effect. This appears to be mediated by an increase in the release of microbicidal molecules, such as nitric oxide and reactive oxygen intermediates (22). Stimulation of mouse macrophages with IFN-γ increases the killing of microbes as diverse as *Mycobacterium, Francisella,* and *Leishmania* (22,23). This has led to the description of IFN-γ as a molecule that enhances resistance to many infectious agents. Moreover, in vivo experiments with neutralizing monoclonal antibodies have suggested a strong role for IFN-γ in the resistance to *Listeria, Mycobacterium,* and *Salmonella,* as well as others (24–26). In numerous systems, IFN-γ appears to act by increasing macrophage function that leads to a decrease in bacterial viability in the tissues. However, as mentioned before, IFN-γ has a wide variety of effects, including an antiproliferative activity on a number of cells, including T cells.

Recent data in a mouse model of *Candida* infections have shown a surprising increase in the susceptibility of mice infected with *Candida* when they are treated with IFN-γ (27); it was suggested that this was related to a diminished proliferative response of systemic T cells. Similarly, a recent surprising finding has shown that *Trypanosoma brucei* infections in mice could be therapeutically improved by neutralizing in vivo IFN-γ, which suggested a negative role for this cytokine in the development of resistance to this organism (28).

It has been known for some time that certain forms of immunity are detrimental to the progression of infections caused by certain classes of microbes or parasites. The better-known example of this is infection with the *Leishmania major* parasite. Infection with this protozoan in resistant mice is followed by a strong TH_1-like response with an important IFN-γ secretion that then enhances macrophage killing of the parasite (29). Conversely, in susceptible hosts, infection appears to induce a TH_2-type response with low levels of IL-2 and IFN-γ and high levels of IL-3, IL-4, and CSF secretion (30). These molecules (IL-4, IL-3) have been shown to interfere with the development of macrophage microbicidal activity presumably via a downregulation of the nitric oxide synthase in these cells. This paradigm of resistance mediated by a TH_1-like response and susceptibility associated with a TH_2-like response has been observed for a number of other infectious agents such as *Listeria* or *Candida* (31,32). However, a TH_2 response with its associated enhanced IgA response and eosinophil infiltration is probably an efficient response that is involved in resistance against a number of gut parasites such as nematodes (33).

Overall, the use of cytokines as immunotherapeutic tools in the setting of infections has given rise to an optimistic view of the use of such reagents. However, it is recognized that there are tremendous problems associated with the eventual use of such material in the human situation. The recent description of IFN-γ as a positive therapeutic agent in the treatment of chronic granulomatous disease is of considerable interest (34). However, other uses for cytokines are probably limited by the inherent toxicity of such material, their unclear pharmacological behavior, and, probably more importantly, by their pleiotropic effects. Our own studies have suggested other potential problems, which are outlined in Chapter 2.

II. INTERACTIONS BETWEEN CYTOKINES AND BACTERIA

It has become apparent that there is a striking and unexpected relationship between mammalian cytokines and successful pathogens. This has been reviewed recently (35) and we will discuss some salient points. In a landmark publication, Mazingue et al. reported that when IL-2 was added to a culture medium of the protozoans of the genus *Leishmania,* the growth of these parasites dramatically increased (36). IL-2 could substitute for a growth factor secreted by the protozoan that was not present when the parasites were continually washed and reincubated in fresh medium. In this work, Mazingue et al. observed that injection of high doses of IL-2 at the inoculation site in resistant mice infected with *Leishmania* rendered these hosts unexpectedly susceptible. Moreover, the immunosuppressive drug, cyclosporin A, rendered susceptible hosts very resistant. Mazingue et al. interpreted these data to suggest that *Leishmania* has adapted to the presence of important cytokines and has evolved to use this material as growth factors; this probably constitutes an important virulence factor.

Kongshavn and Ghadirian observed that incubation of *Trypanosoma musculi* with the cytokine TNF unexpectedly led to a very dramatic increase in the growth of this parasite (37). However, TFN-α–pulsed macrophages were able to kill the parasite. Work by our group showed that fresh clinical strains of *E. coli* were stimulated to grow in the presence of IL-2, GM-CSF, and other cytokines (38). This enhancement of growth was significant only when the bacteria were cultivated in a suboptimal medium such as fresh serum; bacteria growing in an optimal medium, that is, bacterial broth, were not stimulated to grow by cytokines. Porat and colleagues described a similar phenomenon where fresh, clinically important strains of *E. coli* were stimulated in their growth by incubation with the cytokine IL-1 (39). Other findings showed that a specific IL-1–like receptor structure was present at the bacterial surface (39). Cytokine binding and usage appeared not to be limited to fast-growing bacteria inasmuch as recent reports by Shirasutchi et al. as well as our own work showed that acquired immune deficiency syndrome (AIDS)–associated *M. avium* strains were stimulated to grow in the presence of the cytokine IL-6 (40,41). This enhancement of growth occurred in suboptimal medium; enhancement of growth in an optimal medium such as bacterial broth was not significant. Similarly, pulsing of macrophage monolayers with a number of cytokines (IL-6, CSF-1) infected with *M. avium* led to the induced growth of *M. avium* in vitro (42). These findings were interpreted as being relevant to the immunosuppressed state of patients with AIDS who have very high serum levels of IL-6 and probably other cytokines. This may be in part responsible for the unusual susceptibility of patients with AIDS to *M. avium.* Other important papers in this area include the recent work of Barcinski et al. (43). These investigators observed that GM-CSF increased the infectivity of the parasite *Leishmania amazonensis* by protecting the promastigote form of this parasite from heat-induced death. Indeed, it appeared that GM-CSF incubation with *Leishmania* enhanced the resistance of these parasites to heat stress that obviously occurs on infection of mammalian macrophages (43). Luo et al. recently reported that TNF-α binds with high-affinity receptors to the pathogen *Shigella flexneri* (44). The binding of the cytokine was inhibited by trypsin treatment, which indicated a protein component. Interestingly enough, after interaction with TNF-α in solution, *Shigella* organisms became more invasive for epithelial cells in vitro, which suggested that TNF-α–bacteria complexes were interacting with TNF-α receptors present on eukaryotic cells. Our group also showed the presence of high-affinity receptors at the membrane surface of clinical strains of *M. avium.* Scatchard analysis of receptor interaction showed that AIDS-associated strains had a single receptor species with an affinity constant of 50 nM and the number of receptors was approximately 10,000/bacterium (45).

Overall, the data suggest that a range of successful pathogens has adapted to the presence of host cytokines by developing binding structures for these molecules that in the course of an infection may act as a sink for protective cytokines and prevent the activation of cells with these factors. Moreover, the bacteria have adapted to use these cytokines as growth factors and they may become important virulence factors in a number of situations. It remains unclear exactly what mechanism was used by bacteria to acquire this phenotype.

Data in a number of important bacterial strains involved in causing pyelonephritis suggest that bacterial strains that infect humans have recruited mammalian DNA, which may be important in binding to important mucosal surfaces (46). Similarly, receptors for a large number of mammalian molecules at the surface of important pathogens have been described; for example, binding to molecules such as plasminogen and fibrinogen and its related fibrin by-products has been described for *Streptococcus, Haemophilus,* and others (47,48). This type of binding is probably important in preventing an early host response that limits bacterial dissemination and provides a first-line response to these infectious agents. It will be of considerable interest to determine the exact mechanism responsible for both the binding and the growth-enhancing effect of cytokines in bacteria and whether or not bacteria use a signal transduction pathway after the binding of these host molecules.

The exact clinical relevance of these findings is still unclear and probably difficult to ascertain. A number of studies in patients who are receiving infusions of high doses of IL-2 for the treatment of advanced forms of cancer offer some perspective. Indeed, it has become clear that a large fraction of these patients become very susceptible to infections with a number of opportunistic pathogens such as *Escherichia, Klebsiella,* and *Staphylococcus* (49,50). It is unclear if cytokine usage by some of these clinical strains is contributing to this phenomenon. However, infusion with large doses of IL-2 is likely to lead to a large number of manifestations. Indeed, it was reported that patients receiving high doses of IL-2 had a deficient neutrophil chemotactic response, which may be a crucial parameter in determining the susceptibility to bacterial infections (51). The exact relevance of bacterial binding and usage of cytokines will only become clear with further studies in experimental animals as well as in freshly derived clinical strains.

III. IMPLICATIONS FOR CYTOKINE OR ANTICYTOKINE IMMUNOTHERAPY OF BACTERIAL INFECTIONS

Recent data have suggested that a number of cytokines may be involved in downregulating a number of functions that may be important in determining resistance to infections. For example, the TH_2-derived cytokine IL-10 appears to play a pivotal role in downregulating macrophage microbicidal activity. This appears to occur via a diminished nitric oxide synthase activity; IL-10 also blocks the IFN-γ–mediated enhancement of macrophage microbicidal activity (52). Other cytokines have suppressive activities such as molecules in the transforming growth factor beta (TGF-β) family (53). The clear implication of some of these factors in determining susceptibility to infections is shown by the in vivo neutralization of these molecules. IL-10 neutralization in a mouse model of *M. avium* resulted in zero growth of a bacterium that normally grows quite rapidly in the infected tissues (54). In vitro studies showed that as chronic infection developed, there was a progressive decline in splenocyte secretion of IFN-γ and a marked increase in IL-10 release. TGF-β neutralization in a mouse model of *Leishmania* resulted in a dramatically enhanced resistance to this protozoa (55). These data suggest interesting new avenues for immune modulation of bacterial infections.

Another interesting set of observations is the recent description of the capture and the release by viruses of potentially suppressive cytokines. For example, the Epstein-Barr virus, responsible for infectious mononucleosis, codes for and synthesizes the immunosuppressive cytokine IL-10, which may in part be responsible for the observed immunosuppression seen in mononucleosis and possibly other pathologies (56). Similarly, other viruses have adapted to synthesize and release soluble receptors coding for important cytokines such as TNF-α (57). It is probable that approaches based on neutralization of immunosuppressive cytokines in infectious diseases are an area of considerable promise.

In our view, bacterial, viral, and parasitic adaptations to the presence of cytokines pose new and formidable problems for the investigator who is exploring the immunotherapeutic potential of cytokines. Approaches based on cytokine intervention must take these factors into account.

ACKNOWLEDGMENTS

This work was supported by a Medical Research Council-National Health and Welfare Research and Development Program Joint Program on AIDS, the Quebec Lung Association, the Natural Sciences and Engineering Research Council of Canada, and the Fonds de la recherche en santé du Québec.

REFERENCES

1. Balwell FR, Burke F. The cytokine network. Immunol Today 1989; 10:299.
2. Hume DA. Macrophage colony stimulating factor. Life Sci 1990; 2:20.
3. Smith KA. Interleukin-2; inception, impact and implication. Science 1979; 240:1169.
4. Beutler B, Cerami A. Tumor necrosis factor, cachexia, shock and inflammation: A common mediator. Annu Rev Biochem 1988; 57:505.
5. Bone RC. The pathogenesis of sepsis. Ann Intern Med 1991; 115:457.
6. Dinarello CA. Interleukin-1 and its biologically related cytokines. Adv Immunol 1989; 44:153.
7. Van Snick J. Interleukin-6; an overview. Annu Rev Immunol 1990; 8:253.
8. Beutleter B, Cerami A. Cachectin and tumor necrosis factor as two sides of the same biological coin. Nature 1986; 320:584.
9. Wahl SM, McCartney-Francis N, Hunt DA, Smith PD, Wahl LM, Katona IM. Monocyte interleukin-2 receptor gene expression and interleukin-2 augmentation of microbicidal activity. J Immunol 1987; 139:142.
10. Chong KT. Prophylactic administration of interleukin-2 protects mice from lethal challenge with gram-negative bacteria. Infect Immun 1987; 55:668.
11. Weyand C, Goronzy J, Fathman CG, Hanley PO. Administration in vivo of recombinant IL-2 protects mice against septic death. J Clin Invest 1978; 79:1756.
12. Jeevan J, Asherson GL. Recombinant interleukin-2 limits the replication of *Mycobacterium lepraemurium* and BCG in mice. Lymphokine Res 1988; 7:129.
13. Denis M. Cytokine modulation of *Mycobacterium lepraemurium* infection in mice; important involvement of tumor necrosis factor, interleukin-2 and dissociation from macrophage activation. Int J Immunopharm 1991; 13:889.
14. Bermudez LEM, Stevens P, Kolonoski P, Wu P, Young LS. Treatment of experimental disseminated *Mycobacterium avium* infection in mice with recombinant IL-2 and tumor necrosis factor. J Immunol 1988; 143:296.
15. Kaplan G, Kiessling R, Terlemariam S, Hancock G, Sheftel G, Job CK, Converse P, Ottenhoff THM, Becx-Bleumink M, Dietz M, Cohn ZA. The reconstitution of cell-mediated immunity in the cutaneous lesions of lepromatous leprosy by recombinant interleukin-2. J Exp Med 1989; 169:893.

16. Teppler H, Kaplan G, Smith KA, Montana AL, Meyn P, Cohn ZA. Prolonged immunostimulatory effect of low-dose polyethylene glycol interleukin-2 in patients with human immunodeficiency virus type I infection. J Exp Med 1993; 177:483.
17. MTE Vogels, Sweep CGJ, Hermus ADRMM, Van der Meer JWM. Interleukin-1 induced nonspecific resistance to bacterial infection in mice is not mediated by glucocorticoids. Antimicrob Agents Chemother 1992; 36:2785.
18. S Kayashima, Tsuru S, Shinomyia N, Katsura K, Motoyoshi K, Rokutanda M, Nagata N. Effects of macrophage colony stimulating factor on reduction of viable bacteria and survival of mice during *Listeria monocytogenes* infection; characteristics of monocyte subpopulations. Infect Immun 1991; 59:4677.
19. Cenci E, Bartocci A, Puccetti P, Mocci S, Stanley ER, Bistoni F. Macrophage colony-stimulating factor in murine candidiasis; serum and tissue levels during infection and protective effect of exogenous inflammation. Infect Immun 1991; 59:868.
20. Doyle AG, Halliday WJ, Barnett CJ, Dunn TL, Hume DA. Effect of recombinant human macrophage colony stimulating factor 1 on immunopathology of experimental brucellosis in mice. Infect Immuno 1992; 60:1465.
21. Murray HW. Interferon gamma, the activated macrophage, and host defense against microbial challenge. Ann Intern Med 1988; 108:595.
22. Murray HW, Spitalny GL, Nathan CF. Activation of mouse peritoneal macrophages in vitro and in vivo by interferon γ. J Immunol 1985; 135:1619.
23. Denis M. Interferong-treated murine peritoneal macrophages inhibit tubercle bacilli growth via the secretion of reactive nitrogen intermediates. Cell Immunol 1991; 132:150.
24. Denis M. Involvement of cytokines in determining resistance and acquired immunity in murine tuberculosis. J Leukoc Biol 1991; 50:495.
25. Muotiala A, Makela HP. The role of INFg in murine *Salmonella typhimurium* infection. Microb Pathog 1990; 8:135.
26. Buchmeier NA, Schreiber RD. Requirement of endogenous interferon production for resolution of *Listeria* infection. Proc Natl Acad Sci USA 1985; 38:7404.
27. Garner RE, Kuruganti U, Czarniecki CW, Chiu HH, Domer JE. In vivo immune responses to *Candida albicans* modified by treatment with recombinant gamma interferon. Infect Immun 1989; 57:1800.
28. Bakhiet M, Ollsson T, Van der Meide P, Kristensson K. Depletion of $CD8^+$ T cells supresses growth of *Trypanosoma brucei* and interferon γ production in infected rats. Clin Exp Immunol 1990; 81:195.
29. Heinzel FP, Sadick MD, Holaday BJ, Coffmann RL, Locksley RM. Reciprocal expression of interferon γ or interleukin 4 during the resolution or progression of murine leishmaniasis. Evidence for expansion of distinct helper T cell subsets. J Exp Med 1989; 169:59.
30. Heinzel FP, Sadick MS, Mutha SS, Locksley RM. Production of interferon γ, IL-2, IL-4 and IL-10 by $CD4^+$ lymphocytes in vivo during healing and progressive murine leishmaniasis. Proc Natl Acad Sci USA 1991; 88:7071.
31. Romani L, Mencacci A, Ghohmann U, Mocci S, Mosci P, Pucetti P, Bistoni F. Neutralizing antibody to interleukin-4 induces systemic protection of T helper type 1-associated immunity to murine candidiasis. J Exp Med 1992; 176:19.

32. Haak-Frendscho M, Brown JF, Igawa Y, Wagner RD, Czuprynzki CJ. Administration of anti–IL-4 monoclonal antibody 11b11 increases the resistance of mice to *Listeria monocytogenes*. J Immunol 1992; 148:3978.
33. Else KJ, Grencis RK. Cellular immune responses to the murine parasite *Trichuris muris*. Immunology 1991; 72:508.
34. Curnutte JT. Conventional vs interferon-gamma therapy in chronic granulomatous disease. J Infect Dis 1993; 167(Suppl):58–512.
35. Denis M, Campbell D, Gregg EO. Cytokine stimulation of parasitic and microbial growth. Res Microbiol 1991; 142:979.
36. Mazingue C, Cothez-Detoeuf F, Louis J, Kweider M, Auriault C, Capron A. *In vitro* and *in vivo* effects of interleukin 2 on the protozoan parasite *Leishmania*. Eur J Immunol 1989; 19:487.
37. Kongshavn PAL, Ghardirian E. Enhancing and suppressive effects of tumor necrosis factor/cachectin on *Trypanosoma musculi* growth. Parasite Immunol 1989; 10:581.
38. Denis M, Campbell D, Gregg EO. Interleukin-2 and granulocyte-macrophage colony-stimulating factor stimulate growth of a virulent strain of *Escherichia coli*. Infect Immun 1991; 59:1853.
39. Porat R, Clark BD, Wolff SM, Dinarello CA. Enhancement of growth of virulent strains of *Escherichia coli* by interleukin-1. Science 1991; 254:430.
40. Denis M, Gregg EO. Recombinant tumor necrosis factor-alpha decreases whereas recombinant interleukin-6 increases growth of a virulent strain of *Mycobacterium avium* in human macrophages. Immunology 1991; 71:139.
41. Shiratsushi H, Johnson JJ, Ellner JJ. Bidirectional effect of cytokines on the growth of *Mycobacterium avium* within human monocytes. J Immunol 1991; 146:3165.
42. Denis M. Growth of *Mycobacterium avium* in human monocytes: Identification of cytokines which reduce and enhance intracellular microbial growth. Eur J Immunol 1991; 21:391.
43. Barcinski M, Schechtman D, Quintao LG, Costa DA, Soares LRB, Moreira MEC, Charlab R. Granulocyte-macrophage colony-stimulating factor increases the infectivity of *Leishmania amazonensis* by protecting promastigotes from heat-induced death. Infect Immun 1992; 60:3523.
44. Luo G, Niesel DW, Shaban RA, Grimm EA, Kimpel GR. Tumor necrosis factor alpha binding to bacteria; evidence for a high affinity receptor and alteration of bacterial virulence properties. Infect Immun 1993; 61:830.
45. Denis M. Interleukin-6 is used as a growth factor by virulent *Mycobacterium avium*: Presence of specific receptors. Cell Immunol 1992; 141:182.
46. Holmgren A, Branden CI. Crystal structure of chaperone protein *Pap D* reveals an immunoglobulin fold. Nature 1989; 342:248.
47. Ullberg M, Kronvall G, Karlsson I, Wiman B. Receptors for human plasminogen on gram-negative bacteria. Infect Immun 1990; 58:21.
48. Visai L, Speziale P, Bozzini S. Binding of collagens to an enterotoxigenic strain of *Escherichia coli*. Infect Immun 1990; 58:449.
49. Murphy PM, Lane C, Gallin JI, Fauci AS. Marked disparity in incidence of bacterial infections in patients with the acquired immunodeficiency syndrome receiving interleukin-2 or interferon γ. Ann Intern Med 1988; 108:36.

50. Maoleekoonpairoj S, Mittelman A, Sarona S, Ahmed T, Puccio C, Gafney E, Skelos A, Arnold P, Coombe N, Basking P, Arlin Z. Lack of protection against bacterial infections in patients with advanced cancer treated by biologic response modifiers. J Clin Microbiol 1989; 27:2305.
51. Klempner MS, Noring R, Mier JW, Atkins MB. An acquired chemotactic defect in neutrophils from patients receiving interleukin-2 immunotherapy. N Engl J Med 1990; 322:959.
52. Gayzinelli RT, Oswald IP, James SL, Sher A. IL-10 inhibits parasite killing and nitrogen oxide production by IFNγ-activated macrophages. J Immunol 1992; 148:1792.
53. Gazzinelli RT, Oswald IP, Hieny S, James SL, Sher A. The microbicidal activity of interferon-γ treated macrophages against *Trypanosoma cruzi* involves an L-arginine–dependent, nitrogen oxide-mediated mechanism inhibitable by interleukin-10 and transforming growth factor-beta. Eur J Immunol 1992; 22:2501.
54. Denis M, Ghadirian E. IL-10 neutralization augments mouse resistance to systemic *Mycobacterium avium* infections. J Immunol 1993; 151:5425.
55. Barral-Netto M, Barral A, Brownell CE, Skeiky YAW, Ellingsworth LR, Twardzik DR, Reed SG. Transforming growth factor-β in leishmanial infection; a parasite escape mechanism. Science 1992; 257:545.
56. Moore KW, Vieira P, Fiorentino DF, Trounstine ML, Khan TA, Mossmann TR. Homology of the cytokine synthesis inhibitor factor (IL-10) to the Epstein-Barr virus gene BCRF1. Science 1990; 248:1230.
57. Upton C, Jacen JL, Schreiber M, McFadden G. Myxoma virus expresses a secreted protein with homology to the tumor necrosis factor receptor gene family that contributes to virual virulence. Virology 1991; 184:370.

3

Cytokine Profile as a Predisposing or Protective Factor in Immunological Diseases: Present Information, Prospects for Intervention, the Need for Cytokine-Based Epidemiology

Monika Brunner, Nicholas Avrion Mitchison, Joachim Sieper, and Katharina Simon
Deutsches Rheuma-Forschungszentrum, Berlin, Germany

I. INTRODUCTION

A central theme in contemporary immunology is that the activity of regulatory (CD4) T cells is governed by a balance of cytokines secreted by opposing subsets within those cells. Whether or not the original duopoly of TH_1 or TH_2 cells represents an oversimplification, the principle of subpopulations each secreting a distinct profile of cytokines has become firmly over the last 7 years. It has proved particularly valuable in explaining the spectrum of responses seen to such infections as leishmaniasis and leprosy and in accounting for the distribution of susceptibility to autoimmune diseases, again in animal models. It is now being applied intensively in clinical studies of both infections and immunological diseases. This meeting therefore offers an appropriate occasion for reviewing progress in this form of analysis as it has been applied to immunological diseases and looking forward to the next steps.

This article concentrates on inflammatory rheumatic diseases, as this constitutes a group of clinical entities all of which are suspected of having an autoimmune component. Although they are of absorbing interest to us, it must be admitted that inflammatory rheumatic diseases have drawbacks as well as advantages for discussion of the present topic. Their main advantage is that they are sufficiently common for the genetics of inflammatory rheumatic diseases to have been well studied (the commonest among them, rheumatoid arthritis, occurs in the main populations of Europe and the United States at a frequency of approximately 1%). Their main drawback is that inflammatory rheumatic diseases are much less clearly immunological in origin than several of the rare "autoimmune" diseases, such as myasthenia gravis. Although some sort of autoimmune process does seem to be involved, if only because of their strong associations with class II HLA genes, it is also widely believed that inflammatory rheumatic diseases may be triggered by infection, much as is reactive arthritis. If so, this chapter will be closer to the main theme of this book than might otherwise be suspected!

Before moving on to our main theme, it is worth looking back to the origin of the division of CD4 T cells into regulatory subsets. The concept has three roots—one in an activation marker in the rat, another in the cytokines secreted by mouse T-cell clones, and a third in immunosuppressive major histocompatibility complex (MHC) genes. In 1986, the first description of such subsets appeared based on separation of rat CD4 T cells by means of an antibody to the C isoform of CD45 and their differing activity in secreting interleukin-2 (IL-2) (1). Later that year, the division of mouse CD4 T-cell clones into TH_1 and TH_2 types was described at the VIth International Congress of Immunology (2). At the same Congress, CD4 T cells were classified into those restricted by immune-response or immune-suppression class II MHC genes in relation to any given response (3). As it became clearer that the expression of differing CD45 isoforms reflects the activation-quiescence cycle of T cells, it became less likely that this would bear any definite relationship to cytokine secretion profiles, although the matter is still under debate (4). On the other hand, the relationship between cytokine secretion profiles and class II MHC genes has gained increasing credibility and indeed forms the main subject of the present discussion. It is a subject to which we frequently return (5–12).

Our thesis here is that the profile of cytokines secreted by T cells probably varies from one individual to another and between health and disease. It has not yet been studied in a systematic enough way to allow any definitive statement; the point is rather that an effort in that direction is now much needed. The suggestion here is that such a study could be highly rewarding in the context of both infection and autoimmunity for a variety of reasons. It might be possible to warn susceptible individuals to take certain precautions or to avoid certain forms of risk. It could provide new insights into mechanisms of pathogenesis, which in turn could suggest new therapeutic strategies.

The possibility that the T-cell cytokine profile might be predictive of disease susceptibility has already occurred to students of infection. For example, an interesting group of individuals have been identified among those at high risk of infection with human immunodeficiency virus (HIV) (with HIV-positive sexual partners, offspring of HIV-positive mothers, and others) who manifest cell-mediated immunity to HIV peptides without becoming seropositive or developing overt acquired immunodeficiency syndrome (AIDS) (13). It was suggested that these individuals might be predisposed to make a TH_1-type response, which may have enabled them to resist infection, although their immune systems had evidently been exposed to the viral antigens (14). Or to take another example, the lack of an interferon gamma (IFN-γ) response to *Mycobacterium keprae* among persons exposed to this organism identifies those most at risk of developing leprosy (15). Neither of these studies was carried out on a scale large enough to answer the question here posed.

II. PROSPECTS FOR INTERVENTION

It might be argued that the possibility of putting information about cytokines to use is limited because of redundancy in the cytokine network. Cytokines are indeed often pleiotropic, and the same effect can often be brought about by more than one cytokine. It is no doubt for these reasons that so many cytokine knock-out mice tend not to have much of a phenotype, at least on first inspection. On the other hand, therapeutic manipulations with single cytokines have had striking effects in leprosy (16,17), although in this complex disease one needs to be careful about drawing conclusions (18). Even more dramatic was the effect of treatment of rheumatoid arthritis with anti–tumor necrosis factor alpha (anti-TNF-α) in a recent trial (19). The cytokine thickets may not be nearly as impenetrable as had been feared.

III. SUSCEPTIBILITY TO DISEASE AND PROTECTIVE EFFECTS MEDIATED BY CLASS II MHC GENES

A substantial body of information now connects class II MHC genes to disease susceptibility and also to protection against (and during) disease. Within this body of knowledge, we have decided to concentrate on rheumatological aspects, for the reasons given above. We concentrate also on situations where protective as well as the better-known susceptibility effects have been discovered. This is partly because only there can we expect to find a balance between opposing forces, both under MHC control. In addition, the phenomenon of bunching as described below seems to apply only to protective effects.

It is probably for this reason that bias in the cytokine profile has so far been invoked only as a possible mechanism of protection and not of susceptibility. However, it is worth noting that if this mechanism really does operate in protection, then it might conceivably also operate in susceptibility. At present, this would seem a thoroughly heretical hypothesis. To the best of our knowledge, disease susceptibility determined by the MHC is exclusively interpreted in terms of the ability of certain MHC molecules to present chain disease-inducing peptides. We are happy to stick our necks out; at least to the extent of insisting that prior bias not be ruled out without further evidence.

In order to avoid going into too much detail, we have attempted to summarize the relevant information in the form of the four illustrations (Figs. 1–4) all arranged in the same way. The immune system is depicted as pieces moved around on a chess

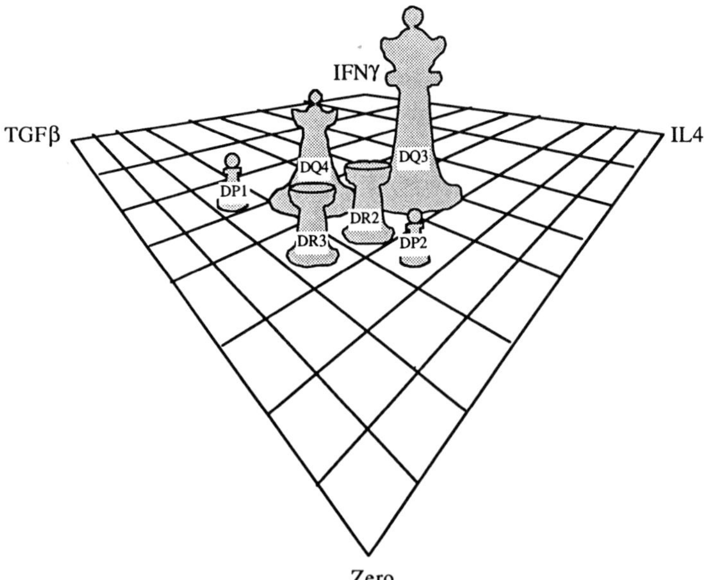

Figure 1 The effect of MHC alleles on the cytokine profile of the immune system depicted as a game of chess. In this representation, the immune system can be pushed in the four directions shown toward synthesis of the three inhibitory cytokines TGF-β, interferon gamma (IFN-γ), and IL-4 and toward zero synthesis. These three are believed to be the chief controlling molecules, although in principle, immuno-chess should be played in *n* dimensions, where *n* is the number of cytokines. The pieces depicted as controlling the location of the immune system on the board (although they are of course not the only factors to do so) are class II MHC molecules shown in approximate order of importance in controlling inhibitory effects and prevention of immunological disease (DQ > DR > DP).

board defined by the four corners TH_1, TH_2, Tm (mucosal T cells), and zero activities. The location within this area of the various diseases and immune responses involves much guesswork, although there are genuine grounds for regarding rheumatoid arthritis as predominantly a TH_1-mediated disease and systemic lupus erythematosus (SLE) as predominantly TH_2 mediated. The MHC is depicted as pushing T cells toward either susceptibility (TH_1 or TH_2 activity) or protection (zero activity). These figures are the best that we can do to illustrate our thesis, and we can only apologize for all the guesswork. Note also that we have already admitted that the TH_1 and TH_2 duopoly is probably a gross oversimplification (for further discussion of this point, see Ref. 21).

Documentation of these illustrations is as follows. Figure 2: collagen-induced arthritis (22), plus our own unpublished data showing suppression of this disease

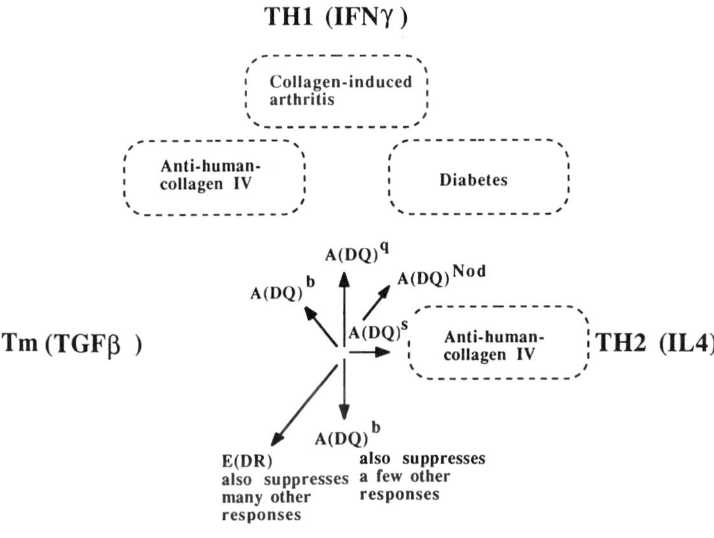

Figure 2 Protective effects and susceptibility to disease mediated by class II MHC genes in the mouse. The immune system is depicted as shown in Figure 1, with the same four corners but also with what are believed to be the main types of cytokine-secreting T cells, Tm (mucosa-associated T cell), TH_1, and TH_2. The figure also shows the location on the board of various model diseases and immune responses. It shows in addition the MHC associations, with susceptibility depicted by arrows pointing toward the TH_1 and TH_2 corners and protection in the opposite direction.

by H-2A^b; anti-human–collagen IV (23); suppression by H-2E (7); suppression by H-2A^b (8,9). Figure 3: rheumatoid arthritis (10); Takayasu arteritis (24); microscopic polyarteritis (25). For discussion of oral tolerance mediated by transforming growth factor beta (TGF-β)–secreting mucosal T cells, see Refs. 20 and 21. Figure 4; (13,14); for possible suppression by TGF-β (another guess!) (27).

Taken together, Figures 2 and 3 confirm that the presence of an appropriate MHC class II molecule can profoundly influence disease susceptibility either by increasing or decreasing it. There is little doubt that these effects apply not only to disease incidence but also to severity of the disease course. That is strikingly confirmed in our work on the impact of the H-2A^b gene on collagen-induced arthritis mentioned above. Furthermore, we have omitted reference to the best-studied instance of protection—that mediated by DR2/DQw1.2 in insulin-dependent diabetes (12).

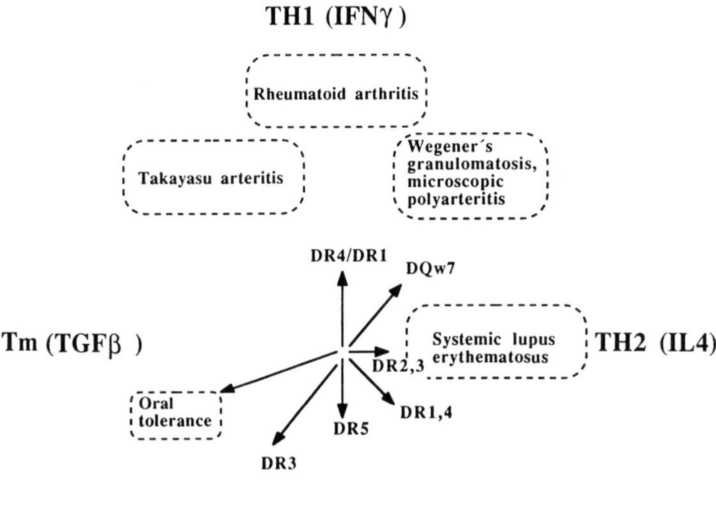

Figure 3 Protective effects and susceptibility to disease mediated by class II MHC genes in inflammatory rheumatic diseases in humans. This illustration has the same corners as before and the diseases and MHC associations are plotted in the same way.

All this is well known, although it is satisfying to note how the evidence for protection continues to grow. The connection with the cytokine profile is much weaker, and as we keep emphasizing involves much guesswork. The evidence in favor of such a connection can be summarized as follows: (1) the lack of any other convincing explanation of the protective effect (6,8,10–12); (2) the evidence from animal models that inducing secretion of at least one inhibitory cytokine (TGF-β) can protect against autoimmune diseases (20); (3) the efficacy of therapeutic intervention with cytokines in leprosy and with an anticytokine in rheumatoid arthritis (cited above); (4) the effect on cytokine profile of the H-2k–H-2d substitution in congenic mice (28,29); and (5) the evidence of cytokine influence in animal models of infection, plus the suggestion that the same influences apply in HIV infection (14). To this list can now be added what is for us the most convincing argument of all; namely, the bunching of protective and immunoinhibitory effects in the mouse. That is illustrated in Figure 1 for H-2E (a gene-locus effect) and for H-2A^b (an allele effect, and therefore more comparable to the HLA effects). Concerning the independence of three independent immunoinhibitory effects (responses to F liver protein alloantigen, to a mycobacterial protein, and to TH_1 alloantigen), the first test in a mouse autoimmune disease also mapped to the same allele. That is what excites us, and that is what this chapter is all about. It must be add-

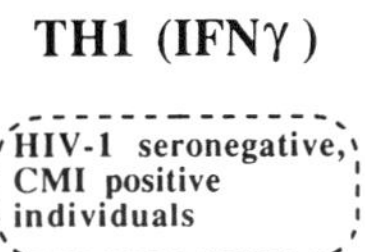

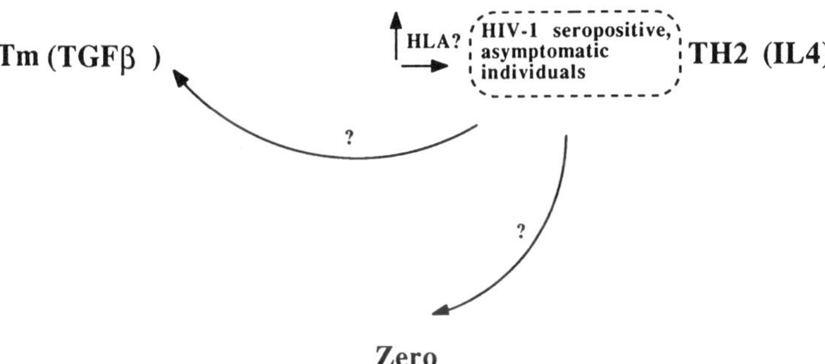

Figure 4 The immune response to HIV. Shearer's hypothesis (13,14) depicted on the same plot as in the previous figures.

ed, however, that protective effects in humans so far show no signs of this kind of bunching except in mapping to a locus, HLA-DQ.

At present, it is entirely unclear whether the immune response is pushed around on the cytokine board by the disease process itself, presumably by the disease-inducing peptides themselves, or whether the pushing occurs prior to the disease. The first possibility can be referred to as the disease- or antigen-specific model and the second as the prior-bias model. Conceivably any prior-bias might be introduced by some special structure in the MHC molecule itself (or in its promoters), but it seems more likely that it would reflect prior manipulation of the immune system by environmental antigens (or possibly, following Irun Cohen's homunculus theory, by self-antigens). In this sense, the two models converge in principle, although they make quite different predictions about the nature of disease predisposition.

IV. THE NEED FOR CYTOKINE EPIDEMIOLOGY

Let us now shift gears and consider what needs to be done in order to test the two linked hypotheses: (1) the cytokine profile influences disease susceptibility, perhaps mainly through protective effects, and (2) that this profile is influenced by HLA polymorphisms. The most obvious first step would be to compare the profile between patients and controls in the response of peripheral blood cells to nonspecific stimulation and to stimulation with disease-related antigen(s). The two types of stimulation are needed so as to distinguish between disease-specific and prior-bias effects. Parallel studies could be carried out in animal models.

Patient-control studies do not distinguish between cause and effect, although they can give valuable information for the next steps. Multiple incidence families could be useful at this stage. Sooner or later one would need to move to studies within the normal population to search for association between HLA genes and the profiles of response to both types of stimulation (again with parallel studies in the mouse, where the genetics are easier to control). Ideally such studies would eventually identify prior risk factors for disease.

We have reviewed methods appropriate for such studies (21). Quantitative down to $\pm 10\%$ of a range of say six cytokines/sample would best be achieved by molecular-biological methods, such as RNAse protection, quantitative polymerase chain reaction (PCR), or in situ hybridization. A year's experience with RNAse protection has left us still uncertain whether this potentially powerful method is workable in practice.

It must be emphasized that such studies might pick up associations of the cytokine profile with either HLA or disease or both. All three possibilities are open, and the epidemiology would need to be designed accordingly.

What else can one say about this need except that it is just the sort of enterprise that the European Economic Community should support.

REFERENCES

1. Arthur RP, Mason D. T cells that help B cell responses to soluble antigens are distinguishable from those producing interleukin 2 on mitogenic or allogeneic stimulation. J Exp Med 1986; 163:774.
2. Mosmann TR, Coffman RL. TH1 and TH2 cells: Different patterns of lymphokine secretion lead to different functional properties. Annu Rev Immunol 1989; 7:145.
3. Mitchison NA, Oliveira DBG. Epirestriction and a specialised subset of T helper cells are key factors in the regulation of T suppressor cells. In: Cinader B, Miller RG, eds. Progress in immunology IV. London: Academic Press, 1986:326.
4. Lightstone EB, Marvel J, Mitchison NA. Hyper-reactivity of mouse CD45RA$^-$ T-cells. Eur J Immunol 1993; 23:2383.
5. Marvel J, Mitchison NA, Oliveira DBG, O'Malley C. The split within the CD4 (helper) T-cell subset, and its implications for immunopathology. Mem Inst Oswaldo Cruz 1987; 82:S260.
6. Mitchison NA. Is genes in the mouse. In: Melchers F, et al., eds., Progress in immunology VII. Berlin: Springer-Verlag, 1989:845.
7. Oliveira DBG, Mitchison NA. Immune suppression genes. Clin Exp Immunol 1989; 75:167.

8. Mitchison NA. The evolution of acquired immunity to parasites. Parasitology 1990; 100:S27.
9. Mitchison NA, Simon K. Dominant rediced responsiveness controlled by H-2(K^b)A^b. A new pattern evoked by Thy-1 antigen and F liver antigen. Immunogenetics 1990; 32:104.
10. Mitchison NA. Genes beneficial in autoimmune disease. Br J Rheumatol 1991; 30:S38.
11. Mitchison NA. Immunoinhibitory genes. Curr Biol 1991; 1:87.
12. Mitchison NA. Specialization, tolerance, memory, competition, latency, and strife among T cells. Ann Rev Immunol 1992; 10:1.
13. Clerici M, Giordi JV, Chou CC, Gudeman VK, Zack JA, Gupta P, Nishanian PG, Berzovsky JA, Shearer GM. Cell mediated immune responses to human immunodeficiency virus (HIV) type 1 in seronegative homosexual men with recent sexual exposure to HIV-1. J Infect Dis 1992; 165:1012.
14. Clerici M, Shearer GM. A TH1 to TH2 switch is a critical step in the etiology of HIV infection. Immunol Today 1993; 14:107.
15. Sampaio EP, Moreira AL, Kaplan G, Alvim MF, Duppre NC, Miranda CF, Sarno EN. Mycobacterium leprae–induced interferon-gamma production by household contacts of leprosy patients: Association with the development of active disease. J Infect Dis 1991; 164:990.
16. Kaplan G, Britton WJ, Hancock GE, Theuvenet WJ, Smith KA, Job CK, Roche PW, Molloy A, Burkhardt RA, Barker J, et al. The systemic influence of recombinant interleukin 2 on the manifestations of lepromatous leprosy. J Exp Med 1991; 173:993.
17. Kaplan G, Walsh G, Guido LS, Meyn P, Burkhardt RA, Abalos RM, Barker J, Findt PA, Fajardo TT, Celona R, et al. Novel responses of human skin to intradermal recombinant granulocyte/macrophage-colony-stimulating factor: Langerhans cell recruitment, keratinocyte growth, and enhanced wound healing. J Exp Med 1992; 175:1717.
18. Sampaio EP, Moreira AL, Sarno EN, Maltra AM, Kaplan G. Prolonged treatment with recombinant interferon gamma induces etythema nodosum leprosum in lepromatous leprosy patients. J Exp Med 1992; 175:1729.
19. Elliott MJ, Maini RN, Feldmann M, Charles P. Treatment of rheumatoid arthritis with climaeric monoclonal antibodies to TNF-α. Safety, clinical efficacy and control of the acute-phase response. Clin Rheumatol 1993; 12:34.
20. Weiner HL, Miller A, Khoury SJ, Zhang ZJ, Al-Sabbagh A, Brod SA, Lider O, Higgins P, Sobel R, Matsui M, Sayegh M, Carpenter C, Eisenbarth G, Nussenblatt RB, Hafler DA. Suppression of organ-specific autoimmune diseases by oral administration of autoantigens. In: Gergely J, ed., Progress in immunology VIII. Berlin: Springer-Verlag, 1989:627.
21. Brunner MC, Caput D, Helbert MR, Mitchison NA, Simon K, Sieper JS, Wu P. Regulation of regulatory T cells. In: Gergely J, ed., Progress in immunology VIII. Berlin: Springer-Verlag, 1989:613.
22. Holmdahl RM, Andersson M, Goldschmidt TJ, Gustafsson K, Jansson L, Mo JA. Type II collagen autoimmunity in animals and provocations leading to arthritis. Immunol Rev 1990; 118:193.
23. Pfeiffer C, Murray J, Madri J, Bottomly K. Selective activation of TH1 and TH2-like cells in vivo. Response to human collagen IV. Immunol Rev 1991; 123:65.

24. Dong RP, Kimura A, Numano F, Yajima N, Hashomoto Y, Kishi Y, Nishimura Y, Sasazuki T. HLA-DP antigen and Takayasu arteritis. Tissue Antigens 1992; 39:106.
25. Spencer AJ, Burns A, Gaskin G, Pusey CD, Rees AJ. HLA class II specificities in vasculitis with antibodies to neutrophil cytoplasmic antigens. Kidney Int 1992; 41:1059.
26. Pappa H, Hadjiyannakos D, Siakotos M, Tarassi K, Nikolopoulo N, Michael S, Kaminis C, Vosnides G, Billis A, Papasteriades C. Antigens and microscopic polyarteritis (MP) with renal involvement. A protective role of HLA-DR3. VIIIth Congress Immunol W-53 (27) 1992.
27. Clerici M, Roilides E, Via CS, Pizzo PA, Shearer GM. A factor from CD8 cells of human immunodeficiency virus-infected patients suppresses HLA self-restricted T helper cell responses. Proc Natl Acad Sci USA 1992; 89:8424.
28. Asherson GL, Dieli F, Gautman Y, Siew LK, Zembala M. Major histocompatibility complex regulation of the class of the immune response. The H-2d haplotype determines poor interferon-gamma response to several antigens. Eur J Immunol 1990; 20:1305.
29. Dieli F, Sireci G, Lio D, Bonano CT, Salerno A. Interleukin 4 is a critical cytokine in delayed Typ hypersensitivity. Eur J Immunol 1993; submitted.

4

Cytokines That Regulate In Vivo Nitric Oxide Production and BCG-Induced Nonspecific Protection

Carol A. Nacy and Shawn J. Green
EntreMed, Incorporated, Rockville, Maryland

Anne H. Fortier
Walter Reed Army Institute of Research, Rockville, Maryland

I. BACTERIAL INTERACTIONS WITH MACROPHAGES

Francisella tularensis is a tiny pleomorphic gram-negative bacterium that can infect mammals through dermal or respiratory routes. It is the etiologic agent of tularemia, or trapper's fever, and the usual method of infection is through inhalation following creation of an aerosol or intradermal inoculation following accidental laceration during the skinning of infected game.

A. Cells Infected by *Francisella* In Vivo

Although *Francisella* can be cultured on synthetic medium without cells, hence its designation as a facultative intracellular pathogen, there are no data to support its extracellular growth in vivo in experimental animals or humans (1,2). Culture or polymerase chain reaction (PCR) analysis of blood collected from *F. tularensis*-infected mice suggests that bacteremia is not a hallmark of disease with this pathogen, and organisms are found only within circulating white blood cells (3). In addition, electron microscopy studies of peritoneal cells from infected mice show *Francisella* only within macrophages. In vivo, then, *Francisella* is an intracellular pathogen, and its principal host cell for intracellular replication is the macrophage (1–4).

B. Macrophages and *Francisella* Interactions In Vitro

Intracellular pathogens develop interesting and diverse strategies for survival in macrophages. Like many obligate intracellular microorganisms, the live vaccine strain (LVS) of *Francisella* does not initiate a respiratory burst during its ingestion by macrophages; like some obligate intracellular pathogens, *Francisella* entry into macrophages is not affected by agents that disrupt microfilaments. However it enters these cells, LVS clearly replicates within a vacuole in macrophages infected in vitro and in vivo. Many fewer LVS-containing vacuoles fuse with prelabeled lysosomes (30%) than *Escherichia coli*-containing vacuoles (70%) (5), although LVS does require an acidic environment for intracellular growth (6). The requirement for low pH may reflect LVS's strict iron requirement. Infective *Francisella* organisms, then, localize in an acidified vesicle, but not necessarily a phagolysosome, within macrophages. Whether they can, like *Mycobacterium* and *Toxoplasma,* actively prevent lysosomal fusion is not yet clear. *Francisella* replicates in both resident and inflammatory macrophages in vitro: bacterial numbers increase 4 to 5 logs in 72 h (2).

Macrophages exposed in vitro to cytokines develop the capacity to kill intracellular *Francisella,* and the predominant cytokine responsible for this antimicrobial activity is interferon gamma (IFN-γ) (2,7). The sensitivity of inflammatory or differentiated macrophages to activation factors differ: as little as 5 U/ml IFN-γ activates inflammatory macrophages, whereas 20 U/ml induces an equivalent amount of intracellular killing in differentiated tissue cells. IFN-γ alone is insufficient to trigger the killing mechanism, however. Tumor necrosis factor alpha (TNF-α) is produced by macrophages infected with LVS, and this TNF-α participates as a necessary second signal for expression of antimicrobial activity (2). Cytokine-activated macrophages have many potential effector activities not present in unstimulated macrophages. Recent studies with inhibitors of nitric oxide (NO) synthesis suggest that this molecule is involved in murine macrophage destruction of *Francisella.* The short-lived reactive nitrogen intermediate is rapidly oxidized to NO_2^-, which can be measured by a simple colorimetric test. NO_2^- in culture fluids, then, is a quantitative index of macrophage activation and *Francisella* killing capacity. Production of NO_2^- correlates with anti-LVS activity in murine macrophages activated in vitro with IFN-γ (2,7,8) or in vivo following infection of mice with *Mycobacterium bovis,* baccilus Calmette-Guerin (BCG) strain (9). A competitive inhibitor of L-arginine–derived NO, N^G-MMLA, blocks IFN-γ–induced anti-LVS activity and production of NO_2^- in all murine macrophage populations.

II. EXPERIMENTAL MODELS OF INFECTION WITH *FRANCISELLA*

The LVS of *F. tularensis* is pathogenic for several different strains of mice and produces infection in these experimental animals similar phenotypically (pathology, symptoms, course of infection) to that seen in humans infected with the wild-type organisms (10). Disease in both human and mouse depends on route of inoculation and ranges from ulceroglandular infection (entry through the skin), to respiratory tularemia (inhalation), to a typhoidal disease, with pneumonia as a common sequela. Peritoneal cells harvested from infected mice harbor up to 10^9 *Francisella* organisms (10), and virtually all of the bacteria are within macrophages. PCR determination of *Francisella* in blood from infected animals confirms that bacteria are not free in the circulation but are found within cells (3).

Bacteria are isolated from all reticuloendothelial organs of experimental animals with 3 days of *Francisella* introduction by any route. Routes of infection that lead to fatal disease in mice result in overwhelming pneumonitis by day 5 and death of the animal shortly thereafter. Intradermal routes of infection that are not lethal show dissemination of organisms to the lung (10,11), but the time of appearance in this tissue is delayed compared with other routes. Complete clearance of *Francisella* from all tissues of mice recovering from intradermal infection occurs within 3 weeks (12,13).

The LD_{50} in mice for the different sites of infection with *F. tularensis* reflect a route-dependent natural resistance: the LD_{50} for intraperitoneal inoculation of LVS is <10 organisms, for intravenous and intranasal inoculation is <100 organisms, but for intradermal inoculation is $>500{,}000$ organisms. The LD_{50} changes from <10 organisms intraperitoneally in C3H/HeN mice to $\gg 10{,}000$ organisms intraperitoneally following recovery from inoculation of LVS in the skin. Both cells and serum passively transfer this immunity to naive recipients (4).

III. BCG PROTECTION AGAINST *FRANCISELLA* INFECTIONS

We explored the nonspecific effects of activated macrophages and NO in vivo using BCG (9). Mice inoculated with BCG and then infected with *Francisella* intraperitoneally 8 days later survive this normally lethal interaction: the *Francisella* LD_{50} changes from <10 organisms to well over a million bacteria. Survival correlates with the secretion of urinary nitric oxides, and macrophages recovered from BCG-treated animals produce NO and actively kill intracellular *Francisella* in vitro (9).

A. Role of Nitric Oxide in Protection

To determine a cause and effect relationship between secretion of urinary nitric oxides and protection from *F. tularensis* infections, we treated mice with the specific inhibitor of nitrogen oxidation of L-arginine, N^GMMLA (9). Mice infected with BCG and exposed to N^GMMLA die of *Francisella* infection and have urinary nitrate levels equivalent to those of control animals who are not injected with BCG. Thus, the in vivo production of NO during BCG infections correlates with BCG-induced nonspecific protection against *Francisella.*

B. Cytokine Regulation of Nitric Oxide Production In Vivo

IFN-γ and TNF-α influence a number of cellular functions in polymorphonuclear neutrophils (PMNs) and macrophages in vitro, but one in which they clearly cooperate is the induction of NO synthase for production of the effector molecule used by activated macrophages to limit intracellular replication of *Francisella.* Treatment of mice with anti–IFN-γ or anti–TFN-α antibodies at the time of BCG inoculation (day 0) blocks the induction of immunity to the mycobacteria, and thus blocks the production of NO required in the intracellular destruction of *Francisella.* The LD_{50} of mice infected with BCG, treated with anti–IFN-γ or anti-TFN monoclonal antibody (MAb), and then exposed to *F. tularensis* is not different than the LD_{50} of mice exposed to LVS alone (<5 bacteria). In contrast, mice infected with BCG and treated with anti–IL-4 MAb survive challenges of up to 10^6 *Francisella* organisms. Treatment of mice with anti–IFN-γ or anti–TNF-α antibodies at the time of *Francisella* infection (day 8) also totally abrogates BCG-induced protection. Mice treated at day 8 after inoculation of BCG develop immunity to the mycobacteria and have high levels of urinary NO; thus, the anticytokine MAb treatment at this time blocks the continuous induction of an effective effector molecule. Macrophages recovered from BCG-infected mice treated with the anticytokine antibodies are not activated to kill *Francisella* in vitro. Thus, the nonspecific protection afforded by BCG, like that of antigen-specific protection, is dependent on IFN-γ and TNF-α. Whether the protective mechanisms induced by specific and nonspecific immunologic events are similar (i.e., both induce activated macrophages that are responsible for elimination of the pathogen) is currently being analyzed.

IV. CONCLUSIONS

IFN-γ and TNF-α are involved in the immune response to mycobacterial infection, and neutralization of these cytokines in vivo results in a population of peritoneal macrophages that is incapable of resolving infection with either *Mycobacterium* or *Francisella.* The immunologic stimulus for induction of cytotoxic levels of NO in vitro by murine macrophages is the synergistic activity between IFN-γ and microbes or microbial products (which stimulate endogenous release of TNF-α by macrophages) or exogenous TNF itself. IFN-γ activates the NO synthase gene and upregulates production of TNF-α stimulated by infectious agents. TNF then acts in an autocrine fashion to amplify the actual synthesis and release of NO by IFN-γ—primed cells. The data with anti–IFN-γ or anti–TNF-α MAb in BCG-infected mice suggest that a similar synergism exists in vivo: either anticytokine MAb introduced by itself reduces the level of NO_3^- excretion by BCG-infected mice and also abolishes protection afforded by BCG.

Injection of anti-TFN MAb interferes with granuloma formation and macrophage antimycobacterial activity. TFN released by macrophages in a developing granuloma acts in a paracrine fashion to promote TNF production, which ultimately influences macrophage antimicrobial activity. It takes several days for antigen-specific α/β T-cells to proliferate and expand for effective protection, so early sources of IFN-γ are likely not to be T cells. TNF production by macrophages infected with *Listeria monocytogenes* is required for IFN-γ production by NK cells obtained from spleens of SCID mice. Exposure to BCG stimulates the early appearance of NK cells, and it is possible that NK cells are a source of IFN-γ that contributes to protection against tularemia during the first few days after BCG infection. TNF, then, can operate at multiple levels in vivo: as a trigger for IFN-γ production by NK cells; as an accessory amplification factor to trigger cytotoxic levels of NO and efficient effector activity; and as a regulatory cytokine for organization of granulomas to isolate infective foci. The cell source(s) and different mechanism(s) of production of IFN-γ and TNF during the initial phase of infection are not yet elucidated, but the central importance of these two cytokines for induction of nonspecific protection and NO synthase activity is clear.

REFERENCES

1. White JD, Rooney JR, Prickett PA, Derrenbacher EB, Beard CW, Griffith WR. Pathogenesis of experimental respiratory tularemia in monkeys. J Infect Dis 1964; 114:277–83.
2. Fortier AH, Polsinelli T, Green SJ, Nacy CA. Activation of macrophages for destruction of *Francisella tularensis*: Identification of cytokines, effector cells, and effector molecules. Infect Immun 1992; 60:817–25.
3. Long GW, Oprandy JJ, Narayanan RB, Fortier AH, Nacy CA. Detection of *Francisella tularensis* in blood by polymerase chain reaction. J Clin Microb 1993; 31:152–4.
4. Fortier AH, Elkins KL, Leiby DL, Green SJ, Crawford RM, Getachew E, Nacy CA. Life and Death of an Intracellular Pathogen: Macrophage interactions with the facultatively intracellular bacterium, *Francisella tularensis*. In: Zwilling B, Eisenstein TK, eds. Macrophages and Intracellular Pathogens. New York: Marcel Dekker, 1992; 349–361.
5. Anthony LSD, Burke RD, Nano FE. Growth of *Francisella* spp. in rodent macrophages. Infect Immun 1991; 59:3291–6.
6. Fortier AH, Leiby DA, Narayanan RB, Toro LA, Asafoadjei E, Crawford RM, Nacy CA, Meltzer MS. Growth of *Francisella tularensis* LVS in macrophages: the acidic intracellular compartment provides essential iron required for growth (submitted).
7. Anthony LSD, Morrissey PJ, Nano FE. Growth inhibition of *Francisella tularensis*

live vaccine strain by IFN-γ-activated macrophages is mediated by reactive nitrogen intermediates derived from L-arginine metabolism. J Immunol 1992; 1829–34.

8. Green SJ, Nacy CA, Meltzer MS. Cytokine-induced synthesis of nitrogen oxides in macrophages: A protective host response to Leishmania and other intracellular pathogens. J Leukoc Biol 1991; 50:93–103.
9. Green SJ, Nacy CA, Granger DL, Schreiber RD, Crawford RM, Meltzer MA, Fortier AH. Administration of monoclonal antibodies to IFN-γ and TNF-α inhibits both production of reactive nitrogen intermediates and protection against *Francisella tularensis* infections in BCG-treated mice. Infect Immun 1993; 61:689–98.
10. Fortier AH, Slayter MV, Ziemba R, Meltzer MS, Nacy CA. Life vaccine strain of *Francisella tularensis*: infection and immunity in mice. Infect Immun 1991; 59:2922–8.
11. Elkins KL, Winegar RK, Nacy CA, Fortier AH. Introduction of *Francisella tularensis* at skin sites induces resistance to infection and generation of protective immunity. Microb Pathog 1992; 13:417–421.
12. Elkins KL, Leiby DA, Winegar RK, Nacy CA, Fortier AH. Rapid generation of specific protective immunity to *Francisella tularensis*. Infect Immun 1992; 60:4571–7.
13. Leiby DA, Fortier AH, Crawford RM, Schreiber RD, Nacy CA. In vivo modulation of the murine immune response to *Francisella tularensis* (LVS) by administration of anti-cytokine antibodies. Infect Immun 1992; 60:84–9.

5

Regulation of Inducible Nitric Oxide Synthase in Macrophages by Cytokines and Microbial Products

Christian Bogdan and Martin Röllinghoff
Universität Erlangen-Nürnberg, Erlangen, Germany

Yoram Vodovotz, Qiao-wen Xie, and Carl Nathan
Cornell University Medical College, New York, New York

I. INTRODUCTION

Cytokines have gained the interest of oncologists and clinical microbiologists, as they appear to be a promising tool for the treatment of malignancies and infectious diseases (1–3). Although these soluble "messengers" are now known to influence the recruitment, interaction, and function of a great variety of cells, there is evidence that effects of cytokines on macrophages are particularly important for their therapeutic potential (4). Cytokine-activated macrophages are endowed with the ability to present antigens and stimulate T lymphocytes as well as to kill tumor cells and microbes (Fig. 1). Several metabolic pathways are enhanced in macrophages after exposure to cytokines, which are particularly important for their antimicrobial activity: (1) the NADPH oxidase, which generates O_2^- and, via the superoxide dismutase and Haber-Weiss reaction, also H_2O_2 and OH· radicals (5); (2) the cytokine-inducible form of nitric oxide synthase (iNOS), which leads to the production of nitric oxide (NO·) (6); and (3) the indoleamine 2,3-dioxygenase, which catabolizes indol derivatives and causes tryptophan depletion (7). While the upregulation of these pathways contributes to the phenotype of the activated macrophage and forms part of the armamentarium of the host organism against tumor cells and microorganisms, the opposite process—termed macrophage deactivation—can be perceived as a regulatory means of our immune system to prevent excessive production of these potentially damaging toxins. On the other hand, suppression of macrophage functions may also result from products of tumor

Macrophage activation

- cytokine production (e.g.,IL-1, IL-6, TNFα)
- production of inorganic toxins (e.g.,NO)
- antigen processing and presentation
- T cell stimulation
- killing of pathogens (bacteria, protozoa, metazoa) and tumor cells

Macrophage deactivation

- regulation of host immune response
- evasion mechanism for parasites/tumors

Figure 1 Modulation of macrophage function.

cells and infectious agents and then serves them as a strategy of survival in the host environment (Fig. 1). In this chapter, we exemplify the concept of macrophage activation and deactivation by focusing on the iNOS pathway in murine macrophages and summarize the current knowledge of its regulation.

II. EXPRESSION AND REGULATION OF iNOS IN VITRO

iNOS is the predominant isoform of nitric oxide synthase in macrophages but also is found in many other cell types, including endothelial cells, hepatocytes, mesangial cells, cardiac myocytes, and keratinocytes. iNOS, like all other forms of nitric oxide synthase, generates NO and citrulline from L-arginine, NADPH, and molecular oxygen in a multistep redox reaction, which has been partially unraveled through studies with the murine macrophage cell line RAW 264.7. NO is a short-lived radical known to react with itself, oxygen, and water to yield $NO_2\cdot$, NO_2^-, and NO_3^-. The activity of iNOS is dependent on a number of cofactors, which include flavinnucleotides (FAD, FMN), heme (8,9), and tetrahydrobiopterin (6,10) (Fig. 2).

iNOS in macrophages differs in several respects from the constitutively expressed isoforms of nitric oxide synthase (cNOS), which, for example, are found in neurons and endothelium (6) (Table 1). First, iNOS activity is not present in resting cells but requires transcriptional induction of the respective gene. Second, once expressed, iNOS is not regulated by Ca^{2+} elevations and subsequent binding of *exogenous* calmodulin. In macrophages, *endogenous* calmodulin is a tightly bound subunit of iNOS, which might explain why the activity of the enzyme is independent of Ca^{2+} fluxes (11). Third, the output of NO by iNOS is characterized by a delayed onset but is large and sustained compared with the amounts of NO released by cNOS. Fourth, although cNOS, at least in the brain, is a strictly cytosolic enzyme, iNOS in macrophages is localized both in the soluble and membrane frac-

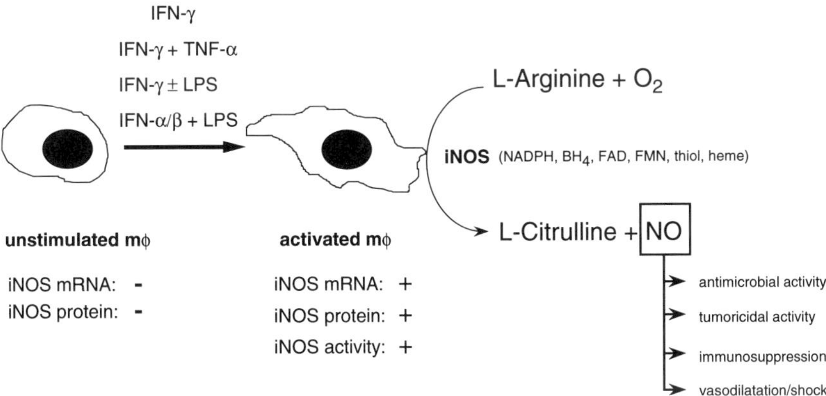

Figure 2 Expression of iNOS in macrophages. iNOS expression in macrophages is a consequence of stimulation with cytokines and/or microbial products (e.g., LPS). The enzymatic activity of iNOS is dependent on several cofactors including tetrahydrobiopterin (BH_4), flavinnucleotides (FMN, FAD), the reduced form of nicotinamide-adenine-dinucleotide-phosphate (NADPH), thiol-groups and ferroprotoporphyrin IX (heme). Some of the functions of NO produced by iNOS are indicated.

Table 1 Comparison of Constitutive and Inducible Nitric Oxide Synthases

Characteristics	cNOS	iNOS
Cellular sources		
Prototypic	Cerebellar cells	Macrophages
Other examples	Endothelial cells	Hepatocytes
	Astrocytes	Tumor cells
	Platelets	Endothelial cells
	Renal macula densa cells	Astrocytes
		Mesangial cells
Mechanism of activation	$\uparrow Ca^{2+} \rightarrow$ binding of calmodulin	1. Transcriptional induction 2. mRNA stabilization/ enhanced translation
Dependent on		
Calcium[a]	Yes	No
Calmodulin[b]	Yes	(No)
THB	Yes	Yes
FAD, FMN	Yes	Yes

[a] At concentrations above those in resting cells.
[b] Added exogenously to purified enzyme.

tion of activated cells. However, the ratio of activities in these compartments differs with the macrophage population (12–17). In the following paragraphs, we discuss the most striking feature of iNOS—its regulation by cytokines.

A. Induction of Macrophage iNOS

High-output generation of NO by macrophages was first observed after stimulation of inflammatory murine peritoneal or splenic macrophages with lipopolysaccharide (LPS) or interferon gamma (IFN-γ) or both (18,19). Later, the macrophage cell line RAW 264.7 also was shown to produce large amounts of NO after treatment with LPS or IFN-γ plus LPS but not with IFN-γ alone (14,20–22). In a detailed study with different populations of primary peritoneal macrophages, IFN-γ was the only cytokine of 11 others tested (IFN-α, IFN-β, tumor necrosis factor alpha [TNF-α], TNF-β, granulocyte-macrophage colony-stimulating factor [GM-CSF], macrophage colony-stimulating factor [M-CSF], interleukin-1β, IL-2, IL-3, IL-4, transforming growth factor beta [TGF-β]) that induced NO synthesis on its own (23). More recently, COS-1 cell-expressed human migration inhibitory factor (MIF) was reported to trigger NO production by inflammatory peritoneal macrophages or J774 cells in the absence of LPS as a cofactor (24), but later the purity of the material used was questioned (24a) and the result could not be reproduced by others with different preparations of MIF (25) (C. Bogdan and S. Stenger, unpublished observations). The list of cytokines that do not seem to affect the iNOS pathway also includes IL-6 (26) and IL-12 (C. Bogdan, et al., unpublished data).

Several cytokines and microbes/microbial products have been described to synergize with IFN-γ for the induction of iNOS in murine macrophages, including LPS (19,23,26,28), muramyl dipeptide (MDP) (26), *Staphylococcus aureus* and staphylococcal toxins (29–31), *Leishmania major* (32–34), TNF-α or TNF-β (23,26), IL-2 (35,36), MIF (24), and IL-7 (37) (summarized in Table 2). In several of these cases the observed synergism with IFN-γ was mediated by the induction of endogenous TNF-α (26,32–35,37). Besides IFN-γ, only IFN-α and IFN-β synergized with LPS to increase substantially the production of NO (22,23,38,38a).

As illustrated in Table 3, the stimulation requirements varied considerably depending on the source of the macrophage population. There are also striking differences in the production of NO between macrophages from various strains of mice (43,44).

Purification (14,45,46) and cloning of iNOS from RAW 264.7 cells (21,47,48) and the generation of anti-iNOS antibodies (21) allowed the detailed analysis of the expression of iNOS in macrophages. In resting primary murine macrophages and unstimulated RAW 264.7 cells, no iNOS mRNA, protein, and enzyme activity are found. After exposure to IFN-γ, IFN-γ/LPS, IFN-γ/TNF-α, or IFN-γ/IL-2, iNOS mRNA and/or protein are readily detectable (16,21,22,35,36,49). In the case of RAW 264.7 cells, increased iNOS gene transcription was measured in response to

Table 2 Activators of iNOS in Macrophages

Substance	Macrophage population[a]	Reference
Cytokines		
IFN-γ	RP, PE	19,23,34
MIF	PE, J774	24
IFN-γ + TNF-α (or TNF-β)	RP, PE, BM	23,26,33
IFN-γ + IL-2	PE, ANA-1	35,36
IFN-γ + MIF	PE	24
IFN-γ + IL-7	PE	37
IFN-γ + TNF-α + IL-10	PE, BM	
Microbes/microbial products		
LPS	PE, BM, RAW 264.7	18–20,22,28,40
Lipid A	BM	28
Lipoteichoic acids	BM	41
Lipopeptide (lipoprotein-derivative)	BM	40
TSST-1	J774.2	30
Cytokines + microbes/microbial products		
IFN-γ + LPS	RP, PE, BM, RAW 264.7	14,19–23,34
IFN-α + LPS	RP, PE, RAW 264.7	22,23,38,38a
IFN-β + LPS		
IFN-β + *S. aureus*	PE	38a
IFN-γ + MDP	PE	26
IFN-γ + *L. major*	RP, PE, BM	32–34
IFN-γ + *S. aureus*	PE	31
IFN-γ + SEB	PE	29,31
IFN-γ + TSST-1		
IFN-γ + LTA		
Metabolites		
picolinic acid	ANA-1	42

[a] ANA-1, J774, and RAW 264.7 are macrophage cell-lines; LTA, SEB, TSST-1 = lipoteichoic acid, enterotoxin B and toxic shock syndrome-toxin 1 (all from *S. aureus*); PE, peritoneal exudate macrophages; RP, resident peritoneal macrophages; BM, bone marrow–derived macrophages.

IFN-γ plus LPS but not to IFN-γ alone (21,22). In contrast, IFN-γ–stimulated peritoneal exudate macrophages expressed strikingly enhanced levels of iNOS mRNA without upregulation of the low-level transcription of the iNOS gene already observed in unstimulated cells (16). It is possible that IFN-γ suppresses iNOS mRNA degradation in this situation and thus causes the accumulation of stable iNOS transcripts (see Table 3).

Table 3 Differences in the Induction and Localization of iNOS in Primary Murine Macrophages and the Murine Macrophate-like Cell Line RAW 264.7

Parameter	Primary macrophage population[a] (Ref.)	RAW 264.7 cells (Ref.)
Inducibility of iNOS by		
IFN-γ	RP: yes (23) PE: yes (19,23) BM: no (34)	No (19,20)
LPS	RP: no (23) PE: yes (19) BM: yes (28,40)	Yes (20,22)
IFN-γ/LPS	Yes (all populations) (19,23,34)	Yes (14,20–22)
Mechanism of iNOS mRNA expression after cytokine stimulation	IFN-γ–stimulated PE: mRNA stabilization? (16) IFN-γ/LPS–stimulated PE: enhanced gene transcription (16)	Enhanced gene transcription after stimulation with LPS or IFN-γ/LPS (21,22)
iNOS mRNA stability	>>3 h (16,50)	~3 h (22)
Localization of iNOS activity	Cytosol ⩽ membrane (16,17)	Cytosol >> particulate (12–15)
Appearance of iNOS in Western blots	Doublet (~135/130 kd) (16)	Single band (~135 kd) (21)

[a] PE = peritoneal exudate macrophages; RP = resident peritoneal macrophages; BM = bone marrow–derived macrophages.

The inducibility of iNOS by IFN-γ, LPS, combinations of IFN-γ with LPS, IL-2, or TNF-α, or by IFN-α/β plus LPS is reflected by the presence of a multitude of consensus sequences in the 1749-bp 5′ promoter region of the iNOS gene, which are known to function as response elements for transcription factors induced by the respective cytokines or LPS. From transfection experiments with RAW 264.7 cells and iNOS promoter/CAT reporter gene constructs, it appears that the upstream portion of the promoter region is absolutely crucial for the responsiveness to IFN-γ (51).

Very little is known so far about the intracellular signaling that leads to the transcription of the iNOS gene after stimulation of macrophages with LPS/cytokines. Cycloheximide potently suppressed the expression of iNOS mRNA demonstrating that the induction of iNOS mRNA depends on de novo protein synthesis (22,51a). More recent studies also indicated that tyrosine phosphorylation is required for maximal transcription of the iNOS gene (51a). Pertussis toxin, which ADP-phosphorylates the α-subunit of certain GTP-binding proteins (52), markedly inhibited the production of NO by LPS in inflammatory murine macrophages suggesting that one or several G-protein(s) participate in the induction of iNOS (53). Finally, LPS-induced iNOS gene transcription requires the translocation of nuclear factor κB from the cystosol to the nucleus (53a).

B. Suppression of Macrophage iNOS In Vitro

So far it is not known what terminates the release of NO by macrophages after the induction of iNOS. Several possible mechanisms are suggested by experimental data, such as autotoxicity of NO for macrophages (19), feedback inhibition of iNOS by NO (54), depletion of substrate/cofactors, or accumulation of endogenously secreted inhibitory cytokines (55). The latter mechanism is particularly attractive, as activated macrophages are known producers of TGF-β (56,57) and IL-10 (58,59), both of which suppress multiple functions of macrophages (60), including the expression of iNOS. In the following we discuss the inhibition of iNOS in murine macrophages by TGF-β, IL-4, IL-10, and other factors (Table 4).

TGF-β was the first cloned cytokine described to suppress NO release. Initially this was demonstrated with naive inflammatory peritoneal macrophages stimulated with IFN-γ or IFN-γ/TFN-α (61) and later confirmed with parasite-infected primary macrophages (62–64) as well as with RAW 264.7 cells (15). More recently, we were able to demonstrate that TFG-β suppresses iNOS activity and protein in IFN-γ–activated inflammatory peritoneal macrophages by (at least) 3 different

Table 4 Inhibitors of iNOS in Macrophages

Substance	Macrophage population[a]	Reference
Cytokines		
TGF-β	RP, PE, RAW 264.7	15,61–64
MDF	PE	61
IL-4	PE, splenic macrophages	50,63,65,66
IL-10	PE, J774	63,64,67–71
IL-13	BM, PE	27,27a
Microbial products		
LPS	PE, J774	49,72–74
Pertussis toxin	PE	53
Pyocyanine	Alveolar macrophages	75
Drugs		
N^{ω}-substituted L-arginines (substrate inhibitors)	RP, PE, RAW 264.7	6,10
Cofactor antagonists	PE, RAW 264.7	6,10
Glucocorticoids	PE	31,76
Antinomycin D	PE, RAW 264.7	21
Cycloheximide	RAW 264.7	
3-Amino-1,2,4-triazole	BM	77
Dithiocarbamates	RAW 264.7, BM	53a,77a
Nifedipine	J774.2	77b

[a] RP = resident peritoneal macrophages; PE = peritoneal exudate macrophages; BM = bone marrow–derived macrophages.

posttranscriptional mechanisms; that is, destabilization of iNOS mRNA, enhanced degradation of iNOS protein, and reduced translation of iNOS mRNA. Most importantly, TGF-β was able to reduce the amount of iNOS protein even after iNOS was fully expressed. No evidence was obtained that TGF-β affects the low level of iNOS gene transcription in IFN-γ–activated macrophages (16). This might explain why TGF-β consistently failed to suppress NO production elicited by IFN-γ plus LPS, which is a situation in which iNOS gene transcription is markedly enhanced (16,21,22).

Several groups have observed that IL-4 also suppresses the production of NO by primary macrophages (63,65,66). However, as shown in a more detailed analysis (50), IL-4 differs strikingly from TGF-β in its ability to modulate the expression of iNOS. First, when added simultaneously with IFN-γ (in the absence of LPS), both IL-4 and TGF-β inhibited the release of NO, but IL-4 was considerably less potent than TGF-β: The maximal suppression afforded by IL-4 was 51% and the 50% inhibitory concentration 45 pM as compared with 90% and 5 pM for TGF-β. Second, when the macrophages were pretreated with IL-4 for 6–8 h, the suppression of IFN-γ–induced NO release reached 83%; under these conditions, IL-4, but not TGF-β, also efficiently blocked the induction of NO by the combination of IFN-γ and lipopolysaccharide. Third, IL-4 strongly reduced iNOS mRNA at 24–72 h of stimulation, but, in contrast to TGF-β, did not alter its stability. The effect of IL-4 might be due to a delayed-onset interference with iNOS transcription. Finally, IL-4 is completely ineffective if added later than 1–3 h after IFN-γ. Consequently, IL-4 does not affect iNOS once it is expressed.

Supernatants from COS 7-cells transfected with the murine IL-10 gene are not only potent inhibitors of TNF-α, IL-1, and H_2O_2 release (58,59,67) but were also reported to inhibit IFN-γ–induced production of NO by primary murine macrophages (63,64,67,68,71) or IFN-γ/LPS–triggered iNOS expression in the murine macrophage cell line J774 (70). However, the studies on the modulation of iNOS by IL-10 yielded results that are far from uniform (Table 5). There is considerable variation as to the extent of suppression observed (37–90%) with COS cell–expressed IL-10. LPS-free *Escherichia coli* expressed murine IL-10 inhibited the production of TNF-α (78) and the killing of *Leishmania major* by peritoneal murine macrophages (A. Gessner, et al., submitted) but did not alter the release of NO (C. Bogdan et al., unpublished data). Recently, COS cell–expressed IL-10 was found to enhance IFN-γ/TNF-α–induced production of NO by murine bone marrow–derived or peritoneal macrophages (39). So far the reason(s) for these discrepant results are not clear. However, there is evidence that IL-10 suppresses iNOS indirectly via its strong inhibitory effect on TNF-α production (69). This suggests that the potency of IL-10 as a suppressor of iNOS could vary with the degree to which endogenous TNF-α contributes to the induction of iNOS by IFN-γ under different culture conditions.

Table 5 Regulation of iNOS in Murine Macrophages by Interleukin-10

Study	Source of IL-10	Macrophage population	Stimulus (h)	Pretreatment with IL-10	NO production
Bogdan et al. (1991) (67)	COS cell supernatant	Peritoneal exudate	IFN-γ (48)	No	Suppressed (37 ± 15%)
			IFN-γ/TNF-α (48)	No	No change
			IFN-γ/LPS (48)	No	No change
Gazzinelli et al. (1992) (64,68)	COS cell supernatant	Peritoneal exudate	IFN-γ (48)	No	Suppressed (80–90%)
			IFN-γ/LPS (48)	No	No change
Oswald et al. (1992) (63,69)	COS cell supernatant	Peritoneal exudate	IFN-γ (48)	No	Suppressed (70–90%)
			IFN-γ/TNF-α (48)	No	No change
Cunha et al. (1992) (70)	COS cell supernatant	J774 cell line	IFN-γ/LPS (24)	No	Suppressed (~45%)
				Yes (18 h)	Suppressed (~70%)
Cenci et al. (1992) (71)	COS cell supernatant	Peritoneal exudate	IFN-γ (22 h) (last 4 with *C. albicans*)	No	Suppressed (~85%)
Corradin et al. (1993) (39)	COS cell supernatant	Peritoneal exudate or bone marrow–drived	IFN-γ/TNF-α (24)	No/yes (1 h)	Enhanced (~100%)
			IFN-γ/TNF-α (48)	No/yes (1 h)	No change
			IFN-γ/LPS (24)	No	Suppressed (54%)
			IFN-γ/LPS (48)	No	No change
Bogdan et al. unpublished	*E. coli* (HPLC) purified)	Peritoneal exudate	IFN-γ (48)	No/yes	No change

In the light of the well-established role of LPS as a (co)inducer of NO production, it was a surprise to learn that LPS can also *suppress* the expression of iNOS. In one set of studies, preexposure of macrophages to LPS inhibited the release of NO after secondary stimulation with LPS or LPS/IFN-γ (72–74), which is reminiscent of classic LPS tolerance or desensitization (79). However, there is threefold evidence that the suppressive activity of LPS does not simply reflect hyporesponsiveness of the macrophages to a second LPS stimulus. First, pretreatment of macrophages with concentrations of LPS, which most potently suppressed the subsequent induction of iNOS, strongly *enhanced* the production of TNF-α elicited by the secondary LPS dose (73). Second, the downregulatory effect of LPS also is observed with secondary stimuli other than LPS; for example, *Staphylococcus aureus,* heat-killed *Listeria monocytogenes,* or IFN-γ (49,73). Third, pretreatment with LPS inhibited the expression of iNOS mRNA at late (≥18 h) but not at early time points (5 h) of stimulation with IFN-γ, which indicates that it does not lead to reduced IFN-γ responsiveness of the macrophages. It is important to note that suppression of NO production was optimal with LPS concentrations in the range of 50–200 pg/ml but was also observed with as little as 10 pg/ml (49). Comparable concentrations of LPS also counteract other macrophage functions (49) and are detected during gram-negative sepsis (80) and therefore might support the survival of the respective bacteria. Inhibition of iNOS by microbial products is likely to be a more general phenomenon, as similar observations were made with *Bordetella pertussis* toxin (53) and pyocyanine from *Pseudomonas aeruginosa* (75).

III. EXPRESSION AND REGULATION OF iNOS IN VIVO

In contrast to the regulation of iNOS in vitro, there is only little known about the expression of macrophage iNOS in vivo. Treatment of rats with LPS or infection of mice with *Mycobacterium bovis* (bacille Calmette-Guerin [BCG] strain) led to a strong increase in the urinary excretion of NO_3^-, a stable endproduct of NO synthesis (18,19,81–83). Macrophages or peritoneal cells isolated from the peritoneal cavity or the spleen of BCG-infected mice produced strikingly enhanced amounts of NO in vitro even without further stimulation with IFN-γ and/or LPS (19,83). Nitrite accumulation in those macrophages populations was associated with a reduced growth of *Francisella tularensis,* a facultative intracellular bacterium. Both the NO production and the antimicrobial activity were suppressed by N_G-monomethyl-L-arginine, a competitive substrate inhibitor of iNOS. In vivo, anti–TNF-α or anti–IFN-γ treatment reduced the urinary NO_3^- levels and abrogated the BCG-mediated protection against *F. tularensis* (83). In a different model, infection with *Leishmania major*, a protozoal parasite, caused a strikingly enhanced excretion of NO_3^- in genetically resistant, but not in susceptible mice (83a).

Although these data already indicate the expression of iNOS in macrophages in vivo, the availability of anti-iNOS antibodies (21,84) and cDNA probes (21,47,48) now allows detailed studies of the tissue distribution of iNOS during infections. Immunohistochemical analysis of rats treated with *Propionibacterium acnes* and LPS revealed the expression of iNOS protein in various organs, including liver, lung, spleen, kidney, and colon. Cells which stained positively for iNOS included macrophages, Kupffer cells, hepatocytes, endothelial cells, neutrophils, and occasionally lymphocytes (85). In a different model, which utilized rats infected with encephalitogenic viruses, iNOS mRNA was detected in the brain when the animals showed clinical signs of encephalitis. In some, but not all, cases was the induction of iNOS mRNA accompanied by the histological appearance of inflammatory infiltrates (81). Finally, in mice infected with *Leishmania major*, iNOS mRNA and protein are found in the skin lesions as well as the draining lymph nodes during the course of the disease. Our preliminary kinetic data indicate a differential upregulation of iNOS in *L. major*–resistant versus susceptible strains. This might reflect the fine balance between macrophage-activating and macrophage-deactivating cytokines that are expressed differently in these mice (S. Stenger et al., in preparation).

IV. FUTURE CHALLENGES

In a large array of in vitro studies, a firm correlation has been established between the antimicrobial activity of macrophages and their capacity to release NO. The list of microorganisms that seem to be attacked by NO now includes metazoa, protozoa, fungi, bacteria (86,87), and even viruses (88). Although a number of molecular targets of NO have been defined (6), there are so far no data that clearly illustrate the microbicidal mechanism of NO within intact infectious agents. Furthermore, NO produced by iNOS of macrophages (as well as other cells) has the potential for adverse activities depending on its concentration and site of release. These include the induction of hypotension and cardiovascular shock (as seen during sepsis) (89) and cytotoxicity toward host cells such as T lymphocytes (90–92) and macrophages themselves (93,94). That these immunosuppressive properties of NO might function during infections in vivo and can promote the survival of microbes was recently demonstrated in murine models of T-cell–dependent resistance to *L. monocytogenes* (95) or *Trypanosoma brucei* (96). Apart from these complexities of NO actions, one major challenge for the future will be the definitive demonstration of iNOS in *human* phagocytes. There is convincing evidence that iNOS is expressed in human tissues (97,98), but there is controversy about the question whether the iNOS pathway is operative in human macrophages (99–103). Clearly, further studies are necessary in order to understand to what degree (macrophage) iNOS contributes to the observed antimicrobial effect of cytokines in patients with infectious diseases (1).

ACKNOWLEDGMENTS

We are grateful to Dr. Steffen Stenger for critical reading of the manuscript. Preparation of the chapter and conduct of some of the studies reviewed were supported by the Deutsche Forschungsgemeinschaft (DFG grant Bo 996/1-1 and Bo 996/2-1) and by the National Institutes of Health (grant CA43610 to C.N. Nathan).

REFERENCES

1. Ezekowitz RAB, ed. Infectious disease management: Advances in adjuvant cytokine therapy. J Infect Dis 1991; 167(Suppl 1):S1.
2. Blankenstein T, Rowley DA, Schreiber H. Cytokines and cancer: Experimental systems. Curr Opin Immunol 1991; 3:694.
3. Oettgen HF. Cytokines in clinical cancer therapy. Curr Opin Immunol 1991; 3:699.
4. Nathan CF. Coordinate actions of growth factors in monocytes/macrophages. In: Sporn MB, Roberts AB, eds. Handbook of experimental pharmacology. Berlin/Heidelberg: Springer Verlag, 1990:427.
5. Klebanoff SJ. Phagocytic cells: products of oxygen metabolism. In: Gallin JI, Goldstein IM, Snyderman R, eds. Inflammation: Basic principles and clinical correlates. New York: Raven Press, 1992:541.
6. Nathan C. Nitric oxide as a secretory product of mammalian cells. FASEB J 1992; 6:3051.
7. Taylor MW, Feng G. Relationship between interferon-γ, indoleamine 2,3-dioxygenase, and tryptophan catabolism. FASEB J 1991; 5:2516.
8. White KA, Marletta MA. Nitric oxide synthase is a cytochrome P-450 type hemoprotein. Biochem J 1992; 31:6627.
9. Stuehr DJ, Ikeda-Saito M. Spectral characterization of brain and macrophage nitric oxide synthase. J Biol Chem 1992; 267:20547.
10. Stuehr DJ, Griffith OW. Mammalian nitric oxide synthases. Adv Enzymol Relat Areas Mol Biol 1992; 65:287.
11. Cho HJ, Xie Q-w, Calaycay J, Mumford RA, Swiderek KM, Lee TD, Nathan C. Calmodulin is a subunit of nitric oxide synthase from macrophages. J Exp Med 1992; 176:599.
12. Marletta MA, Yoon PS, Iyengar R, Leaf CD, Wishnok JS. Macrophage oxidation of L-arginine to nitrite and nitrate: Nitric oxide as an intermediate. Biochemistry 1988; 27:8706.
13. Stuehr DS, Kown NS, Gross SS, Thiel BA, Levi R, Nathan CF. Synthesis of nitrogen oxides from L-arginine by macrophage cytosol: requirement for inducible and constitutive components. Biochem Biophy Res Commun 1989; 161:420.
14. Stuehr DJ, Cho HJ, Kwon NS, Weise MF, Nathan CF. Purification and characterization of the cytokine-induced macrophage nitric oxide synthase: an FAD- and FMN-containing flavoprotein. Proc Natl Acad Sci USA 1991; 88:7773.
15. Förstermann U, Schmidt HHHW, Kohlhaas KL, Murad F. Induced RAW 264.7 macrophages express soluble and particulate nitric oxide synthase: inhibition by transforming growth factor-β. Eur J Pharmacol 1992; 225:161.

16. Vodovotz V, Bogdan C, Paik J, Xie Q-w, Nathan C. Mechanisms of suppression of macrophage nitric oxide release by transforming growth factor-β. J Exp Med 1993; 178:605.
17. Vodovotz Y. Unpublished observations, 1993.
18. Stuehr D, Marletta MA. Mammalian nitrite biosynthesis: Mouse macrophages produce nitrite and nitrate in response to *Escherichia coli* lipopolysaccharide. Proc Natl Acad Sci USA 1985; 82:7738.
19. Stuehr DJ, Marletta MA. Induction of nitrite/nitrate synthesis in murine macrophages by BCG infection, lymphokines or interferon-γ. J Immunol 1987; 139:518.
20. Stuehr DJ, Marletta MA. Synthesis of nitrite and nitrate in murine macrophage cell-lines. Cancer Res 1987; 47:5590.
21. Xie Q-w, Cho HJ, Calaycay J, Mumford RA, Swiderek KM, Lee TD, Ding A, Troso T, Nathan C. Cloning and characterization of inducible nitric oxide synthase from mouse macrophages. Science 1992; 256:225.
22. Lorsbach RB, Murphy WJ, Lowenstein CJ, Snyder SH, Russell SW. Expression of the nitric oxide synthase gene in mouse macrophages activated for tumor cell killing. Molecular basis for the synergy between interferon-γ and lipopolysaccharide. J Biol Chem 1993; 268:1908.
23. Ding AH, Nathan CF, Stuehr DJ. Release of reactive nitrogen intermediates and reactive oxygen intermediates from mouse peritoneal macrophages. Comparison of activating cytokines and evidence for independent production. J Immunol 1988; 141:2407.
24. Cunha FQ, Weiser WY, David JR, Moss DW, Moncada S, Liew FL. Recombinant migration inhibitory factor induces nitric oxide synthase in murine macrophage. J Immunol 1993; 150:1908.
24a. David JR. Erratum. Important note on recombinant MIF. J Immunol 1993; 151:5114.
25. Herriott MJ, Jiang H, Stewart CA, Fast DJ, Leu RW. Mechanistic differences between migration inhibitory factor (MIF) and IFN-γ for macrophage activation. J Immunol 1993; 150:4524.
26. Drapier J-C, Wietzerbin J, HIbbs JB, Jr. Interferon-γ and tumor necrosis factor induce the L-arginine-dependent cytotoxic effector mechanism in murine macrophages. Eur J Immunol 1988; 18:1587.
27. Doherty TM, Kastelein R, Menon S, Andrade S, Coffman RL. Modulation of murine macrophage function by IL-13. J Immunol 1993; 151:7151.
27a. Doyle AG, Herbein G, Montaner LJ, Minty AJ, Caput D, Ferrara P, Gordon S. Interleukin 13 alters the activation state of murine macrophages in vitro in an interleukin-4-like manner. Eur J Immunol 1994 (in press).
28. Keller R, Gehrl R, Keist R. The interaction of macrophages and bacteria: *Escherichia coli* species, bacterial lipopolysaccharide, and lipid A differ in their ability to induce tumoricidal activity and the secretion of reactive nitrogen intermediates in macrophages. Cell Immunol 1992; 141:47.
29. Fast DJ, Shannon BJ, Herriott MJ, Kennedy MJ, Rummage JA, Leu RW. Staphylococcal exotoxins stimulate nitric oxide-dependent murine macrophage tumoricidal activity. Infect Immun 1991; 59:2987.
30. Zembowicz A, Vane JR. Induction of nitric oxide synthase activity by toxic shock syndrome toxin 1 in a macrophage-monocyte cell line. Proc Natl Acad Sci USA 1992; 89:2051.

31. Cunha FQ, Moss DW, Leal LMCC, Moncada S, Liew FY. Induction of macrophage parasiticidal activity by *Staphylococcus aureus* and exotoxins through the nitric oxide synthesis pathway. Immunology 1993; 78:563.
32. Green SJ, Crawford RM, Hockmeyer JT, Meltzer MS, Nacy CN. *Leishmania major* amastigotes initiate the L-arginine-dependent killing mechanism in IFN-γ stimulated macrophages by induction of tumor necrosis factor-α. J Immunol 1990; 145:4290.
33. Betz-Corradin S, Buchmüller-Rouiller Y, Mauël J. Phagocytosis enhances murine macrophage activation by interferon-γ and tumor necrosis factor-α. Eur J Immunol 1991; 21:2553.
34. Betz-Corradin S, Mauël J. Phagocytosis of *Leishmania* enhances macrophage activation by IFN-γ and lipopolysaccharide. J Immunol 1991; 146:279.
35. Cox GW, Melillo G, Chattopadhyay U, Mullet D, Fertel RH, Varesio L. Tumor necrosis factor-α–dependent production of reactive nitrogen intermediates mediates IFN-γ plus IL-2–induced murine macrophage tumoricidal activity. J Immunol 1992; 149:3290.
36. Deng W, Thiel B, Tannenbaum CS, Hamilton TA, Stuehr DJ. Synergistic cooperation between T cell lymphokines for induction of the nitric oxide synthase gene in murine peritoneal macrophages. J Immunol 151:322.
37. Gessner A, Vieth M, Will A, Schröppel K, Röllinghoff M. Interleukin 7 enhances antimicrobial activity against *Leishmania major* in murine macrophages. Infect Immun 1993; 61:4008.
38. Kamuo R, Shapiro D, Le J, Huang S, Aguet M, Vilcek M. Generation of nitric oxide and induction of major histocompatibility complex class II antigen in macrophages from mice lacking the interferon γ receptors. Proc Natl Acad Sci USA 1993; 90:6626.
38a. Zhang X, Alley EW, Russell SW, Morrison DC. Necessity and sufficiency of beta interferon for nitric oxide production in mouse peritoneal macrophages. Infect Immun 1994; 62:33.
39. Betz-Corradin S, Fasel N, Buchmüller-Rouiller Y, Ransijn A, Smith J, Mauël J. Induction of macrophage nitric oxide production by interferon-γ and tumor necrosis factor-α is enhanced by interleukin-10. Eur J Immunol 1993; 23:2045.
40. Hauschildt S, Bassenge E, Bessler W, Busse R, Mülsch A. L-arginine-dependent nitric oxide formation and nitrite release on bone marrow-derived macrophages stimulated with bacterial lipopeptide and lipopolysaccharide. Immunology 1990; 70:332.
41. Keller R, Fischer W, Keist R, Bassetti S. Macrophage response to bacteria: induction of marked secretory and cellular activities by lipoteichoic acid. Infect Immun 1992; 60:3664.
42. Melillo G, Cox GW, Radzioch D, Varesio L. Picolinic acid, a catabolite of L-tryptophan, is a co-stimulus for the induction of reactive nitrogen intermediate production in murine macrophages. J Immunol 1993; 150:4031.
43. Liew FY, Li Y, Moss D, Parkinson C, Rogers MV, Moncada S. Resistance to *Leishmania major* infection correlates with the induction of nitric oxide synthase in murine macrophages. Eur J Immunol 1991; 21:3009.
44. Oswald IP, Afroun S, Bray D, Petit J-F, Lemaire G. Low response of BALB/c macrophages to priming and activating signals. J Leukoc Biol 1992; 52:315.
45. Hevel JM, White KA, Marletta MA. Purification of the inducible murine macrophage

nitric oxide synthase. Identification as a flavoprotein. J Biol Chem 1991; 266:22789.
46. Yui Y, Hattori R, Kosuga K, Eizawa H, Hiki K, Kawai C. Purification of nitric oxide synthase from rat macrophages. J Biol Chem 1991; 266:12544.
47. Lyons CR, Orloff GJ, Cunningham JM. Molecular cloning and functional expression of an inducible nitric oxide synthase from a murine macrophage cell-line. J Biol Chem 1992; 267:6370.
48. Lowenstein CJ, Glatt CS, Bredt DS, Snyder SH. Cloned and expressed macrophage nitric oxide synthase contrasts with the brain synthase. Proc Natl Acad Sci USA 1992; 89:6711.
49. Bogdan C, Vodovotz Y, Paik J, Xie Q-w, Nathan C. Traces of bacterial lipopolysaccharide suppress IFN-γ–induced nitric oxide synthase gene expression in primary mouse macrophages. J Immunol 1993; 151:301.
50. Bogdan C, Vodovotz Y, Paik J, Xie Q-w, Nathan C. Mechanism of suppression of nitric oxide synthase expression by interleukin-4 in primary mouse macrophages. J Leukoc Biol 1994; 55:227.
51. Xie Q-w, Whisnant R, Nathan C. Promoter of the mouse gene encoding calcium-independent nitric oxide synthase confers inducibility by interferon-γ and bacterial lipopolysaccharide. J Exp Med 1993; 177:1779.
51a. Xie Q-w, Nathan C. Promotor of the mouse gene encoding calcium-independent nitric oxide synthase confers inducibility by interferon-γ and bacterial lipopolysaccharide. Trans. of the Assoc. Am. Phys. (Cadmus Journal Services) 1993; 106:1.
52. Neer EJ, Clapham DE. Roles of G protein subunits in transmembrane signalling. Nature 1988; 333:129.
53. Zhang X, Morrison DC. Pertussis-toxin-sensitive factor differentially regulates lipopolysaccharide-induced tumor necrosis factor-α and nitric oxide production in mouse peritoneal macrophage. J Immunol 1993; 150:1011.
53a. Xie Q-w, Kashiwabara Y, Nathan C. Role of transcription factor NF-κB/Rel in induction of nitric oxide synthases. J Biol Chem 1994; 269:4705.
54. Assreuy J, Cunha FQ, Liew FY, Moncada S. Feedback inhibition of nitric oxide synthase by nitric oxide. Br J Pharmacol 1993; 108:833.
55. Vodovotz Y, Kwon NS, Pospischil M, Manning J, Paik J, Nathan C. Inactivation of nitric oxide synthase following prolonged incubation of mouse macrophages with IFN-gamma and bacterial lipopolysaccharide. J Immunol 1994 (in press).
56. Assoian RK, Fleurdelys BE, Stevenson HC, Miller PJ, Madtes DK, Raines EW, Ross R, Sporn MB. Expression and secretion of type β transforming growth factor by activated human macrophages. Proc Natl Acad Sci 1987; 84:6020.
57. Twardzik DR, Mikovits JA, Ranchalis JE, Purchio AF, Ellingsworth L, Ruscetti FW. γ-Interferon-induced activation of latent transforming growth factor-β by human monocytes. Ann NY Acad Sci 1990; 593:713.
58. de Waal Malefyt R, Abrams J, Bennett B, Figdor CG, de Vries JE. Interleukin 10 (IL-10) inhibits cytokine synthesis by human monocytes: an autoregulatory role of IL-10 produced by monocytes. J Exp Med 1991; 174:1209.
59. Fiorentino DF, Zlotnik A, Mosmann TR, Howard M, O'Garra A. IL-10 inhibits cytokine production by activated macrophages. J Immunol 1991; 147:3815.
60. Bogdan C, Nathan C. Modulation of macrophage function by transforming growth factor-β, interleukin 4 and interleukin 10. Ann NY Acad Sci 1993; 685:713.

61. Ding A, Nathan CF, Graycar J, Derynck R, Stuehr DJ, Srimal S. Macrophage deactivating factor and transforming growth factors-β1, -β2, and -β3 inhibit induction of macrophage nitrogen oxide synthesis by IFN-γ. J Immunol 1990; 145;940.
62. Nelson BJ, Ralph P, Green SJ, Nacy CA. Differential susceptibility of activated macrophage cytotoxic reactions to the suppressive effects of transforming growth factor-β1. J Immunol 1991; 146:1849.
63. Oswald IP, Gazzinelli RT, Sher A, James SL. IL-10 synergizes with IL-4 and transforming growth factor-β to inhibit macrophage cytotoxic activity. J Immunol 1992; 148:3578.
64. Gazzinelli RT, Oswald IP, Hieny S, James SL, Sher A. The microbicidal activity of interferon-γ–treated macrophages against *Trypanosoma cruzi* involves an L-arginine-dependent, nitrogen oxide-mediated mechanism inhibitable by interleukin 10 and transforming growth factor-β. Eur J Immunol 1992; 22:2501.
65. Liew FY, Li Y, Severn A, Millott S, Schmidt J, Salter M, Moncada S. A possible novel pathway of regulation by murine T helper type-2 (Th2) cells of a Th1 cell activity via the modulation of the induction of nitric oxide synthase in macrophages. Eur J Immunol 1991; 21:2489.
66. Al-Ramadi BK, Meissler JJ, Jr, Huang D, Eisenstein TK. Immunosuppression induced by nitric oxide and its inhibition by interleukin-4. Eur J Immunol 1992; 22:2249.
67. Bogdan C, Vodovotz Y, Nathan C. Macrophage deactivation by IL-10. J Exp Med 1991; 174:1549.
68. Gazzinelli RT, Oswald IP, James SL, Sher A. IL-10 inhibits parasite killing and nitrogen oxide production by IFN-γ activated macrophages. J Immunol 1992; 148:1792.
69. Oswald IP, Wynn TA, Sher A, James SL. Interleukin 10 inhibits macrophage microbicidal activity by blocking the endogenous production of tumor necrosis factor α required as a costimulatory factor for interferon γ-induced activation. Proc Natl Acad Sci USA 89:8676.
70. Cunha FQ, Moncada S, Liew FY. Interleukin-10 (IL-10) inhibits the induction of nitric oxide synthase by interferon-γ in murine macrophages. Biochem Biophys Res Commun 1992; 182:1155.
71. Cenci E, Romani L, Mencacci A, Spaccapelo E, Schiaffella E, Puccetti P, Bistoni F. Interleukin-4 and interleukin-10 inhibit nitric oxide-dependent macrophage killing of *Candida albicans*. Eur J Immunol 1993; 23:1034.
72. Lorsbach RB, Russell SW. A specific sequence of stimulation is required to induce synthesis of the antimicrobial molecule nitric oxide by mouse macrophages. Infect Immun 1992; 60:2133.
73. Zhang X, Morrison DC. Lipopolysaccharide-induced selective priming effects on tumor necrosis factor α and nitric oxide production in mouse peritoneal macrophages. J Exp Med 1993; 177:511.
74. Severn A, Xu D, Doyle J, Leal LMC, O'Donnell CA, Brett SJ, Mooss DW, Liew FY. Preexposure of murine macrophages to lipopolysaccharide inhibits the induction of nitric oxide synthase and reduces leishmanicidal activity. Eur J Immunol 1993; 23:1711.
75. Shellito J, Nelson S, Sorensen RU. Effect of pyocyannine, a pigment of *Pseudomonas aeruginosa,* on production of reactive nitrogen intermediates by murine alveolar macrophages. Infect Immun 1992; 60:3913.

76. DiRosa M, Radomski M, Carnuccio R, Moncada S. Glucocorticoids inhibit the induction of nitric oxide synthase in macrophages. Biochem Biophys Res Commun 1990; 172:1246.
77. Buchmüller-Rouiller Y, Scheider P, Corradin SB, Smith J, Mauël J. 3-Amino-1,2,4-triazole inhibits macrophage NO synthase. Biochem Biophys Res Commun 1992; 183:150.
77a. Mülsch A, Schray-Utz B, Mordvintcev PI, Hauschildt S, Busse R. Diethyldithiocarbamate inhibits induction of macrophage NO synthase. FEBS Lett. 1993; 321:215.
77b. Szabo C, Mitchell JA, Gross SS, Thiemermann C, Vane JR. Nifedipine inhibits the induction of nitric oxide synthase by bacterial lipopolysaccharide. J Pharmacol Exp Ther 1993; 265:674.
78. Bogdan C, Paik J, Vodovotz Y, Nathan C. Contrasting mechanisms for suppression of macrophage cytokine release by transforming growth factor-β and interleukin-10. J Biol Chem 1992; 267:23301.
79. Mathison JC, Virca GD, Wolfson E, Tobias PS, Glaser K, Ulevitch RJ. Adaptation to bacterial lipopolysaccharide controls lipopolysaccharide-induced tumor necrosis factor production in rabbit macrophages. Clin Invest 1990; 85:1108.
80. Redl H, Bahrami S, Schlag G, Traber DL. Clinical detection of LPS and animal models of endotoxemia. Immunobiol 1993; 187:330.
81. Koprowski H, Zheng YM, Heber-Katz E, Fraser N, Rorke L, Fu ZF, Hanlon C, Dietzschold B. *In vivo* expression of inducible nitric oxide synthase in experimentally induced neurologic diseases. Proc Natl Acad Sci USA 1993; 90:3024.
82. Granger D, Hibbs JB, Jr, Broadnax LM. Urinary nitrate excretion in relation to murine macrophage activation. Influence of dietary L-arginine and oral N-monomethyl-L-arginine. J Immunol 1991; 146:1294.
83. Green SJ, Nacy CA, Schreiber RD, Granger DL, Crawford RM, Meltzer MS, Fortier AH. Neutralization of gamma interferon and tumor necrosis factor alpha blocks in vivo synthesis of nitrogen oxides from L-arginine and protection against *Francisella tularensis* infection in *Mycobacterium bovix* BCG-treated mice. Infect Immun 1993; 61:689.
83a. Evans TG, Thai L, Granger DL, Hibbs JB, Jr. Effect of in vivo inhibition of nitric oxide production in murine leishmaniasis. J Immunol 1993; 151:907.
84. Ohshima H, Brouet I, Bandaletova TY, Adachi H, Oguchi S, Iida S, Kurashima Y, Morishita Y, Sugimura T, Esumi H. Polyclonal antibody against an inducible form of nitric oxide synthase purified from the liver of rats treated with *Propionibacterium acnes* and lipopolysaccharide. Biochem Biophys Res Commun 1992; 187:1291.
85. Bandaletova T, Brouet I, Bartsch H, Sugimura T, Esumi H, Ohshima H. Immunohistochemical localization of an inducible form of nitric oxide synthase in various organs of rats treated with *Propionibacterium acnes* and lipopolysaccharides. Acta Pathol Microbiol Immunol Scand 1993; 101:330.
86. Green SJ, Nacy CA. Antimicrobial and immunopathologic effects of cytokine-induced nitric oxide synthesis. Curr Opin Infect Dis 1993; 6:384.
87. Nathan C, Hibbs JB, Jr. Role of nitric oxide synthesis in macrophage antimicrobial activity. Curr Opin Immunol 1991; 3:65.
88. Karupiah G, Xie Q-w, Buller ML, Nathan C, Duarte C, MacMicking JD. Inhibition of viral replication by interferon-γ-induced nitric oxide synthase. Science 1993; 261:1445.

89. Kilbourn RG, Griffith OW. Overproduction of nitric oxide in cytokine-mediated and septic shock. J Natl Cancer Inst 1992; 84:827.
90. Hoffman RA, Langrehr JM, Billiar TR, Curran RD, Simmons RL. Alloantigen-induced activation of a rat splenocytes is regulated by the oxidative metabolism of L-arginine. J Immunol 1990; 145:2220.
91. Albina JE, Abate JA, Henry WL, Jr. Nitric oxide production is required for murine resident peritoneal macrophages to suppress mitogen-stimulated T cell proliferation. Role of IFN-γ in the induction of the nitric oxide-synthesizing pathway. J Immunol 1991; 147:144.
92. Mills CD. Molecular basis of "suppressor" macrophages. J Immunol 1991; 146:2719.
93. Albina JE, Cui S, Mateo RB, Reichner JS. Nitric oxide-mediated apoptosis in murine peritoneal macrophages. J Immunol 1993; 150:5080.
94. Albina JE, Mills CD, Henry WL, Jr, Caldwell MD. Regulation of macrophage physiology by L-arginine: Role of the oxidative L-arginine deiminase pathway. J Immunol 1989; 143:3641.
95. Gregory SH, Wing EJ, Hoffman RA, Simmons RL. Reactive nitrogen intermediates suppress the primary immunologic response to *Listeria*. J Immunol 1993; 150:2901.
96. Sternberg J, McGuigan F. Nitric oxide mediates suppression of T cell responses in murine *Trypanosoma brucei* infection. 1992; 22:2741.
97. Hibbs JB, Jr, Westenfelder C, Taintor R, Vavrin Z, Kablitz C, Baranowski RL, Ward JH, Menlove RL, McMurray MP, Kushnr JP, Samlowski WE. Evidence for cytokine-inducible nitric oxide synthesis from L-arginine in patients receiving interleukin-2 therapy. J Clin INvest 1992; 89:867.
98. Geller DA, Lowenstein CJ, Shapiro RA, Nussler AK, Di Silvio M, Wang SC, Nakayama DK, Simmons RL, Snyder SH, Billiar TR. Molecular cloning and expression of inducible nitric oxide synthase from human hepatocytes. Proc Natl Acad Sci USA 1993; 90:3491.
99. Denis M. Tumor necrosis factor and granulocyte macrophage-colony stimulating factor stimulate human macrophages to restrict growth of virulent *Mycobacterium avium* and to kill avirulent *M. avium*: Killing effector mechanism depends on the generation of reactive nitrogen intermediates. J Leukoc Biol 1991; 49:380.
100. Sherman MP, Loro ML, WOng VZ, Tashkin DP. Cytokine- and *Pneumocystis carinii*–induced L-arginine oxidation by murine and human pulmonary alveolar macrophages. J Protozool 1991: 38:234.
101. Murray WH, Teitelbaum RF. L-arginine–dependent reactive nitrogen intermediates and the antimicrobial effect of activated human mononuclear phagocytes. J Infect Dis 1992; 165:513.
102. Schneeman M, Schoedon G, Hofer S, Blau N, Guerrero L, Schaffner A. Nitric oxide synthase is not a constituent of the antimicrobial armature of human mononuclear phagocytes. J Infect Dis 1993; 167:1358.
103. Bermudez LE. Differential mechanisms of intracellular killing of *Mycobacterium avium* and *Listeria monocytogenes* by activated human and murine macrophages. The role of nitric oxide. Clin Exp Immunol 1993; 91:277.

6

ST 789 Modulates In Vitro and In Vivo Production of Interleukin-6 and Interleukin-2

Claudio De Simone and Giuseppe Famularo
University of L'Aquila, L'Aquila, Italy

Antonio Rosi, Claudio Albertoni, and Pietro Foresta
Sigma Tau, Pomezia, Italy

Sonia Tzantzoglou and Vito Trinchieri
University of Rome, Rome, Italy

I. INTRODUCTION

Hypoxanthine derivatives have well-established immunomodulant properties (1). Furthermore, the addition of arginine to these molecules strongly enhances their immunomodulant activity (1).

ST 789 is a synthetic compound that belongs to a new family of hypoxanthine derivatives exhibiting a terminal arginine at the N-9 position of the purine ring, as found in tutfsin, thymic hormones, and substance P (2). Owing to the ability of arginine to enhance the immune functions both in vitro and in vivo (3–10), the recently reported immunomodulant properties of ST 789 (11–18) are likely to be mostly dependent on arginine residual in the moiety. Notably, ST 789 has been shown to enhance the lymphocyte responsiveness to mitogens (12) as well as to stimulate the natural killer cell progenitor differentiation (13). In addition, the putative antitumor activity of ST 789 is probably mediated via the boosting of natural immunity (14). Moreover, ST 789 proved to be a strong inducer of interleukin-6 (IL-6) production by mitogen-driven peripheral blood mononuclear cells (PBMCs) (15) from humans. However, ST 789 is not able per se to induce cytokine release (15,16). Finally, the immunoenhancing activity of ST 789 has been further pointed out by the demonstration that in vivo administered ST 789 is able to restore the immune response in immunodepressed hosts and to protect against experimental infections with both bacterial and fungal pathogens (17,18).

The aim of this study was to evaluate in vitro the effect of ST 789 on IL-2 and IL-6 production in cultures of PBMCs from healthy humans as well as to assay the production of IL-2 and IL-6 in ST 789–treated cultures of murine splenocytes and peritoneal macrophages, respectively. In addition, we tested the ability of the in vivo treatment with ST 789 to affect the production of IL-2 and IL-6 by splenocytes and peritoneal macrophages, respectively, obtained from both normal and cyclophosphamide (Cy)–immunodepressed mice.

II. MATERIALS AND METHODS

A. Animals

Male C57B1/6, B6D2F1 and Balb/c mice were purchased from Iffa-Credo (Lyon, France). Both mice (10 weeks of age) were housed in groups of five and were allowed food and water ad libitum. They were kept under specific pathogen-free conditions and were randomly assigned to receive different treatment schedules.

B. Drugs and Chemicals

The ST 789 hypoxanthine derivative was synthesized by Sigma-Tau Chemical Laboratories (Pomezia, Italy). Recombinant murine IL-2 and IL-6 were both purchased from Genzyme (Boston, MA) and Rat T-STIM–conditioned medium was from Collaborative Research Inc. (Bedford, MA). Cy was purchased from Sigma (St. Louis, MO). Mitogens were phytohemoagglutinin (PHA, Flow Laboratories, Scotland), concanavalin A (ConA) (Sigma), and lipopolysaccharide (LPS) (DIFCO, Detroit, MI). All chemicals were dissolved in pyrogen-free physiologic saline solution just prior to use. Tritriated thymidine ([^{3}H]TdR), 5 Ci/mmol) was from Amersham (Aylesbury, Buckingamshire, England).

C. Cell Lines and Tissue Culture Media

The murine IL-2–dependent T-cell line (CTLL-2) was grown in RPMI 1640 medium containing 25 mM HEPES, 2 mM L-glutamine (Biochrom KG, Berlin, Germany), 5×10^{-5} M 2-mercaptoethanol (Sigma), 50 μg/ml gentamicin (Sigma), and supplemented with 10% heat-inactivated Myoclone FCS (fetal calf serum) (GIBCO, Grand Island, NY) and 5% Rat T-STIM. The murine hybridoma cell line B9, which is strictly dependent on IL-6 for growth, was maintained in the above culture medium conditioned with 10 U/ml recombinant murine IL-6 instead of Rat T-STIM.

D. Separation Techniques of PBMCs from Humans

PBMCs obtained from healthy volunteers by the standard Ficoll-Hypaque (Pharmacia Fine Chemicals, Uppsala, Sweden) density gradient technique were washed three times in Hank's Balanced Salt Solution (Flow Laboratories) and suspended at a final concentration of 1 × 10^6 cells/ml in RPMI 1640 (Flow Laboratories) containing 10% heat-inactivated fetal calf serum (FCS), 1% L-glutamine (GIBCO), 100 U/ml penicillin, 200 μg/ml streptomycin, and 100 μg/ml gentamicin.

E. Separation Techniques of Peritoneal Exudate Cells from Mice

Briefly, exudate peritoneal cells were obtained by washing the peritoneal activity of experimental animals with RPMI 1640. The harvested cells were centrifuged at 400*g* for 10 min at 4°C and density adjusted to 2 × 10^6 cells/ml.

F. PBMC Cultures

PBMCs at a density of 2 × 10^6 cells/ml were stimulated with PHA (10 μg/ml) and cultured for 48 h at 37°C in a 5% CO_2 atmosphere with or without ST 789. ST 789 was used at concentrations ranging from 1 to 100 μg/ml. At the end of the culture period, the supernatants were harvested for measuring IL-2 and IL-6. Then the supernatants were filtered (0.22-μm pore size filter) and stored at –80°C until tested.

G. Splenocyte Cultures

Briefly, in order to promote IL-2 production, splenocytes (1 × 10^7 cells/ml in a 24-well plate) were cultured in RPMI 1640 supplemented with 10% FCS and 5 × 10^{-5} M beta-mercaptoethanol) in the presence of ConA (2 μg/ml) at 37°C for 24 h. Cell-free supernatants were then collected and assayed for IL-2 activity.

H. Macrophage Cultures

Peritoneal exudate cells at a density of 1.5 × 10^6 cells/ml were seeded in a 24-well plate and incubated for 2 h at 37°C in a 5% CO_2 atmosphere to allow them to adhere. Then the nonadherent cells were removed by washing three times with warm RPMI 1640. This procedure resulted in an adherent cell population enriched in macrophages (85–95%), as determined by morphology and esterase staining. The macrophage monolayers were then incubated with either ST 789 alone or in combination with LPS (2 μg/ml) in complete medium for 24 h at 37°C in a 5% CO_2 atmosphere.

I. Bioassay for IL-2

This bioassay was used for measuring IL-2 in supernatants obtained from cultures of splenocytes from mice. Serial supernatant dilutions were added (100 μl/ml) to

CTLL-2 cells (5 × 10^3 cells/well) and incubated for 24 h at 37°C in a 5% CO_2 atmosphere. Six hours before harvesting, the cells were pulsed with [^{3}H]TdR (0.5 μCi/well). Afterwards, the cells were harvested onto glass fiber filtermats (Pharmacia, Turku, Finland) by using an automated cell harvester (Pharmacia), and [^{3}H]TdR incorporation (an index of cell proliferation) was assessed by a beta-plate counter (Pharmacia). The CTLL-2 proliferation data were converted into IL-2 units by direct comparison with the radioactivity incorporation in a standard culture with known concentrations of recombinant murine IL-2.

J. Bioassay for IL-6

This bioassay was used for measuring IL-6 in cultures of splenocytes and peritoneal macrophages from mice. Briefly, the levels of IL-6 in splenocyte or macrophage culture supernatants were measured by their ability to induce the proliferation of the murine B-cell B9 hybridoma line. B9 cells were washed twice prior to use in the assay to remove the exogenous recombinant murine IL-6 additioned to sustain their growth. Briefly, twofold serial dilutions of supernatants in RPMI 1640 supplemented with 10% FCS and 5 × 10^{-5} M 2-mercaptoethanol were added to flat-bottomed 96-well microtiter plates containing 5 × 10^3 cells per well. Then the cells were incubated for 96 h at 37°C in a 5% CO_2 atmosphere and pulsed with 0.5 μCi/well of [^{3}H]TdR during the last 18 h of incubation. Afterwards, the cells were harvested onto glass fiber filtermats (Pharmacia) by using an automated cell harvester (Pharmacia), and the [^{3}H]TdR incorporation (an index of cell proliferation) was assessed by a beta-plate counter (Pharmacia). The IL-6 activity is expressed as B9 hybridoma growth units per milliliter, and 1 U is defined as the reciprocal of the sample dilution required to produce 0.5 maximal proliferation. Values for each sample were calculated by interpolating the sample counts per minute of four to six dilutions by using linear regression analysis.

K. Enzyme-linked Immunosorbant Assays (ELISA) for IL-2 and IL-6

The levels of IL-2 and IL-6 in supernatants of PBMCs from humans were measured by using commercially available ELISA test kits (Intertest-2 and Intertest-6, Genzyme, for IL-2 and IL-6, respectively). These assays were based on the dual immunometric sandwich principle and were performed according to the manufacturer's instructions, with minor modifications, as previously reported (19,20).

L. Statistical Analysis

The statistical analysis was performed by using Student's *t*-test.

III. RESULTS

A. Effects of ST 789 on In Vitro Production of IL-6 and IL-2

In our experiments, ST 789 was able strongly to enhance the IL-6 production by mitogen-driven PBMCs from humans (Table 1). No dose-dependent effect of ST 789 was found for concentrations ranging from 50 to 100 μg/ml.

However, we did not find any increase of IL-6 levels in supernatants of murine splenocytes cultured in the presence of both mitogens and ST 789 as compared with cell cultures driven by the mitogen alone (Table 2).

In addition, ST 789 resulted in a strong increase in the production of IL-6 by murine peritoneal macrophages not driven by LPS (Table 3). Notably, ST 789 enhanced the IL-6 production by LPS-stimulated peritoneal macrophages to a significantly lesser degree with respect to cells not stimulated by endotoxin (Table 3). A negative relationship between the concentration of ST 789 and the ability of LPS-driven macrophages to produce IL-6 was established, since the highest IL-6 levels were found in culture media of cells cultured in the presence of the lowest concentration (1 μg/ml) of ST 789 (Table 3).

On the other hand, ST 789 did not affect the production of IL-2 by both PBMCs from humans and splenocytes from mice (data not shown).

Table 1 In Vitro Effect of ST 789 on IL-6 Produced by PHA-Driven Human PBMCs[a]

Experimental sample	IL-6 (ng/ml)
PHA	<1
PHA plus ST 789 (50 μg/ml)	12.4 ± 5.3[b]
PHA plus ST 789 (100 μg/ml)	15.6 ± 4.2[b]

[a] The results are expressed as the mean IL-6 levels ± SE of PBMC cultures from five healthy individuals. PHA was used at the concentration of 10 μg/ml.
[b] $P < .001$ as compared with PHA alone.

Table 2 In Vitro Effect of ST 789 on IL-6 Produced by ConA-Driven Murine Splenocytes[a]

Experimental sample	IL-6 (U/ml)
ConA	275.16 ± 58.1
ConA plus ST 789 (100 μg/ml)	280.65 ± 70.1

[a] The results are expressed as the mean IL-6 levels ± SE of splenocyte cultures from five B6D2F1 mice. ConA was used at the concentration of 2 μg/ml.

Table 3 In Vitro Effect of ST 789 on IL-6 Produced by Murine Peritoneal Macrophages[a]

Experimental sample	IL-6 (U/ml)
Medium	17.5
LPS	870.0
ST 789 (10 μg/ml)	484.4
ST 789 (100 μg/ml)	443.7
LPS plus ST 789 (1 μg/ml)	1211.7
LPS plus ST 789 (10 μg/ml)	849.6
LPS plus ST 789 (100 μg/ml)	510.3

[a] The results are expressed as the mean IL-6 levels ± SE of pooled macrophage cultures from five B6D2F1 mice. LPS was used at the concentration of 2 μg/ml.

B. Effects of In Vivo Treatment with ST 789 on IL-6 Production by Peritoneal Macrophages

The treatment of Balb/c mice with ST 789 at two different dosages (0.25 and 25 mg/kg) resulted in a strongly enhanced IL-6 production by peritoneal macrophages both constitutively and following LPS stimulation (Table 4).

In Cy-immunodepressed mice we found somewhat surprising results (Table 5). In fact, the Cy treatment resulted unexpectedly in strong enhancement of the IL-6 production by peritoneal macrophages as compared with control animals. Moreover, the treatment with ST 789 of Cy-immunodepressed mice did not in turn affect the IL-6 production by peritoneal macrophages.

Table 4 Effect of In Vivo Treatment with ST 789 on IL-6 Produced by Murine Peritoneal Macrophages[a]

Treatment	LPS	IL-6 (U/ml)
Control	−	81.9
Control	+	16384
ST 789 (0.25 mg/kg/day)	−	131.7
ST 789 (25 mg/kg/day)	+	28509

[a] The results are expressed as the mean IL-6 levels ± SE of pooled macrophage cultures from five Balb/c mice. LPS was used at the concentration of 2 μg/ml. ST 789 was administered intraperitoneally from day −5 to day −1 and peritoneal macrophages collected at day 0.

Table 5 Effect of In Vivo Treatment with ST 789 on IL-6 Produced by Peritoneal Macrophages from Cy-Immunodepressed Mice[a]

Treatment	IL-6 (U/ml)
Control	8430 ± 1904
Cy	28191 ± 2987[b]
Cy plus ST 789	17115 ± 5935[b]

[a] The results are expressed as the mean IL-6 levels ± SE of macrophage cultures from five B6D2F1 mice. ST 789 (25 mg/kg/day) was administered intraperitoneally from day −5 to day −1 and Cy (100 mg/kg) was injected intraperitoneally at day −2. Peritoneal macrophages were collected at day 0.

[b] $p < .001$ as compared with control group.

C. Effects of In Vivo Treatment with ST 789 on IL-6 Production by Splenocytes

The treatment of Balb/c mice with ST 789 at three different dosages (0.25, 2.5, and 25 mg/kg) did not affect the IL-6 production by splenocytes (Table 6).

The immunosuppressive treatment with Cy significantly impaired the IL-6 production by mitogen-driven splenocytes and ST 789 was able, at least to a certain degree, to restore the production of the cytokine (Table 7).

D. Effects of In Vivo Treatment with ST 789 on IL-2 Production

The treatment with ST 789 strongly increased the ability of splenocytes from B6D2F1 mice to produce IL-2 on mitogen stimulation (Table 8). Furthermore, no

Table 6 Effect of In Vivo Treatment with ST 789 on IL-6 Produced by ConA-Driven Murine Splenocytes[a]

Treatment	IL-6 (U/ml)
Control	762.6 ± 102.6
ST 789 (0.25 mg/kg/day)	646.5 ± 56.9
ST 789 (2.5 mg/kg/day)	677.9 ± 71.6
ST 789 (25 mg/kg/day)	994.6 ± 168.6

[a] The results are expressed as the mean IL-6 levels ± SE of splenocyte cultures from five Balb/c mice. ST 789 (0.25, 2.5, and 25 mg/kg/day) was administered intraperitoneally from day −5 to day −1 and the spleen removed at day 0.

Table 7 Effect of In Vivo Treatment with ST 789 on IL-6 Produced by ConA-Driven Splenocytes from Cy-Immunodepressed Mice[a]

Treatment	IL-6 (U/ml)
Control	304.3 ± 12.2
Cy	42.3 ± 8.5[b]
Cy plus ST 789	104.1 ± 31.9[c]

[a] The results are expressed as the mean IL-6 levels ± SE of splenocyte cultures from five B6D2F1 mice. ST 789 (25 mg/kg/day) was administered intraperitoneally from day −5 to day −1 and Cy (100 mg/Kg) was injected intraperitoneally at day −2. The spleen was removed at day 0.
[b] $p < .001$ as compared with control group.
[c] $p < .001$ as compared with Cy group.

Table 8 Effect of In Vivo Treatment with ST 789 on IL-2 Produced by ConA-Driven Murine Splenocytes[a]

Treatment	Spleen weight (mg)	Splenocytes ($\times 10^7$)	IL-2 (U/ml)
Control	83.4 ± 8.4	7.2 ± 0.9	27.1 ± 2.1
ST 789	74.2 ± 2.6	6.9 ± 0.3	39.7 ± 4.6[b]

[a] The results are expressed as the mean IL-2 levels ± SE of splenocyte cultures from five B6D2F1 mice. ST 789 (25 mg/kg/day) was administered intraperitoneally from day −5 to day −1 and the spleen removed at day 0.
[b] $p < .05$ as compared with control group.

difference was found between control and ST 789–receiving mice in terms of both spleen weight and the number of recovered splenocytes (Table 8).

IV. DISCUSSION

Our experiments indicate that ST 789 could act as a strong inducer of IL-6 production both in vitro and in vivo.

We found that ST 789 was able to increase in vitro the production of IL-6 by PBMCs from healthy individuals and peritoneal macrophages from mice. Our data indicating that ST 789 exerts the maximum IL-6–enhancing effect on peritoneal exudate cells points out that cells of monocyte-macrophage linkage are likely to be the major candidates as targets for the IL-6–enhancing effect of ST 789. The finding that only low concentrations (1 μg/ml) of ST 789 further enhanced the IL-6 production by LPS-driven macrophages, whereas higher concentrations (10 and 100 μg/ml) did not, possibly indicates that maximally activated cells are not susceptible to further stimulation.

However, a direct effect of ST 789 on IL-6 production by T lymphocytes could not be ruled out. A selective action of ST 789 on different T-lymphocyte subsets might be hypothesized as well as ST 789 could mediate a complex modulation of the cellular networks regulating the production of the cytokine. Currently, further studies are in progress in our laboratory to establish the role of T cells in accounting for the enhanced IL-6 production by ST 789–treated PBMCs from humans.

Our data pointing out no effect of ST 789 on the in vitro production of IL-2 confirms previously published results by us and by others (15,16).

The in vivo treatment of immunocompetent mice with ST 789 proved to enhance strongly the ability of peritoneal macrophages to produce IL-6 both constitutively and following endotoxin stimulation, which suggests that in vivo ST 789 also has a major effect on cells of the monocyte-macrophage lineage.

In our study, peritoneal macrophages from Cy-immunodepressed mice produced significantly more IL-6 as compared with normal animals and ST 789 was not able to further augment the production of IL-6. This finding could not be easily explained. Notably, strongly increased production of IL-6 by PBMCs has been recently shown in patients with common variable hypogammaglobulinemia (21) suggesting that the overproduction of IL-6 could be a common pathway shared by several disorders with immunodeficiency in spite of different pathogenesis.

The effects of Cy on the production of cytokines, most notably IL-6, are complex, as pointed out by our finding that splenocytes from Cy-treated mice produce significantly less IL-6 with respect to control animals. In addition, ST 789 treatment proved to restore, at least to some degree, the IL-6 production by splenocytes from Cy-treated mice. Overall, our results indicate that Cy could exhibit either enhancing or suppressive effects on IL-6 production as far as different cell targets are evaluated, such as peritoneal macrophages and splenic lymphocytes. Notably, at least in this latter model of experimental immunosuppression, ST 789 proved able to counteract the effects of Cy on IL-6 production.

The increased production of IL-2 by splenocytes from mice treated with ST 789, in spite of no IL-2–enhancing effect exhibited in vitro by ST 789, indicates that ST 789 could be useful in vivo as an inducer of IL-2. However, the mechanisms accounting for the discrepancy in the ST 789–driven production of IL-2 between in vitro and in vivo experiments remain to be fully established. In addition, since no difference was found between control and ST 789–receiving mice in terms of both spleen weight and the number of recovered splenocytes, our data suggest that ST 789 could exert its IL-2–enhancing activity at the single cell level.

In conclusion, at least in our experimental models, ST 789 acts both in vitro and in vivo as a strong inducer of the production of cytokines, such as IL-6 and, although to a lesser degree, IL-2. Since cytokines are required to mount the immune response, ST 789 could prove helpful in humans in the combination therapy of both infections and cancer.

REFERENCES

1. Cornaglia-Ferraris P, Coffey RG, Hadden JW. Substituted purines as immunomodulators: speculative considerations. EOS J Immunol Immunopharmacol 1984; 2:134.
2. De Simone C, Arrigoni Martelli E. ST 789: A new synthetic immunomodulator. Thymus 1992; 19:S1.
3. Daly JM, Reynolds JV, Thon AK, Kinsley L, Dietrich-Gallagher M, Shou J, Ruggieri B. Immune and metabolic effects of arginine in the surgical patient. Am Surg 1988; 208:512.
4. Daly JM, Reynolds JV, Signal RK, Shou J, LIberman MD. Effect of dietary protein and aminoacids on immune function. Crit Care Med 1990; 18:86.
5. Lowell JA, Parnes HL, Blackborn GL. Dietary immunomodulation: Beneficial effects on oncogenesis and tumor growth. Crit Care Med 1990; 18:145.
6. Reynolds JV, Daly JM, Zhang SM, Ziegler MM, Nasi A. Arginine, protein malnutrition and cancer. J Surg Res 1988; 45:513.
7. Reynolds JV, Daly JM, Shou J, Sigal R, Ziegler MM, Naji A. Immunological effects of arginine supplementation in tumor bearing and non tumor-bearing hosts. Ann Surg 1990; 211:202.
8. Albina JE, Caldwell MD, Henry WL, Mills CD. Regulation of macrophage functions by L-arginine. J Exp Med 1989; 169:1021.
9. Keller R, Geiges M, Keist R. L-arginine–dependent reactive intermediates as mediator of tumor cell killing by activated macrophages. Cancer Res 1990; 50:1421.
10. Fabris N, Mocchegiani E. Arginine-containing compounds and thymic endocrine activity. Thymus 1992; 19:S21.
11. De Simone C, Arrigoni Martelli E, Famularo G, Foresta P, Ruggiero V, Giacomelli R, Tonietti G, Jirillo E. ST 789, a candidate for combination therapies of infections and cancer. In: Goldstein AL, Garaci E, eds. Combination Therapies. New York: Plenum Press, 1992:223.
12. Giovarelli M, Arione R, Jemma C, Musso T, Benetton G, Forni G, Cornaglia-Ferraris P. In vitro and in vivo immunomodulatory activity of an N-9 arginyl hypoxanthine derivative (PCF-39). Int J Immunopharmacol 1987; 9:659.
13. Riccardi C, Migliorati G. Enhancement of murine NK cell activity generation by ST 789. Thymus 1992; 19:S109.
14. Jemma C, Giovarelli M, Vai S, Cavallo F, Forni G. In vitro and in vivo immunomodulatory activity of ST 789 on two human lymphoblastic T cell lines. Thymus 1992; 19:S79.
15. Famularo G, Giacomelli R, Sacchetti S, Valdarnini D, De Simone C, Tonietti G. Modulation by ST 789 of in vitro lymphocyte activation and cytokine production. Thymus 1992; 19:S71.
16. Munno I, Pellegrino NM, Marcuccio C, De Simone C, Jirillo E. Influence of ST 789 on interleukin-1, interleukin-2 and interferon-gamma production from normal human peripheral blood mononuclear cells Thymus 1992; 19:S63.
17. Foresta P, Riccardi C, Arrigoni Martelli E. Immunorestorative properties of ST 789 on experimentally immunosuppressed and infected mice. Thymus 1992; 19:S97.

18. De Simone C, Arrigoni Martelli E, Foresta P, Famularo G, Giacomelli R, Jirillo E, Ruggiero V, Tonietti G. Hypoxanthine derivatives in experimental infections. Can J Infect Dis 1992; 3:106B.
19. Famularo G, Procopio A, Giacomelli R, Danese C, Sacchetti S, Perego MA, Santoni A, Tonietti G. Soluble interleukin-2 receptor, interleukin-2 and interleukin-4 in sera and supernatants from patients with progressive systemic sclerosis. Clin Exp Immunol 1990; 81:368.
20. Famularo G, Giacomelli R, Di Giovanni S, Sacchetti S, Tonietti G. Cytokine production in patients with monoclonal gammapathies. J Clin Lab Immunol 1991; 34:63.
21. Oliva A, Pandolfi F, Paganelli R, Quinti I, Guerra E, Sacco G, Fanales-Belasio E, Giovannetti A, Aiuti F. Ruolo delle citochine nella patogenesi dell'ipogammaglobulinemia comune variabile, Abstract Book of XII Congresso Societa' Italiana di Immunologia ed Immunopatologia, Giardini Naxos, Italia, 1992:316.

7

Effect of Interferon-Gamma Injections on Murine Visceral (*Leishmania donovani*) Leishmaniasis

Albrecht F. Kiderlen
Robert Koch-Institute, Berlin, Germany

Marie-Luise Lohmann-Matthes
Fraunhofer-Institute of Toxicology, Hannover, Germany

I. INTRODUCTION

Microorganisms of the genus *Leishmania* are flagellated protozoa that cause a wide spectrum of diseases in humans. These include inapparent infestations, cutaneous leishmaniasis manifesting as self-healing lesions, and visceral leishmaniasis (VL), or kala-azar, a nonhealing, systemic form of infection. The different clinical manifestations usually, but not necessarily, correspond to different *Leishmania* species (1). Inbred mice are widely used as animal models for most forms of leishmaniasis. These are especially interesting, as mouse strains vary in terms of early resistance or susceptibility to the different species of *Leishmania* and also in terms of late self-healing or nonhealing capacity (2,3).

The concept for the experiments shown here is based on three major findings.

1. In their mammalian host, all *Leishmania* species are obligate intracellular pathogens of the monocyte-macrophage system.
2. It has been extensively shown in vitro that these obligate host cells can be transformed into cytotoxic effector cells with a capacity to destroy their intracellular parasites by various protocols of immunologic activation (4). The cytokine interferon gamma (IFN-γ), usually given together with bacterial endotoxin (lipopolysaccharide, LPS) as a necessary second signal, is the classic activating factor, MAF, of macrophage (MΦ) cytotoxicity in vitro (5).

3. Recombinant murine IFN-γ (rIFN-γ), produced in *Escherichia coli* (6), is biologically active in vivo. This could be shown for instance by activating peritoneal MΦ in situ with intraperitoneal injections of rIFN-γ and subsequently testing their potential for killing P815-tumor cells in vitro (7). Also, rIFN-γ protects mice from local or systemic infection with the bacterial pathogen *Listeria monocytogenes* (7).

These findings led to the assumption that leishmaniasis should be an ideal candidate for IFN-γ therapy.

II. MATERIALS AND METHODS

A. Animals

Six- to 10-week-old female C57BL/6 mice from Charles River Wiga GmbH, Sulzfeld, Germany, were used.

B. *L. donovani*

The isolate LRC-LD51, from a case of kala-azar in India in 1954, was received from Dr. Frank Ebert, Bernhard-Nocht-Institute, Hamburg, Germany. It has since been passaged in hamster (*Mesocricetus auratus*) and later in cotton rats (*Sigmodon hispidus*) every 2 months. Amastigote (AM) *Leishmania* organisms were isolated from the spleen of animals infected for approximately 3 months as described for hamster by Channon et al. (8). Promastigote (PM) *Leishmania* organisms were grown as suspension culture in a 1:1 mixture of RPMI 1640 (GIBCO BRL, Glasgow, Scotland) and Hosmem II medium (9) supplemented with 5% fetal calf serum (FCS), referred herein as growth medium (GM). The cultures were incubated at 25°C in humidified normal atmosphere containing 6% CO_2.

C. rIFN-γ

Recombinant murine and human IFN-γ, expressed in *E. coli*, were produced by Genentech (San Francisco, CA) and kindly provided by Boehringer Ingelheim (Ingelheim, Germany). Recombinant IFN-γ was quantitated and regularly controlled via its ability to protect murine L929 fibroblasts from the cytopathic effect of human vesicular stomatitis virus (VSV) [10].

D. Monitoring Systemic Infection with *L. donovani* in Mice

Approximately 2×10^7 PM or 5×10^6 AM *L. donovani* organisms were injected intravenously into a lateral tail vein (day 0). Recombinant IFN-γ was either added to this inoculum or injected intravenously separately as indicated in the Results section. We waited at least 14 days after the end of rIFN-γ treatment before quantitating standard infection parameters, thereby excluding purely short-term effects of this treatment. At the times indicated, the mice were killed, their spleens aseptically removed, and spleen weights determined. For better comparison between individual experiments, the data are presented as percent-values of the mean spleen weight of the infected, untreated control group (= 100%). The relative amount of viable *L. donovani* organisms/spleen was quantitated radiometrically: The organs were homogenized in 9.5 ml phosphate-buffered saline (PBS) with a Teflon homogenizer (Schütt Labortechnik, Göttingen, Germany). The homogenates were washed twice in PBS (12 min, 2000 *g*), resuspended in GM, and incubated in Petri dishes at 25°C. When viable PM *Leishmania* organisms appeared in cultures from untreated control mice and began to multiply (3–5 days of culture), triplicate 100 μl aliquots from all cultures were cultivated in 96-well microtiter plates for another 18 h in the presence of 1 μCi/well tritium-labeled thymidine ([^{3}H]ThR). Under these conditions, *Leishmania* PM were the only cells to show significant specific incorporation of [^{3}H]ThR. The data are presented as percent-values of the mean specific [^{3}H]ThR-incorporation (= cpm [infected mice] minus cpm [noninfected controls]) of the infected, untreated control group (= 100%).

E. Intraperitoneal Implantation of Osmotic Pumps

Mice were anesthetized by injecting intraperitoneally 60 mg Na-pentobarbital/kg mouse (Narcoren, Iffa Merieux, Laupheim, Germany). After 5 min, mice were in deep narcosis, which held for over 30 min. The lower half of the ventral surface was disinfected by repeated swabbing with 70% ethanol. The osmotic pumps (Alzet mini-osmotic pump, 200, Alza, Palo Alto, CA) were gently inserted into the peritoneal cavity through an approximately 1.5 cm latitudinal incision. These pumps are designed to release their content of 100 μl at a constant flow rate over 7 days (11). Peritoneum and skin were closed with three stitches of catgut or polyester, respectively. Until the mice had completely regained motility, they were kept warm under an infrared lamp and their eyes were moistened repeatedly with sterile saline. After 11 days, the osmotic pumps were removed in a similar operation, as recommended.

III. RESULTS

We examined the protective effects of rIFN-γ injections on leishmaniasis in mouse models. In this study, we injected intravenously *L. donovani,* a causative agent of fatal visceral leishmaniasis in humans, into susceptible, late-healing C57BL/6 mice. In the first sets of experiments, we tested single and repeated injections of different quantities of rIFN-γ at different stages of infection: before, simultaneous, early, and late. Standard parameters of infection were determined at least 14 days after the last injection of rIFN-γ. In summary, single injections of rIFN-γ were not effective in this murine VL model. Repeated injections led to considerable protection only when treatment began shortly before or within the first days of infection. With lower doses of rIFN-γ (10^3–10^4 U/injection), the protective effect was dose dependent. The protective effect could not be significantly improved using higher doses of rIFN-γ (10^5–10^6 U/injection; data not shown). In the representative experiment shown here (Fig. 1a and 1b), mice were infected with 2×10^7 PM and received repeated injections of 10^3 or 10^4 U rIFN-γ/injection at 3-day intervals as indicated, either before [day −5, −3, −1 postinjection (p.i.)], early (day 0, 3, 6, 9 p.i.) or late (day 14, 17, 20, 23 p.i.) during the course of infection. Splenomegaly, one of the diagnostic hallmarks of VL, was compared between groups of infected controls, and rIFN-γ–treated mice day 35 p.i. (Fig. 1a). As illustrated, 10^3 U rIFN-γ injected 4 times at 3-day intervals had no significant protective effect. A 10-fold higher dose gave a mean of 40% protection when given before and marginal protection when treatment was started simultaneously with infection. Next, we compared the relative amount of viable *L. donovani* organisms in whole spleen cultures expressed as percent specific [^{3}H]ThR incorporation as described in the Materials and Methods section (Fig. 1b). As shown, protection expressed by this direct parameter of infection had the same tendencies as above but was more pronounced. The effect was specific for murine rIFN-γ as an identically produced preparation of human rIFN-γ was completely ineffective (data not shown).

In another set of experiments, we tested whether more repeats at shorter (24 h) intervals would enhance the protective effect of low-dose rIFN-γ. Again mice were infected with 2×10^7 PM. Groups received daily injections of either PBS or 1×10^3 U rIFN-γ/mouse for 9 consecutive days beginning 4 h preinfection. Splenomegaly was compared day 33 p.i. (Fig. 2a). As illustrated, rIFNγ had only little beneficial effect on splenomegaly (23% mean specific reduction of spleen weight). Next we compared the relative amount of viable *L. donovani* organisms in whole spleen cultures (Fig. 2b). In this representative experiment, the mean specific [^{3}H]ThR incorporation in spleen cultures of rIFN-γ–treated mice was 79% lower than that in the untreated control group. Furthermore, after incubating for 7 days, two of five of these cultures still did not show any viable parasites (Fig. 2b). In comparison, quantitation of viable parasites per organ appeared to be a more sensitive and direct monitoring method than determining spleen weights alone. As shown here, repeated sham treatment (sterile PBS) sometimes led to enhanced infection parameters, possibly in response to inflammatory stimuli. It must be noted that often mice repeatedly treated with rIFN-γ but none of the mice injected with PBS alone showed pronounced tissue destruction at the sites of injection. These localized ulcers were free of detectable *Leishmania* parasites. In this particular experiment, tissue destruction led to loss of tail segments in two of five cases.

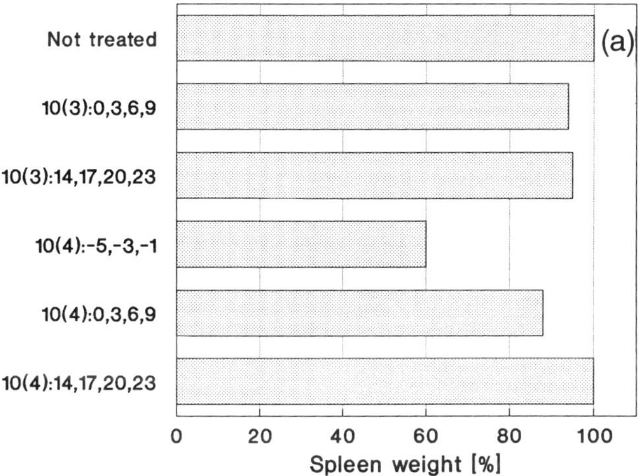

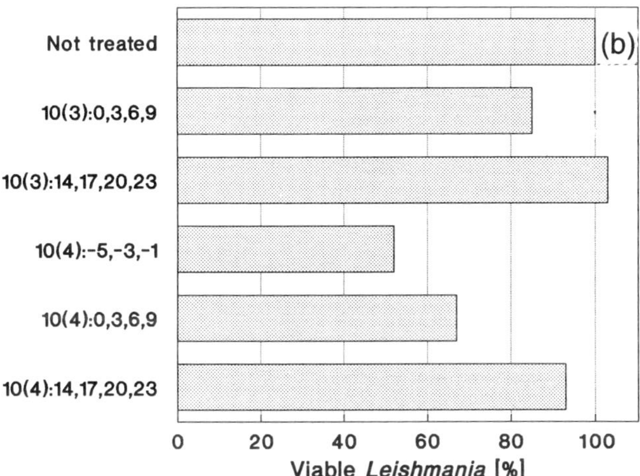

Figure 1 Repeated injections of rIFN-γ reduced splenomegaly (a) and viable parasites/spleen (b) depending on the dose and the time injected. C57BL/6 mice were infected with 2 × 107 PM *L. donovani.* At the days p.i. indicated (negative numbers indicate treatment before infection), mice received either 1×10^3 or 1×10^4 U rIFN-γ in PBS/injection. Day-35 p.i. spleen weights (a) and the relative amount of viable Leishmania parasites/spleen (b) were determined. Bars indicate mean values of five mice/group. SD: 10–17% (a); 7–12% (b).

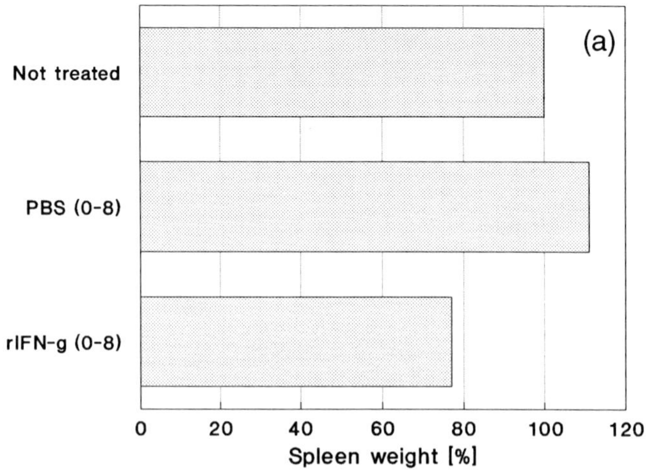

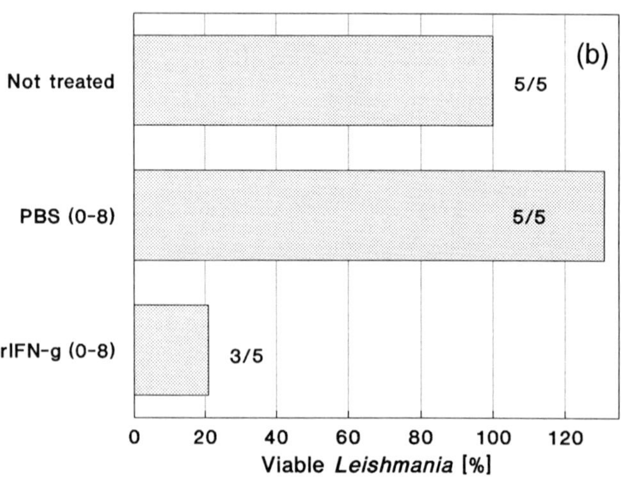

Figure 2 Repeated daily injections of low-dose rIFN-γ marginally prevented splenomegaly (a) and strongly reduced viable parasites/spleen (b) in *L. donovani*-infected C57BL/6 mice. Mice were infected i.v. with 2×10^7 PM *L. donovani.* Beginning 4 h preinfection, mice received daily i.v. injections of 1×10^3 U rIFN-γ/mouse in 200 mL PBS or PBS alone. Day-33 p.i. spleen weights (a) and the relative amount of viable parasites/spleen (b) were determined. Numbers/5 indicate the spleen homogenates with viable PM after 7 days of culture. Bars indicate the mean values of five mice/group. SD: 6–28% (a); 18–22% (b).

In an attempt to ensure a more constant supply of rIFN-γ and to reduce its momentary local concentration, we implanted osmotic pumps filled with 1×10^6 U rIFN-γ in 0.2 ml PBS plus 0.3% bovine serum albumin (BSA)/pump or with PBS plus BSA alone into the peritoneal cavity of mice. The osmotic pumps are designed to release their content at a constant rate over 7 days. In the experiment shown here, the pumps were implanted or the sham operations performed 4 h before mice were infected with intravenous injections of 5×10^6 *L. donovani* AM, and the mice were killed day 35 p.i. Again, the development of splenomegaly was only marginally reduced compared with untreated, sham-operated, or PBS-treated controls (Fig. 3a). In terms of viable parasites/spleen, however, this protocol gave appreciable protection from VL (Fig. 3b). In this particular experiment, the infection was highly pathogenic. Granulomalike structures were apparent on all infected mice except those that had been treated with rIFN-γ. However, all spleen cultures, including those from the rIFN-γ–treated group, showed abundant viable *L. donovani* PM after 7 days of incubation. Again, PBS treatment possibly led to some enhancement of viable parasites/spleen (Fig. 3b). When the osmotic pumps were surgically removed on day −11 postimplantation as recommended, the pumps that had delivered rIFN-γ were surrounded by hemorrhagic tissue and one had even been turned front-to-rear as an obvious irritant. On the other hand, those pumps dispensing PBS alone had been completely embedded in healthy tissue. Again, the long-term confrontation of tissue with rIFN-γ seemed to provoke reactions leading to necrosis of this tissue.

IV. DISCUSSION

In the vertebrate host, *Leishmania* are obligate intracellular parasites in cells of the monocyte-macrophage system. With the exception of the natural infection via the bite of *Phlemotomus* sandflies during which the parasite is in its promastigote stage, and during the invasion of new host cells as amastigotes, these parasites can therefore not be reached by factors of the humoral immune response. Instead, it has been repeatedly shown that the cellular immune response mechanisms are critical for spontaneous healing from leishmaniasis with the activation of MΦ from permissive host cells to cytotoxic effector cells as the decisive event. As generally accepted, IFN-γ, a glycopeptide released by activated T-cell populations and natural killer cells, together with a second signal that, at least in vitro is generally supplied by LPS, is a major MΦ-activating factor in vitro and in vivo. On the other hand, malfunctions of these mechanisms are responsible for the chronic and pathogenic aspects of established forms of leishmaniasis. Indeed, it seems as if clinical forms of leishmaniasis do not result only from lack of MΦ activation but also from immune response patterns actually enhancing host cell capacities.

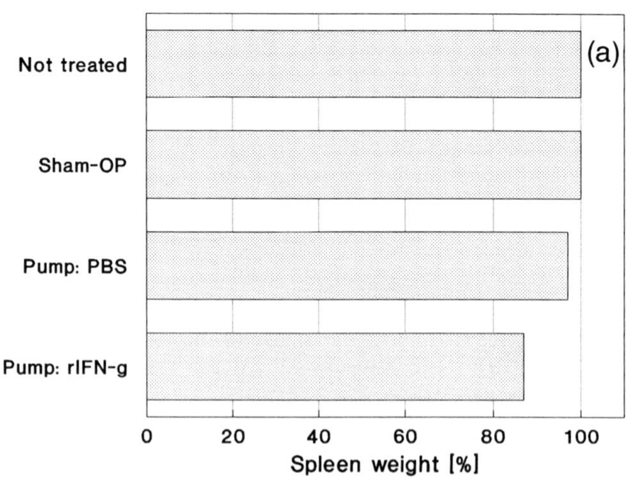

Treatment:

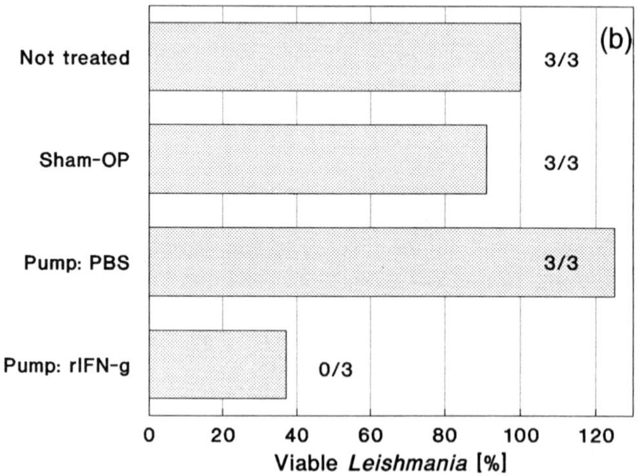

Figure 3 Continuous i.p. release of rIFN-γ via osmotic pumps implanted into the peritoneal cavity marginally prevented splenomegaly (a) and reduced viable parasites/spleen (b) in *L. donovani*-infected C57BL/6 mice. Mice were infected i.v. with 5×10^6 AM L. donovani organisms. Osmotic pumps were implanted or sham operations were performed 4 h preinfection. Pumps were filled with 1×10^6 U rIFN-γ in 0.2 ml PBS + 0.3% BSA or with 0.2 ml PBS + BSA alone, and released their content continuously over the next 7 days. Day-35 p.i. spleen weights (a) and the relative amount of viable parasites/spleen (b) were determined. All spleen homogenates of infected animals contained viable PM after 7 days of culture. Numbers/3 indicate granulomalike structures on surface of livers; bars indicate mean values of three mice/group. SD: 11–18% (a); 15–18% (b).

With this background, it seemed plausible that treatment with rIFN-γ should help resolve or protect from infection with *Leishmania* organisms. It was our hope to thereby circumvent the problems of inadequate immune response in patients with leishmaniasis, especially in cases of general immunosuppression as well as the rising problems of resistance to the available chemotherapeutics and their innate toxic side effects. In this series of experiments, we infected susceptible, late-healing, or nonhealing mice with *L. donovani,* the causative agent of lethal human visceral leishmaniasis, and treated randomly chosen groups with injections of murine rIFN-γ at different stages of infection and with various rIFN-γ doses and repeats. The results indicate: (1) Treatment of VL with intravenous injections of rIFN-γ is exquisitely time dependent. The highest relative level of reduction of viable parasites in the spleen was achieved when rIFN-γ was injected immediately before or very early after infection. Any curative effect of rIFN-γ injections rapidly decreases with the establishment of visceral infection. (2) Treatment effects also depend on the individual dose of rIFN-γ and the number of repeats. Best effects could be achieved with many repeats of relatively low doses of rIFN-γ injected intravenously or via osmotic pumps implanted into the peritoneal cavity. (3) Splenomegaly and amount of viable parasites per spleen do not fully correlate. The direct quantitatition of viable parasites is more sensitive to protective or curative effects of rIFN-γ. (4) Sterile cure of *Leishmania* parasites could possibly be achieved but only in individual cases. (5) Repeated or continuous treatment with rIFN-γ intravenously or intraperitoneally caused histotoxic side effects (necrosis, hemorrhages) at the sites of rIFN-γ release. When treatment was terminated, the tissue healed. Even with the most effective application protocol, this form of treatment with unmodified rIFN-γ must therefore be handled with caution.

It is unclear at this time why treatment with rIFN-γ, which is so effective in vitro or very early during infection, is so ineffective at later disease stages. One could speculate on altered host cell characteristics or the accumulation of suppressive factors. On the other hand, we could repeatedly show in vitro that suboptimal activation of *Leishmania*-parasitized Mφ with rIFN-γ may actually enhance intracellular survival/multiplication of these organisms (5). The implications of these findings for the in vivo situation remain speculative. Although sterile protection possibly cannot be achieved, a reduction of the parasite load might already prove advantageous. For instance, the combination of rIFN-γ with standard chemotherapeutics might reduce the dose requirements for both components decisively. In activating not only cytotoxic effector capacity but also other immunologic functions such as major histocompatibility (MHC) class II expression for antigen presentation rIFN-γ might help break down resistance to chemotherapeutics and lessen the chances of relapse. Encapsulation in liposomes should be a sensible way of achieving higher levels of in vivo activity by targeting the material for phagocytes and reducing its nonspecific degradation. In reducing the effective dose and enhancing its dispersal, encapsulation of rIFN-γ in liposomes should also minimize the histotoxic side effects of rIFN-γ injections observed in the experiments reported here.

REFERENCES

1. Control of the leishmaniases: report of a WHO expert committee. Who Technical Report Series 793, WHO, Geneva, 1990.
2. Bradley DJ. Regulation of Leishmania populations within the host: II. Genetic control of acute susceptibility of mice to *Leishmania donovani* infection. Clin Exp Immunol 1977; 30:130.
3. Blackwell JM, Roberts B, Alexander J. Response of Balb/c mice to leishmanial infection. Curr Tops Microbiol Immunol 1985; 122:97.
4. Adams DO, Hamilton TA. The cell biology of macrophage activation. Ann Rev Immunol 1984; 2:283.
5. Roach TIA, Kiderlen AF, Blackwell JM. Role of inorganic nitrogen oxides and tumor necrosis factor alpha in killing *Leishmania donovani* amastigotes in gamma interferon-lipopolysaccharide–activated macrophages from Lsh^s and Lsh^r congenic mouse strains. Infect Immun 1991; 59:3935.
6. Gray PW, Goeddel DV. Cloning and expression of murine interferon cDNA. Proc Natl Acad Sci USA 1983; 80:5842.
7. Kiderlen AF, Kaufmann SHE, Lohman-Matthes M-L. Protection of mice against the intracellular bacterium *Listeria monocytogenes* by recombinant immune interferon. Eur J Immunol 1984; 14:964.
8. Channon JY, Roberts MB, Blackwell JM. A study of the differential respiratory burst activity elicited by promastigotes and amastigotes of *Leishmania donovani* in murine resident peritoneal macrophages. Immunology 1984; 53:245.
9. Berens RL, Marr JJ. An easily prepared defined medium for cultivation of *Leishmania donovani* promastigotes. J Parasitol 1978; 64:160.
10. Marcucci F, Klein B, Kirchner H, Zawatzky R. Production of high titers of interferon gamma by prestimulated spleen cells. Eur J Immunol 1982; 12:787.
11. Fara J. Parenteral fundamentals. Recent advances in parenteral drug delivery systems. J Parenteral Sci Technol 1983; 37:20.

8

Investigation into the Mode of Action of Interleukin-1 in the Arthritic Condition

Rainer Joachim Box
Robert Koch-Institut, Berlin, Germany

I. INVOLVEMENT OF INTERLEUKIN-1 IN INFLAMMATORY PROCESSES

Arthritic inflammations of the joint are typified by a massive infiltration of the synovial membrane with activated macrophages, T and B cells, and plasma cells and proliferation of the synovial fibroblasts. The connective tissues—cartilage, tendon, bone—are destroyed in the longer term by degradation of the extracellular matrix and the failure of repair mechanisms. The resident mesenchymal cells of the joint and the invading cells produce a variety of cytokines, of which interleukin-1 (IL-1), in concert with tumor necrosis factor alpha (TNF) and interferon gamma, seems to play an important role in the pathogenesis of rheumatoid arthritis (1). In vivo IL-1 induces the synthesis of prostaglandin E_2 (PGE_2), nitric oxide, and the neutral proteinase collagenase in articular chondrocytes and synovial fibroblasts (2). Studies in vitro and in animal models have shown that IL-1, in association with TNF, induces bone degradation in osteoblastic cell cultures (3), and that it initiates erosive arthritis in rabbit joints (4). However, the intracellular events that follow IL-1 binding to its receptor are not yet understood sufficiently. The signaling mechanism apparently comprises activation of a G-protein and protein phosphorylation (5), but there are conflicting reports as to which messenger system is involved (cyclic adenosine monophosphate [cAMP], calcium ions, or inositolphosphates). Although a recent report excludes protein kinase C involvement in IL-1 signaling, at least in articular chondrocytes (6), the relationship between actions of IL-1 and intracellular cAMP is still unclear (7). With the variety of biologic events known to be triggered by IL-1, the mode of action may well depend on the cell types examined and the particular biologic activity of IL-1 studied.

II. DEFINITION OF THE MODEL SYSTEM

This study was conducted in vitro with monolayer cultures of synovial fibroblasts (SFs) from healthy young rabbits. In these primary cells, IL-1 induces collagenase (EC 3.4.24.7) gene expression, and therefore this cell system may be useful in studying events during the genesis of rheumatoid arthritis. The SF cells obtained are described here in terms of IL-1 binding, hormone-induced cAMP accumulation, and the biologic effects of the recombinant human IL-1 receptor antagonist protein (Synergen Inc., Boulder, CO). Rabbit SFs were prepared weekly; that is, fresh knee synovial membranes were gently disrupted into a single-cell suspension with a bacterial collagenase and placed into culture flasks containing Dulbecco's modified Eagle's medium (DMEM), 10% fetal calf serum, penicillin, and streptomycin (8). For the experiments, 3- to 5-week old cells were transferred after trypsinization from 600-ml culture flasks to six-well plates (about 8×10^5 cells per well in 3 ml of medium). After 24 h, cells were washed repeatedly and rendered quiescent in serum-free DMEM containing 0.2% lactalbumin hydrolysate. After further 24 h, this medium was renewed and experiments started (e.g., 1.67 U/ml recombinant human IL-1β (Boehringer Ingelheim, Germany) were added to induce collagenase activity).

III. EXPERIMENTAL DATA

SF cells displayed a single class of high-affinity binding sites for iodinated recombinant human IL-1β (Amersham International, Plc, Slough, England). The receptor number and the dissociation constant (Kd) was calculated from a Scatchard plot and values for both (9) were in the range given for chondrocytes (10).

Hormonal stimulation of SF cells lead to a significant rise in intracellular cAMP measured over 10 min at 37°C in the presence of 10^{-5} M isobutylme thylxanthine (IBMX), an inhibitor of phosphodiesterase activity (Fig. 1). IBMX is added routinely in such studies, since the induced cAMP levels are transient or prove difficult to detect in some cell types. In general, however, IBMX amplifies the response and, as was verified again here, does not cause qualitative changes. Histamine (HIS) at a final concentration of 10^{-5} M did not increase cAMP over basal levels (30 pmol/10^6 cells). Isoproterenol (IPR, 10^{-6} M), a potent analogue of epinephrine, PGE (10^{-6} M), or forskolin (For, 10^{-6} M), respectively, induced an 8- to 10-fold increase in cAMP. When forskolin, which acts directly on adenylyl cyclase and not via a specific receptor, was used at 10^{-4} M, a concentration regarded as indicative for the capacity of the enzyme in a given cell type, the cAMP values were around 3000 pmol/10^{-6} cells. These values are uncommonly high in comparison to suspension

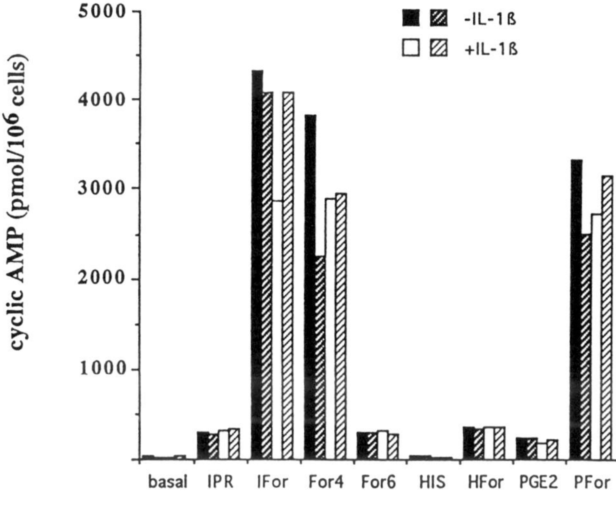

Figure 1 Effect of IL-1β on hormone-induced accumulation of cAMP in synovial SFs. Cells were incubated for 10 min at 37°C in the presence (white and light-shaded areas) or absence (black and dark-shaded areas) of 1.67 U/ml IL-1β with the following hormones (final concentrations given in parentheses): from left—basal, no addition; IPR, isoproterenol (1 μM); IFor, isoproterenol and forskolin (both 1 μM); For4, forskolin (100 μM); For6, forskolin (1 μM); HIS, histamine (10 μM); HFor, histamine and forskolin (10 μM and 1 μM, respectively); PGE_2, prostaglandin E_2 (1 μM); PFor, prostaglandin E_2 and forskolin (both 1 μM). The hormone addition directly followed preincubation (10 min at 37°C) with the phosphodiesterase inhibitor isobutylmethylxanthine (10 μM) without change of serum-free DMEM.

cultures of murine (11) and human lymphoid cells, clones, or cell lines (12). Their basal values are often lower than 5 pmol/10^6 cells, and forskolin at 10^{-4} induces only up to a few hundred pmol/10^6 cells. The high cAMP response was confirmed in separate experiments using four different cell preparations, and it appears therefore that a very effective adenylyl cyclase is present in rabbit SF cells. A combination of the lower dose of forskolin (10^{-6} M) with IPR or PGE_2 induced a response of equal high amplitude, whereas with HIS, a similar amount was found as with 10^{-6} M forskolin alone. The synergistic response of forskolin with IPR or PGE_2 is also found in other cell types (e.g., murine T-cell lines but not B-cell lines [11]). The lack of the response to HIS may be due to SF cells not having histamine receptors or carrying H_1 histamine receptors that are not linked to adenylyl cyclase. The simultaneous addition of IL-1β with forskolin, hormone, or the combination of both did not show any significant change in the cAMP response (Fig. 1). Incubation with IL-1β for 48 h resulted typically in a collagenase activity of about 130 μg collagen type I degraded per minute and per milligram DNA of SF cells (W. Ehrlich, personal communication). In order to see whether cAMP had any effect on IL-1–induced collagenase, SF cells were incubated with both IL-1β and 10^{-5} M IBMX. After 48 h, cells contained about 330 pmoles and the culture medium 450 pmoles cAMP (780 moles in total; control levels with IL-1β alone were negligible), but no difference in collagenase expression was noted (7).

The anti-inflammatory properties of the recombinant human IL-1 receptor antagonist were tested and quantified in mice treated with potassium peroxochromate. In SF cells, the antagonist reduced IL-1β mediated collagenase expression in a dose-dependent fashion (9). A significant decrease in the specific binding of IL-1β was noted to occur in parallel.

ACKNOWLEDGMENTS

I thank Fr. H. Wohlert, Dr. H. Huser, Dr. W. Ehrlich, and Prof. H. Kröger for their support.

REFERENCES

1. Chin J, Hatfield C, Krzesicki R, Herblin W. Interactions between interleukin-1 and basic fibroblast growth factor on articular chondrocytes. Arthritis Rheum 1991; 34:314.
2. Stadler J, Stefanovic M, Billiar T, Curran R, McIntyre L, Georgescu H, Simmons R, Evans C. Articular chondrocytes synthesize nitric oxide in response to cytokines and lipopolysaccharide. J Immunol 1991; 147:3915.
3. Thomas B, Mundy G, Chambers T. Tumour necrosis factor α and β induce osteoblastic cells to stimulate osteoclast bone resorption. J Immunol 1987; 138:775.
4. Pettipher E, HIggs G, Henderson B. Interleukin 1 induces leucocyte infiltration and cartilage degradation in the synovial joint. Proc Natl Acad Sci 1986; 83:8749.
5. O'Neill L, Bird T, Gearing A, Saklatvala J. Interleukin 1 signal transduction. J Biol Chem 1990; 265:3146.
6. Hulkower K, Georgescu H, Evans C. Evidence that responses of articular chondrocytes to interleukin-1 and basic fibroblast growth factor are not mediated by protein kinase C. Biochem J 1991; 276:157.
7. Takahashi S, Ito A, Nagino M, Mori Y, Xie B, Nagase H. Cyclic AMP suppresses interleukin 1 induced synthesis of matrix metalloproteases but not of tissue inhibitor of metalloproteinases in human uterine cervical fibroblasts. J Biol Chem 1991; 266:19894.
8. Brinckerhoff C, Benoit M, Culp W. Autoregulation of collagenase production by a protein synthesized and secreted by synovial fibroblasts: A cellular mechanism for control of collagen degradation. Proc Natl Acad Sci 1985; 82:1916.

9. Box R, Miesel R, Ehrlich W, Wohlert H. Suppression of collagenase activity and inflammatory effects of interleukin 1 by a potent antagonist of the interleukin 1 receptor: Study of the receptor requirement. Submitted.
10. Chin J, Horuk R. Interleukin 1 receptors on rabbit articular chondrocytes: relationship between biological activity and receptor binding kinetics. FASEB J 1990; 4:1481.
11. Box R, Portenier M, Staehelin M. Similarities in cAMP responses between murine lymphoid cell lines and subsets of mouse lymphocytes. J Leukoc Biol 1987; 42:144.
12. Box R, Simmer B. Investigation into the correlation between cell surface phenotype and hormonal response patterns in human lymphocyte subsets, clones and lymphoid cell lines (submitted).

9

Induction of Cytokines by Yeast Copper/Zinc Superoxide Dismutase

Zvetanka H. Stefanova, Valentina H. Valeva, and Hristo O. Neychev
Bulgarian Academy of Sciences, Sofia, Bulgaria

Ivan G. Mitov
Medical Academy, Sofia, Bulgaria

I. INTRODUCTION

Superoxide radicals are suspected of initiating or participating in several pathologic states such as aging, inflammation processes, and actions of drugs and chemicals. Superoxide dismutase (SOD) detoxifies these radicals and plays an important role in the biologic defense against oxidative stress. There are some reports concerning the application of exogenous SOD in chronic inflammation and radiation damage (1,2). In our previous investigations, we found that yeast superoxide dismutase (SOD) administered intraperitoneally to mice enhanced the natural killer (NK) cell activity (3). In vitro application of SOD did not influence the NK cell activity. It is important to find whether the NK cell activation after in vivo SOD treatment is mediated by cytokines.

The present study was performed to evaluate the effect of exogenous copper/zinc (Cu/Zn) SOD on the cytokine induction in vitro and in vivo.

II. MATERIALS AND METHODS

A. Cu/Zn SOD

The enzyme is isolated from *Kluyveromyces marxianus*. It is a blue-green water-soluble substance, a glycoprotein homodimer with molecular weight of 36 kd. An oligosaccharide moiety of 1600 daltons is associated with the protein and consists of 0.68 M xylose; 1 M mannose; 4.90 M galactose and 2.79 M glucose. The specific activity of the preparation is 3000 U/mg protein determined by the cytochrome *c* method. In some experiments, SOD from bovine erythrocytes (Serva) was included for comparison.

B. Pyrogen Assay

Outbred New Zealand white rabbits received an intravenous injection of SOD (25, 50, 500 μg or 2 mg/kg body weight) and temperature was monitored for 24 h.

C. Thymocyte Comitogenic Assay for Interleukin-1 (IL-1)

Thymocytes were cultured for 3 days with test samples (pMϕ supernatants) in the presence of suboptimal concentration of phytohemagglutinin (PHA). Cultures were pulsed with 0.25 μCi/well [^{3}H]thymidine for the last 18 h. Cells were harvested on glass fiber filters and thymidine incorporation was measured in a liquid scintillation counter.

D. Cytokine Induction in Human Whole Blood Cultures

Heparinized blood was diluted 1:3 with complete RPMI, peripheral blood mononuclear cells (PBMCs) were counted, adjusted to 1×10^6/ml, and stimulated with SOD (10, 100, and 200 μg/ml) for 24 h. Supernatants were used for cytokine determinations.

E. Bioassays for Tumor Necrosis Factor Alpha (TNF-α)

TNF in serum samples from SOD-treated mice and in PBMC supernatants was detected by its cytolytic effect on the sensitive L929 cell line in the presence of actinomycin D. After staining with crystal violet the percentage of lysed cells was measured in an enzyme-linked immunosorbent assay (ELISA)-reader.

F. Assay for Interferon Gamma (IFN-γ)

A monoclonal antibody–based ELISA was used for determination of IFN-γ in supernatants from human PBMCs (4).

G. IL-6 Bioassay

IL-6 levels in supernatants were measured using the hybridoma cell line B9 (5). The growth of the IL-6–dependent cells after cultivation with test samples was determined by 3- (4,5-dimethylthiazol-2-yl) 2,5-diphenyltetrazolium bromide (MTT) colorimetric assay.

III. RESULTS

A. Pyrogenic Effect of SOD

Stimulation of cytokine release by yeast SOD became apparent as pyrogenic effect. Within the concentration range from 25 μg to 2 mg/kg body weight, a dose-dependent effect was observed (Fig. 1). A maximum increase of the body temperature occurred between the third and sixth hour after the intravenous SOD injection.

B. IL-1 Release

In vitro cultivation of mouse pMϕ in the presence of 10, 100, and 1000 U/ml yeast SOD for 24 h resulted in an IL-1 release at all concentrations tested (Fig. 2). When pMϕ were triggered to IL-1 secretion by lipopolysaccharide (LPS), the presence of yeast SOD in the culture (10 U/ml) decreased the amount of released IL-1.

C. TNF-α Induction

TNF was detected in serum samples of mice injected intraperitoneally with 2 and 10 mg/kg yeast and bovine SOD (Fig. 3). Serum TNF levels in yeast SOD–treated

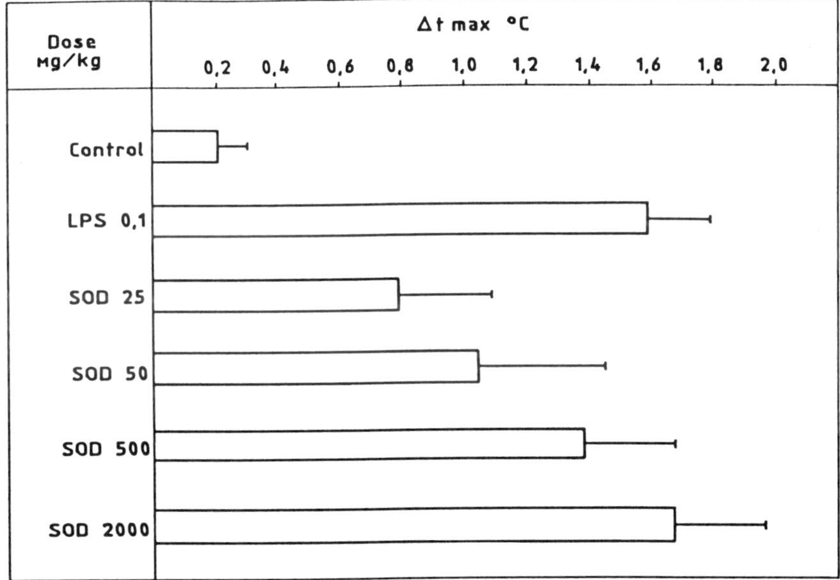

Figure 1 Pyrogenic effect of SOD in rabbits after a single intravenous injection. Maximum increases in the body temperatures are presented as means ±SD. Negative control—nonpyrogenic PBS; positive control—LPS.

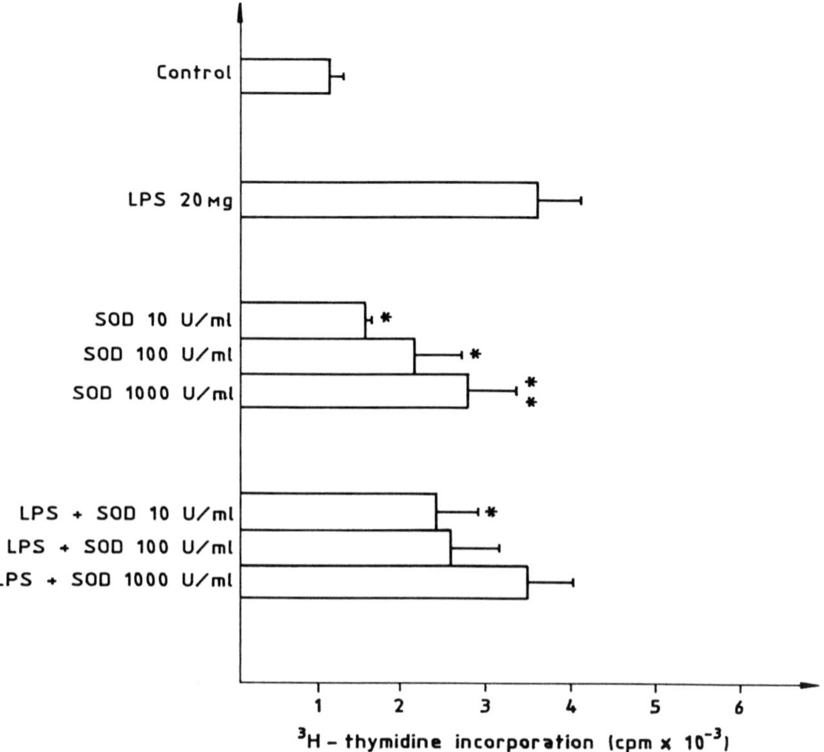

Figure 2 IL-1 release from mouse pMϕ after in vitro SOD stimulation. 1×10^6 pMϕ were cultured for 24 h in the presence of indicated concentrations of SOD with or without LPS. Supernatants were tested in a thymocyte comitogenic assay. Means ±SD. Significant differences are indicated versus the negative control (for SOD cultures) and versus the positive control (for SOD LPS cultures). *P < .05; ***P* < .02.

mice were raised 3 h after the administration and remained elevated till the third day. Bovine SOD showed some delay in the TNF induction with a maximum at 24 h when TNF levels were the highest. Cultivation of human PBMCs in the presence of 10, 100, and 200 μg/ml yeast SOD did not result in detectable TNF release (Table 1). When LPS at a concentration of 1 μg/ml was added to the cultures, a slight potentiation in the LPS-induced TNF production was detected.

D. IFN-γ and IL-6 Induction

The yeast-derived SOD caused a significant induction of IFN-γ by human PBMCs cultured in the presence of 10, 100, and 200 μg/ml SOD for 24 h (see Table 1). IL-6 was not detected under the same conditions.

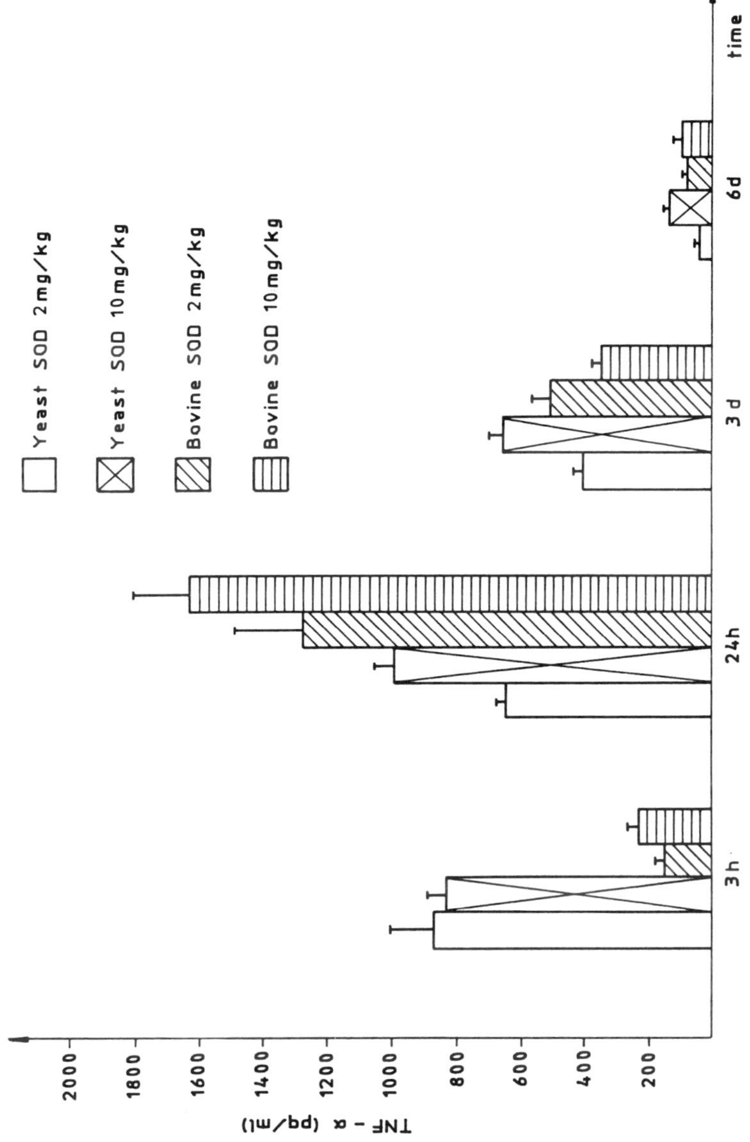

Figure 3 TNF-α levels in sera from mice intraperitoneally injected with SOD. Sera were pooled and diluted 1:40. Values represent means ±SD of triplicate determinations.

Table 1 TNF-α and IFN-γ Release from SOD-Stimulated Human PBMCs

LPS (μg/ml)	SOD (μg/ml)	Cytokine: TNF-α (pg/ml)	Cytokine: IFN-γ (ng/ml)
	10	0	1.85 ± 0.14
–	100	0	2.65 ± 0.31
	200	0	2.81 ± 0.33
1	–	140.2 ± 13.7[a]	5.26 ± 0.44
	10	174.8 ± 14.7[b]	2.90 ± 1.71
1	100	198.0 ± 21.6[c]	3.98 ± 1.23
	200	174.9 ± 11.4[b]	3.79 ± 2.25

[a] Mean ±SD in a representative experiment.
[b] $P < .02$.
[c] $P < .05$.

IV. DISCUSSION

Evidence exists about the stimulating effect of cytokines, and TNF-α in particular, on SOD synthesis (6,7). The opposite side of the relationship between SOD and the immune system has been poorly investigated. A few works have been published on the influence of exogenous and endogenous SOD on the cytokine levels. Helgestat et al. (8) has proved the stimulating effect of SOD on the granulocyte-macrophage colony-stimulating factor (GM-CSF), and this fact in addition to the radical scavenger functions explains the well-known radioprotective properties of the enzyme.

The present results showed that yeast Cu/Zn SOD influenced the cytokine release both in vitro and in vivo. The induction of IL-1 and IFN-γ explained the enhancement of NK cell activity for a prolonged period of time reported in our previous work (3). The IFN-γ induction was significant in cultures of human PBMCs and in sera of mice treated with yeast SOD (data not shown). SOD influenced other parameters of the cellular host defense system as well. After a single intraperitoneal injection, an enhancement of the phagocytic activity of mouse pMϕ and polymorphonuclears was observed with a maximum at the third day (9). These findings corresponded well with the elevated serum TNF-α levels detected until the third day after the treatment.

The comparison between the yeast and bovine erythrocyte Cu/Zn SOD showed differences in their effect on NK cell activation, in vivo IL-1 (3), and in vivo TNF-α induction. These differences may reflect the presence of oligosaccharide moiety in the yeast enzyme molecule unlike SOD from mammals.

The cytokine induction together with our previous results on the NK cell and pMϕ activation suggests some additional immunologic mechanisms of action of exogenous SOD as a part of the biologic defense against oxidative stress. The role of SOD in some pathologic states needs further elucidation.

ACKNOWLEDGMENT

This work was supported by a Project grant No. LI/221/92 from the Bulgarian National Foundation for Scientific Research.

REFERENCES

1. Flohe L. Superoxide dismutase for therapeutic use: Clinical experience, dead ends and hopes. Mol Cell Biochem 1988; 84:123.
2. Petkau A. Role of superoxide dismutase in modification of radiation injury. Br J Cancer 1987; 55(Suppl. VIII):87.
3. Neychev H, Stefanova Z, Toschkova R, Kujumdjieva A, Savov V, Kenarova B. Enhancement of natural killer (NK) cell activity by superoxide dismutase (SOD). 17th International Congress of Chemotherapy, Berlin, 1991 (Abstr).
4. Gallati H, Pracht I, Schmidt J, Haring P, Garota G. A simple, rapid and large capacity ELISA for biologically active native and recombinant human IFN-γ. J Biol Regul Homeost Agents 1987; 1:109.
5. Aarden LA, de Groot ER, Shaap OL, Landsdorp PM. Production of hybridoma growth factor by monocytes. Eur J Immunol 1987; 17:1411.
6. Kawaguchi T, Takeyasu A, Matsunobu K, Uda T, Ishizawa M, Suzuki K, Nishiura T, Ishikawa M, Taniguchi N. Stimulation of Mn-superoxide dismutase expression by tumor necrosis factor-α: Quantitative determination of Mn-SOD protein levels in TNF-resistant and sensitive cells by ELISA. Biochem Biophys Res Commun 1990; 171:1378.
7. Visner GA, Dougall WC, Wilson JM, Burr A, Nick HS. Regulation of manganese superoxide dismutase by lipopolysaccharide, interleukin-1 and tumor necrosis factor. J Biol Chem 1990; 265:2856.
8. Helgestad J, Storm-Mathisen I, Lie SO. Endogenous inhibitors of human granulocyte/macrophage colony forming cells in vitro are inactivated by free radical scavengers. Scand J Haematol 1986; 37:395.
9. Stefanova Z, Mizinska-Boevska J, Janchev I, Neychev H. Influence of yeast superoxide dismutase on phagocytic activity of mouse peritoneal exudate cells (PEC). CR Acad Bulg Sci 1993; 46:117.

10

Tumor Necrosis Factor-Alpha in Combination with Riminophenazines Enhances Oxidative Metabolism of Neutrophils from Patients with Leprosy

Marta M. Krajewska
Medical University of Southern Africa, Medunsa, South Africa

Ronald Anderson
University of Pretoria, Pretoria, South Africa

I. INTRODUCTION

The treatment of leprosy, which is caused by the obligate intracellular bacterium *Mycobacterium leprae* multiplying within the mononuclear phagocytes, is long and a risk of developing a resistance to the drugs by *M. leprae* remains a problem. The survival of mycobacteria inside the phagocytes is of fundamental importance in the pathogenesis of leprosy and tuberculosis (1).

Activated phagocytes undergo a chain of oxidative metabolic reactions called the respiratory burst (RB), as a result of which strongly antimicrobial reactive oxidants (RO) are formed. The products of RB are essential for killing invading microorganisms (2). These reactive oxygen compounds (superoxide [$O_2^{\cdot -}$], hydroxyl radical [OH·], hydrogen peroxide [H_2O_2], hypochlorous acid [HOCl] and chloramines [R-NHC1]) provide most of the microbicidal oxidative activity within the phagosome and extracellular environment (3). Furthermore, susceptibility of mycobacteria to H_2O_2 appears to be inversely related to their virulence (4). Therefore, enhancement of the oxidative reactions of the immune response should improve the treatment of leprosy.

Tumor necrosis factor alpha (TNF-α) has been reported to have stimulatory effects on neutrophils (increased phagocytosis, antibody-dependent cell-mediated cytotoxicity, respiratory burst, and degranulation), the products of which may mediate the cytostatic and antimicrobial activity of TNF-α (7–9).

Clofazimine, the antileprosy drug, shows strong antimycobacterial properties and is now recommended by the World Health Organization (WHO) as part of a combination therapy for all cases of leprosy (5). Furthermore, clofazimine has been reported to enhance oxidative metabolism of neutrophils (6).

In the present study, we tested the effects of combination treatment of TNF-α with clofazimine or its analogue (the halogen-free cycloalkylimino compound B669) on neutrophils from patients with leprosy. Neutrophils from four untreated patients with leprosy were assayed for myeloperoxidase (MPO)–mediated iodination, cytochrome *c* reduction assay for superoxide generation, and luminol-enhanced chemiluminescence (LECL).

II. MATERIALS AND METHODS

Unless indicated, all chemicals and reagents were obtained from the Sigma Chemical Corp. (St. Louis, MO).

A. Chemicals and Reagents

Recombinant human TNF-α (rhTNF-α) was supplied by Amersham Laboratories (Aylesbury, England). Clofazimine (3-[p-chloro-anilino]-10-[p-chlorophenyl]-2,10-dihydro-2-[isopropylimino]-phenazine) and its analogue B669 were synthesized by Dr. J.F. O'Sullivan, the Health Research Board Laboratories, Department of Chemistry, University College Dublin, Republic of Ireland. The agents were selected on the basis of previously published data (10).

B. Solubilization of Agents

rhTNF-α was diluted in appropriate buffer, as recommended by the manufacturer, and further diluted in Hanks' Balanced Salt Solution (HBSS) just before the assay to the final concentration of 10 ng/ml, which is in the therapeutic range (12).

Two milligrams of clofazimine or B669 were dissolved in 1 ml of dimethyl sulfoxide (DMSO) and further diluted in HBSS, buffered with 4.2 mM N-2-hydroxy-ethyl-piperazine-N′-2-ethane sulfonic acid (HEPES), pH 7.4. The agents were further diluted in HBSS to the final concentration of 0.5 μg/ml immediately prior to the assay. Yawalkar and Vischer (11) reported peak serum levels of clofazimine 0.7–1.0 μg/ml after oral dose of 200 mg clofazimine daily.

C. Cell Preparation

Leukocytes were prepared from heparinized venous blood obtained from four patients with leprosy (newly diagnosed and untreated) in Vac-U-Test tubes containing lithium heparin. Three cases were classified as borderline (BB) and one as borderline lepromatous (BL) form of leprosy according to the clinical and histopathological classification of Ridley and Jopling (13). Polymorphonuclear leukocytes (PMNLs) were separated on Ficoll (Pharmacia, Uppsala, Sweden) – metrizoate (Nyegaard & Co., Oslo, Norway) cushions as previously described (14).

D. Myeloperoxidase (MPO)-Mediated Iodination Assay

This was determined by a modified method of Klebanoff and Clark (15). Briefly, the cells (10^7/ml in HBSS) were incubated for 15 min at 37°C with bovine serum albumin (10 mg/ml), 40 nmol of carrier sodium iodide, tested agents, and 0.6 μCi of ^{125}I-labeled sodium iodide. Thereafter N-formyl-L-methionyl-L-leucyl-L-phenylalanine (FMLP) was added, and 30 min later the reaction was terminated by the addition of ice-cold 10% TCA. The incorporation of ^{125}I into acid-precipitable protein was determined by solid scintillation counting of the resulting precipitate. The results were expressed as nanomoles of ^{125}I contained in the protein precipitate/ 10^6 PMNLs/30 min.

E. Superoxide Generation Assay

This was measured by superoxide dismutase (SOD)–inhibitable reduction of ferricytochrome *c* (cyt *c* type VI) as described elsewhere (16). The cells were stimulated with FMLP at a predetermined optimal concentration (1 μM). In this study, superoxide ($O_2^{\cdot -}$) release was measured by an endpoint assay (fixed time measurements). The reaction was terminated by chilling and the cells removed by centrifugation prior to measuring the difference in the optical density in a known aliquot of the cell-free supernatant at 500 nm in a Pye Unicam ultraviolet double-beam spectrophotometer (model SP 1700, Unicam, England). The amount of reduced cyt *c* was calculated using an extinction coefficient of 21.1 mM at 550 nm (17).

F. Luminol-Enhanced Chemiluminescence (LECL) Assay

Experimental mixtures containing PMNLs, tested agent(s), and luminol (final concentration 0.2 mM) were preincubated for 5 min. Thereafter FMLP (1 μM) was added and LECL measured at 37°C in a LKB Wallac 1251 luminometer (Turku, Finland). The results were expressed as peak responses in millivolts (mV) per second.

G. Expression and Analysis of Results

The results were expressed as a mean ±SE of the mean (SEM) of triplicate assay.

III. RESULTS

A. Effects of TNF-α in Combination with Clofazimine and B669 on PMNL MPO-Mediated Iodination

These results are shown in Tables 1–4. TNF-α and riminophenazines used alone primed PMNLs for increased iodination, especially in response to stimulation with FMLP. Furthermore, the effects of combinations of TNF-α with clofazimine or B669 were more potent than those of any agent used individually.

Table 1 Patient No. 1. Effects of TNF-α (10.0 ng/ml) Alone or in Combination with Clofazimine (clof) and B669 (0.5 μg/ml) on PMNL Iodination and Superoxide ($O_2^{\cdot -}$) Generation

	Iodination[a] nmol $^{125}I/5 \times 10^5$ PMNL		$O_2^{\cdot -}$ generation[a] nmol red. cytochrome $c/5 \times 10^5$ PMNLs	
Experiment	no FMLP	FMLP 1 μM	no FMLP	FMLP 1 μM
Control	0.38 ± 0.01	0.68 ± 0.05	2.05 ± 0.68	5.61 ± 0.49
TNF-α	0.43 ± 0.05	1.26 ± 0.07[c]	4.10 ± 0.49[c]	9.80 ± 0.13[b]
clof	0.39 ± 0.03	0.83 ± 0.01[d]	2.92 ± 0.14	8.13 ± 0.49[b]
B669	0.47 ± 0.04	0.83 ± 0.04[b]	2.61 ± 0.48	9.08 ± 0.60[d]
clof + TNF-α	0.47 ± 0.02[d]	1.39 ± 0.22[d]	3.87 ± 0.59	10.35 ± 0.27[c]
B669 + TNF-α	0.47 ± 0.05	1.46 ± 0.12[c]	3.24 ± 0.36	10.90 ± 0.24[b]

[a] Results expressed as a mean ± SEM of triplicate assay, PMNL concentration 5×10^5/ml.
[b] $p < .005$.
[c] $p < .01$.
[d] $p < .05$.

Table 2 Patient No. 2. Effects of TNF-α (10.0 ng/ml) Alone or in Combination with Clofazimine (clof) and B669 (0.5 μg/ml) on PMNL Iodination and Superoxide ($O_2^{\cdot -}$) Generation

	Iodination[a] nmol $^{125}I/3 \times 10^5$ PMNL		$O_2^{\cdot -}$ generation[a] nmol red. cytochrome $c/3 \times 10^5$ PMNLs	
Experiment	no FMLP	FMLP 1 μM	no FMLP	FMLP 1 μM
Control	0.62 ± 0.05	0.90 ± 0.06	0.39 ± 0.13	4.43 ± 0.49
TNF-α	0.74 ± 0.12	0.95 ± 0.10	1.19 ± 0	6.87 ± 0.48[d]
clof	0.80 ± 0.10	1.00 ± 0.03[d]	1.10 ± 0.13	6.32 ± 0.76[d]
B669	0.79 ± 0.05	1.12 ± 0.10	0.95 ± 0[d]	6.79 ± 0.83[d]
clof + TNF-α	0.90 ± 0.08[c]	1.49 ± 0.12[c]	1.89 ± 0.82	6.56 ± 0.73[d]
B669 + TNF-α	0.97 ± 0.09[d]	1.42 ± 0.06[c]	3.32 ± 0.63[d]	10.59 ± 0.49[b]

[a] Results expressed as a means ± SEM of triplicate assay, PMNL concentration 3×10^5/ml.
[b] $p < .001$.
[c] $p < .01$.
[d] $p < .05$.

Table 3 Patient No. 3. Effects of TNF-α (10.0 ng/ml) Alone or in Combination with Clofazimine (clof) and B669 (0.5 μg/ml) on PMNL Iodination and Superoxide ($O_2^{\cdot -}$) Generation

	Iodination[a] nmol ^{125}I/3 × 10^5 PMNL		$O_2^{\cdot -}$ generation[a] nmol red. cytochrome *c*/3 × 10^5 PMNLs	
Experiment	no FMLP	FMLP 1 μM	no FMLP	FMLP 1 μM
Control	0.44 ± 0.11	0.73 ± 0.05	1.19 ± 0.41	6.56 ± 1.07
TNF-α	0.46 ± 0.02	1.16 ± 0.03[c]	1.58 ± 0.37	8.61 ± 0.76[d]
clof	0.49 ± 0.05	1.11 ± 0.02[c]	1.74 ± 0.77	8.61 ± 0.36
B669	0.52 ± 0.03	0.96 ± 0.04[d]	2.69 ± 0.83[d]	9.48 ± 0.24[d]
clof + TNF-α	0.53 ± 0.08	1.35 ± 0.10[c]	2.69 ± 0.36[c]	9.87 ± 1.54
B669 ± TNF-α	0.55 ± 0.01	1.08 ± 0.08[c]	3.40 ± 0.77[b]	11.69 ± 0.96[c]

[a] Results expressed as a means ±SEM of triplicate assay, PMNL concentration 3 × 10^5/ml.
[b] $p < .001$. [c] $p < .005$. [d] $p < .05$.

Table 4 Patient No. 4. Effects of TNF-α (10.0 ng/ml) Alone or in Combination with Clofazimine (clof) and B669 (0.5 μg/ml) on PMNL Iodination and Superoxide ($O_2^{\cdot -}$) Generation

	Iodination[a] nmol ^{125}I/10^6 PMNL		$O_2^{\cdot -}$ generation[a] nmol red. cytochrome *c*/10^6 PMNLs	
Experiment	no FMLP	FMLP 1 μM	no FMLP	FMLP 1 μM
Control	0.30 ± 0.04	0.44 ± 0.04	0.87 ± 0.27	3.55 ± 0.48
TNF-α	0.35 ± 0.01	1.22 ± 0.07[c]	1.82 ± 0.36[d]	9.32 ± 0.59[b]
clof	0.30 ± 0.03	0.63 ± 0.01[d]	1.66 ± 0.41	4.97 ± 0.41
B669	0.33 ± 0.01	0.68 ± 0.07[d]	2.69 ± 0.83	8.14 ± 0.60[d]
clof + TNF-α	0.40 ± 0.02	1.93 ± 0.15[c]	2.61 ± 0.63[d]	11.85 ± 0.63[b]
B669 + TNF-α	0.43 ± 0.05[d]	2.00 ± 0.33[d]	3.16 ± 0.59[d]	12.72 ± 0.36[c]

[a] Results expressed as a means ±SEM of triplicate assay, PMNL concentration 3 × 10^5/ml.
[b] $p < .001$. [c] $p < .005$. [d] $p < .05$.

B. Effects of TNF-α in Combination with Clofazimine and B669 on PMNL Superoxide Generation

These results are presented in Tables 1–4. TNF-α, clofazimine, and B669 caused enhanced $O_2^{\cdot -}$ generation by PMNLs from patients with leprosy, especially in FMLP-stimulated cells. Combination treatment of these cells with TNF-α and clofazimine or B669 caused a significant enhancement of $O_2^{\cdot -}$ generation in comparison to the effects of the agents used alone.

C. Effects of TNF-α in Combination with Clofazimine and B669 on PMNL LECL

These results appear in Figs. 1–4 and are expressed as a mean of duplicate readings. TNF-α and riminophenazines used alone were stimulatory for LECL of PMNL. Moreover, when used in combination, they caused at least additive effects in LECL assays in comparison with the effects of individual agents. Interestingly, this effect was in all four cases time dependent, especially in the case of PMNLs from patient no. 2 (Fig. 2).

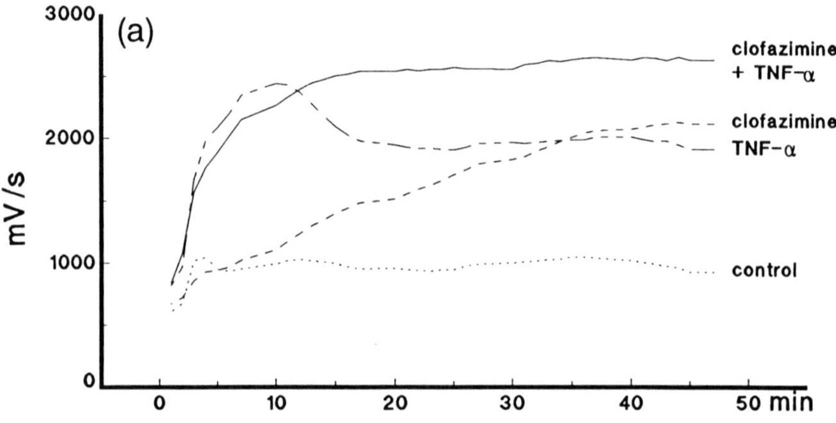

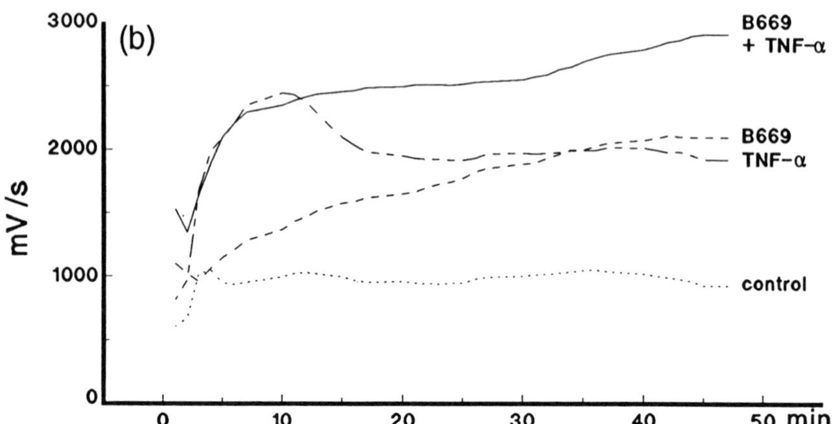

Figure 1 Patient no. 1. Effects of TNF-α (10 ng/ml) alone or in combination with (a) clofazimine or (b) B669 (0.5 μg/ml) on luminol-enhanced chemiluminescence of FMLP-stimulated PMNLs.

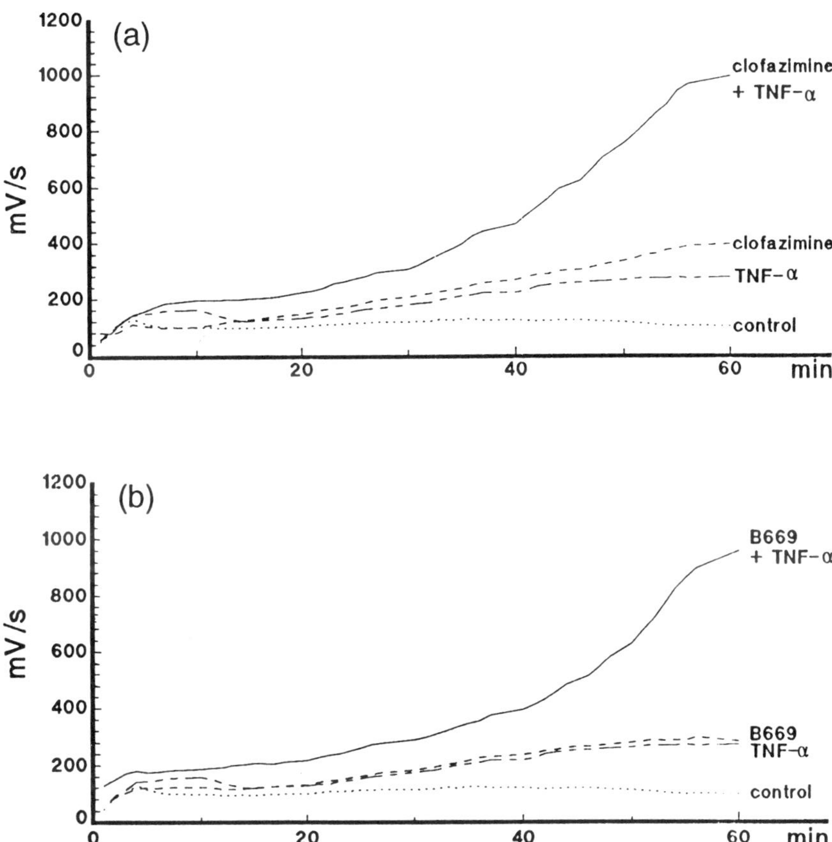

Figure 2 Patient no. 2. Effects of TNF-α (10 ng/ml) alone or in combination with (a) clofazimine or (b) B669 (0.5 μg/ml) on luminol-enhanced chemiluminescence of FMLP-stimulated PMNLs.

IV. DISCUSSION

Studies on the effects of combinations of cytokines with pharmacologic agents are emerging as important areas of immunologic research. Synergistic or additive interactions may allow reduction of the doses and side effects of the agents and allowing at the same time maximal beneficial effects. There are known synergistic interactions of TNF-α in cancer therapy, including combinations with interferon gamma (IFN-γ) (18) or interleukin-2 (IL-2) (19).

TNF-α was originally described by Carswell et al. (20) as a serum factor secreted by activated monocytes and macrophages that is able to cause necrosis of certain tumors. However, many biologic activities involving various cells have been since described. TNF-α appears to be both a primary mediator in the pathogene-

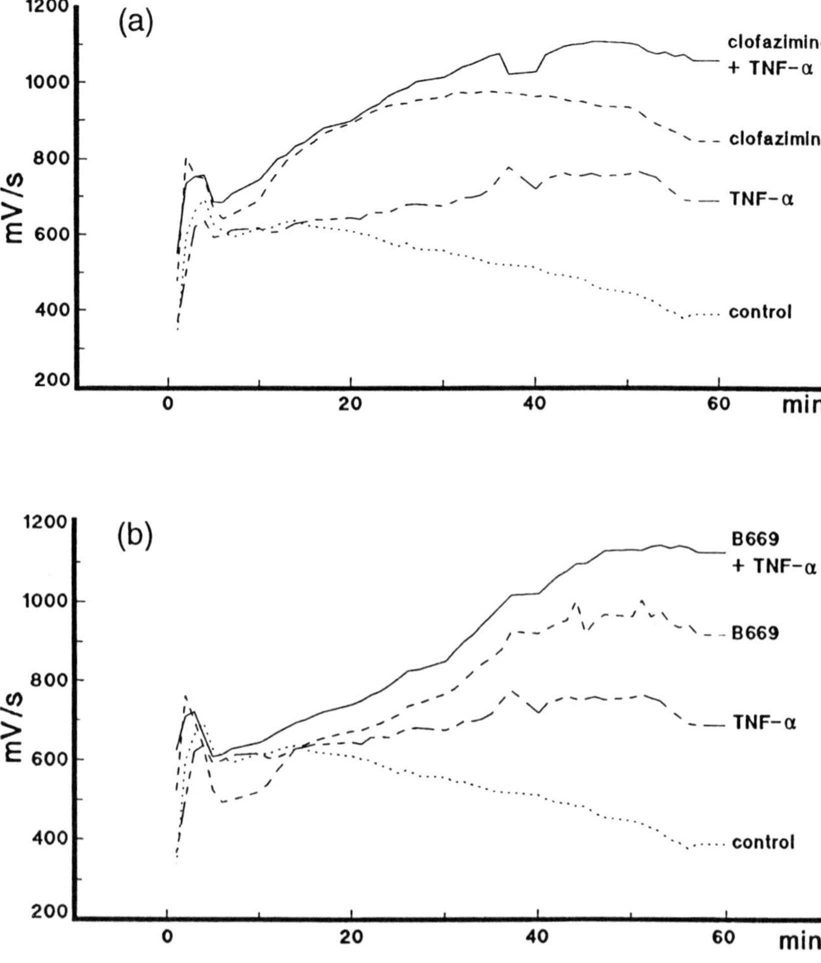

Figure 3 Patient no. 3. Effects of TNF-α (10 ng/ml) alone or in combination with (a) clofazimine or (b) B669 (0.5 μg/ml) on luminol-enhanced chemiluminescence of FMLP-stimulated PMNLs.

sis of infection, injury, and inflammation, whereas at the same time plays a role in beneficial processes of host defence and tissue homeostasis (21,22). When lesser amounts of TNF-α are released in tissues, the beneficial effects predominate leading to enhanced host defence against pathogens. TNF-α also has some antimicrobial properties, enhancing neutrophil-mediated killing of *Plasmodium falciparum* (23) and increasing eosinophil ability to kill schistosomula (24). TNF-α enhances oxidative burst, $O_2^{\cdot -}$ production and arachidonic acid metabolism in PMNLs

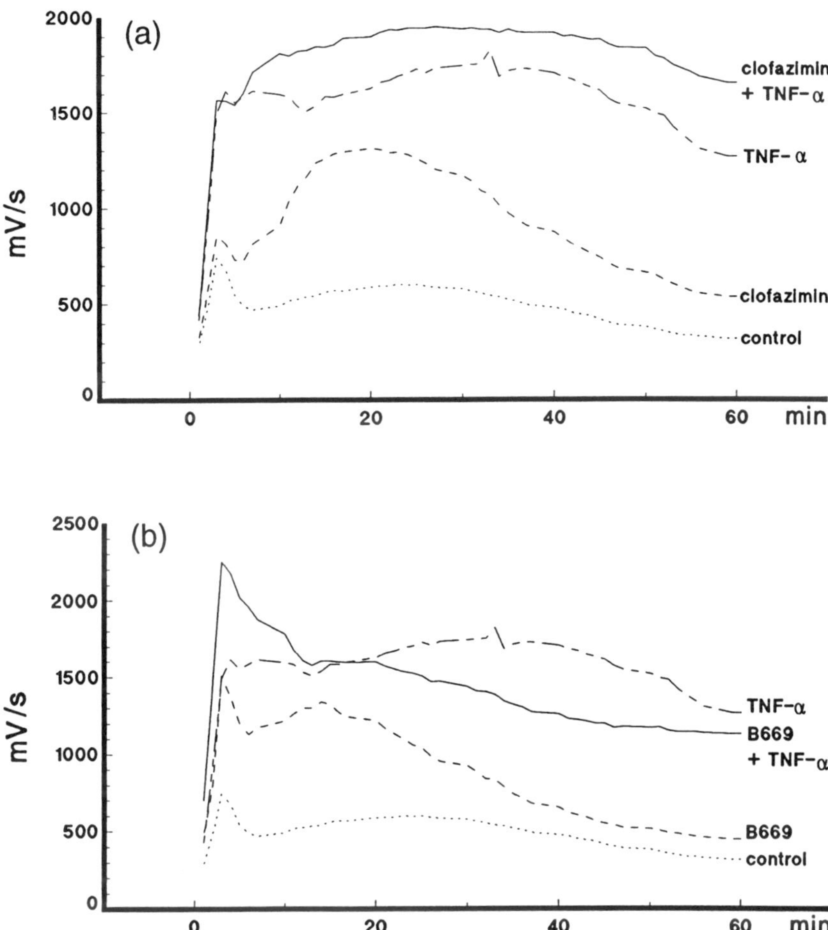

Figure 4 Patient no. 4. Effects of TNF-α (10 ng/ml) alone or in combination with (a) clofazimine or (b) B669 (0.5 μg/ml) on luminol-enhanced chemiluminescence of FMLP-stimulated PMNLs.

(8,25–27). Furthermore, TNF-α primes PMNLs for increased H_2O_2 generation, MPO-mediated iodination, and degranulation of PMNLs (7). Shau (9) postulated that H_2O_2 produced by TNF-α–treated PMNLs is the mediator of cytostatic activity of TNF-α.

Clofazimine has been reported to have strong antimycobacterial properties (28). Clofazimine is active not only against *M. tuberculosis* and *M. leprae* (29,30) but also against many other species such as *M. fortuitum, M. chelonae* (31), and *M.*

avium complex (32,33). Moreover, it has been successfully used in the treatment of other nonmycobacterial intracellular infections with pathogens that interfere with phagocyte microbicidal functions; for example, *Brucella abortus* (34) or *Legionella pneumophila* (35). Furthermore, clofazimine has been useful in the treatment of fistulous withers in horses, which is caused by *Brucella abortus, Actinomycoses bovis,* or *Streptococcus zooepidemicus* (36).

Clofazimine at different concentrations (0.5–10.0 μg/ml) was capable of enhancing the spontaneous production of H_2O_2 and intracellular killing ability of phagocytes. This may reverse the inhibitory effect of mycobacteria on intracellular killing mechanisms of phagocytes (37). Clofazimine also potentiates the reactivity of the MPO/H_2O_2/halide system as measured by LECL and oxygen consumption (6). Clofazimine-mediated dose-dependent (0.1–5.0 μg/ml) potentiation of $O_2^{\cdot-}$ generation by neutrophils is stimulus nonspecific (6).

Phagocytes of patients with leprosy showed both decreased phagocytes and diminished chemotaxis. Monocytes from patients with lepromatous leprosy secrete subnormal quantities of H_2O_2 after stimulation with phorbol myristate acetate (PMA) and *M. leprae* fails to stimulate significant $O_2^{\cdot-}$ release from human monocytes and neutrophils (38). Furthermore, neutrophils from untreated patients with leprosy have reduced production of $O_2^{\cdot-}$ and OH· as well as increased SOD activity in comparison to healthy individuals (39). Interestingly, *M. leprae* itself contains a large amount of SOD, which is probably active. Combined SOD activity, both of human and bacterial origin, would further diminish the concentration of $O_2^{\cdot-}$ and OH·.

In this study, we report superior effects of combinations of TNF-α with clofazimine or B669 on oxidative metabolism of PMNLs from patients with leprosy in comparison with the effects of individual agents. The effects of the combinations were especially potent on FMLP-stimulated PMNLs. Measurements of LECL were taken over a period of time and thus the results reflected the peak values, whereas the cyt *c* reduction assay was read at a fixed time and reflected the cumulative production of $O_2^{\cdot-}$. LECL appears to be more sensitive than the assays for the specific oxygen intermediates (40). Furthermore, these enhancing effects in LECL assays were clearly time dependent, which might be of significance in vivo. This may also apply to the treatment of other nonmycobacterial infections caused by microorganisms multiplying intracellularly. Since oxygen radicals are highly toxic and antimicrobial, such enhancement of host defenses may significantly improve the effects of therapy.

V. CONCLUSIONS

TNF-α enhances the generation of cytostatic and antimicrobial products of neutrophils. Clofazimine, a riminophenazine, and its analogue B669 also potentiate oxidative metabolism of neutrophils. Combinations of these agents were tested for possible synergistic enhancement in the generation of strongly antimicrobial reactive oxidants by phagocytes of patients with leprosy. This may improve the effect of treatment of infections caused by intracellular pathogens. Neutrophils from four patients with untreated leprosy were exposed to the combinations of TNF-α and clofazimine or B669 and assayed for myeloperoxidase-mediated iodination, superoxide generation, and LECL. Combination treatment with TNF-α and clofazimine or B669 was more stimulatory for reactive oxidant generation by neutrophils than these agents used individually. Furthermore, in LECL assays a marked time-dependent response was recorded, which might be of significance in vivo.

ACKNOWLEDGMENTS

The authors would like to thank Dr. P.W. Mars of the Department of Dermatology at Medunsa and Mrs. Koch of Westford Hospital in Pretoria for help in obtaining samples of blood from patients with leprosy. We are also grateful to Mrs. M.E.P. Petzer of the Audiovisual Department at Medunsa for production of the figures.

REFERENCES

1. Lowrie DB. Mononuclear phagocyte-mycobacterium interaction. In: Ratledge C, Stanford J, eds. The biology of the mycobacteria. Vol. 2: Immunological and Environmental Aspects. London: Academic Press, 1983:235.
2. Babior BM. Oxidants from phagocytes: Agents of defence and destruction. A review. Blood 1984; 64:959.
3. Baggiolini M, Wymann MP. Turning on the respiratory burst. TIBS 1990; 15:69.
4. Edwards CK, Hedegaard HB, Zlotnik A, Gangadharam PR, Johnston RB, Jr, Pabst MJ. Chronic infection due to *Mycobacterium intracellulare* in mice: Association with macrophage release of prostaglandin E_2 and reversal by injection of indomethacin, muramyl dipeptide or interferon gamma. J Immunol 1986; 136:1820.
5. World Health Organization, Chemotherapy of leprosy for control programmes. Report of a WHO study group, WHO, Geneva, World Health Organization Technical Report Series. 1982; 675:7.
6. Anderson R, Zeis BM, Anderson IF. Clofazimine-mediated enhancement of reactive oxidant production by human phagocytes as a possible therapeutic mechanism. Dermatologica 1988; 176:234.
7. Klebanoff SJ, Vadas MA, Harlan JM, Sparks LH, Gamble JR, Agosti JM, Waltersdorph AM. Stimulation of neutrophils by tumor necrosis factor. J Immunol 1986; 136:4220.
8. Nathan CF. Neutrophil activation on biological surfaces. Massive secretion of hydrogen peroxide in response to products of macrophages and lymphocytes. J Clin Invest 1987; 80:1550.
9. Shau H. Characteristics and mechanism of neutrophil-mediated cytostasis induced by tumor necrosis factor. J Immunol 1988; 141:234.
10. Savage JE, O'Sullivan JF, Zeis BM, Anderson R. Investigation of the structural prop-

erties of dihydrophenazines which contribute to their pro-oxidative interactions with human phagocytes. J Antimicrob Chemother 1989; 23:691.

11. Yawalkar SJ, Vischer W. Lamprene (clofazimine) in leprosy. Lepr Rev 1979; 50:135.
12. Sheron N, Lau JN, Hofmann J, Williams R, Alexander GJM. Dose-dependent increase in plasma interleukin-6 after recombinant tumour necrosis factor infusion in humans. Clin Exp Immunol 1990; 82:427.
13. Ridley DS, Jopling WH. Classification of leprosy according to immunity: a five-group system. Int J Lepr 1966; 34:255.
14. Anderson R. Enhancement by clofazimine and inhibition by dapsone of production of prostaglandin E_2 by human polymorphonuclear leucocytes in vitro. Antimicrob Agents Chemother 1985; 27:257.
15. Klebanoff SJ, Clark RA. Iodination by human polymorphonuclear leukocytes: A re-evaluation. J Lab Clin Med 1977; 89:675.
16. Clifford DP, Repine JE. Measurement of oxidizing radicals by polymorphonuclear leucocytes. Meth Enzymol 1984; 105:393.
17. Van Gelder BF, Slater EC. The extinction coefficient of cytochrome *c*. Biochim Biophys Acta 1962; 58:593.
18. Balkwill FR, Ward BG, Moodie E, Fiers W. Therapeutic potential of tumour necrosis factor-alpha and gamma-interferon in experimental human ovarian cancer. Cancer Res 1987; 47:4755.
19. Nishimura T, Ohta S, Sato N, Togashi Y, Goto M, Hashimoto Y. Combination tumour-immunotherapy with recombinant tumour necrosis factor and recombinant interleukin 2 in mice. Int J Cancer 1987; 40:255.
20. Carswell EA, Old LJ, Kassel RL, Green S, Fiore N, Williamson B. An endotoxin-induced serum factor that causes necrosis of tumors. Proc Natl Acad Sci USA 1975; 72:3666.
21. Havell EA. Evidence that tumor necrosis factor has an important role in antibacterial resistance. J Immunol 1989; 143:2894.
22. Tracey KJ, Vlassara H, Cerami A. Cachectin/tumour necrosis factor. Lancet 1989; 1:1122.
23. Kumaratilake LM, Ferrante A, Rzepczyk CM. Tumor necrosis factor enhances neutrophil-mediated killing of *Plasmodium falciparum*. Infect Immun 1990; 58:788.
24. Silberstein DS, David JR. Tumor necrosis factor enhances eosinophil toxicity to *Schistosoma mansoni* larvae. Proc Natl Acad Sci USA 1986; 83:1055.
25. Berkow RL, Dodson MR. Biochemical mechanisms involved in the priming of neutrophils by tumor necrosis factor. J Leukoc Biol 1988; 44:345.
26. She Z-W, Wewers MD, Herzyk DJ, Sagone AL, Davis WB. Tumor necrosis factor primes neutrophils for hypochlorous acid production. Am J Physiol 1989; 257:L338.
27. Test ST. Effect of tumor necrosis factor on the generation of chlorinated oxidants by adherent human neutrophils. J Leukoc Biol 1991; 50:131.
28. Barry VC, Belton JG, Conalty ML, Denneny JM, Edward DW, O'Sullivan JF, Twomey D, Winder F. A new series of phenazines (rimino-compounds) with high antituberculous activity. Nature 1957; 179:1013.
29. Barry VC, Conalty ML. The antimycobacterial activity of B663. Lepr Rev 1965; 36:3.
30. Levy L. Activity of four clofazimine analogues against *Mycobacterium leprae*. Lepr Rev 1981; 52:23.

31. Ausina V, Condom MJ, Mirelis B, Luquin M, Coll P, Prats G. *In vitro* activity of clofazimine against rapidly growing nonchromogenic mycobacteria. Antimicrob Agents Chemother 1986; 29:951.
32. Agins BD, Berman DS, Spicehandler D, El-Sadr W, Simberfoff MS, Rahal JJ. Effect of combined therapy with ansamycin, clofazimine, ethambutol and isonazid for *Mycobacterium avium* infection in patients with AIDS. J Infect Dis 1989; 159:784.
33. Centers for Disease Control, Diagnosis and management of mycobacterial infection and disease in persons with human T-lymphotropic virus type III/lymphadenopathy associated virus infection. MMWR 1986; 35:448.
34. Canning PC, Roth JA, Tabatabai LB, Deyoe BL. Isolation of components of *Brucella abortus* responsible for inhibition of function in bovine neutrophils. J Infect Dis 1985; 152:913.
35. Gabay JE, Horwitz MA. Isolation and characterization of the cytoplasmic and outer membranes of the legionnaires' disease bacterium (*Legionella pneumophila*). J Exp Med 1985; 161:409.
36. Knottenbelt DC, Hill FWG, Morton DJ. Clofazimine for the treatment of fistulous withers in three horses. Vet Rec 1989; 125:509.
37. Wadee AA, Anderson R, Rabson AR. Clofazimine reverses the inhibitory effect of *Mycobacterium tuberculosis* derived factors on phagocyte intracellular killing mechanisms. J Antimicrob Chemother 1988; 21:65.
38. Santos DO, Salgado JL, Moreira AL, Sarno EN. Reactive oxygen intermediates in the phagocytes from leprosy patients: Correlation with reactional states and variation during treatment. Int J Lepr 1992; 60:92.
39. Niwa Y, Sakane T, Miyachi Y, Ozaki M. Oxygen metabolism in phagocytes of leprotic patients: Enhanced endogenous superoxide dismutase activity and hydroxyl radical generation by clofazimine. J Clin Microbiol 1984; 20:837.
40. Sample AK, Czuprynski CJ. Priming and stimulation of bovine neutrophils by recombinant human interleukin-1 alpha and tumor necrosis factor alpha. J Leukoc Biol 1991; 49:107.

II

SYNTHETIC IMMUNOMODULATORS

11

Immunological Profile of Methyl Inosine Monophosphate

Marina Sosa, Anutosh R. Saha, Elba M. Hadden, and John W. Hadden
University of South Florida Medical College, Tampa, Florida

I. INTRODUCTION

Immunostimulants, including chemically defined compounds and biological substances, have been successfully employed in the prevention and treatment of human diseases (1). An increasing interest in immunotherapy over recent decades has resulted in a large and diverse collection of these agents. One group of immunologically active compounds has been derived from purine structures such as inosine and hypoxanthine. These compounds have been classified as "thymomimetic drugs" (2) in that they stimulate the immune system by actions primarily on thymus-derived T lymphocytes, although they may have actions on other cells involved in immune responses. Prior work has documented the thymomimetic and immunotherapeutic activity of isoprinosine, a purine molecule containing inosine (3). Synthesis of a series of new molecular structures has yielded a methylated form of inosine monophosphate (MIMP) as a third-generation compound of this type (Fig. 1). Like the other purines, in vitro MIMP induces in human prothymocytes the expression of T-lymphocyte differentiation markers (CD3, CD4, CD8) and interleukin-2 (IL-2) receptor (CD25) indicating a thymic hormone–like action (4). Based on such results, it has been postulated that T-cell precursors bear a purine receptor for these inosinelike compounds that leads via a yet to be defined signaling pathway to differentiation of T cells (5) and that this action is central to the designation "thymomimetic."

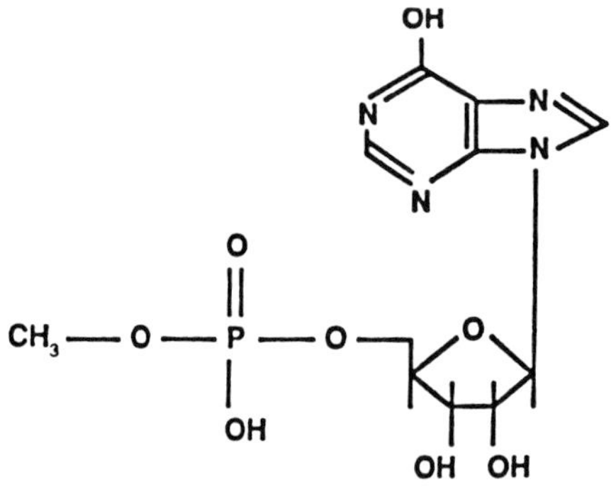

Figure 1 Methyl inosine monophosphate.

Secondary immunodeficiency involving the thymus-dependent T-lymphocyte system is common to many infirmities. It is evident in cancer and in human immunodeficiency virus (HIV) and other viral infections, severe physical stress, and aging. In each of these conditions, although lymphopenia may occur, suppressive influences extrinsic to the T lymphocyte have been documented suggesting that therapy designed to reverse immunosuppressive influences might ameliorate the immunodeficiency. The current studies were undertaken to evaluate the ability of MIMP to reverse various immunosuppressive influences in human and murine lymphocytes in vitro and to examine the effect of MIMP on the immune responses of normal mice.

The chemical and immunopharmacological properties of MIMP have been reviewed (6). Its structure was confirmed by nuclear magnetic resonance (NMR) and high-performance liquid chromatography (HPLC). After double crystallization, the purity of MIMP was found >99%, with a sharp melting point (145°C). MIMP was assessed for toxicity in vivo, and the lethal dose for half the mice (LD_{50}) exceeded 500 mg/kg of body weight by both the parenteral and oral routes (7).

In vitro and in vivo experiments have been carried out in order to evaluate the effect of MIMP on different immunologic aspects.

II. IN VITRO EFFECT OF MIMP ON HUMAN AND MICE LYMPHOCYTES

Peripheral blood lymphocytes obtained from normal donors (human peripheral blood lymphocytes, HPBLs) and mice splenocytes were stimulated with phytohemagglutinin (PHA) and concanavalin A (ConA) in the presence of MIMP at concentrations of 0.1–100 μg/ml (8). The results showed that MIMP is able to enhance mitogen-driven lymphocyte activation and proliferation in both human and mice models in a dose-dependent manner. At optimal mitogenic dose of PHA or ConA, MIMP showed its maximum effect at 100 μg/ml; however, MIMP did not show any effect in the absence of the mitogen (8). Since the PHA or ConA responses reflect mainly T-lymphocyte proliferation, the observations suggest that the immunoenhancing properties of MIMP are mainly via T-cell activation.

IL-2 acts within the immune system as a protein hormone that is necessary for the differentiation or activation of thymus-dependent lymphoid cells which bear specific receptors on their membrane (IL-2 receptor complex, IL-2R (9)). Binding of IL-2 to IL-2R on antigen-stimulated T cells leads to cell division, interferon gamma (IFN-γ) production, and enhancement of killer cell activity. IL-2 also reacts with other cells unrelated to the immune system such as natural killer cells and modifies their biological behavior (10). Many conditions are associated with defective responses to IL-2. The effect of MIMP to induce IL-2 production or substitute for IL-2 was studied in a normal murine IL-2–dependent cell line (CTLL). MIMP alone at 1–100 μg/ml had no effect and did not augment the effect of a saturating dose of IL-2; on the other hand, a CTLL line that had become unresponsive to IL-2 was preincubated with MIMP at 10 μg/ml for 3 days and was subsequently washed to remove the MIMP. When incubated for 48 h with MIMP at 6–50 μg/ml, the cells showed significant dose-dependent proliferation when compared with cells that were not preincubated with MIMP. When the defective cells were preincubated for 3 days with MIMP at 10, 25, and 50 μg/ml, washed, and then incubated for 48 h with 10 Us/ml of IL-2 in the absence of MIMP, the cells showed significant dose-dependent proliferation compared with those that had not been preincubated with MIMP (11). MIMP restored the deficient IL-2–induced proliferative response of the CTLL line as measured by thymidine incorporation.

A review on the effect of MIMP on HIV infection has been published (12). The effect of MIMP to reverse the immunosuppression induced by an HIV-derived synthetic peptide was tested. This 17–amino acid synthetic peptide (AA 571–587), which is homologous to the immunosuppressive site of GP 41 segment of HIV (13), showed inhibition in a dose-dependent manner of the PHA response. MIMP reversed the suppression induced by the peptide when it was mild or moderate. When the suppression was severe (>50%), MIMP showed no significant effect (8). Ruegg et al. (13) have published that the inhibitory effect of this peptide is linked to the inhibition of protein kinase C (PKC) in lymphocytes; thus MIMP may have effects related to PKC (14).

Other studies were carried out in order to assess the activity of MIMP under circumstances of HIV infection. Many of the pathological aspects of this disease are not directly due to infection by HIV; rather they are secondary effects that accompany the infection. Virus products, chronic stress, and poor nutrition all may result in detrimental effects on the general health of HIV-infected individuals and eventually contribute to the pace of the loss of helper $CD4^+$ T lymphocytes. Since

MIMP improves the PHA response under different conditions, and since MIMP reversed the suppression by AA 571–584 peptide, we used MIMP in vitro in the range of 0.1–100 μg/ml with lymphocytes obtained both from eight patients with acquired immunodeficiency disease (AIDS) and eight pre-AIDS (AIDS-related complex, ARC) patients. As control, eight normal donors were used. The CD4 counts were estimated as follows: for normal individuals, the count was approximately 1200, for ARC patients it was 544, and for the patients with AIDS it was 40. Insignificant effects of MIMP were observed on the PHA responses of lymphocytes from patients with AIDS, since these responses were $<10\%$ of control. Since in patients with AIDS, the CD4 counts are so low, the MIMP did not have any target on which to act. On the other hand, MIMP at 0.1–100 μg/ml augmented progressively the depressed PHA responses of ARC lymphocytes with maximal response close to those of control lymphocytes.

Normal lymphocytes can be suppressed because of different conditions such as viral infections, stress, inflammation, and aging. We have tested the effect of MIMP under these conditions (8). We found in all the measured parameters that MIMP could reverse the suppression of T-cell responses to PHA to close to normal values when the suppression was moderate ($>50\%$).

One study was carried out in murine splenocytes with hydrocortisone (HC), which is one of the mediators released during stress. Hydrocortisone progressively inhibited over a concentration range of 10^{-8} to 10^{-6} M both PHA and ConA responses of murine splenocytes. MIMP abrogated the suppressive effects of HC (10^{-8} to 10^{-7} M) on the proliferative response to PHA by murine splenocytes (14).

During inflammatory processes, as in rheumatoid arthritis, different kinds of mediators are released, among them prostaglandins (prostaglandin E_2, PGE_2). Prostaglandins are implicated in the regulation of immune response and can exert both positive and negative effects. Most in vitro studies have demonstrated that PGE_2 inhibits a wide variety of immune functions such as proliferation of T cells activated by plant lectins or monoclonal antibodies to T-cell receptor (TCR)/CD3 complex (15,16). This inhibition is thought to be related to changes in intracellular levels of cyclic adenosine monophosphate (AMP) induced by PGE_2 leading to inhibition of IL-2 production and IL-2 receptor expression. We found that MIMP, at concentrations of 1–100 μg/ml, reverses the inhibition of PGE_2 (10^{-5} M) on PHA response of normal HPBLs. When the response of PHA was more severely depressed, MIMP could not reverse the suppressive effect.

During viral infections, suppressor factors may be produced by lymphocytes, macrophages, or infected cells other than those of the immune system. Among these factors are the interferons (IFNs). Although IFNs are antiviral factors, under appropriate circumstances, they can also be immunosuppressive. IFN-α decreases the in vitro lymphoproliferative responses to PHA. MIMP used at concentrations of 1–100 μg/ml reversed almost to normal the inhibitory effect produced by IFN-α at a range of 10–100 Us/ml.

During the aging process, it has been shown that the response to PHA is more variable and generally depressed. Studies were carried out with 16 aged donors (84± years) and 22 controls (20–50 years). MIMP at 1–100 μg/ml returned the in vitro responses to PHA of the aged to normal.

III. IN VIVO EFFECT OF MIMP ON MICE

To determine whether MIMP has the same potency in vivo, experiments were carried out to investigate the effect of MIMP on humoral and cellular immune responses in mice (17). Intraperitoneal and oral routes of administration were tested. The doses of MIMP from 0.1–100 mg/kg were used in both assays.

Plaque-forming cell (PFC) assay was used to study the effect of MIMP on humoral immune response. Balb/c mice were immunized with 1×10^8 sheep red blood cells (SRBCs) by intraperitoneal (i.p.) injection. MIMP was given either intraperitoneally or orally by gavage (p.o.). Measurement of PFC showed that MIMP both intraperitoneally and orally significantly stimulates the PFC response, with the highest response being at 50 mg/kg.

Delayed-type hypersensitivity (DTH) response was used to determine the effect of MIMP on cellular immune response. Mice were immunized injecting 1×10^7 SRBCs intraperitoneally. After 4 days, the mice were challenged by injecting 1×10^8 SRBCs in the hind footpad. The results showed optimal effect of MIMP on DTH at 1 mg/kg i.p. This stimulatory effect of MIMP on DTH was observed on the sensitization rather than on the expression phase (18).

As these results show, MIMP was able to stimulate T-cell–dependent antibody response and cellular immunity as measured by PFC and DTH assays, respectively.

To find the effect on mitogen-induced spleen cell proliferation in vitro, the spleen cells of MIMP-treated (i.p.) mice were cultured with PHA and ConA. The splenocytes showed a progressive increase of proliferative responses to PHA and ConA. These results show clearly that MIMP augments T-lymphocyte proliferation in vivo.

Our studies in vitro demonstrated that MIMP enhances the proliferative response to PHA in both ARC PBLs and suppression induced in normal PBLs by a synthetic retroviral peptide. In order to demonstrate a possible clinical usefulness of MIMP, Balb/c mice were infected with Friend leukemia virus (FLV) as a murine model of AIDS.

Three days after being infected with FLV, mice received 1 mg/kg of MIMP by both intraperitoneal and oral routes and treated for 10 days (18). We found that MIMP delayed the mean day of death in mice infected with FLV by both routes. In control groups, the mean survival time was 39 ± 1 day, whereas the treated mice had a mean survival time of 46 ± 2 days. The mechanism by which MIMP prolongs survival is unknown.

IV. CONCLUSIONS

MIMP is nontoxic in vitro and in vivo and is active both intraperitoneally and orally. MIMP is effective in augmenting the responses of normal human T lymphocytes and murine splenocytes to PHA and ConA in vitro as well as in vivo in mice. MIMP also augments impaired mitogen responses as observed in aging, with defective IL-2 response in a cell line, and following immunosuppression with an HIV-derived peptide, IFN-α or PGE_2. MIMP has a potent effect on T-cell–dependent antibody response and cellular immunity as measured by PFC and DTH response.

REFERENCES

1. Hadden JW. Immunotherapy of human immunodeficiency virus (HIV). TIPS 1991; 12:107.
2. Hadden JW. Thymomimetic drugs. In: Miescher PA, Bolis L, Ghione M, eds. Serono symposium on immunopharmacology. New York: Raven Press, 1985:183–92.
3. Hadden JW, Englard A, Sadlick JR, Hadden EM. The comparative effects of isoprinosine, levamisole, muramyl dipeptide and SM1213 on lymphocyte and macrophage proliferation and activation in vitro. Int J Immunopharmacol 1979; 1:17.
4. Touraine J-L, Sandhadji K, deBouteiller O, Hadden JW. In vitro effects of inosine monophosphate methyl (MeIMP) on human prothymocyte differentiation. Int J Immunopharmacol 1991; 13:761.
5. Hadden JW, Specter S, Galy A, Touraine J-L, Hadden EM. Thymic hormones, interleukins, endotoxin and thymomimetic drugs in T lymphocyte ontogeny. In: Chedid L, Hadden JW, Spreafico F, Dukor P, Willoughby D, eds. Advances in immunopharmacology 3. Oxford, England: Pergamon Press, 1986:487–97.
6. Sosa M, Saha AR, Giner-Sorolla A, Hadden EM, Hadden JW. Immunopharmacological properties of MIMP. In: Gorgiev V, Yamaguchi H, eds. Immunomodulatory drugs. New York Academy of Science, New York, New York, 1993:458–63.
7. Sosa M, Saha AR, Wang Y, Wadsworth T, Hadden EM, Hadden JW. Methyl inosine monophosphate (MIMP) promotes immune response in mice. Int J Immunopharmacol 1991; 13:761.
8. Hadden EM, Sosa M, Strand M, Coffey RG, Hadden JW. Methyl inosine monophosphate (MIMP) restores depressed lymphoproliferative responses of normal human and murine T lymphocytes. Int J Immunopharmacol 1991; 13:762.
9. Smith KA. The two chain structure of high affinity IL-2 receptors. Immunol Today 1987; 8:11–13.
10. Fink S, Fimasz M, Sterin-Borda L, Borda E, Bracco MME de. Stimulation of heart contractility by supernatants from lectin-activated lymphocytes. Role of IL-2. Int J Immunopharmacol 1989; 11:367–70.
11. Hadden EM, Wang Y, Sosa M, Coffey RG, Giner-Sorolla A, Hadden JW. Methyl inosine monophosphate (MIMP) augments T lymphoproliferative responses and reverses various immunosuppressants. Int J Immunopharmacol 1991; 13:762.
12. Hadden JW, Giner-Sorolla A, Hadden EM. Methyl inosine monophosphate (MIMP).

a new purine immunomodulator for HIV infection. Int J Immunopharmacol 1991; 13(S1):49–54.

13. Ruegg C-L, Strand M. Inhibition of protein kinase C and anti-CD_3-induced Ca^{2+} influx in jurkat T cells by a synthetic peptide with sequence identify to HIV-1 gp41°. J Immunol 1990; 144:3928–35.
14. Hadden JW, Hadden EM. Therapy of T cell immunodeficiency with biological substances and drugs. In: Faist E, Meakings T, Schildberg W, eds. Host defense dysfunction in trauma, shock and sepsis. Berlin: Springer-Verlag, 1993:1097–1107.
15. Goodwin JS, Bankhurst AD, Messner RP. Suppression of human T-cell mitogenesis by prostaglandin. Existence of prostaglandin-production suppressor cell. J Exp Med 1977; 146:1719–34.
16. Chouaib S, Bertoglio JH. Prostaglandins E as modulators of the immune response. Lymphokine Res 1988; 7:237–45.
17. Sosa M, Saha AR, Wang Y, Coto JA, Giner-Sorolla A, Hadden EM, Hadden JW. Potentiation of immune responses in mice by a new inosine derivative—Methyl inosine monophosphate (MIMP). Int J Immunopharmacol 1992; 14(7):1259–66.
18. Hadden JW, Ongradi J, Specter S, Nelson R, Sosa M, Monell C, Strand M, Giner-Sorolla A, Hadden EM. Methyl inosine monophosphate: A potential immunotherapeutic for early human immunodeficiency virus (HIV) infection. Int J Immunopharmacol 1992; 14(4):555.

12

Enhancement of the Immune Response to Hepatitis B Surface Antigen by Methyl Inosine Monophosphate

Tatiana A. Semenenko and M.I. Mikhailov
Academy of Medical Science, Moscow, Russia

John W. Hadden
The University of South Florida College of Medicine, Tampa, Florida

I. INTRODUCTION

Enhancement of the immunogenicity of vaccine preparations against hepatitis B is still an important public health problem, since nearly 2–15% of those vaccinated have been found not to produce antibodies to the hepatitis B virus surface antigen (anti-HBs) and are thus nonprotected against this infection (1). Hereafter, these individuals will be referred to as nonresponders. They are most frequently among the patients undergoing chronic hemodialysis treatment (2,3), those with human immunodeficiency virus (HIV) infection, and other immunocompromised patients (4). All of these patients belong to groups having high risk of infection with hepatitis B virus. Those who become involved in these groups and have not previously had contact with hepatitis B virus face the highest risk and, therefore, need rapid and maximal protection from the infection.

One possible way of increasing the efficacy of prophylaxis is the enhancement of the immunogenicity of the vaccines by using biologically active substances possessing adjuvant properties.

The results obtained in combined application of hepatitis B vaccine and various immunomodulators point to the expediency of this approach, since combination treatment can reduce the number of persons who either do not completely respond or respond with low levels of anti-HBs (5,6). The encouraging data obtained by us point to the necessity of expanded search on immunomodulators that promote rapid induction of maximal levels of specific antibodies and increase protection in immunocompromised hosts.

One of the possible candidates for this role is methyl inosine monophosphate (MIMP), a new purine immunomodulator synthesized at the University of South Florida by a group of scientists headed by Dr. J.W. Hadden (7). In previous reports, MIMP was found to have effects in vitro on T-cell proliferative responses greater than on B-cell responses and to induce T-cell differentiation (8,9), and it can thus be classified as a "thymomimetic" drug (10). Under optimal conditions, MIMP shows small effects to modulate the activity of interleukin-2 (IL-2), but under unusual circumstances of a deviated CTLI line, MIMP had marked activity on pathways related to IL-2 action (11).

Studies with murine peritoneal exudate and alveolar macrophages indicate that the macrophage is also a target of MIMP action (12). Also, recent in vivo results in mice indicate that MIMP possesses adjuvant activity for both humoral and cellular immune responses (13).

The present study was designed to evaluate the effect of MIMP on immune response to surface HBsAg in mice.

II. MATERIALS AND METHODS

A. Hepatitis B Virus Surface Antigen (HBsAg)

HBsAg (subtypes ad and ay) was purified from the blood plasma of the antigen carriers according to the following scheme:

> HBsAg containing blood plasma initially diluted with two volumes of physiological solution was heated in a water bath for 60 min at 80°C. The clot that formed was mechanically disrupted and the denatured proteins removed by centrifugation.
>
> HBsAg was reprecipitated with polyethylene glycol 6000 M at a final concentration of 15%.
>
> HBsAg containing precipitate was dialysed against 0.9 M NaCl and treated with pepsin (100 mg/ml) and Tween-80 (final concentration 2%).
>
> HBsAg solution was twice ultracentrifuged in a linear sucrose gradient.
>
> The purified HBsAg preparation was dialysed against 0.9 M NaCl, aliquoted in 1 ml volumes, and frozen at 60°C.

After 50- to 100-fold concentration, the resultant preparation had the following characteristics: It consisted only of spherical particles 18–25 nm in diameter, was devoid of Dane particles or filamentous particles of HBsAg; negative results were obtained in tests for HBeAg (EIA, Abbott, Chicago, IL); DNA-polymerase; HBV-

DNA by the method of directed amplification; and the presence of proteins characteristic of normal human blood serum.

The preparation obtained served as the basis for elaboration of experimental lots of hepatitis B vaccine (14).

B. Animals

Male mice (line DBA/2) weighing 16–18 g were obtained from the Animal Laboratory, Academy of Medical Science, Moscow, Russia. This line of mice was chosen as poor responders of hepatitis B vaccine.

Experimental mice were irradiated at a dose rate of 25 rad/min (1 rad = 0.01 Gy) using a ^{60}CO γ beam source for 4 min. Data represent the average of 10 animals in each group.

C. Immunoassay Procedures

Serum specimens from immunized mice were tested for the presence of anti-HBs in enzyme immunoassay using the diagnostic kits of Roche Diagnostica. Anti-HBs EIA (Roche), with the sensitivity of <10 IU/L.

Antibody concentration was determined with the help of calibration curves designed on the basis of the results of anti-HBs detection in standard serum panel (Roche Diagnostica) with the concentrations as follows: 10 IU/L, 50 IU/L, 100 IU/L, and 150 IU/L. The results of the two experiments were used for the calculation of mean values.

D. Schemes and Doses of MIMP and HBsAg Administration

HBsAg was administered at a dose of 16 mg/mouse in 0.2 ml PBS (pH 7.4) twice with an interval of 2 weeks.

MIMP was introduced both orally or intraperitoneally at a dose of 50 mg/kg. Three schemes of MIMP administration were used:

Scheme 1: HBsAg and MIMP orally by gavage 30 min prior to the antigen introduction

Scheme 2: HBsAg and MIMP intraperitoneally

Scheme 3: HBsAg and MIMP intraperitoneally followed by oral administration for 4 days.

E. Statistical Analysis

Data were analyzed for statistical significance using Student's *t*-test.

Table 1 Effect of MIMP on Anti-HBs Production in DBA/2 Mice

Scheme No.	Days After Injection	
	Day 14	Day 21
HBsAg	< 10 IU/L	260 IU/L ± 16.3
Scheme 1	< 10 IU/L	370 IU/L ± 18.5[a]
Scheme 2	100 IU/L ± 12.6[b]	610 IU/L ± 24.3[b]
Scheme 3	100 IU/L ± 12.6[b]	420 IU/L ± 20.6[a]

[a] $0.01 < p < .05$.
[b] $p < .01$.

III. RESULTS

Previous studies were helpful in the proper selection of mouse line DBA/2 as poor responders in antibody production to HBsAg (15).

The results of anti-HBs detection in the control group of mice treated with HBsAg alone and in the groups treated with the combination of HBsAg and MIMP are presented in Table 1.

Antibody responses obtained with combined administration of HBsAg and MIMP according to the various schemes exceeded the antibody level in the control group of animals.

In a second set of experiments, the effect of MIMP on the induction of a specific immune response to HBsAg in immunocompromised DBA/2 mice subjected to ionizing radiation was analyzed.

The results of anti-HBs detection in the irradiated mice of the control group (injected only with HBsAg) and those treated with the combination (HBsAG & MIMP using schemes 2 and 3) are presented in Table 2.

Table 2 Effect of MIMP on Anti-HBs Responses in Immunocompromised Irradiated DBA/2 Mice

Scheme No.	Days After Injection	
	Day 14	Day 21
Only HBsAg	< 10 IU/L	< 10 IU/L
Scheme 2	< 10 IU/L	100 IU/L ± 10.8[a]
Scheme 3	< 10 IU/L	100 IU/L ± 10.0[a]

[a] $p < .01$.

The data indicate that in the case of schemes 2 and 3, anti-HBs were significantly increased on day 21 indicating a restoring effect of MIMP upon the immune system of these irradiated animals.

IV. DISCUSSION

It is well known that adjuvants are able to stimulate antibody formation in response to heterologous antigens. In vaccine preparations with hepatitis B, HBsAg is generally absorbed onto aluminum hydroxide to enhance the immunogenic effect in order to achieve a protective titer of anti-HBs (>10 MU/L) that prevents infection.

Various schemes of multiple administrations of the antigen preparation are recommended to induce intense immunity against hepatitis B; however, in some vaccinated persons, it is impossible to create protective immunity owing in part to peculiarities of the immune response determined at the level of the major histocompatibility complex (MHC) (14,15). It has been suggested that "nonresponders" lack a dominant gene of the immune response in the MHC and, as a result, synthesis of anti-HBs occurs, at most, at low levels that can barely be detected by currently applied methods of detection (16). Similarly, immune response genes associated with the murine major histocompatibility complex (H-2) have been shown to control cellular and humoral responses to determinants on numerous T-cell–dependent antigens, including HBsAg. Different strains of mice, coded by Pre-S2 zone S-gene HBV on their immune responses for proteins, place them in the following haplotype order (17): $H\text{-}2^b > H\text{-}2^d > H\text{-}2^s > H\text{-}2^k > H\text{-}2^f$. DBA/2 mice do not produce antibodies to HBsAg at a high level when treated with antigen alone.

The results presented here demonstrate that MIMP is able to influence the development of humoral immune response to HBsAg. It was shown that MIMP as a single intraperitoneal as well as a single oral administration dose of 50 mg/kg can induce a significant increase in the level of antibodies to HBsAg. After intraperitoneal administration with antigen, additional oral administration of MIMP did not increase its activity.

The mechanism of adjuvant effect of MIMP is not clear and needs further study; however, one may speculate about it. Nonresponsiveness to hepatitis B vaccine has been observed in hemodialysis patients (3) and in immunocompromised individuals (4). In some instances, larger vaccine doses or an increased number of doses have resulted in seroconversion (15). Soluble interleukin-2–receptor levels have been observed in these patients and the resulting impairment of interleukin-2 action may have an effect on response to hepatitis B vaccine (3,6). Meyer et al. (6) injected 40-mg doses of the vaccine followed by $1.2\ \mu\ 10^5$ U of natural interleukin-2 to 10 hemodialysis patients who were nonresponders. Four weeks later, 6 of the 10 patients showed seroconversion, although antibody levels were still below normal.

MHC-linked unresponsiveness to protein or peptide antigens also can be overcome by administration of IL-2.

The results of our investigation in relation to knowledge about the immunopharmacology of MIMP allow the suggestion that MIMP, as an adjuvant, may be realized via T cells and may involve interleukin-2.

REFERENCES

1. Deinhardt F. Aspects of vaccination against hepatitis B, passive-active immunization schedules and vaccination responses in different age groups. Scand J Infect Dis 1983; 38(Suppl):17.
2. Ferguson M. Hepatitis vaccines. Curr Opin Infect Dis 1990; 3:367.
3. Walz G, Kenzendorf U, Haller H, Keller F, Offerman G, Josimovic-Alasevic O, Diamantstein T. Factors influencing the response to hepatitis B vaccination of haemodialysis patients. Nephron 1989; 51:474.
4. Hess G, Rossol S, Voth R, Cheatham-Speth D, Clemens R, Mayer ZUM, Buscenfelde K-H. Active immunization of homosexual men using a recombinant hepatitis B vaccine. J Med Virol 1989; 29:229.
5. Celis E, Abraham KG, Miller RW. Modulation of the immune response to hepatitis B virus by antibodies. Hepatology (Baltimore) 1987; 7:563.
6. Meuer SC, Dumann H, Meyer ZUm, Buschenfeld K-H, Kohler H. Low dose interleukin-2 induces immune response against HBsAg in immunodeficient nonresponders to hepatitis B vaccination. Lancet 1989; 1:15.
7. Hadden JW, Giner-Sorolla A, Hadden EM. Methyl inosine monophosphate (MIMP), a new purine immunomodulator for HIV infection. Int J Immunopharmacol 1992; 14:555.
8. Hadden JW, Hadden EM, Wang Y, et al. Methyl inosine monophosphate (MIMP)—A new purine immunomodulator. Int J Immunopharmacol 1991; 13:761.
9. Touraine J-L, Sanhadju K, Bouteiller O, Hadden JW. *In vitro* effects of inosine 5'-methyl monophosphate (MeIMP) on human prothymocyte differentiation. Int J Immunopharmacol 1991; 13:761.
10. Hadden JW. Thymomimetic drugs. In: Miescher PA, Bolis L, Ghione M, eds. New York: Raven Press, 1985:183.
11. Hadden EM, Sosa M, Strand M, et al. Methyl inosine monophosphate (MIMP) restores depressed lymphoproliferative response of normal human and murine T lymphocytes. Int J Immunopharmacol 1991; 13:762.
12. Sosa M, Hadden EM, Hadden JW, et al. Methyl inosine monophosphate (MIMP) promotes immune responses in mice. Int J Immunopharmacol 1991; 13:762.
13. Sosa M, Saha AR, Wang Y, et al. Potentiation of immune response in mice by a new inosine derivative-methyl inosine monophosphate (MIMP). Int J Immunopharmacol 1992; 14:1259.
14. Alper CA, Kruscall MD, Marcus-Bagley D, Graven DE, Katz AJ, Deinstag JL, Awdeh Z, Yunis EJ. Genetic prediction of nonresponse to hepatitis B vaccine. N Engl J Med 1989; 321:708.

15. Walker M, Szmuness W, Stevens C, Rubinstein P. Genetics of anti-HBs responsiveness. I. HLA-DR7 and nonresponsiveness to hepatitis vaccination. Transfusion (Philadelphia) 1981; 21a:601.
16. Thomson G, Bodmer W. The genetic analysis of HLA and disease. In: Dausset J, Sneigaard A, eds. HLA and disease. Copenhagen: Munksgaard, 1977:84–93.
17. Benaceraf B, McDevitt HO. Histocompatibility-linked immune response genes. Science 1972; 175:273.
18. Recommendations of the Immunization Practices Advisory Committee (ACIP). Protection aginst viral hepatitis. MMWR 39, 1990.

13

Combination Therapies in Human Immunodeficiency Virus Infection

Antonio Mastino, Cartesio Favalli, Cartesio D'Agostini, Maria Staglianò, Sandro Grelli, and Enrico Garaci
University of Rome, Tor Vergata, Rome, Italy

Luigi Perroni and Fabrizio Soscia
St. Maria Goretti Hospital, Latina, Italy

I. INTRODUCTION

To date, treatment strategies against human immunodeficiency virus (HIV) infection have focused on the use of antiretroviral agents based on the assumption that viral proliferation is the unique factor involved in the progression of acquired immunodeficiency syndrome (AIDS). The availability of zidovudine (azidothymidine, AZT) has reduced risk of disease progression and has increased the median survival of HIV-infected patients. But AZT, even in combination with other drugs active on opportunistic infections, does not appear to provide long-term efficacy and has partial and transient effect on the decline of immunologic functions (35). In addition, AZT therapy seems to be correlated with the appearance of resistant HIV strains (4,17). On the other hand, immune mechanisms other than a direct cytotoxic effect of the virus on target cells seem to be involved in the damage of the immune system caused by HIV. As a consequence, the use of agents able to affect the immune system should be considered in combination with antiretroviral agents. Here we discuss these two major points; that is (1) mechanisms leading to immunodepression during HIV infection and (2) the possibility to utilize immunotherapeutic drugs in combination with antiretroviral agents in controlling HIV infection, and report our personal experience in these fields of research.

II. MECHANISMS BY WHICH HIV CAUSES IMPAIRMENT OF THE IMMUNE SYSTEM

The exact mechanisms by which HIV infection impairs the immune system of the host are not yet understood. The severe and selective depletion of $CD4^+$ T lymphocytes remains the most specific and visible damage associated with HIV infection. Until recently, it was considered to be exclusively due to the cytopathic effect of the virus. There is now growing evidence to suggest that many other complex mechanisms could be responsible for the depletion and immune impairment of $CD4^+$ cells. Among these autoimmune responses by antibodies or by cytotoxic major histocompatibility complex (MHC) restricted and nonrestricted cells could play an important role (32). Moreover, independently by the mechanisms implicated, some investigators have clearly shown that the loss of immune functions precedes the significant reduction in $CD4^+$ cell number by at least 12 months (19,32), and it seems to be correlated with the reduction of interleukin-2 (IL-2) production (25). The decline of cytokine production of HIV-infected patients seems to affect IL-2 and interferon gamma (INF-γ) differently in comparison with IL-2 and IL-6. As a consequence, a sort of switching from TH_1-type cellular response toward TH_2 type has been recently proposed as one of the mechanisms leading to the severe dysfunction of the immune system during the progression of the HIV infection (5). On the basis of recent results, it has been also hypothesized that $CD4^+$ cell depletion could be induced by an unidentified HIV superantigen (15,16,20). In addition, $CD4^+$ T cells are not the exclusive target of HIV. In fact, several investigators have focused their attention on the role of the HIV infection of "reservoir" cells with antigen-presenting functions (APCs). Among them dendritic cells (21,27,33) or macrophages (13,22,29) whose functions, particularly their antigen-presenting capacity, seem to be strongly affected by the virus (17). In any case, it is clear that important pathogenetic events associated with the progression to AIDS occur at lymph node and probably thymus levels in HIV-infected patients (21,26,31).

Apoptosis of lymphocytes is considered one of the most probable mechanisms leading to the destruction of the immune system associated to HIV infection (1). This hypothesis, even if sustained by some experimental evidence (14,18), has not yet been demonstrated directly. Recently, it has been shown that gp120 primes T lymphocytes for activation-induced apoptosis by the T-cell receptor (TCR) through cross linking CD4 molecules, whereas neither cross linking of gp120 nor TCR alone did significantly increase apoptosis level over background (2). Many researchers have recently focused their attention on the study of apoptosis of immune cells; however, the exact mechanisms and the physiologic mediators involved in this phenomenon are still unknown (6). We have recently demonstrated that prostaglandin E_2 (PGE_2) administration induces thymocyte apoptosis in mice; thus suggesting that the endogenous production of this substance could be involved in the process of natural apoptosis of immune cells in vivo (23). We have thus investigated the possibility that the intercellular contacts between HIV-infected APCs and lymphocytes could induce apoptosis in the latter population and that PGE_2 production by HIV-infected APCs could play a role in the hypothesized phenomenon.

III. INDUCTION OF LYMPHOCYTE APOPTOSIS BY HIV-INFECTED MACROPHAGES

Peripheral blood lymphocytes (PBLs) collected from healthy donors and frozen, were thawed and analyzed for apoptosis after 24-h coculture with autologous macrophages (PBMΦ) collected from the same donors and previously infected in vitro with a monocytotropic strain (HTLV-$III_{Ba\text{-}L}$) of HIV, as described (28–30), or not. PBLs and PBMΦ were separated from the same donor by countercurrent centrifugal elutriation, as previously described (28). In parallel experiments, apoptosis was detected after culture of PBLs with cell-free supernatants collected from the PBMΦ cultures whether infected with HIV or not. The results of a representative experiment are reported in Figure 1. When either HIV-infected

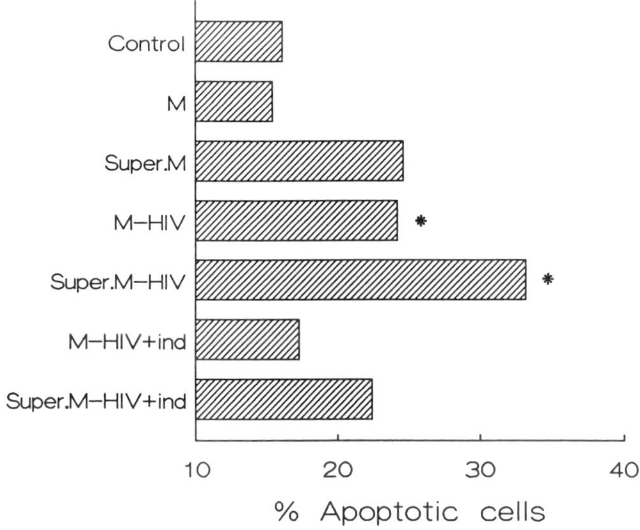

Figure 1 Induction of lymphocyte apoptosis by HIV-infected macrophages. Peripheral blood lymphocytes from healthy donors were cultured for 24 h in the absence (control) or in the presence of autologous uninfected macrophages (M) and HIV-infected macrophages (M-HIV) or of cell-free supernatants from uninfected (Super.M) and HIV-infected macrophage (Super.M-HIV) cultures. Indomethacin (ind) also was added to some cultures. Data are expressed as percentage of apoptotic cells evaluated by microscopic analysis of morphologic features and positivity for the enzyme tissue transglutaminase, as detected by immunohistochemical staining with a specific antibody. Percentages are referred to counts obtained in duplicate slides from the same culture (400 cells each). $^*P < .01$ against corresponding value with uninfected macrophages on their supernatant by χ^2 test.

PBMΦ or supernatants from their cultures were added to the autologous lymphocyte cultures, the percentage of apoptotic lymphocytes was significantly enhanced, at the χ^2 test, with respect to values obtained with uninfected PBMΦ or with supernatants from uninfected PBMΦ, respectively. Higher values of apoptosis obtained in PBMΦ-free cultures could be due to the lack of apoptotic lymphocyte–specific removal by macrophages in the cultures exposed to cell-free supernatants rather than to a real difference in apoptosis induction. When the cyclooxygenase inhibitor indomethacin was added to the HIV-infected PBMΦ cultures, the percentage of apoptotic lymphocytes was reduced at levels similar to those observed after cocultivation with uninfected PBMΦ or exposure to supernatants from uninfected PBMΦ. This fact indicates that PGE_2 production is implicated in lymphocyte apoptosis induced by HIV-infected PBMΦ. Lymphocyte cultures infected by direct exposure to HIV lymphocytotropic or monocytotropic strains did not show any increase over the control percentage of apoptotic lymphocytes 24 h after the infection (data not shown). In addition, preapoptotic and apoptotic lymphocytes after direct PGE_2 treatment in vitro were found at PGE_2 concentrations of 10^{-5} and 10^{-6} M. On the contrary, at the higher concentration of PGE_2 tested (10^{-4}), morphologic evidence of necrosis was observed (data not shown). The production of PGE_2 by mock-infected or HIV-infected PBMΦ was estimated by competitive binding radioimmunoassay in supernatants of the cultures. The results are reported in Figure 2. As shown, PGE_2 was promptly produced by HIV-infected macrophages, with very high values of release starting from the first hours following the infection. Anyhow, the larger amounts of PGE_2 production by HIV-infected PBMΦ with respect to uninfected cells also was maintained during the successive weeks of culture. Thus, an increased level of PGE_2 production seems to be a characteristic strictly connected with the typical chronic infection of macrophages by HIV and not only a phenomenon associated with nonspecific stimulation during the early phase of the infection, as could be suggested by previous results on PGE_2 production following gp120 stimulation or HIV-infection in vitro (9,34). The results obtained in our work clarify the kinetics of this production and indicate a mechanism that could link PGE_2 production by HIV-infected APCs with the immunopathogenesis of AIDS. In fact, considering our results, it is possible to hypothesize that endogenous PGE_2 released by HIV-infected APCs in the lymph nodes and/or the thymus during strict intercellular contacts could locally reach concentrations active to induce cell death of lymphocytes by apoptosis. This phenomenon could act synergistically with other events as an additional mechanism in inducing T-cell loss. Our study on lymphocyte apoptosis induction by HIV-infected APCs and PGE_2 could thus contribute to explain some of the mechanisms involved in the dramatic alterations of the immune system induced by HIV. The results of this study have been recently reported by us in detail (24).

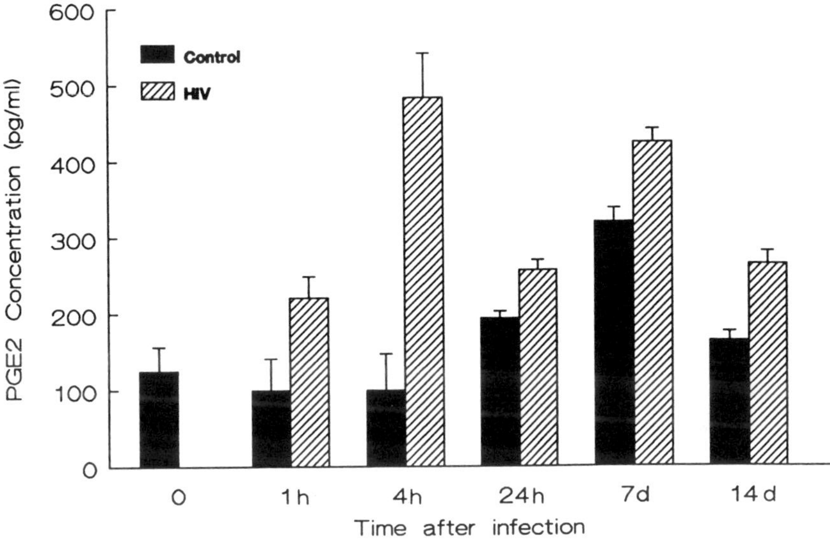

Figure 2 Comparative dosage of PGE_2 concentration in supernatants from mock-infected (control) or HIV-infected (HIV) primary culture of human monocyte/macrophages from peripheral blood. A highly enriched monocyte/macrophage population obtained from healthy adult donors by countercurrent centrifugal elutriation was infected in vitro with the monocytotropic HTLV-III Ba-L strain. At the indicated times, supernatants were completely aspirated for PGE_2 dosage and medium was replenished. Results represent mean values derived from duplicate samples of triplicate cultures ±SD $P < .02$ in respect to control cultures at each time tested.

IV. USE OF THYMOSIN α_1 IN COMBINATION WITH INTERFERON AND ZIDOVUDINE IN HIV INFECTION

Since our previous studies demonstrated the high immunopotentiating efficacy of combination therapy with thymosin α_1 ($T\alpha_1$) plus interferon (IFN) in animal models of immunodepression (7,8,11,12), we performed a series of in vitro experiments in order to verify the possibility of transferring a combination therapy with $T\alpha_1$ plus IFN associated with AZT in HIV-positive patients. Figure 3 illustrates the results of an in vitro experiment in which PBLs from healthy donors and HIV-infected donors were treated with single or combined therapies in order to study the effects of cytotoxicity. After 24 h of culture, we observed an impairment of cytotoxicity against natural killer (NK)–sensitive target cells in PBLs from HIV-infected patients in comparison with those from healthy donors. A similar effect had been previously obtained by other investigators using fresh PBLs (3). Pretreatment with $T\alpha_1$

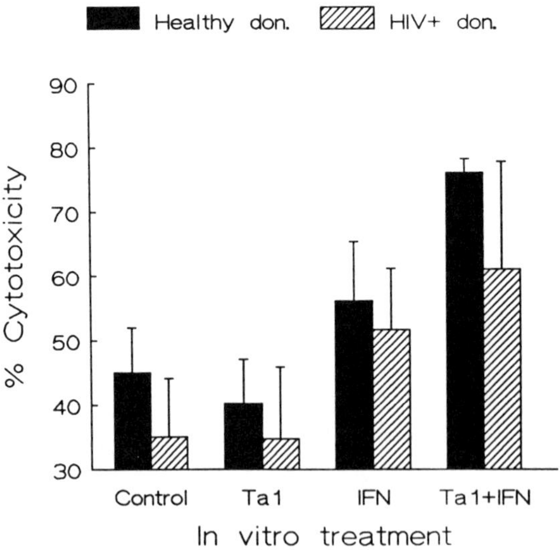

Figure 3 Effect of combination treatment in vitro with thymosin $\alpha 1$ and interferon on cytotoxicity in PBLs from healthy donors or HIV-seropositive volunteer donors. Peripheral blood mononuclear cells from 10 HIV-seronegative or HIV-seropositive volunteers were cultured in complete culture medium alone (control) or in complete medium with addition of thymosin $\alpha 1$ (T$\alpha 1$), 100 ng/ml), interferon alpha (IFN, 500 U/ml) or both (T$\alpha 1$ + IFN). Cytotoxicity against K-562 cells was tested after 24 h of culture with a 4 h ^{51}Cr release assay. Results are expressed as mean values per group calculated from individual values of cytotoxicity obtained for each patient from quadruplicate wells at effector/target ratio of 50:1 ± SD. Statistics by paired Student's *t*-test. Group HIV$^-$: IFN, $P = .0025$ against control; T$\alpha 1$ + IFN, $P = .0018$ against control and $P = .0169$ against IFN. Group HIV$^+$: IFN, $P = .0001$ against control; T$\alpha 1$ + IFN, $P = .0001$ against control and $P = .0156$ against IFN.

significantly potentiated the cytotoxic activity of IFN-α–stimulated PBLs collected from HIV-positive patients. Cytotoxicity values in this group were higher than those observed in PBLs from healthy donors when stimulated by IFN-α alone showing complete restoration of NK cell response to this cytokine. PBLs from healthy controls also responded to Tα_1 pretreatment. Treatment with Tα_1 alone had no effect on cytotoxicity in either group. In contrast, treatment with IFN alone stimulated the cytotoxic activity of cells collected from healthy donors and, even if at a lower level, from HIV-positive patients. In summary, these experiments demonstrated a synergistic effect of Tα_1 and IFN in PBLs from HIV-infected patients, which was similar to that previously obtained by us in animal models of immunosuppression. Other experiments showed that combination treatment with Tα_1 in vitro did not potentiate the direct antiviral activity of IFN-α at a suboptimal dose whether or not associated with AZT. On the other hand, no evident negative interference with the direct antiviral activity of AZT was observed following $T\alpha_1$ and IFN-α treatment when used alone or in combination; thus opening the way to their potential use in association in vivo in HIV-infected patients. As a consequence, we then performed a pilot 1-year clinical study to examine the safety and the potential beneficial effects of this combination therapy in HIV-infected patients. Forty subjects (23 men and 17 women, aged 22–43 years) who were HIV seropositive were enrolled in a not-blinded clinical study between January and June 1991. Of the 40 patients, 27 were drug abusers, 9 were homosexuals, and 4 were heterosexuals. Of the four heterosexuals, two had sexual partners who were drug abusers. Criteria used for eligibility were positive test for HIV, made by ELISA and confirmed by Western blot; age 18 years or older; $CD4^+$ lymphocytes ranging from 200 to 500/mm^3; Centers for Disease Control (CDC) classes II or III; Karnofsky >70; no previous antiretroviral or immunomodulator treatment; normal hematologic, hepatic, renal, and coagulation functions. Written informed consent was obtained from each patient. The subjects were randomly assigned to four groups that were homogeneous in terms of age, risk factor, and sex. The first group received AZT (The Wellcome Foundation Ltd., London, England), 500 mg/day in two administrations; the second group received AZT plus natural human lymphoblastoid IFN-α (The Wellcome Foundation Ltd.), 2 MU i.m. twice weekly; the third group received AZT plus $T\alpha_1$ (Sclavo Pharmaceuticals, Siena, Italy), 1 mg s.c. twice weekly; the fourth group received AZT plus $T\alpha_1$ plus IFN-α. After 1-year, 28 patients (7 patients in each group) had completed the cycle of treatment. None of the patients withdrew because of side effects, disease progression, or death but rather because of either admission to therapeutic communities or their own lack of compliance. Particularly, no adverse effects or major toxicity was observed in any patient in any of the groups, and none of the volunteers had to interrupt the treatment or reduce the dosage of drugs they were receiving, as shown by monthly laboratory testing that included complete blood count with differential, platelet count, glucose, electrolytes, blood urea nitrogen, liver function tests, cholesterol, triglycerides, coagulation profile, and β_2-microglobulin assayed by turbidimetric immunoassay. Patients also were monitored monthly for $CD4^+$ cell counts by monoclonal antibodies and flow cytometry. Results of part of these data are shown in Figure 4 and indicate the mean absolute $CD4^+$ cell counts from patients who completed the 1-year treatment plan at time 0 and after 3, 6, 9, and 12 months of treatment. $CD4^+$ lymphocytes in the AZT group averaged from 382 ± 140, and in the group receiving AZT and $T\alpha_1$ the cell count changed from 411 ± 84 to 446 ± 115. However, in all these groups, $CD4^+$ values at time 0 and after 12 months were not significantly different when tested by the paired Student's *t*-test. In contrast, in the group treated with SZT, $T\alpha_1$, and IFN-α, the $CD4^+$ cell number averaged from 309 ± 77 before treatment to 496 ± 230 after 12 months, with a statistically significant difference (P value $= .029$ at paired Student's *t*-test). In addition to this parameter, the proliferative response to *Candida albicans* or tetanus toxoid antigens and the polymerase chain reaction (PCR) analysis

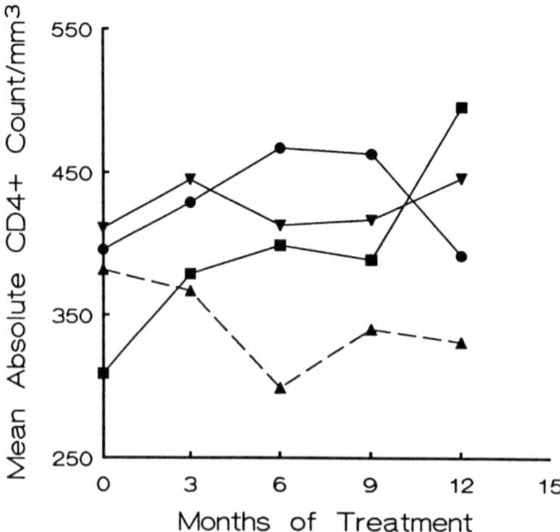

Figure 4 Effect of combination therapy with AZT + Tα1+ IFN-α on $CD4^+$ cell count. Solid lines, mean values obtained in patients treated with AZT + IFN-α (•), AZT + Tα1 (▼), or AZT + Tα1 + IFN--α (■); dashed line, mean values obtained in patients treated with AZT alone.

of HIV-DNA were studied. The proliferative responses to both recall antigens in the patients of all four groups in the study were significantly decreased as compared with those observed in healthy donors either at enrollment or after 1 year of treatment. However, the group receiving $T\alpha_1$ plus IFN-α in combination with AZT showed an increased response recall antigens after 1 year of treatment, with a difference that was statistically significant in respect to values obtained in the other groups of patients after 1 year of treatment and versus those obtained in the same group on entry. The HIV-DNA analysis by PCR, performed only in patients belonging to the groups that received AZT alone or AZT plus combination treatment with $T\alpha_1$ and IFN-α, demonstrated that the calculated number of viral copies per $CD4^+$ cells, expressed as a percentage with respect to the total $CD4^+$ cell population, was statistically significantly lower in the patients receiving combined treatment ($0.61 \pm 0.16\%$ vs. 1.24 ± 0.55, $P = .013$ by Student's two-sided t-test). A smaller cohort of patients has continued to be treated and observed after 1 year of treatment. Results of the $CD4^+$ cell counts at 18 and 24 months, restricted to this cohort of patients, confirmed the trend previously observed with a difference among patients receiving AZT plus $T\alpha_1$ and IFN-α and those receiving all the other treatments that was even more evident than that observed after 12 months of therapy.

In conclusion, the results of immunologic monitoring indicated that the combination of AZT plus $T\alpha_1$ and IFN was more effective than AZT or AZT with either single agent alone in increasing the number of circulating $CD4^+$ lymphocytes.

During the first 6 months of treatment with AZT and IFN-α, an increase in CD4$^+$ cells was demonstrated, but this declined over the following 6 months. Other researchers have similarly observed that the duration of the beneficial effects of AZT therapy starts to decrease after 6 months and seems linked to the emergence of HIV-1–resistant strains (4). Regarding the mechanism of action of combination immunotherapy, no evident in vitro antiviral activity was induced by either Tα_1 or by the combination of Tα_1 plus IFN-α in respect to IFN alone; thus suggesting an indirect effect. Tα_1 may induce thymic maturation of the CD4$^+$ population and increase production of IL-2 and IFN-α from CD4$^+$ mature cells. Tα_1 could also protect CD4$^+$ cells from, at present, not-well identified damages caused by mechanisms other than a direct cytopathic action of the HIV. Moreover, combination treatment could induce the reduction of CD4$^+$-infected cells by increasing cytotoxic activities. Anyhow, the increase in the CD4$^+$ cell number and the restoration of some immune functions in the group treated with AZT plus Tα_1 and IFN-α is encouraging and suggests a possible impact on the disease. The results of this study (12a) formed the theoretical basis for a controlled, double-blind multicenter clinical trial in a larger cohort of patients that is now in progress with the approval and the support of the Ministry of Italian Health.

ACKNOWLEDGMENTS

Research reported in this chapter was supported by National Research Council (C.N.R.), Projects "Prevention and Control of Disease" and "Aging" and by the Italian Health Ministry "AIDS Research Project"; we are grateful to Dr. E. Di Macco, Prof. G. Novelli, and Prof. B. Dallapiccola for HIV-DNA analysis by PCR.

REFERENCES

1. Ameisen JC. Programmed cell death and AIDS: From hypothesis to experiment. Immunol Today 1992; 13:388–91.
2. Banda NKD, Bernier J, Kurahara DK, Kurrle R, Haigwood N, Sekaly RP, Finkel TH. Crosslinking CD4 by human immunodeficiency virus gp120 primes T cells for activation-induced apoptosis. J Exp Med 1992; 176:1099–1106.
3. Bonovida D, Katz J, Gottlieb M. Mechanism of defective NK cell activity in patients with acquired immunodeficiency syndrome (AIDS) and AIDS related complex. J Immunol 1988; 137:1157–63.
4. Boucher CA, O'Sullivan E, Mulder JW, Ramautarsing C, Kellam P. Ordered appearance of zidovudine resistance mutations during treatment of 18 human immunodeficiency virus-positive subjects. J Infect Dis 1992; 165:105–10.
5. Clerici M, Shearer GM. A Th1-Th2 switch is a critical step in the etiology of HIV infection. Immunol Today 1993; 14:107–11.
6. Cohen JJ. Apoptosis. Immunol Today 1993; 14:126–130.
7. Favalli C, Jezzi T, Mastino A, Rinaldi-Garaci C, Riccardi C, Garaci E. Modulation

of natural killer activity by thymosin alpha 1 and interferon. Cancer Immunol Immunother 1985; 20:189–92.

8. Favalli C, Mastino A, Jezzi T, Grelli S, Goldstein AL, Garaci E. Synergic effect of thymosin alpha 1 and alpha-beta interferon on Nk activity in tumor-bearing mice. Int J Immunopharmacol 1989; 11:443–50.
9. Foley P, Kazazi F, Biti R, Sorrell TC, Cunningham AL. HIV infection of monocytes inhibits the T-lymphocyte proliferative response to recall antigens, via production of eicosanoids. Immunology 1992; 75:391–7.
10. Fox CH, Cottler-Fox M. The pathobiology of HIV infection. Immunol Today 1992; 13:353–6.
11. Garaci E, Mastino A, Pica F, Favalli C. Combination treatment using Thymosin α 1 and interferon after cyclophosphamide is able to cure Lewis lung carcinoma in mice. Cancer Immunol Immunother 1990; 32:154–60.
12. Garaci E, Pica F, Mastino A, Palamara AT, Belardelli F, Favalli C. Antitumor effect of thymosin α1/interleukin-2 or thymosin α1 interferon α,β following cyclophosphamide in mice injected with highly metastatic Friend erythroleukemia cells. J Immunother 1993; 13:7–17.

12a. Garaci E, Rocclin G, Perroni L, D'Agostini C, Sascia F, Grelli S, Mastino A, Favalli C. Combination treatment with zidovudine, thymosin α1 and interferon-γ in HIV infection. J Clin Lab Res (in press).

13. Gartner S, Markovits P, Markovits DM, Kaplan MH, Gallo RC, Popovic M. The role of mononuclear phagocytes in HTLV-III/LAV infection. Science 1986; 233:215–19.
14. Groux H, Torpier G, Monte D, Mouton Y, Capron A, Ameisen JC. Activation-induced death by apoptosis in CD4+ T cells from human immunodeficiency virus infected asymptomatic individuals. J Exp Med 1992; 175:331–40.
15. Imberti L, Sottini A, Bettinardi A, Puoti M, Primi D. Selective depletion in HIV infection of T cells that bear specific T cell receptor V beta sequences. Science 1991; 254:860–62.
16. Kanagawa O, Nussrallah BA, Wiebenga ME, Murphy KM, Morse HC, Carbone FR. Murine AIDS superantigen reactivity of the T cells bearing V beta 5 T cell antigen receptor. J Immunol 1992; 149:9–16.
17. Larder BA, Coates KE, Kemp SD. Zidovudine-resistant human immunodeficiency virus selected by passage in cell culture. J Virol 1991; 65:5232–6.
18. Laurent-Crawford AG, Krust B, Muller S, Riviere Y, Rey-Cuille MA, Bechet JM, Montagnier L, Hovanessian AG. The cytopathic effect of HIV is associated with apoptosis. Virology 1991: 185:829–39.
19. Lucey DR, Melcher GP, Hendrix CW, Zajac RA, Goetz DW. Butzin CA, Clerici M, Warner RD, Abbadessa S, Hall K, et al. Human immunodeficiency virus infection in the US Air Force: Seroconversion assay to predict change in CD4+ T cell counts. J Infect Dis 1991; 164:631–7.
20. Lyerly HK, Matthews TJ, Langlois AJ, Bolognesi DP, Weinhold KJ. Human T-cell lymphotropic virus IIIB glycoprotein (gp120) bound to CD4 determinants on normal lymphocytes and expressed by infected cells serves as target for immune attack. Proc Natl Acad Sci USA 1987; 84:4601–5.
21. Macatonia SE, Lau R, Patterson S, Pinching AJ, Knight SC. Dendritic cell infection, depletion and dysfunction in HIV-infected individuals. Immunology 1990; 71:38–45.

22. Mann DL, Gartner S, LeSane F, Buchow F, Popovic M. HIV-1 transmission and function of virus-infected monocytes/macrophages. J Immunol 1990; 144:2152–8.
23. Mastino A, Piacentini M, Grelli S, Favalli C, Autori F, Tentori L, Oliverio S, Garaci E. Induction of apoptosis in thymocytes by prostaglandin E2 in vivo. Dev Immunol 1992; 2:263–71.
24. Mastino A, Grelli S, Piacentini M, Oliverio S, Favalli C, Perno CF, Garaci E. Correlation between induction of lymphocyte apoptosis and prostaglandin E2 production by macrophages infected with HIV. Cell Immunol 1993; 152:120–30.
25. Murray HW, Rubbin BY, Masur H, et al. Impaired production of lymphokines and immune (gamma) interferon in the acquired immunodeficiency syndrome. J Engl J Med 1988; 310:883–9.
26. Pantaleo G, Graziosi C, Fauci AS. New concepts in the immunopathogenesis of human immunodeficiency virus infection. N Engl J Med 328:327–35.
27. Patterson S, Knight SC. Susceptibility of human peripheral blood dendritic cells to infection by human immunodeficiency virus. J Gen Virol 1987; 68:1177–81.
28. Perno CF, Yarchoan R, Cooney DA, Hartman NR, Gartner S, Popovic M, Hao Z, Gerrard TL, Wilson YA, Johns DG, Broder S. Inhibition of human immunodeficiency virus (HIV-1/HTLV-IIIBa-L) replication in fresh and cultured human peripheral blood monocytes/macrophages by azidothymidine and related 2′,3′-dideoxynucleosides. J Exp Med 1988; 168:1111–25.
29. Perno CF, Yarchoan R, Cooney DA, Hartman NR, Webb DSA, Hao K, Mitsuya H, Johns DG, Broder S. Replication of human immunodeficiency virus in monocytes. Granulocyte/macrophage colony-stimulating factor (GM-CSF) potentiates viral production yet enhances the antiviral effect mediated by 3αazido-2′3′-dideoxythymidine (AZT) and other dideoxynucleoside congeners of thymidine. J Exp Med 1989; 169:933–51.
30. Perno CF, Baseler MW, Broder S, Yarchoan R. Infection of monocytes by human immunodeficiency virus type 1 blocked by inhibitors of CD4-gp120 binding, even in the presence of enhancing antibodies. J Exp Med 1990; 171:1043–56.
31. Schnittman SM, Denning SM, Greenhouse JJ, Justement JS, Baseler, M, Kurtzberg J, Hynes BF, Fauci AS. Evidence for susceptibility of intrathymic T-cell precursors and their progeny carrying T-cell antigen receptor phenotypes TCR alpha beta+ and TCR gamma delta+ to human immunodeficiency virus infection: A mechanism for CD4+ (T4) lymphocyte depletion. Proc Natl Acad Sci USA 1990; 87:7727–31.
32. Shearer GM, Clerici M. Early T-helper cell defects in HIV-1 infection. AIDS 1991; 5:245–53.
33. Via CS, Shearer GM. Autoimmunity and the acquired immune deficiency syndrome. Curr Opin Immunol 1989; 1:753–6.
34. Wahl LM, Corcoran ML, Pyle SW, Arthur LO, Harel-Bellan A, Farrar WL. Human immunodeficiency virus glycoprotein (gp120) induction of monocyte arachidonic acid metabolite and interleukin-1. Proc Natl Acad Sci 1989; 86:621–5.
35. Yarchoan R, Mitsuya H, Broder S. Challenges in the therapy of HIV infection. Immunol Today 1993; 14:303–6.

14

The Use of Thymic Peptides for Patients with Undetermined Human Immunodeficiency Virus Test Results

V.G. Morozov, K.S. Ivanov, V.V. Sheyback, and A.D. Kasyanov
Military-Medical Academy, St. Petersburg, Russia

I. INTRODUCTION

Disturbances of the functions of immune system are often discovered during screening research for human immunodeficiency virus (HIV) infection of blood donors and patients with various diseases. The main difficulty relates to interpretation of enzymoimmunoassay positive blood serum test results when negative and especially doubtful characteristics of immunoblotting are obtained (detection of antibodies to certain protein core or HIV ferments, but not to its protein membrane—gp41, gp120, gp160) (Table 1).

According to the results obtained in our laboratory, screening research for HIV infection reveals about 3.5% of primary enzyme-linked immunosorbent assay (ELISA)-positive results (1).

The discovery of ELISA-positive results may relate to the presence of antibodies to HIV as well as to its cross character of immunological reactions of transitory or constant nature that in turn proves some disturbances in the immune system, although the mechanism of this process is not yet known.

Thymic medications are designed to correct the disturbances in the immune system because thymic peptides control differentiation processes of T lymphocytes that take part in the process of antigen detection (2,3).

Thymalin is one of the thymic medications that has been used in clinical practice for prevention and treatment of secondary immunodeficiency since 1977. It con-

Table 1 The Interpretation of Immunoblotting Results When Obtaining Antibodies to HIV

Interpretation
Positive—antibodies to the HIV protein membrane (gp41, gp120, gp160) are discovered
Undetermined—antibodies to either HIV protein core or ferments but not to its protein membrane are discovered
Negative—antibodies to HIV proteins are not found

tains a complex of immunoregulating peptides isolated from calf thymus by means of acid extraction (4). According to its biological activity and clinical effectiveness, thymalin can be compared with thymosin, thymostimulin, and other known thymic medications.

The study discussed in this chapter shows the influence of thymic peptides on the immune system of patients with ELISA-positive results to HIV infection.

II. METHODS

The quantity of T and B lymphocytes was determined by means of immunofluorometric assay (5) with monoclonal antibodies (Ortho Diagnostic Systems, Inc., USA). Functional activity of lymphocytes was estimated in capillary modification of the leukocyte migration inhibition test (MIF assay) with concanavalin A (ConA) (6) and by means of detection of immunoglobulins (M, G, A) in blood serum (7). The content of immune complexes was estimated by spectrophotometric assay after blood serum treatment with polyethylene glycol (8). ELISA was used to detect antibodies to HIV. As a confirming test an immunoblotting method (Western blot) was applied using test systems produced by the Du Pont (USA) and Organon (USA) companies. Mathematical processing of the results obtained was carried out on the IBM PC/AT produced by the Spec Co. (Korea). The validity of the results was estimated in regard to Student's t test.

III. RESULTS

Repeated and confirming tests of 2133 patients (Fig. 1) allowed us to distinguish the following: group I—patients whose ELISA-positive results have not been confirmed (60.3%); group II—persons with ELISA-positive results while having negative characteristics of immunoblotting (27.7%); and group III—patients with ELISA-positive results and undetermined characteristics of immunoblotting (12.7%). Most often antibodies to p18/17 (63.7%), p15 (31.9%), p24 (31.9%), p55 (24.8%), and p-34-31 (10.6%) were found (1). Clinical research of 56.7–65.8% patients revealed infectious diseases (78.6%), diseases of the digestive organs (14.1%),

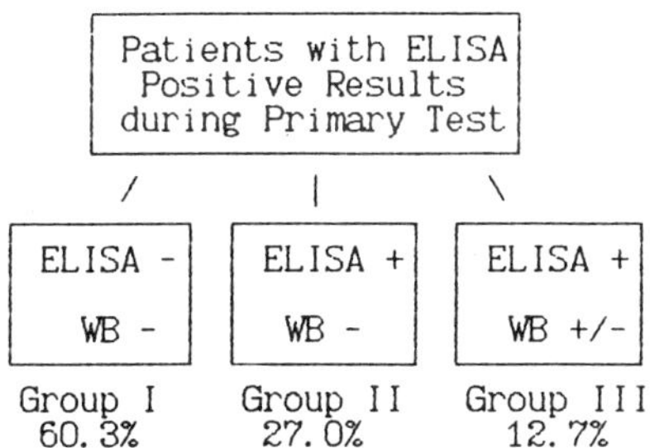

Figure 1 Group distribution of patients with ELISA-positive test results to HIV infection (n = 2133).

autoimmune diseases (14.0%), cardiovascular diseases (11.3%), and oncological (3.6%) and other diseases. Frequency and degree of pathology were highest in group III.

Immunotherapy with thymalin was carried out by means of intramuscular injection of 10 mg daily over 5 days. For this purpose 38 individuals from group II and 19 persons from group III were selected. These patients suffered both from acute infections (e.g., respiratory infectious diseases, flu, hepatitis A, pneumonia, infectious mononucleosis) and chronic infections (e.g., chronic tonsillitis, toxoplasmosis, cholecystitis). Group II also included patients with pathology of liver and autoimmune, cardiovascular, and oncologic diseases. Sixty-two patients of similar age from group I (with ELISA-negative results in regard to HIV infection) served as a control group. Thirty-five apparently healthy donors were tested in order to define normal immunological characteristics.

The results of the immunological investigation are given in Table 2. Patients with ELISA-positive results to HIV infection (groups II and III), an increase of $CD8^+$ cells, and disturbance of the relation of $CD4^+/CD8^+$ cells were revealed. Distinct changes also were shown in the humoral chain of the immune system. The patients tested revealed the increase of B cells and the content of immune complexes in serum, the concentration of immunoglobulins G and A was also changing which proved the development of autoimmune process. The distinct disturbance of the functional activity of lymphocytes in comparison with the control group was shown. The patients in group III also showed a valid increase of the total number of lymphocytes in the blood. It is also worth mentioning that the differences revealed in regard to immunologic characteristics between the groups of patients correlated with the results of the screening test to HIV infection in ELISA.

Thymalin immunotherapy of patients with ELISA-positive results led to an improvement in the characteristics of immune status in 70.2% of cases. A decrease of immune complexes and a tendency toward normalization of the relation between

Table 2 Immunological Parameters for Patients with Different Results in Regard to HIV Infection

Index	Normal (n = 35)	I group (n = 62)	II group (n = 38)	III group (n = 19)
Lymphocytes ($\times 10^9$/L)	1.96 ± 0.06	2.02 ± 0.15	2.10 ± 0.11	2.40 ± 0.13
$CD3^+$ cells ($\times 10^9$/L)	1.09 ± 0.08	1.48 ± 0.19	1.36 ± 0.17	1.27 ± 0.20
$CD4^+$ cells ($\times 10^9$/L)	0.65 ± 0.05	0.69 ± 0.09	0.67 ± 0.08	0.54 ± 0.12
$CD8^+$ cells ($\times 10^9$/L)	0.41 ± 0.03	0.57 ± 0.14	0.62 ± 0.11^a	0.77 ± 0.13^a
$CD4^+/CD8^+$	1.64 ± 0.12	1.21 ± 0.12^a	1.10 ± 0.07^a	0.72 ± $0.17^{a,b}$
MIF assay (ConA) (%)	59.1 ± 1.7	58.4 ± 2.4	68.3 ± 6.7^b	72.1 ± $2.9^{a,b}$
B Ig^+ cells ($\times 10^9$/L)	0.29 ± 0.02	0.37 ± 0.02	0.41 ± 0.05^a	0.46 ± 0.01^a
IgM (g/L)	1.15 ± 0.06	1.07 ± 0.21	1.28 ± 0.25	1.15 ± 0.29
IgG (g/L)	11.50 ± 0.50	10.48 ± 0.36	11.70 ± 0.58^b	12.48 ± $0.42^{a,b}$
IgA (g/L)	1.90 ± 0.08	1.91 ± 0.18	1.56 ± $0.16^{a,b}$	1.86 ± 0.20
Immune complexes (%)	92.6 ± 1.8	89.8 ± 1.8	77.3 ± $2.0^{a,b}$	69.8 ± $2.8^{a,b}$

[a] Distinct difference in regard to norm ($P < .05$).
[b] Distinct difference in regard to parameters I group ($P < .05$).

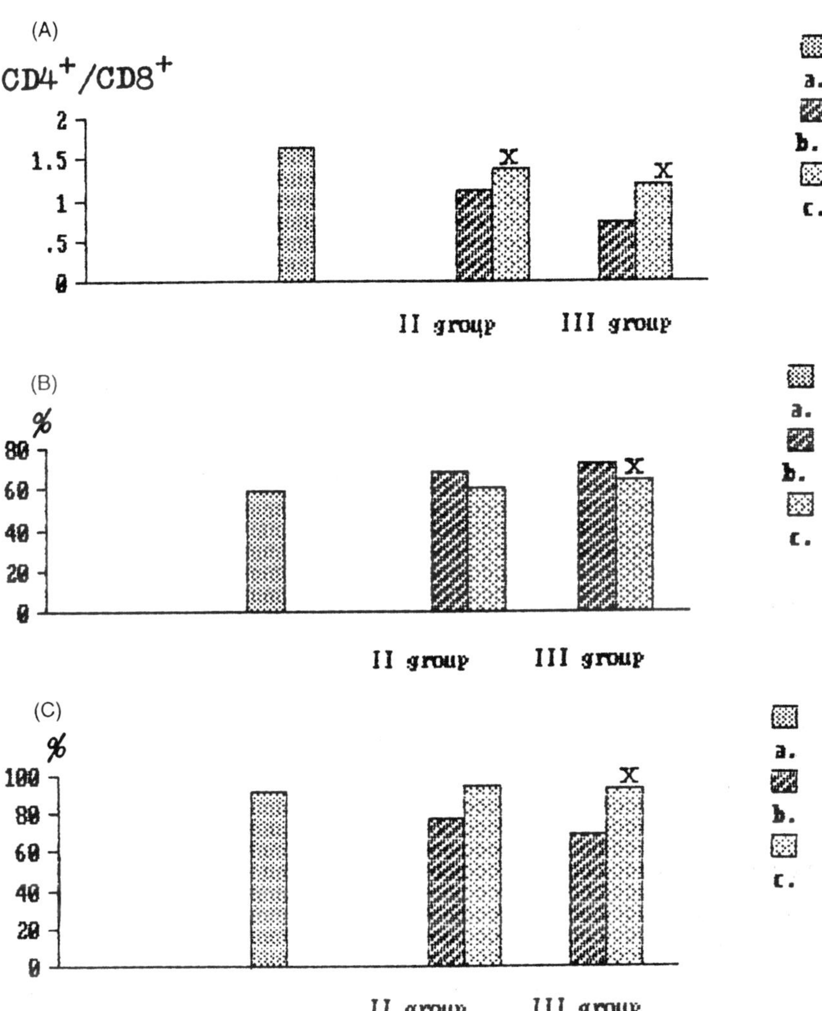

Figure 2 Peripheral blood CD4+/CD8+ ratio (A), MIF-assay with ConA (B), and content of immune complexes (C) in patients before and after thymalin treatment. (a, normal, b, before treatment, c, after treatment); x, distinct difference ($P < .05$).

T lymphocytes and MIF assay was noted (Fig. 2). Also, after using thymalin, the patients from group II demonstrated decrease in the antibodies to HIV, extinction characteristics being as follows: (control 0.211 $\pm$ 0.018) before treatment and 0.180 $\pm$ 0.016 (control 0.207 $\pm$ 0.014) after treatment. However, despite improvement status in group III, significant changes of antibodies to HIV were not discovered $\pm$ 0.099 (control 0.250 $\pm$ 0.020) and 0.539 $\pm$ 0.107 (control 0.248 $\pm$ 0.017), respectively.

IV. CONCLUSIONS

The appearance of ELISA-positive cross reactions among patients tested may on the one hand be explained by activation of latent infection, the pathogens of which have similar ferments or structural components to those of HIV and on the other hand may be due to the autoimmune nature of ELISA-positive reactions related to the disturbance of distinguishing homologous HIV tissue structures of different diseases. Thymic peptide correction of disturbances of the immune system influence the degree of distinction of cross immunological reactions that are registered in ELISA.

According to the results obtained, one can suggest that the response to antigens of the patients tested depends on the thymus and develops in regard to interleukin-4, -5, and -6 dependent pathway with participation of $CD4^{+}TH_{2}$ cells. Disturbances of immunoregulation may take place on the level of restriction of activated $CD4^{+}TH_{0}$ cells. Although in order to obtain a more definite answer it is necessary to check the cytokine profile, particularly the content of interleukins 2, 4, 5, 6, 9, and 10, interferon-gamma, and tumor necrosis factor-beta or their receptors.

REFERENCES

1. Morozov VG, Sheybak VV, Selivanov AA. Diagnostic value of different test-systems to screening detection of antibodies to HIV. J Microbiol Epidemiol Immunobiol 1992; 4:17 (in Russian).
2. Cohen GH, Hooper JA, Goldstein AL. Thymosin induced differentiation of murine thymocytes in allogenic mixed lymphocyte cultures. Ann NY Acad Sci 1975; 249:145.
3. Scheid MP, Goldstein G, Boyse EA. The generation and regulation of lymphocyte populations. Evidence from differentiative induction systems in vitro. J Exp Med 1978; 147:1727.
4. Morozov VG, Khavinson VKh. Thymalin and its immunobiological activity. In: Grinevich, Chebotarev VF, eds. Immunobiology of thymic hormones. Zdorovja, Kiev, 1989:125.
5. Reinhertz EL, Kung PC, Goldstein G, Schlossman SF. Separation of functional subsets of human T-cells by a monoclonal antibody. Proc Natl Acad Sci USA 1979; 76:4061.

6. Soborg M, Bendixen G. Human lymphocyte migration as a parameter of hypersensitivity. Acta Med Scand 1967; 181:247.
7. Mancini G, Carbonara AO, Heremans JF. Immunochemical quantitation of antigen by single radial immunodiffusion. Immunochemistry 1965; 2:235.
8. Grinevich JuA, Alferov An. Detection of immune complexes in oncological patients' blood. Lab Delo 1981; 8:4493 (in Russian).

15

Site-Specific Differentiation Induction on T Cells In Vivo by Synthetic Thymic Peptides to Fight Microbial Infections and Cancer

Christian Birr
Heidelberg University and ORPEGEN, Med.-Molekularbiologische Forschungsgesellschaft mbH, Heidelberg, Germany

C. Thomas Nebe and Gerhard Becker
Heidelberg University, Heidelberg, Germany

In previous contributions (1–3), we had reported on in vitro bioactivities of synthetic thymic peptides derived from thymosin α_1 (4). Findings of immunoactivity in both mitogen costimulation and mixed lymphocyte culture assays as well as in α-amanitine-inhibited E rosette tests on human peripheral lymphocytes directed our search for more specific methods for in vitro investigations. By application of the azathioprine-inhibition assay on the rosetting phenomenon with mouse spleen cells, we had identified a fundamental signal character (5) in thymosin α_1–related sequences to induce cytotoxic lymphocytes. The parent active polypeptide thymosin α_1 shows peculiar sequence characteristics among its 28 residues. Several of them occur in pairs resulting in a unique sequence pattern of basic, acidic, polar, and lipophilic amino acids. By comparison of our findings with the literature data, we had identified a structure-related in vitro activity pattern that we were able to condense into a peptide sequence motif like basic-acidic-lipophilic-acidic. Since in that assay spleen cells were applied known to represent a pool of cytotoxic T cells, we nowadays interpret our earlier findings on the specific sequence pattern as an indication for a molecular major histocompatibility complex (MHC) restriction motif (6) for thymic selection and priming of cytotoxic lymphocytes from a pool of immature $CD4^+CD8^+$ thymocytes (7).

Several of these thymic peptides then were shown to exhibit extraordinary efficacy in animal models as reported in the previous proceedings of these symposia series and elsewhere (8). Thus, the spread of lung metastases could be suppressed in mice inoculated with B16 melanoma cells. Furthermore, there was a significant enhancement of the survival rate of mice that had been immunosuppressed by 5-fluorouracil (215 mg/kg/day) or x-ray and then challenged with P388 or L1210 leukemia cells. It was shown in those studies that the protective effect by synthetic small thymic peptides was both sequence and dose dependent. Some of the peptides tested showed the same activity in mice when administered at 100 times higher dosage with food.

These data are in support of our view on a molecular peptide signal motif for the in vivo induction of activated cytotoxic T lymphocytes. On these grounds, we became interested in learning more about the regulatory hierarchy in the differentiation induction of T lymphocytes in vivo in bone marrow, in thymus, and in the peripheral immune system by synthetic peptide signal molecules. Mice were treated subcutaneously daily for 15 days with thymic peptides (30 nmol/kg, five animals/peptide). Nineteen different fragments of thymosin α_1, encoded T2–T29, were investigated this way and compared with thymopoietin pentapeptide (TP-5, TIMUNOX Cilag, Zürich), thymosin α_1, and 15 nmol/kg of twin α_1 (9) (N,N′-succinoyl-bis-thymosin α_1), respectively. From day 6 to 10, the animals received in addition a sublethal daily dose of 25 mg/kg of 5-FU for general immune suppression caused by this cancer chemotherapeutic agent. On day 15, the animals were sacrificed and dissected. Cell preparations from bone marrow, thymus, spleen, lymph nodes, and peripheral blood were investigated by flow cytometry side by side, each with the same set of fluorescent labeled monoclonal antibodies specific for lymphocyte differentiation states as shown in Figures 1–4 in comparison with control preparations. Control group 1 was a blank group receiving only 200 µl/day of 0.9% saline for 15 days (PL^-). In addition to saline, group 2 received the 5-FU treatment from days 6 to 10, as described above (PL^+).

The treatment with 5-FU from days 6 to 10 in mice on day 15 results in a significant increase of T lymphocytes not only in bone marrow but also in thymus, spleen (data not shown), and peripheral blood. In lymph nodes, we do not see this phenomenon (data not shown). The well-known immunosuppressive effect of this cancer chemotherapeutic agent is demonstrated by decreased numbers both of early B marked leukocytes in bone marrow and of B cells in peripheral blood as compared with the blank control. Therefore, we have to take into account these elevated levels of induction at individual differentiation stages of lymphocytes already achieved by 5-FU treatment. Several of our thymic peptides result in statistically significant inductions of lymphocyte differentiation at various sites of the immune system over the 5-FU effect, although applied systemically in very low dosage. Thymosin α_1 (THY) and its N,N′-dimeric twin α_1 version (TWI) are more tissue specific and mainly active in the thymus as expected for a homing factor. In comparing the responses from Figure 1 with those of Figures 2–4, one can generate a better understanding on distinct site specificities of several of the thymic peptides;

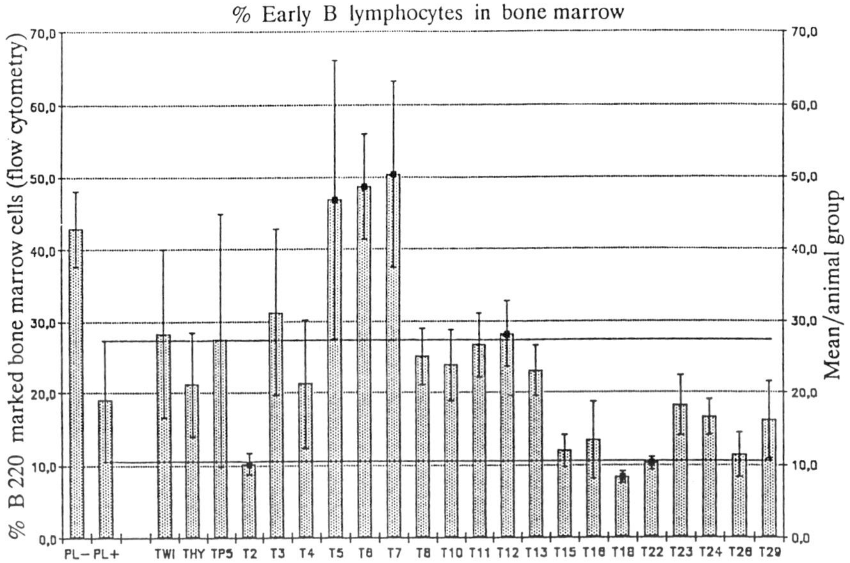

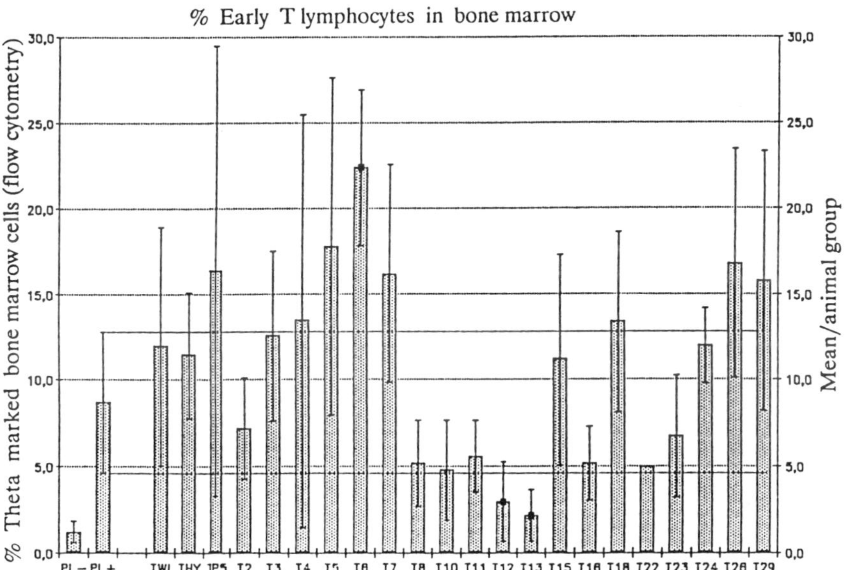

Figure 1 Site-specific in vivo induction of early B- and T-precursor cells in bone marrow of 5-FU–suppressed mice by synthetic thymic peptides, s.c.; significance <0.05.

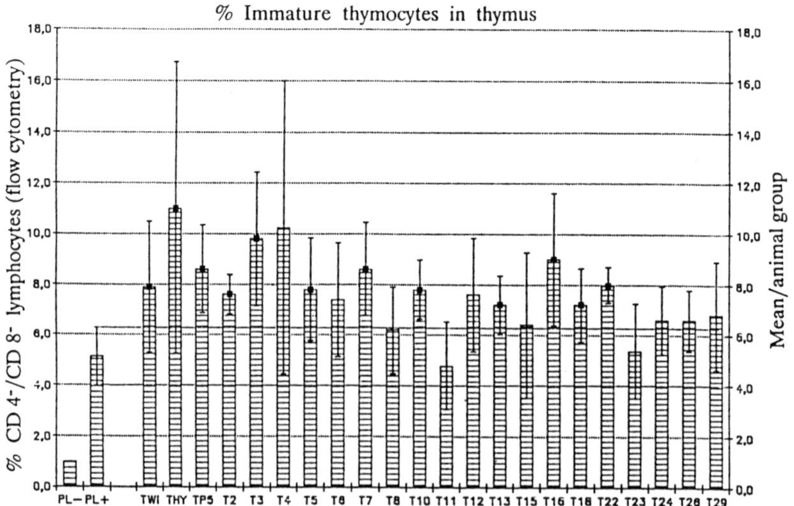

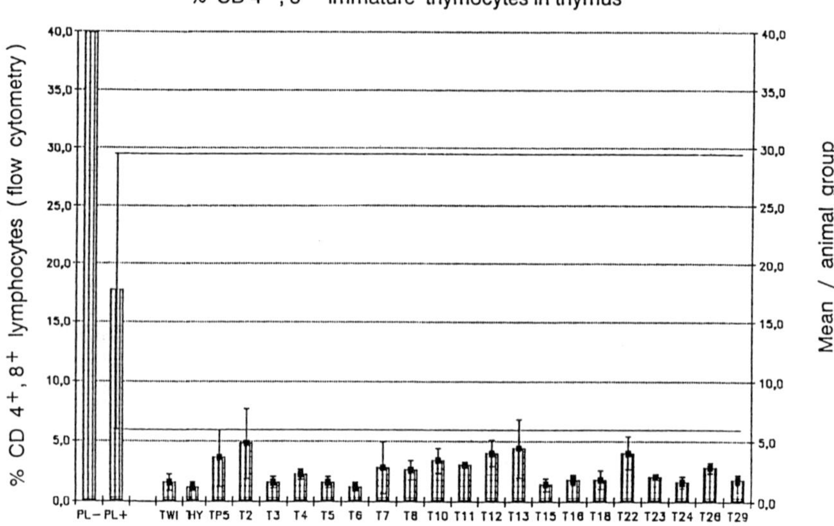

Figure 2 Site-specific in vivo induction in thymus of immature double negative thymocytes and of the depletion from double positive lymphocytes by signal-dependent advanced selection for maturation of T helper and T suppressor cells; significance <0.05. Model as in Figure 1.

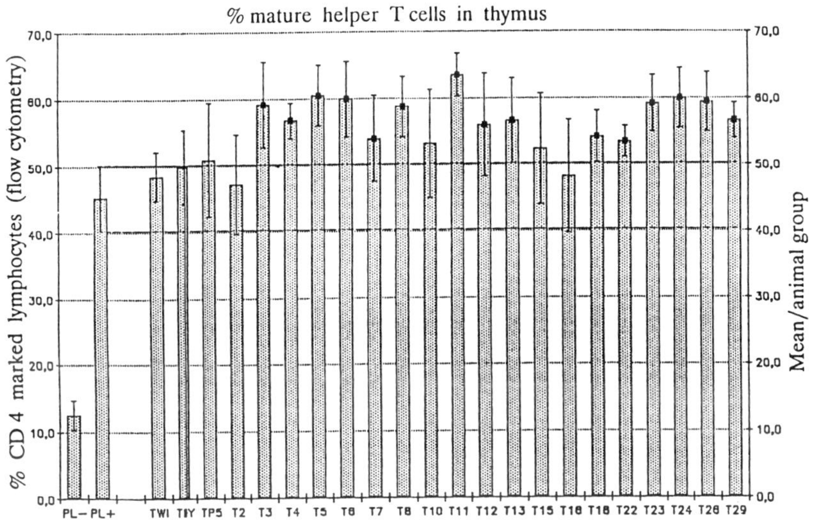

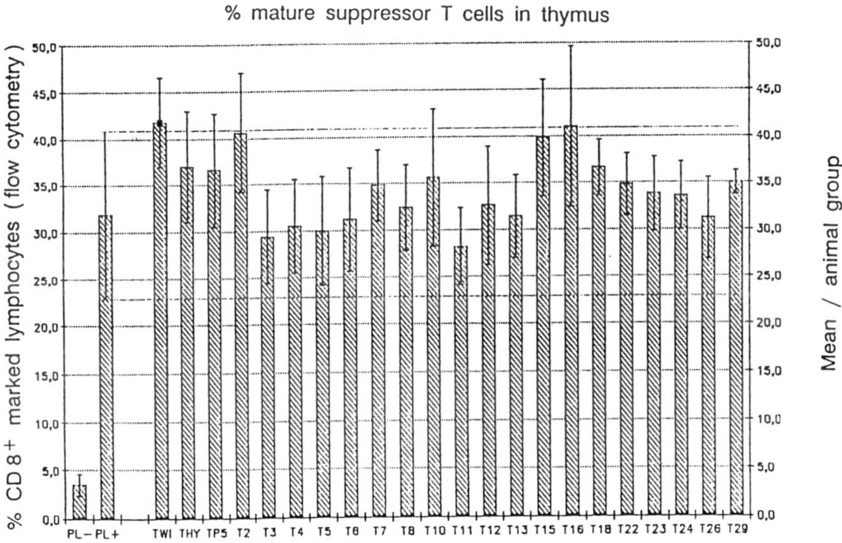

Figure 3 Site-specific in vivo induction in thymus of mature T helper and T suppressor/cytotoxic lymphocytes; significance <0.05. Model as in Figure 1.

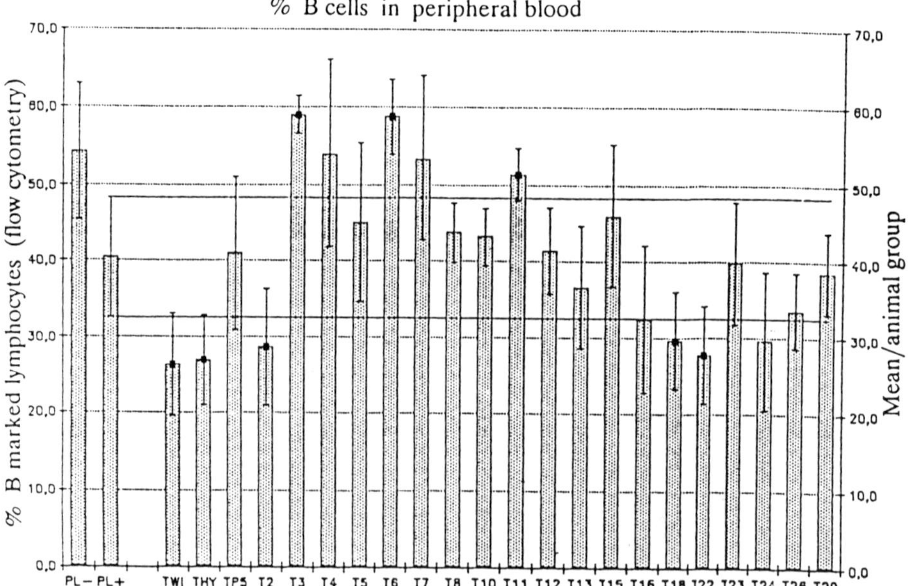

Figure 4 Site-specific in vivo induction in peripheral blood of B lymphocytes and suppressor/cytotoxic T cells; significance <0.05. Model as in Figure 1.

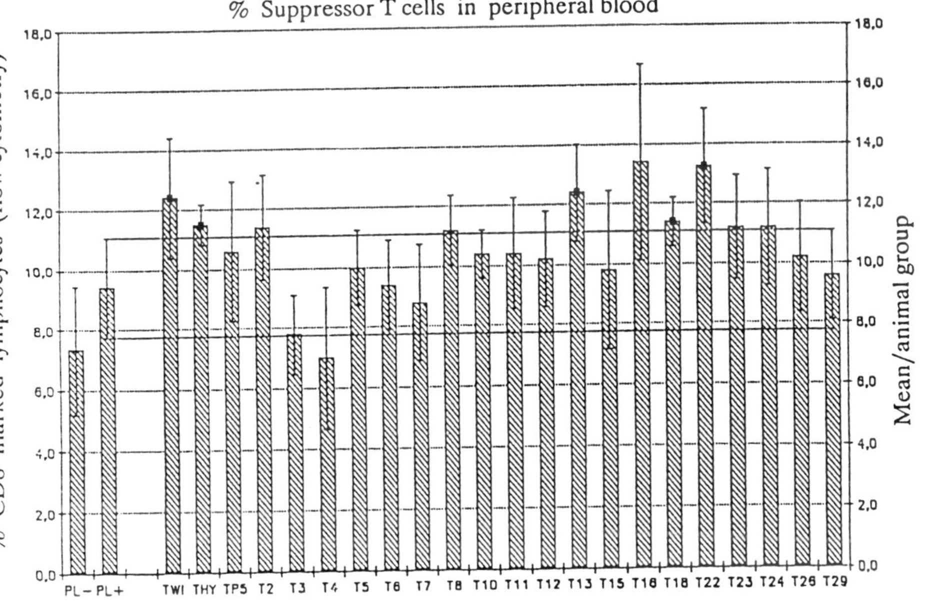
% Suppressor T cells in peripheral blood
% CD8 marked lymphocytes (flow cytometry)
Mean/animal group
18,0
16,0
14,0
12,0
10,0
8,0
6,0
4,0
2,0
0,0
PL− PL+ TWI THY TP5 T2 T3 T4 T5 T6 T7 T8 T10 T11 T12 T13 T15 T16 T18 T22 T23 T24 T26 T29

for example, in bone marrow depleted from early B lymphocytes by the action of 5-FU, four of our peptides can overcompensate this deficiency significantly, which can also be seen in Figure 4 as an increased count of B cells in peripheral blood. Thymosin α_1–related peptides T3–T7 in bone marrow induce the increase of early theta markers in stem cell differentiation. Consequently, under the influence of these peptides, we can demonstrate a significant increase of homing of immature thymocytes in the thymus and maturation of T helper and suppressor cells from this compartment as presented in Figures 2 and 3. In thymus, about all of the peptides investigated exhibit significant activity in the site-specific increase of the number of immature $CD4^-CD8^-$ thymocytes. It is most essential also to note from Figure 2 that the natural large pool of double-positive thymocytes ($CD4^+CD8^+$) becomes completely depleted under the influence of the thymic peptides tested. For comparison, the very left bar in the graph (blank control, no 5-FU, no thymic peptides, PL^-) represents the normal count for $CD4^+CD8^+$, >40% of all cells measured.

From this large pool, an advanced selection both for $CD4^+$ mature T helper and $CD8^+$ cytotoxic lymphocytes occurs under the influence of our thymic peptide signal molecules. Most probably the process is MHC restricted and calls for the specific structural motifs mentioned above. As a consequence of the advanced pool depletion, an increased number of T helper and suppressor cytotoxic cells are maturated in and released from the thymic environment (Fig. 3). In the circulation (Fig. 4), thymosin α_1 as well as its dimeric version and the shorter fragments T_3, T_{18}, and T_{22} significantly increase the number of suppressor/cytotoxic T-cells. This finding should be appreciated as a hint for further pharmaceutical investigations with specific thymic peptides. Thymopoietin pentapeptide (TP5), the only synthetic immunopharmaceutical currently on the market, which we ran for comparison with our thymosin α_1 peptides, did not exert any superiority but mean activity in all sites investigated.

Our findings on site-specific activities of thymosin α_1 peptides administered systemically may further support our working hypothesis on the regulatory signal properties of short immunopeptides, not only generated synthetically but also probably by metabolic fragmentation of tissue-specific factors like thymosin α_1 and thymopoietin and also of lymphokines and cytokines (10). It is well experienced by clinicians that therapeutic application of interferons or by interleukin-2 (IL-2) systemically is accompanied by severe side effects. Thymic signal peptides, however, are not toxic at all and can induce activation of both $CD4^+$ helper and $CD8^+$ cytotoxic T cells. By flow cytometric measurements, we were able to detect expression of IL-2 receptors as a sign for T-cell activation at different sites when the synthetic signal peptides were administered systemically. In this way, site-specific release of lymphokines should occur as a second signal for local cell/cell communication. In other words, synthetically available peptide signal molecules exhibit the immunopharmacologic property of a site-specific trigger of the lymphokine communication network for regulation of leukocyte activation in the periphery. This concept should be suitable for therapeutic applications.

REFERENCES

1. Birr C, Stollenwerk U, Werner I, Manke HG, Brodner O. Synthesis and properties of immunostimulating peptides. In: Brunfeldt K, ed. Peptides 1980. Copenhagen: Scriptor Press, 1981:420–7.
2. Ciardelli TL, Krueck I, Birr C, Brodner O. Synthesis and properties of immunostimulating peptides. II. Peptides, Synthesis-Structure-Function. In: Rich D, Gross E, eds. Rockford, IL: Pierce Chemical Co. Press, 1981:541–4.
3. Birr C, Krueck I, Stollenwerk U, Ciardelli TL, Brodner O. Synthesis and properties of immunostimulating peptides. III. In Rich D, Gross E, eds. Peptides, Synthesis-Structure-Function. Rockford, IL: Pierce Chemical Co. Press, 1981:545–8.
4. Goldstein AL, Low TLK, McAdoo M, McClure J, Thurman GB, Rossio J, Lai CY, Wang SS, Havey C, Ramel AH, Meienhofer J. Thymosin-α_1: Isolation and sequence analysis of an immunologically active polypeptide. Proc Natl Acad Sci USA 1977; 74:725–9.
5. Ciardelli TL, Incefy GS, Birr C. The activity of synthetic thymosin-α_1 C-terminal peptides in the azathioprine E-rosette inhibition assay. Biochemistry 1982: 21:4233–7.
6. Van Ewijk WA. T-cell differentiation is influenced by thymic microenvironments. Annu Rev Immunol 1991; 9:591–615.
7. Vukmanovic S, Grandea AG III, Faas SJ, Knowless BG, Bevan MJ. Positive selection of T-lymphocytes induced by intrathymic injection of a thymic epithelial cell line. Nature 1992; 359:729–32.
8. Bohn B, Nebe CT, Birr C. Pharmacological potential of thymosin-α_1 fragments. Adv Biosci 1988; 68:405–11.
9. Birr C, Chiardelli TL, Nebe CT, Heinzel W, Young M, Ho AD, Stehle B. Twin-α_1, the head-to-head analog of thymosin-α_1, is a more potent immunoregulatory polypeptide drug than the parent. In: Deber CM, Hruby VJ, Kopple KD, eds. Peptides—Structure & Function. Rockford, IL: Pierce Chemical Co. Press, 1985:79–82.
10. Birr C. Synthetic small thymic peptides—An immunoregulatory concept. In: Goldstein AL, ed. Thymic Hormones and Lymphokines. New York: Plenum, 1984:97–109.

16

Immunomodulation with Thymosin Fraction 5 Reduced the Number of *Trichuris muris* Adult Worms Developed in the Intestine of Nonresponder Mice

J. Hermanek
Institute of Parasitology, Ceske Budejovice, Czech Republic

D. Wakelin
University of Nottingham, Nottingham, England

I. INTRODUCTION

The nematode parasite *Trichuris muris,* which inhabits the cecum and proximal colon of mice, is closely related to *Trichuris trichiura* causing chronic trichuriasis in humans. It has been shown previously that immunocompetent mice of most inbred strains (with exception of the B10.BR and AKR strains) expel primary infections with *T. muris* before reaching patency (i.e., before day 32 postinfection [1]) and that this expulsion is usually associated with the TH_2 type of immune response (2–4). On the other hand, chronic infections have been described in athymic nude (5) or neonatally thymectomized mice (6), as well as in mice immunosuppressed with antithymocyte serum (6) or corticosteriods (7,8). Consequently, the functional T-cell–mediated immunity seems to be essential for the immune expulsion of worms (9). At least, the hydrocortisone-treated Balb/K mice developed the same TH_1 type of immune response as the primary nonresponder B10.BR mice (10).

Thymosins represent a wide group of various immunomodulatory polypeptides isolated from bovine thymus, which are involved in the differentiation of precursors of T lymphocytes into immunocompetent cells (11). Numerous different thymic preparations have been isolated in the last two decades, and some of them are widely used for immunocorrection in different immunodeficient conditions (12). More recently, a beneficial effect of immunomodulation with thymic preparations on the resistance to some experimental parasitic infections has been described (13–15). Moreover, Somali children treated with thymopentin have been recently reported to be more resistant to reinfection with *T. trichiura* when compared with nontreated children (16).

Regarding the crucial role of T-cell–mediated immune response in the output of primary infections in murine trichuriasis (5,6), the present study was designed to investigate the principal possibility to influence the resistance of nonresponder mice (mice of the B10.BR strain and hydrocortisone-treated CBA mice) to infection with *T. muris* by immunomodulation with thymosin fraction 5.

II. MATERIALS AND METHODS

A. Mice

Specific pathogens free female mice of the CBA/Ca and B10.BR strains (HARLAN OLAC), aged 6–8 weeks at the moment of infection, were used throughout the experiments.

B. Parasite

The maintenance of *T. muris* and the preparation of infective eggs were as described previously (1). Experimental mice were infected perorally with 400, 300, or 200 embryonated eggs in 0.2 ml of phosphate-buffered saline (PBS) at DPI (day post-infection) 0. Mice were killed by an overdose of chloroform on given days and the cecum and first 4 cm of the colon were removed from each mouse, placed in Hank's buffer, and frozen at –20°C. The larval stages were counted after scraping of mucosa from the thawed opened gut in PBS, and the adult worms were counted as they were removed individually from the gut mucosa.

C. Treatment

Hydrocortisone-treated CBA mice were injected subcutaneously with 1.25 mg of hydrocortisone acetate (HCA, Sigma) in 0.1 ml of PBS on DPI 7, 9, and 11, which regimen has been described as suitable for the establishment of chronic infection (8). Thymosin fraction 5 was injected intraperitoneally as a 100 μg/dose in 0.2 ml of PBS. The time schedules of treatment and design of individual experiments are shown in the presented table footnotes and figure legends.

D. Antigens

The adult worms (DPI 35) were washed several times and homogenized in sterile PBS using a glass tissue homogenizer. After centrifugation (30 min, 1500 *g*) the resulting supernatant was dialysed overnight against 3 × 5 L of PBS, sterilized by filtration through 0.22 μm Millipore filters, analyzed for protein content using Lowry assay, aliquoted, and stored at –20°C.

E. Parasite-Specific Serum Antibodies

Mice killed by an overdose of chloroform were bled by cardiac puncture and individual sera were collected and stored at –20°C until used. Levels of parasite-specific immunoglobulins (Ig) GAM, Gl, and G2a in serum samples were determined using indirect enzyme-linked immunosorbent assays (ELISA) as described previously (1). Microplates (Falcon) were coated with 0.5 μg/well of adult worm homogenate (AWH) and individual sera diluted to 1:100 in PBS-Tween were tested in triplicate. Alkaline phosphatase–conjugated goat anti-mouse IgGAM (Sigma), anti-mouse IgGl (Sigma), or anti-mouse IgG2a (Binding Site), all diluted 1:500, were used for the detection of bound antibodies.

F. In Vitro Lymphocyte Proliferation Assay

The evaluation of the in vitro proliferation of spleen and mesenteric lymph node (MLN) cells (pooled within experimental groups) was carried out as described previously (17). Briefly, 2×10^5 cells in 200 μl of RPMI 1640 media (Gibco) supplemented with 10% of fetal calf serum (FCS, Gibco) were placed into the wells of 96-well microplates (Falcon). Mitogens—concanavalin A (ConA, Sigma), lipopolysaccharide (LPS, Sigma), and phytohemagglutinine (PHA, Sigma)—were added in the volume of 20 μl/well so the final concentration was 5 μg/ml. The final concentrations of *T. muris* AWH used were 4–50 μg/ml (four wells for each concentration tested). The plates were incubated for 4 days, the wells were pulsed with 0.25 μCi of ^{3}H-labeled thymidine ([^{3}H]TdR, Amersham) for 24 h. The results are expressed as mean differences between the [^{3}H]TdR uptake (counts per minute, cpm) of stimulated and the mean uptake of nonstimulated cells: diff.cpm = mean ($cpm_{stimulated}$ – mean $cpm_{nonstimulated}$).

G. Evaluation of Cytokine Production In Vitro

Single cell suspensions of lymphoid cells were diluted in 10% FCS-supplemented RPMI 1640 up to 5×10^6 cells/ml. Two milliliters of the final suspension (i.e., 10^7 cells) were placed into each well of 24-well plate (Costar) and incubated with 5 μg/ml of ConA for 48 h. After centrifugation, the supernatants were collected, aliquoted and stored at –20°C until used.

Interferon gamma (IFN-γ), interleukin-4 (IL-4), and interleukin-5 (IL-5) were measured by a sandwich ELISA method using a protocol developed by Else and Grencis (2). The limit of detectability in our assays was 1 U/ml for IFN-γ and 4 U/ml for IL-5.

H. Statistical Analysis

The significance of differences in the experimental data between different groups were compared using the Mann-Whitney U-test. A value of $P < .05$ was considered significant.

III. RESULTS

A. Immunomodulation with Thymosin in B10.BR Mice

The influence of immunomodulation with thymosin fraction 5 on the development of *T. muris* in the large intestine of B10.BR mice (three independent experiments) is summarized in Table 1. These results clearly demonstrate a significant protective effect (21.6–61.9% reduction in the worm burden) of thymosin, which strongly depended on the timing of thymosin administration and—to some extent—on the infection dose used. Control nontreated B10.BR female mice were unable to expel primary infection with 400 or 300 eggs of *T. muris* and carried more than 200 adult worms on DPI 35. Repeated intraperitoneal injections of 100 μg/mouse of thymosin fraction 5 caused only slight reduction in the worm burden (21.6% in comparison with the nontreated control group) when administered at about the moment of infection; however, a highly significant protection (44.9% respective 59.6%) was achieved by the administration within the second week after infection (experiments 1 and 2). The similar protective effect was observed after a prolonged administration of thymosin at the later stages of infection (experiment 3).

Only minor, nonsignificant differences in humoral immune responses (*T. muris*-specific IgGl, IgG2a, IgGAM serum antibodies) between immunomodulated and nontreated control B10.BR mice have been found in our experiments (data not

Table 1 Mean Worm Burdens in Nonresponder B10.BR Mice Treated with Thymosin

Experiment	Infection dose (eggs/mouse)	Thymosin treatment (100 μg/mouse on DPI)	No. of worms on DPI 35 (mean ± SD)
1	400	—	265 ± 38[a]
	400	7,9,11	146 ± 47[a]
2	400	—	208 ± 82[b]
	400	−3, −1,1,3	163 ± 41
	400	7,9,11,13	84 ± 49[b]
3	300	—	202 ± 37[c]
	300	18,21,23,25,28,30,32	77 ± 21[c]

Female mice of the B10.BR strain (five to eight mice per group) were infected with embryonated eggs of *T. muris* on DPI 0 and injected intraperitoneally with 100 μg/mouse of thymosin fraction 5 on given days after infection (experiments 1–3). The number of adult worms developed in the large intestine (cecum and proximal colon) was assessed on DPI 35.

[a] $P < .01$.

[b] $P < .05$.

[c] $P < .005$.

shown). Similarly, there were no significant differences in the in vitro proliferative responses of spleen or MLN cells to different mitogens (ConA, PHA, LPS) assessed on DPI 35 (data not shown). However, highly significant differences in the in vitro antigen-induced proliferative responses of lymphoid cells to AWH antigens were detected in most of our experiments (representative data from experiment 1 are shown in Fig. 1). The AWH revealed a noticeable mitogenic ac-

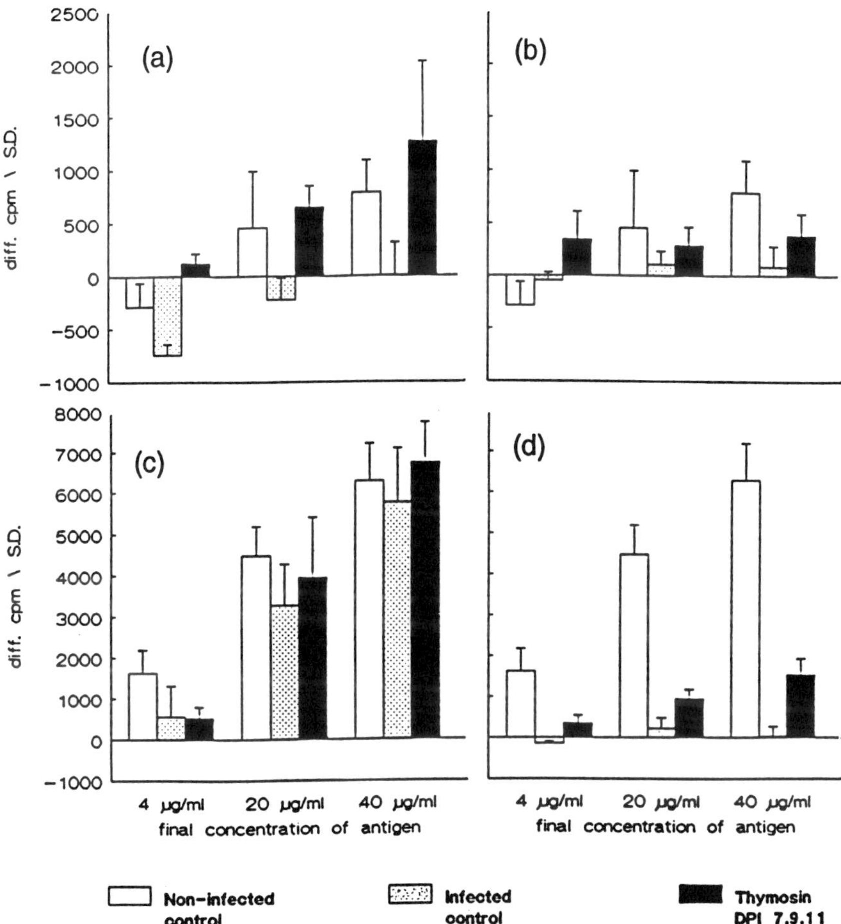

Figure 1 The in vitro proliferative response of spleen (a,b) and MLN (c,d) cells from female B10.BR mice to *T. muris* antigens (AWH) assessed on DPI 14 (a,c) and 35 (b,d)—experiment 1. 2×10^5 lymphoid cells were incubated with 4, 20, or 40 µg/ml of AWH for 4 days and then pulsed for 24 h with 0.25 µCi of tritium-labeled thymidine. The results are expressed as differences in the uptake of stimulated and nonstimulated cells (diff. cpm).

tivity on spleen and MLN cells from noninfected B10.BR mice. Surprisingly, proliferative responses of lymphoid cells from infected control animals were, in general, significantly lower when compared with naive controls. However, immunomodulation with thymosin restored and even elevated antigen-induced proliferation of spleen and MLN cells in infected mice.

MLN cells from B10.BR mice infected with *T. muris* produced large amounts of IFN-γ after an in vitro stimulation with ConA—especially when the production was assessed on DPI 14 (Fig. 2a). In contrast, only minute amounts of IL-5 were produced by MLN cells at this time (Fig. 2b). By DPI 35, the IFN-γ production significantly declined and production of IL-5 raised. A similar pattern of IFN-γ production was observed in spleen cells (>307 U/ml on DPI 14 and 16.4 $\pm$ 1.5 U/ml on DPI 35), but production of IL-5 was much lower in comparison with MLN cells. However, only insignificant differences in the production of measured cytokines were noticed between immunomodulated and control mice in our experiments (Fig. 2).

B. Immunocorrection by Thymosin in Hydrocortisone-Treated CBA Mice Infected with *T. muris*

Influence of the administration of thymosin on the worm burden in HCA-treated CBA mice (experiments 4–6) is summarized in Table 2. Control HCA-treated mice

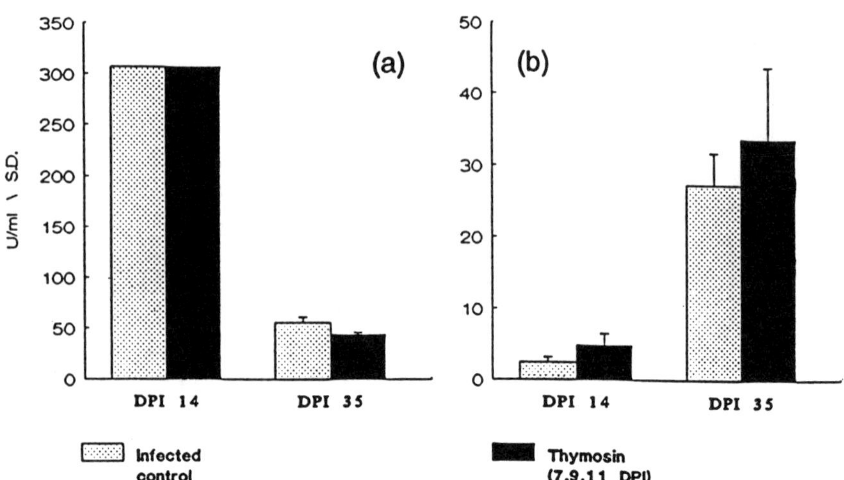

Figure 2 The in vitro production of IFN-γ (a) and IL-5 (b) by MLN cells from B10.BR mice infected with *T. muris* (experiment 1). 5×10^6 cell/ml were incubated with 5 μg/ml of ConA for 48 h and the concentration of individual cytokines in resulting supernatants was measured by a sandwich ELISA method. The limit of detectability in our experiments was 1 U/ml for IFN-γ and 4 U/ml for IL-5.

Table 2 Mean Worm Burdens in HCA-Treated CBA Mice After Administration of Thymosin

Experiment	Infection dose (eggs/mouse)	Thymosin treatment (100 μg/mouse on DPI)	No. of worms on DPI 35 (mean ±SD)	Reduction (%)
4	400	—	256 ± 30	—
	400	13,15,17	169 ± 29	34.0
5	300	—	175 ± 55	—
	300	13,15,17,19,21	130 ± 77	25.7
6	200	—	192 ± 58[a]	—
	200	13,15,17,19	62 ± 98[a]	67.7

Female mice of the CBA strain (five to eight mice per group) were infected with embryonated eggs of *T. muris* on DPI 0 and injected subcutaneously with 1.25 μg/mouse of HCA on DPI 7, 9, and 11. Thymosin fraction 5, 100 μg/mouse, was administered intraperitoneally on given days after infection and the number of adult worms developed in the large intestine (cecum and proximal colon) was assessed on DPI 35.
[a] $P < .01$.

infected with different numbers (200–400/mouse) of *T. muris* eggs on DPI 0 retained a significant portion of worms (58.3–96.0% of infection dose) on DPI 35. Repeated intraperitoneal injections of 100 μg/mouse of thymosin fraction 5 given after the termination of the HCA treatment significantly reduced the number of *T. muris* worms in the large intestine (25.7–67.7% reduction in comparison with control group). The maximal protective effect of thymosin was observed when experimental mice were infected with the lowest infection dose (200 eggs/mouse)—67.7%.

The level of serum parasite-specific antibodies (IgGAM, IgGl, IgG2a) assessed on DPI 35 did not differ significantly between control and thymosin-treated mice in experiments 4 and 5 (data not shown). In experiment 6, the total level of parasite-specific antibodies (IgGAM) was remarkably lower in the immunomodulated mice when compared with control animals, but these mice produced more IgGl antibodies (data not shown).

The in vitro proliferative response of spleen and MLN cells to various mitogens (ConA, PHA, LPS) assessed on DPI 35 did not differ significantly between the immunomodulated and control groups (data not shown). However, there were highly significant differences in the proliferative response of MLN and especially spleen cells to AWH between these groups (the representative data from experiment 6 are shown on Fig. 3). Whereas spleen cells from control HCA-treated mice did not respond to *Trichuris*-derived antigens at all, spleen cells from thymosin-treated mice were able to display a significant response (Fig. 3a).

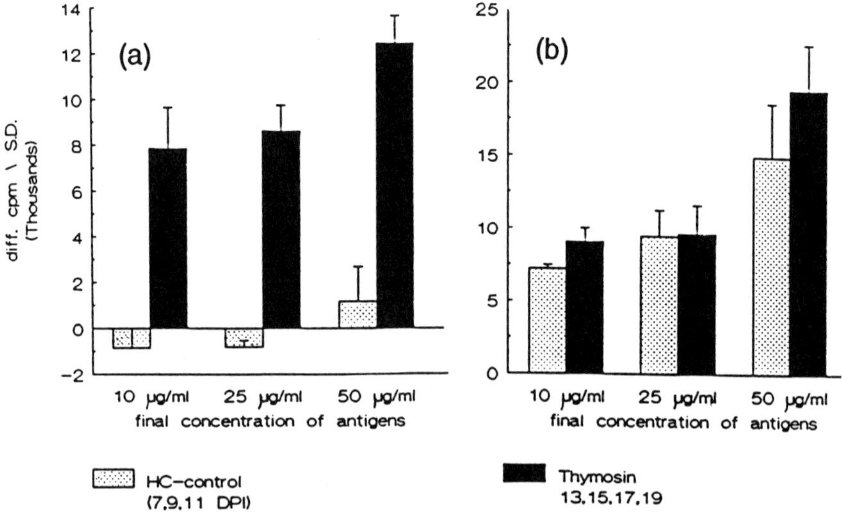

Figure 3 Antigen-induced proliferative responses of spleen (a) and MLN (b) cells from HCA-treated CBA mice on DPI 35 (experiment 6). 2×10^5 lymphoid cells were incubated with 10, 25, or 50 μg/ml of AWH for 5 days and processed as described in Figure 1.

The high in vitro ConA-induced production of IFN-γ by spleen cells from immunosuppressed CBA mice was only slightly reduced by the treatment with thymosin (data not shown); however, this treatment resulted in a significant increase of IL-5 production by spleen cells when assessed on DPI 20 and 35 (Fig. 4).

IV. DISCUSSION

Chronic parasitic infections are frequently associated with altered immunological responsiveness of the host (18). Regarding the crucial role of T-cell–mediated immunity in the outcome (expulsion or chronic infection) of many parasitic infections, thymic factors involved in the differentiation of T-lymphocytes (11) represent perspective immunomodulatory substances for therapy of chronic parasitoses.

In the present study, we demonstrated for the first time the principal possibility to influence the outcome of infection with *T. muris* in primary nonresponder B10.BR mice by a systemic administration of thymosin fraction 5, which is a crude polypeptide extract prepared from bovine thymus. In our experiments, the number of adult worms developed in the large intestine of mice repeatedly injected with 100 μg of thymosin within week 2 or weeks 4–5 after infection (but not around the moment of infection) was significantly lower in comparison with nontreated control mice. This reduction in the worm burden was not accompanied by any significant changes

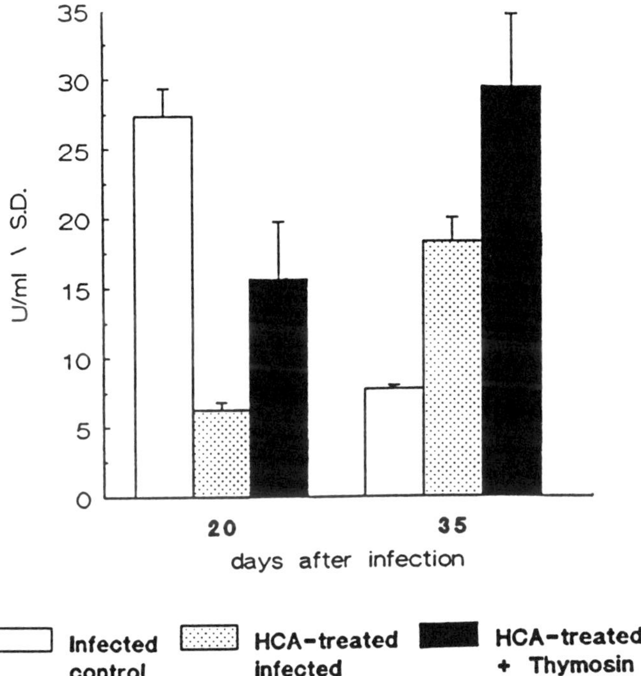

Figure 4 The in vitro production of IL-5 by spleen cells from HCA-treated female CBA/Ca mice infected with *T. muris* assessed on DPI 35 (experiment 4). The preparation of supernatants and the measurement of cytokines were the same as described in Figure 2.

in the production of parasite-specific serum antibodies (IgGl, IgG2a, IgGAM). Once again, this observation clearly demonstrated that the humoral immune response probably plays only a limited role in the control of infection in mouse trichuriasis. It has been shown previously that mice of the primary nonresponder B10.BR strain produced a comparable or even higher level of parasite-specific polyclonal antibodies when compared with mice of the resistant NIH mice (17).

In contrast, highly significant correlations between the reduction of the worm burden and the in vitro antigen-induced proliferative responses of lymphoid cells were observed in immunomodulated mice. Whereas lymphoid cells from control infected mice did not respond to any concentration of AWH, lymphoid cells from thymosin-treated mice showed proliferative responses comparable with naive noninfected mice. This restoration was antigen-specific, because there were no significant differences in mitogen-induced (ConA, PHA, LPS) proliferative responses between control and treated mice. Similarly, in our previous experiments, we noticed highly significant differences in the ability of lymphoid cells from infected mice of the NIH (resistant) and B10.BR (susceptible) strains to recognize *Trichuris* antigens in vitro, whereas cells of both strains were able to respond to various mitogens (17). Once again, these facts demonstrated that the altered cellular responsiveness observed in mice of the B10.BR strain chronically infected with *T. muris* represents rather parasite-specific unresponsiveness than some kind of general immunodeficiency.

It has been proved that resistance in mouse model of trichuriasis is associated with the TH_2 type of immune response, whereas mice of the susceptible B10.BR strain display TH_1 response when infected with *T. muris* (2,4). However, B10.BR mice previously immunized with *Trichuris* antigens (9), as well as mice concurrently infected with *Trichinella spiralis* (19), were able to display a protective TH_2 type of response. Moreover, we were not able to demonstrate any significant differences in the in vitro production of basic TH_1 (IFN-y)– and TH_2 (IL-5)–associated cytokines between control infected and immunomodulated mice in our experiments with thymosin. Consequently, the administration of thymosin did not directly influence this regulatory mechanism in chronic mouse trichuriasis.

It has been shown previously that high doses of HCA damaged proliferation and differentiation of T lymphocytes and that thymic factors could accelerate the recovery of the immune system when administered after the cessation of HCA (20,21). Moreover, repeated injections of thymic factors into mice previously immunosuppressed with HCA resulted in the restoration of the resistance to some experimental parasitoses (13). Similarly, in the present study, the administration of thymosin to CBA mice infected with *T. muris* embryonated eggs and then immunosuppressed by a short treatment with high doses of HCA resulted in a significant immunorestoration. This protective effect of thymosin was associated (like in B10.BR mice) with the restoration of the ability of spleen cells to proliferate in vitro in the presence of *Trichuris*-derived antigens. However, the significantly higher level of serum parasite-specific IgGl antibodies and in vitro production of IL-5 by spleen cells in thymosin-treated mice reflected at least partial involvement of TH_1-TH_2 shift in the protective effect of thymosin in this model of chronic trichuriasis.

Unfortunately, only limited information concerning the possible in vivo influence of thymic factors on the production of main regulatory cytokines is available at the present time (22,23). It has been shown that treatment with thymopentin, an immunoregulatory pentapeptide from bovine thymus, reduced the susceptibility of aged mice to cutaneous leishmaniasis by upregulating the TH_1 type of immune response (24); however because of the complexity of many thymic preparations, it is very difficult to extrapolate this finding. More likely, different thymic preparations in different model parasitic infections could influence the TH_1-TH_2 balance in different ways. Consequently, the exact mechanism of the protective effect of thymosin in our models of chronic trichuriasis remains unclear.

V. CONCLUSIONS

Immunocompetent mice of most inbred strains expel primary infections with *T. muris* before patency, whereas mice of the B10.BR strain and corticosteroid-treated mice of other strains (nonresponder mice) develop chronic infections and retain adult worms in the large intestine. In the present study, we investigated the possible influence of nonspecific immunomodulation with thymosin fraction 5, a polypeptidic immunomodulatory preparation, on the development of immune responses and worm establishment in B10.BR and hydrocortisone-treated CBA mice. Administration of 100 μg/dose of thymosin on DPI 7, 9, 11, 13 as well as on DPI 18-32 (but not on DPI -3, -1, 1, 3) significantly reduced the number of *T. muris* adult worms in the cecum and colon of female B10.BR mice assessed on DP 35. A similar protective effect of thymosin (administered on DPI 13, 15, 17, 19) was observed in hydocotisone-treated female CBA mice. In both models, the reduction in worm number was accompanied by a significant increase of the in vitro proliferative response of lymphoid cells from immunomodulated mice to *T. muris* antigens in comparison with nontreated animals. On the other hand, significant differences in the in vitro production of cytokines characteristic for TH_1 (IFN-γ) or TH_2 (IL-5) immune responses were found only in hydrocortisone-treated CBA but not in B10.BR mice.

ACKNOWLEDGMENTS

J.H. was supported by a Wellcome Trust postdoctoral fellowship. The authors are also very grateful to Dr. K.J. Else and Dr. R.K. Grencis (Department of Structural Biology, Medical School, University of Manchester) who provided all reagents for cytokine assays. Dr. V. Kral (Immunological Department of Regional Hygienic Station, Usti nad Labem, Czech Republic) kindly provided thymosin fraction 5 for our experiments.

REFERENCES

1. Else KJ, Wakelin D. The effects of H-2 and non-H-2 genes on the expulsion of the nematode *Trichuris muris* from inbred and congenic mice. Parasitology 1988; 96:543.
2. Else KJ, Grencis RK. Cellular immune responses to the murine nematode parasite *Trichuris muris*. I. Differential cytokine production during acute or chronic infection. Immunology 1991; 72:508.
3. Else KJ, Grencis RK. Helper T-cell subsets in mouse trichuriasis. Parasitol Today 1991; 7:313.
4. Else KJ, Hultner L, Grencis RK. Cellular immune responses to the murine nematode parasite *Trichuris muris*. II. Differential induction of T_H-cell subsets in resistant versus susceptible mice. Immunology 1992; 75:232.

5. Ito Y. The absence of resistance in congenitally athymic nude mice toward infection with the intestinal nematode *Trichuris muris*: Resistance restored by lymphoid cell transfer. Int J Parasitol 1991; 21:65.
6. Tanabe M, Hosaka Y, Ito Y, Senda F, Okuda T. Effect of neonatal thymectomy and administration of immunosuppressants on the acquired resistance of mice to infection with the nematode, *Trichuris muris*. Jpn J Parasitol 1979; 28:339.
7. Lee TDG, Wakelin D. Cortisone-induced immunotolerance to nematode infection in CBA/Ca mice. I. Investigations of the deffect in the protective response. Immunology 1982; 47:227.
8. Lee TDG, Wakelin D. Cortisone-induced immunotolerance to nematode infection in CBA/Ca mice. II. A model for human chronic trichuriasis. Immunology 1983; 48:571.
9. Lee TDG, Wakelin D, Grencis RK. Cellular mechanisms of immunity to the nematode *Trichuris muris*. Int J Parasitol 1983; 13:349.
10. Else KJ, Hultner L, Grencis RK. Modulation of cytokine production and response phenotypes in murine trichuriasis. Parasite Immunol 1992; 14:441.
11. Oates KK, Sztein MB, Goldstein AI. Mechanisms of action of the thymosins: Modulation of lymphokines, receptors and T-cell differentiation antigens. Immunol Ser 1989; 45:273.
12. Bryom NA, Hobbs JR, eds. Thymic factor therapy. New York: Raven Press, 1984.
13. Hermanek J. The nonspecific immunomodulation influences the resistance of mice to the experimental infection with *Mesocestoides corti* (Hoeppli, 1925) and *Ascaris suum* (Goeze, 1782). J Helminthol 1991; 65:121.
14. Hermanek J, Koudela B. Influence of the administration of thymic preparations on the course of infection and efficacy of chemotherapy in neonatal mice infected with *Cryptosporidium parvum*. Ther Infect Dis 1991; 6:327.
15. Kociecka W, Rauhut W, Gustowska I. The effect of thymus factor X (TFX Polfa) on *Trichinella spiralis* muscle invasion in mice. Parasitological and histological study. Wiadom Parazytol 1989; 35:413.
16. Shiddo S, Ilardi I, Mussa c, Mohamud MA, Aceti A, Leone F, Sebastiani A, Laghi, Amiconi G. Reinfection of Somali children with *Trichuris trichiura* after chemotherapy: Relevance of immunostimulation. Trans Soc Trop Med Hyg 1990; 84:832.
17. Hermanek J, Wakelin D. Dynamics of the *in vitro* antigen-induced proliferative responses of spleen and mesenteric lymph node cells in resistant and susceptible strains of mice infected with *Trichuris muris*. In 1993; in preparation.
18. Behnke JM, Barnard CJ, Wakelin D. Understanding chronic infections: Evolutionary considerations, current hypotheses and the way forward. Int J Parasitol 1992; 22:861.
19. Hermanek J, Goyal PK, Wakelin D. Lymphocyte, antibody and cytokine responses during concurrent infections between helminths that selectively promote T-helper-1 or T-helper-2 activity. Parasite Immunol (in press).
20. Hadden EM, Malec P, Sosa M, Hadden JW. Mixed interleukins and thymosin fraction 5 synergistically induce lymphocyte-T development in hydrocortisone-treated aged mice. Cell Immunol 1992; 144:228.
21. Thurman GB, Rossio JL, Goldstein AI. Thymosin-induced recovery of murine T-cell functions following treatment with hydrocortisone acetate. Transplant Proceed 1977; 9:1201.

22. Cillari E, Milano S, Perego R, Gromo G, D'Agostino R, Arcoleo F, Dieli M. Modulation of IL-2, IFN-y, TNF-α and IL-4 production in mice of different ages by thymopentin. Int J Immunopharmacol 1992; 14:1029.
23. Mastino A, Favalli C, Grelli S, Garaci E. Thymic hormones and cytokines. Int J Immunopathol Pharmacol 1992; 5:77.
24. Cillari E, Milano S, Dieli M, Arcoleo F, Perego R, Leoni F, Gromo G, Severn A, Liew FY. Thymopentin reduces the susceptibility of aged mice to cutaneous leishmaniasis by modulating CD4 T-cell subsets. Immunology 1992; 76:362.

17

Carbapenems and Potential Immunomodulating Properties

N.A. Carlone, A.M. Cuffini, V. Tullio, and G. Cavallo
Institute of Microbiology, University of Turin, Turin, Italy

I. INTRODUCTION

The failure of antibiotic therapy is often due to an inability of the patient's immune system to provide the support that antibiotics need for the eradication of the infection. Thus the modern trend of therapy requires that drugs with a stimulatory effect rather than an immunosuppressive effect on the cells of the immune system be used. It follows that the immunological side effects of antibiotics should always be taken into account besides the various potential side effects of antibiotics. The literature reports evidence suggesting that many antimicrobial drugs can modulate host defenses in different ways (1–4). Some antibiotics are concentrated in the phagocytes where they enhance intracellular killing, some lead to alterations the bacteria that render them more susceptible to phagocytosis, some directly enhance the phagocytic and/or microbial activities of the host cells, some depress bacterial uptake by phagocytes, and still others become inert following intracellular uptake. General absence of phagocytic function impairment and synergism with phagocytes has been shown for cell wall–acting antibiotics (5,6). Moreover, their great efficacy exhibited in the treatment of clinical infections, which substantially exceeds any expectations based on their activity and pharmacokinetic behavior in vitro (7), seems to confirm the ability of these antibiotics to "manipulate" the body immune defenses improving the final outcome of infection. This investigation delineates the effects of two cell wall inhibitors, imipenem and meropenem. The cell wall inhibitors possess the broadest antibacterial spectrum of currently available β-lactam antibiotics

(8,9) on the activities of human macrophages toward *Staphylococcus aureus,* the intracellular survival of which has been associated with therapeutic failures and recurrences of treated infections (10).

II. MATERIALS AND METHODS

A. Bacteria

A clinical isolate of *S. aureus* was cultured to mid-exponential base in brain heart infusion broth (BHI; Oxoid, London, England) concentrated 10 times in BHI broth containing 30% glycerol, quick frozen in dry ice-ethanol, and stored at −70°C.

B. Antibiotic

Imipenem and meropenem were kindly provided by Merck Sharpe & Dohme S.P.A., Rome, Italy, and by Zeneca S.P.A., Milan, Italy, respectively. Antibiotic solutions were prepared freshly for each experiment. The MICs for the strain of *S. aureus* were determined by a dilution technique in Mueller-Hinton broth (Oxoid) with an inoculum of 2×10^7 CFU (colony-forming units)/ml.

C. Macrophages

Human macrophages were collected from inflamed appendices and gallbladders of hospitalized patients who had not previously received antibiotics. The cells were harvested by repeated washings of the collected appendices and gallbladders with a total volume of 15 ml of cold Earle's balanced salt solution. The fluids were pooled and the cells washed three times by centrifugation at 100*g* for 10 min at 4°C. After suspending the pellets in RPMI 1640 (GIBCO Laboratories, Grand Island, NY) supplemented with 10% fetal calf serum (GIBCO), the cell suspensions were placed in plastic culture plates (30 × 8 mm) (Nunc, Copenhagen, Denmark) and incubated for 2 h at 37°C in 5% CO_2. The nonadherent cells were eliminated by rinsing with Earle's balanced salt solution and fresh RPMI 1640 was added to the monolayers. The number of macrophages per plate, as assessed by microscopy, was approximately 10^6. Cell counts (both total and differential) and viability were determined by examining stained smears (Diff Quick; Harleco, Gibbstown, NY) and by counting latex bead–ingesting cells following incubation, respectively. Only preparations containing 90% of macrophagelike cells, confirmed by both criteria, were used.

D. Phagocytosis Assay

Experiments were initiated by transferring 500 μl of the frozen bacterial culture into fresh BHI broth containing 150 μCi of [^{3}H]uracil (specific activity: 26 Ci/mmol; Sorin, Saluggia, Italy). After incubation for 4 h at 37°C, the radiolabeled bacteria were washed several times by centrifugation with cold BHI broth and the pellet was resuspended in fresh medium to yield a final suspension of 2×10^7 CFU/ml confirmed by viable counts that were performed in triplicate (11).

Aliquots of 1 ml of BHI broth containing labeled staphylococci (2×10^7 CFU) were added to the macrophage monolayers (10^6 cells/plate). The plates were incubated on a rocking platform at 37°C in a moist air containing 5% CO_2. After incubating for periods varying from 30 min to 24 h, the monolayers were rinsed with Earle's buffer to eliminate free, extracellular staphylococci and were then treated with lysostaphin (1 μg/ml; Sigma; St. Louis, MO) at 4°C to remove attached, noningested, macrophage-associated bacteria. After two further washings, the adherent cell monolayers were resuspended in 1 ml of sterile distilled water for 5 min; 100-μl samples of this suspension were then placed in a scintillation fluid (Atomlight; Nen, Dreieich, England) and the radioactivity was measured by liquid scintillation spectrometry. Radioactivity was expressed as counts per minute (cpm)/sample. Percentage phagocytosis at a given sampling time was calculated according to the following formula (12):

$$\%\ \text{phagocytosis} = \frac{\text{cpm in macrophage pellet}}{\text{cpm in total bacterial pellet}} \times 100$$

E. Intracellular Killing Assay

Staphylococci in 1 ml of BHI broth containing 2×10^7 CFU were added to the macrophage layers (10^6 cells/plate) and incubated for 30 min at 37°C to allow phagocytosis to proceed. The cells were washed vigorously several times with Earle's buffer to remove free, extracellular bacteria and were then treated with lysostaphin (1 μg/ml) at 4°C to remove adherent, noningested staphylococci. A sample of the bacteria-containing macrophages was withdrawn and the cells were lysed by the addition of sterile water; the numbers of liberated, viable staphylococci were then determined (Time_0). Other plates containing macrophage monolayers and staphylococci were incubated for periods that varied from 30 min up to 24 h (Time_x) after which the numbers of surviving intracellular staphylococci were determined in the same way. Macrophage killing was expressed as the survival index (SI), which was calculated by adding the number of surviving bacteria at T_0 to the number of survivors at T_x and dividing by the number of survivors at T_0 (13). According to this formula, if bacterial killing was 100% effective, the SI would be 1.

F. Effect of Carbapenems on Macrophage Phagocytic and Bactericidal Functions

The effects of both imipenem and meropenem on the phagocytosis and intracellular killing of *S. aureus* by human macrophages were investigated by incubating the bacteria and macrophage monolayers at 37°C for periods ranging from 30 min to 24 h in the presence of antibiotics at concentrations of 0.5 MIC; antibiotic-free controls were also included. Phagocytosis and intracellular killing were assessed according to the methods described above.

In order to differentiate between the effects of carbapenems on *S. aureus* from those on the macrophages, the phagocytic and bactericidal activities of the macrophages against *S. aureus* were assessed following 1 h preexposure of the macrophages and bacteria individually to subinhibitory concentrations of the antibiotics. Preexposed staphylococci were added to untreated monolayers and untreated staphylococci were added to preexposed monolayers and the plates were incubated at 37°C for periods varying from 30 min to 24 h. The phagocytic and bactericidal activities of the macrophages were determined as described above.

All experiments were carried out in quadruplicate with appropriate controls. Results were expressed in terms of the mean and standard errors of the mean of the four separate experiments.

G. Prophylactic Activity of Carbapenems in Experimental Infections

Either imipenem or meropenem (20 mg/kg) was given intravenously twice a day to Swiss male mice (Morini; San Polo d'Enza, Italy), weighing 25–30 g for 3 days before intravenous challenge with the carbapenem-resistant pathogen *Candida albicans*. The challenge was 10^9 CFU/mouse, and it killed mice within 3 days. The mortality rate of the animals (infected carbapenem treated) was recorded within 2 weeks since the infection.

H. Statistical Analysis

Differences between test and control results were compared by an analysis of variance (ANOVA) with Tukey's test.

III. RESULTS AND DISCUSSION

Table 1 shows that when either imipenem or meropenem in subinhibitory concentrations were incubated in the presence of both bacteria and macrophages, the susceptibility of *S. aureus* to phagocytosis was enhanced. After periods of incubation of 60 min to 2 h, a significantly higher percentage of staphylococci was ingested compared with antibiotic-free controls ($P < .01$); there was no difference in phagocytosis following incubation for periods >3 h. Under the same experimental conditions, the presence of both carbapenems also led to greater intracellular killing of staphylococci (Table 2); although there was a progressive increase in the numbers of viable intracellular bacteria in the control macrophages (SI values were consistently >2), the numbers of organisms that survived after

Table 1 Effect of Subinhibitory Concentration of Imipenem and Meropenem on the Phagocytosis (%) of *S. aureus* by Human Macrophages

	Mean (±SEM)% phagocytosis		
Time	Control	Imipenem	Meropenem
30 min	24.0 ± 1.6	32.6 ± 1.3	31.5[b] ± 2.6
60 min	29.1 ± 2.1	43.7[a] ± 4.8	44.2[a] ± 1.6
90 min	33.5 ± 3.5	51.8[a] ± 2.2	54.2[a] ± 1.5
2 h	24.2 ± 1.8	44.2[a] ± 2.0	39.6[a] ± 2.4
3 h	9.2 ± 0.82	10.7 ± 0.1	11.2 ± 1.1
6 h	4.4 ± 0.4	4.1 ± 1.7	4.5 ± 0.09
24 h	4.5 ± 0.43	3.9 ± 0.6	4 ± 0.1

[a] Significantly different ($P < .01$) from the controls.
[b] Significantly different ($P < .05$) from the controls.

Table 2 Effects of a Subinhibitory Concentration of Imipenem and Meropenem on the Killing of *S. aureus* by Human Macrophages

	Survival Index ±SEM				
Time	Controls	Imipenem	(% Killed[a])	Meropenem	(% Killed[a])
30 min	1.89 ± 0.09	1.5[b] ± 0.31	(50)	1.62[b] ± 0.02	(38)
60 min	>2	1.29[b] ± 0.07	(71)	1.41[b] ± 0.09	(59)
90 min	>2	1.28[b] ± 0.01	(72)	1.27[b] ± 0.03	(73)
2 h	>2	1.15[b] ± 0.04	(85)	1.11[b] ± 0.01	(89)
3 h	>2	1.09[b] ± 0.01	(91)	1.08[b] ± 0.02	(92)
6 h	>2	1.03[b] ± 0.01	(97)	1.06[b] ± 0.03	(94)
24 h	>2	1.27[b] ± 0.4	(73)	1.04[b] ± 0.02	(96)

[a] Percentage of initial bacterial population killed by macrophages in presence of carbapenems.
[b] Significantly different ($P < .01$) from the controls.

incubation in the presence of the antibiotics decreased significantly with time (killing of 70–95% of the initial bacterial population; $P < .01$). These results are at variance with those of Adinolfi and Bonventre (14), who demonstrated that imipenem in high concentrations (10 and 40 × MBC) exerted little effect on the intracellular activities of phagocytes against *Escherichia coli.* The 24-h exposure of staphylococci and human macrophages to half the MIC of meropenem resulted in a higher enhancement of microbicidal activity against phagocytosed staphylococci as compared with the imipenem containing systems (SI = 1.04 vs 1.07). The in vitro susceptibilities of extracellular bacteria and viable organisms recovered from lysed macrophages were indistinguishable indicating that surviving bacteria had not acquired resistance to carbapenems following exposure to the antibiotic (data not shown). Such enhanced macrophage activity as we observed is probably related to carbapenem-induced phenotypic and biochemical changes in *S. aureus*, which may have facilitated the phagocytosis and altered susceptibility to microbicidal macrophage systems.

It was also of interest to study separately the effect of both carbapenems on staphylococci and macrophages. Preincubation of staphylococci with subinhibitory concentrations of imipenem and meropenem made the bacteria more susceptible to phagocytosis (Table 3; A) and intracellular killing by human macrophages (Fig. 1) than organisms that had not undergone previous exposure. These results are consistent with those of Adinolfi and Bonventre (15) and of Mandell and Afnam (16) for imipenem, but strongly conflict with Easmon's data (17) showing that both imipenem and meropenem had any significant effect on the uptake or killing of *S. aureus* exposed to varying drug concentrations either for 2 h before or during the assay or for the whole experiment, even at concentrations equivalent to 8 × MBC.

The mechanisms by which sub-MICs of some antibiotics, including carbapenems, modify bacterial susceptibility to the phagocytic and bactericidal activities of macrophages remain poorly understood, although direct damage to the bacterium by the antibiotic may, at least in part, be responsible. Electron microscopic studies demonstrated that staphylococci that were neither totally killed nor inhibited by half the MIC of imipenem had undergone ultrastructural morphological alterations that probably involved subtle physiological changes both in the synthesis of peptidoglycan and to antiphagocytic surface factors (18); this explanation has been proposed by other investigators (19,20).

Following preexposure of macrophages to subinhibitory concentrations of both carbapenems, there was a significant increase in the phagocytosis (Table 3; B) and intracellular killing of staphylococci (Fig. 1) compared with the controls confirming the ability of imipenem (21) and meropenem (22) to cross biological membranes, bind to it, and remain active within the cells. Hand and King-Thompson (21) demonstrated that although imipenem binds to phagocytes, the concentration of cell-associated drug declined steadily during an incubation period of 1 h. Our results do not support this observation, since both phagocytic and killing activities continued to increase after 6 and 24 h. Following preexposure of *S. aureus* or macrophages to imipenem bactericidal concentration, a similar pattern has been observed (23): imipenem potentiated both the phagocytic and microbicidal activities of the macrophages.

That meropenem has excellent antibacterial activity in vivo has been demonstrated (24) in both normal and immunocompromised animals and in some models of infection is superior to imipenem. The in vivo models of experimental infections assessed in this study indicate that prophylactic treatment with meropenem enhanced the survival of Swiss mice infected by carbapenem-resistant *Candida albicans* (Table 4). When imipenem (without cilastatin) was given to a group of 30 animals, there was no significant influence on the mortality rate of mice compared with that of the control infected mice (93.3 vs 96.6%). Prophylactic treatment with meropenem

Table 3 Effects of Phagocytosis Following Preexposure of *S. aureus* (A) or Macrophages (B) to Sub-MIC Carbapenems

	Mean (±SEM)% phagocytosis				
Time of exposure	Controls	A Imipenem	A Meropenem	B Imipenem	B Meropenem
30 min	24.0 ± 1.6	31.1 ± 0.2	30.5[b] ± 0.16	25.9 ± 2.2	19.5 ± 0.1
60 min	29.1 ± 2.1	78.2[a] ± 1.81	73.2[a] ± 1.39	46.5[a] ± 1.1	34.8[b] ± 0.13
90 min	33.5 ± 3.45	59.3[a] ± 0.9	51.9[a] ± 1.1	52.5[a] ± 3.5	47.2[a] ± 0.3
2 h	24.2 ± 1.8	55[a] ± 1.2	45.4[a] ± 0.9	50.9[a] ± 3.2	53.3[a] ± 0.41
3 h	9.2 ± 0.8	8.2 ± 0.4	8.1 ± 0.3	19.1[a] ± 0.2	16.8[b] ± 0.2
6 h	4.4 ± 0.4	6.7 ± 0.7	5.4 ± 0.2	8.1[a] ± 0.2	4.7 ± 0.1
24 h	4.5 ± 0.43	5.7 ± 0.3	5.9 ± 0.05	5.4 ± 0.13	3.4 ± 0.22

[a] Significantly different ($P < .01$) from the controls.
[b] Significantly different ($P < .05$) from the controls.

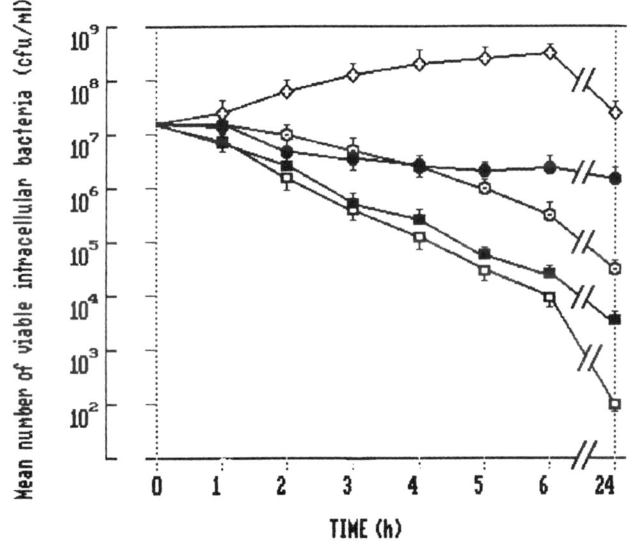

Figure 1 Influence of imipenem and meropenem exposure on the survival of intracellular *S. aureus* in human macrophages. Results are expressed as the mean ±SEM of four experiments. ◇ Control (untreated *S. aureus* and macrophages); ● 0.5 MIC meropenem pretreated *S. aureus;* ⊙ 0.5 MIC imipenem pretreated *S. aureus;* ■ 0.5 MIC meropenem pretreated macrophages; □ 0.5 MIC imipenem pretreated macrophages.

Table 4 Influence of Prophylactic Treatment with Carbapenems of Mice Infected with *C. albicans*

3 Days treatment and dosage	No. of mice	i.v. challenge *C. albicans* (10^9 CFU)	No. at 15 days dead/ no. tested	Mortality rate (%)
Placebo control	30	30	29/30	96.6
Imipenem	30	30	28/30	93.3
Meropenem	30	30	2/30	6.6

reduced the mortality rate from 96.6 to 6.6% in a group of 30 animals indicating a protective effect of meropenem given prophylactically. The mechanisms by which the drug exerts its immunoenhancing properties is still unknown.

IV. CONCLUSIONS

From the results reported in this study, it emerges that imipenem and meropenem possess interesting immunomodulatory properties. Both carbapenems at subinhibitory concentrations are able to modulate the interactions between *S. aureus* and host defense mechanisms, possibly by causing alterations to the bacterium and thereby increasing its susceptibility to the phagocytic and microbicidal activities of human macrophages. Once both imipenem and meropenem had been concentrated intracellularly, they could then act more effectively on the replicating bacteria: this may have therapeutic implications in the context of treating infections caused by microorganisms that have the capacity to remain viable within phagocytes.

Finally, the in vivo immunomodulatory properties shown by meropenem, makes this antibiotic more suitable for the therapy of patients with defects of the phagocytic components of the immune system. The therapeutic potential of these latter observations requires further investigation.

ACKNOWLEDGMENTS

This work was supported in part by grants from the Italian MURST 60%, 40% "Immunodiagnosi" and from C.N.R. (91.00540.CTO), Rome, Italy.

REFERENCES

1. Van der Auwera P. The immunomodulating effects of antibiotics. Curr Opin Infect Dis 1988; 1:363.
2. van den Broek PJ. Antimicrobial drugs, microorganisms and phagocytes. Rev Infect Dis 1989; II:213.

3. Miler I, Borte M, Hejzlar M, Braun W. The effect of antibiotics on the function of the phagocytic system. Pad Grenz 1990; 29:377.
4. Carlone NA, Cuffini AM, Tullio V, Cavallo G. Interactions of antibiotics with phagocytes in vitro. J Chemother 1991 3(Suppl 1):98.
5. Lingaas E, Midvedt T. The influence of cefoperazone, cefotaxime, ceftazidime and aztreonam on phagocytosis by human neutrophils in vitro. J Antimicrob Chemother 1989; 23:701.
6. Cuffini AM, Carlone NA, Tullio V, Cavallo G. Cell wall inhibitors and bacterial susceptibility to phagocytosis. In: Faist E, Meakins JL, Schildberg FW, eds. Host defense dysfunction in trauma, shock and sepsis. Berlin: Springer-Verlag, 1993:979.
7. O'Grady F. Antibiotics and host defences. In: Greenwood D, O'Grady F, eds. The scientific basis of antimicrobial chemotherapy. Cambridge, England: Cambridge University Press, 1985:341.
8. Jones RN. Review of the *in vitro* spectrum of activity of imipenem. Am J Med 1985; 78(Suppl 6A):22.
9. Jones RN, Barry AL, Thornsberry C. In vitro studies of meropenem. J Antimicrob Chemother 1989; 24(Suppl A):9.
10. Van der Auwera P, Husson M, Fruhling J. Influence of various antibiotics on phagocytosis of *Staphylococcus aureus* by human polymorphonuclear leucocytes. J Antimicrob Chemother 1987; 20:399.
11. Cuffini AM, Tullio V, Fazari S, Paizis G, Carlone NA. The effects of sub-MICs of cefonicid on the interaction of human macrophages with *Klebsiella pneumoniae.* J Antimicrob Chemother 1991; 28:933.
12. Carlone NA, Cuffini AM, Ferrero M, Tullio V, Avetta G. Cellular uptake, and intracellular bactericidal activity of teicoplanin in human macrophages. J Antimicrob Chemother 1989; 23:849.
13. Cuffini AM, Tullio V, Fazari S, Allocco A, Carlone NA. Pefloxacin and immunity: cellular uptake, potentiation of macrophage phagocytosis and intracellular bioactivity for *Klebsiella pneumoniae.* Int J Tissue React 1992; XIV(3):131.
14. Adinolfi LE, Bonventre PF. Intraphagocytic activity of imipenem and piperacillin. J Antimicrob Chemother 1988; 21:508.
15. Adinolfi LE, Bonventre PF. Enhanced phagocytosis, killing, and serum sensitivity of *Escherichia coli* and *Staphylococcus aureus* treated with sub-MICs of imipenem. Antimicrob Agents Chemother 1988; 32:1012.
16. Mandell AL. Afnan M. Mechanisms of interaction among subinhibitory concentrations of antibiotics, human polymorphonuclear neutrophils and Gram-negative bacilli. Antimicrob Agents Chemother 1991; 35:1291.
17. Easmon CSF. Interaction of meropenem with humoral and phagocytic defences. J Antimicrob Chemother 1989; 24(Suppl A):259.
18. Carlone NA, Cuffini AM, Tullio V. Imipenem: Valutazione degli effetti ultrastrutturali e dell'attività battericida. Giorn Ital Chemioter 1988; 35(Suppl 5):61.
19. Veringa EM, Verhoef J. Influence of subinhibitory concentrations of clindamyoin on opsonophagocytosis of *Staphylococcus aureus,* a protein-A-dependent process. Antimicrob Agents Chemother 1986; 30:796.
20. Tripodi MF, Adinolfi LE, Utili R, Marrons A, Ruggiero G. Influence of subinhibitory

concentrations of loracarbef (LY 163892) and daptomycin (LY 146032) on bacterial phagocytosis, killing and serum sensitivity. J Antimicrob Chemother 1990; 26:491.

21. Hand WL, King-Thompson NL. The entry of antibiotics into human monocytes. J Antimicrob Chemother 1989; 23:681.
22. Cuffini AM, Tullio V, Allocco A, Giachino P, Fazari S, Carlone NA. The entry of meropenem into human macrophages and its immunomodulating activity. J Antimicrob Chemother 1993; 32.
23. Cuffini AM, Tullio V, Allocco A, Fazari S, Giachino P, Carlone NA. Enhanced *Staphylococcus aureus* susceptibility to immunodefenses induced by subinhibitory and bactericidal concentrations of imipenem. J Antimicrob Chemother 1993; 31:559.
24. Edwards JP, Williams S, Nairn K. Therapeutic activity of meropenem in experimental infections. J Antimicrob Chemother Suppl A 1989; 24:279.

18

Immunomodulatory Action of Lysine Derivatives of Cinnamic Acid

N. Ivanovska, Z. Stefanova, V. Valeva, V. Dimov, V. Bankova, and S. Popov
Bulgarian Academy of Sciences, Sofia, Bulgaria

I. INTRODUCTION

The biological activity of many polyphenols is well known (1), but their low solubility limits their practical application. In previous works, we reported on the immunomodulatory activity of propolis, a bee product, composed mainly of polyphenols. Recently, we synthesized complexes of propolis with basic amino acids (lysine and arginine) and by this way we achieved full solubility in water. These complexes also possessed an improved immunoadjuvant activity (2). In order to investigate this phenomenon, we synthesized lysine derivatives of individual propolis constituents and estimated their ability to influence the immune response. The complexes of one simple aromatic acid, cinnamic acid, formed with different amounts of lysine are the object of the present study.

II. MATERIALS AND METHODS

A. Cinnamic Acid Complexes

The complexes of lysine (Ly) and cinnamic acid (CN) were obtained by dissolving corresponding amounts of both compounds in water and evaporating under vacuum. The molar ratios Ly:CN were as follows: CN1, 1:1; CN2, 2:1, CN3, 3:1, CN6, 6:1; and CN8, 8:1.

B. Treatment

Solutions of CN2, CN8, or Ly in a dose of 10 mg/kg were administered intraperitoneally to ICR mice, either once or in three consecutive days. The concentration of cinnamic acid was approximately 3 mg/kg in CN2 complex and 1 mg/kg in CN8 complex.

C. Complement Assays

Classical pathway (CP) and alternative pathway (AP) inhibition of complement activity in normal human serum (NHS) was determined by the method of Klerx et al. (3). For determination of AP activity, serial dilutions of individual mouse sera were made and hemolysis of rabbit erythrocytes was measured using the same method. C3 hemolytic activity in mouse serum was determined by estimating the ability of sera from treated animals to restore the AP activity of C3-deficient mouse serum.

D. Total Protein Secretion

Peritoneal macrophages (pMϕ) at a concentration of 2.10^6 cells/ml were cultured in leucine-deficient minimum essential medium Eagle (MEM with the addition of [^{3}H]leucine (2 μCi/ml). Labeled proteins in the supernatants were coprecipitated in the presence of 1% bovine serum albumin (BSA) with 10% trichlor acetic acid (TAA).

E. In Vivo and In Vitro Determination of Antibody Response to Sheep Red Blood Cells

Mice were treated with CN2, CN8, or Ly in 3 days as described. On the next day, they were immunized intraperitoneally with sheep red blood cells (SRBC) (10^6 cells). At day 7 mice were bled and hemagglutination titers in individual sera were determined.

For in vitro determination of anti-SRBC response, spleen cells from treated animals were received on the next day after the last application of complexes or Ly and cultivated in the presence of SRBC (10^7 splenocytes: 5.10^6 SRBC). Cells were collected starting from day 3, washed, and mixed with 5% suspension of SRBC for 2 h at 37°C. After that guinea pig serum (1:8) was added for 30 min and hemolysis in the supernatants was measured at 405 nm in a enzyme-linked immunosorbent assay (ELISA)-reader.

III. RESULTS AND DISCUSSION

It is well known that Ly inhibits CP and AP complement activity in vitro (4,5). We supposed that the lysine content would be of importance in complement-mediated reactions. It was established that complexes inhibited in vitro complement activity in NHS at a concentration interval between 5×10^{-4} and 2×10^{-4} M of cinnamic acid (Table 1). Only CN1 complex exhibited a lower effect. Lysine itself

Table 1 Inhibition of CP and AP Complement Activity in NHS by Different Lysine Complexes of Cinnamic Acid

	5×10^{-5} M[a]		10^{-4} M		2.5×10^{-4} M		5×10^{-4} M	
	CP[b]	AP[b]	CP	AP	CP	AP	CP	AP
CN1[c]	0	0	10	10	35–40	10	35–45	10
CN2	10	10–14	25–40	12	75–85	50	90–100	70–80
CN3	5–12	10	12–20	11	65–70	25–35	70–80	30–40
CN6	10	10	14–24	10–16	80–90	50	90–100	65–70
CN8	10–18	10–20	35–40	15–20	90–95	60–70	100	100
Ly	0	0	10	0	40–50	50	70–80	Lysis

[a] A final concentration of cinnamic acid in the assay.
[b] Percentage inhibition calculated to the controls where buffer instead of substances was added. Data are mean from five independent determinations.
[c] Index giving the molar ratio of Ly to CN.

possessed strong suppressive activity at a concentration higher than 20 mM. Such concentration is achieved only in 5×10^{-4} M CN8. Evidently, the anticomplementary action of the complexes with low lysine content should not be attributed to lysine. In further experiments the effects of CN2 and CN8 complexes after in vivo application were compared with those caused by free lysine. The complexes and Ly modulated complement activity in vivo and the results indicated that the C3 component was one of the targets of their action (Table 2). It is important to note that 24 h after their application, the complexes, but not Ly, suppressed AP and particularly C3 functional activities in mouse serum. This fact coincided with the previous results showing that the best effect against some bacterial and fungal infections was achieved in the case of inoculation 24 h after the intraperitoneal injection of propolis lysine derivative (2). The C3 component plays a major role in the processes of opsonization and presentation of the antigens. Its functional inactivation may result in a longer persistence of antigen and augmentation of the humoral immune response. Such possibility is supported by the results from determining anti-SRBC response in vivo and in vitro. The antibody production began earlier and was strongly enhanced in the cultures of splenocytes from treated animals (Fig. 1). At day 7 after in vivo immunization with SRBC, an elevation of antibody titers was not observed (Fig. 2). Evidently, the peak of antibody synthesis was missed. The substances also increased the number of pMϕ and protein secretion at the first hours after their application (Table 3).

The present results confirmed our hypothesis that the adjuvant activity of propolis lysine derivative might be attributed to its phenolic constituents. At molar ratio 1:1 a salt between the basic NH_2 group of lysine and COOH group of cinnamic acid is formed, and the increase of Ly leads to formation of complexes that may

Table 2 AP and C3 Hemolytic Complement Activity in Mouse Serum

		1 h[a]	6 h[a]	24 h[b]
CN2	AP	−23.0[c]	+56.0[c]	−29.0[c]
	C3	+18.0[c]	+10.0	−50.0[c]
CN8	AP	−35.5[c]	−11.9[c]	−24.0[c]
	C3	−18.0	+34.2[c]	−50.0[c]
Ly	AP	−31.1[c]	−100[c]	0
	C3	+7.0	+11.2	0

[a] Hours after i.p. application of substances in a single dose of 10 mg/kg when serum samples were collected.

[b] Hours after the last application of substances in 3 consecutive days in a dose of 10 mg/kg daily.

[c] Values are expressed as percentages of corresponding levels in saline-treated animals. $P < .05$, Student's *t*-test, n = 5.

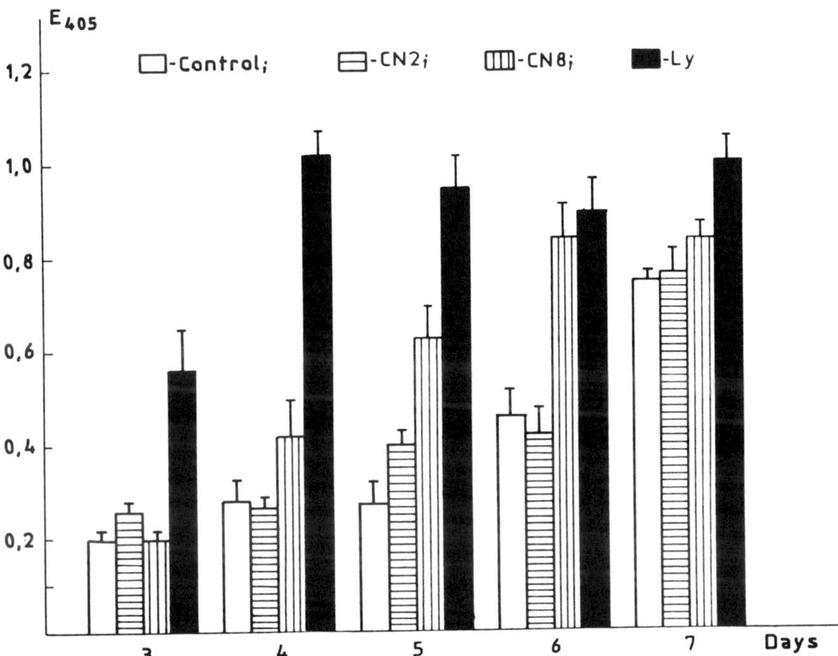

Figure 1 Changes in anti-SRBC response in cultures of spleen cells. Abscissa: Days of cultivation. Ordinate: Hemolysis given by antibody-producing cells. The values are means ±SD of five determinations in two studies.

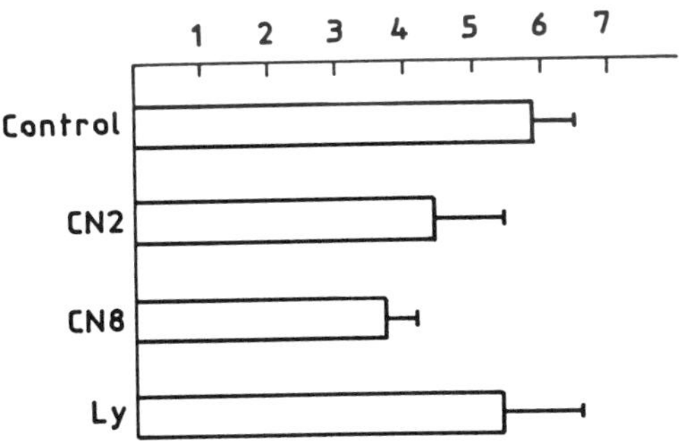

Figure 2 In vivo anti-SRBC response.

Table 3 Influence of Lysine Complexes on the Number and TP Secretion of pMϕ

		1 h[a]	6 h[a]	24 h[b]
	pMϕ	+46.8[c]	+9.6	+20.2
CN2	TP	1376 ± 343[d]	2267 ± 64	1944 ± 258
		(0.75)[e]	(1.27)	(0.84)
	pMϕ	+64.7	+26.7	+33.3
CN8	TP	1517 ± 179	4480 ± 366	2222 ± 127
		(0.83)	(2.52)	(0.94)
	pMϕ	+59.1	+32.0	−10.1
Ly	TP	1378 ± 195	4261 ± 997	2470 ± 437
		(0.76)	(2.39)	(1.05)

[a,b] Time after treatment as described in Table 2.
[c] Counted as a mean number of cells yielded per group.
[d] Estimated in counts/min.
[e] Values in parenthesis are ratios calculated to control animals.

differ in their immunomodulatory activity. A better understanding of the mechanism of action of lysine complexes should prove highly relevant in developing new drugs for immunotherapy.

ACKNOWLEDGMENT

This work has been completed with the financial support of the National Foundation for Scientific Research under contract MY-3.

REFERENCES

1. Havsteen B. Flavonoids, a class of natural products of high pharmacological potency. Biochem Pharmacol 1983; 32:1141.
2. Dimov V, Ivanovska N, Bankova V, Popov S. Immunomodulatory action of propolis: IV. Prophylactic activity against gram-negative infections and adjuvant effect of the water-soluble derivative. Vaccine 1992; 12:817.
3. Klerx JPAM, Beukelman CJ, Van Dijk H, Willers JMN. Microassay for colorimetric estimation of complement activity in guinea pig, human and mouse serum. J Immunol Methods 1983; 63:215.
4. Takada Y, Arimoto Y, Mineda H, Takada A. Inhibition of the classical and alternative pathways by amino acids and their derivatives. Immunology 1978; 34:509.
5. Shirahama S, Sugawara Y, Takada T, Takada A. Inhibition by lysine and arginine of the conversion of C3 and B in the serum and a purified system. Int J Immunopharmacol 1986; 8:269.

19

The Role of Major Histocompatibility Complex Class II Antigen Expression in the Pathogenesis of Adjuvant Arthritis: Action of Immunomodulatory Drugs

Lilia Marinova-Mutafchieva and Ivan Goranov
Bulgarian Academy of Science, Sofia, Bulgaria

I. INTRODUCTION

Cyclophosphamide (CY), which has a poor immunosuppressive effect on adjuvant arthritis (AA) in rats, evokes an aggravating form of the disease when applied at a single low dose before rats' sensitization with complete Freund's adjuvant (CFA) (1). The aggravating effect of CY on AA development was accompanied with an aberrant expression of major histocompatibility complex (MHC) class II antigens in articular tissues (2).

The aim of the present study is to examine whether indomethacin (IND), which has anti-inflammatory action and is widely used in the treatment of arthritis (3), also could exert an aggravating effect on AA in rats.

The results demonstrate that IND, like CY, provokes an aggravating form of AA and a hyperexpression of MHC class II antigens in articular tissues.

II. MATERIALS AND METHODS

A. Animals

Male Wistar rats, 6–8 weeks old, from the Breeding Centre of the Bulgarian Academy of Sciences were used. The rats were kept under routine laboratory conditions.

B. Induction and Clinical Assessment of AA

Adjuvant arthritis was induced in Wistar rats by intradermal injection at the base of the tail of 0.1 ml CFA (DIFCO, Detroit, MI) containing 10 mg/ml dessicated, heat-killed *Mycobacterium tuberculosis* (H37Ra). The clinical severity of AA was estimated in arthritic units (AU) as previously described (1).

C. Drug Treatment

IND (Sigma) was administered intradermally at a dose of 0.5 mg/kg, given daily during the experiment up to day 21 after CFA sensitization. The animals were divided into four groups according to the treatment (Table 1).

D. Statistical Analysis

Student's *t*-test was used to determine the significant difference between the groups. The results are presented as mean values ±SD.

E. Preparation of the Tissues and Cryostat Sections

The rats were killed on day 21 after CFA sensitization. Both knee joints were dissected in toto and skin and superficial layers were removed. The joints were embedded in an aqueous solution of 8% (w/v) gelatin white (Sigma), frozen slowly in liquid nitrogen, and stored at –80°C. The frozen whole knee joints were cut on 10-μm thick sections at a cabinet temperature of –25°C and were subsequently fixed with waterproof adhesive tape onto slides.

Table 1 Clinical Severity of Adjuvant Arthritis

Animal group (n)	Treatment		Incidence of AA	Clinical manifestation of AA on day 21 (expressed in AU)
	CFA	Indomethacin		
Control (9)	–	–	0	0
I (20)	+	–	8	9.8 ± 1.3
II (20)	+	+	9	14.3 ± 1.6[a]
III (20)	+	+	8	7.2 ± 1.1[a]

Groups: Control, untreated; I, sensitized with CFA; II, indomethacin treatment commenced on day 5 before CFA; III, indomethacin administration began together with CFA or on day 5 after CFA. Indomethacin was administered i.d. at a dose of 0.5 mg/kg given daily during the experiment up to day 21 after CFA sensitization. Data are mean ±SD.

[a] $P < .05$ compared with group I by using Student's *t*-test.

F. Immunoperoxidase Staining

The sections were briefly fixed in acetone, air dried, and processed for peroxidase-antiperoxidase staining (2). The monoclonal antibodies employed as primary probes in the staining were OX-6 (Sera-Lab), an antibody that recognized a nonpolymorphic determinant on the Ia molecule (MHC, RT-1B) common to all rat strains (4); W3/25 (Serotec), a surface marker positive on T helper lymphocytes and macrophages (5); OX-8 (Serotec), which identifies T suppressor/cytotoxic lymphocytes (5).

III. RESULTS

The clinical severity of AA in the experimental groups is presented in Table 1. In group I, the severity of AA on day 21 after sensitization of the animals with CFA according to the manifestations of the clinical signs was 9.8 AU. In the other two experimental groups (II and III) with induced AA and treated with IND, the clinical manifestations of disease were markedly changed. In group II, where the IND application commenced 5 days before CFA sensitization, the AU reached a value of 14.3, whereas in group III, where IND treatment began on day 0 or 5 after CFA application, the AU decreased to 7.2.

Immunohistochemical data also show significant differences between the experimental groups. The results are summarized in Table 2 and represent the average pattern of the amount and location of the various cell types in the knee joint.

In control (untreated) joints, both the synovial membranes and the joint spaces are virtually free of OX6-, W3/25-, and OX8-positive cells. In arthritic joints, however, a considerable number of OX6 (Ia)–positive cells were found dispersed throughout the inflammatory tissues. The amount Ia-positive cells was markedly higher in group II, where IND was administered before CFA sensitization. The accumulation of Ia-positive cells in clusters around blood vessels within the synovial tissue was striking. Two cell types could be recognized—round cells with a ringlike membrane staining which were probably lymphocytes, and larger cells, staining very strongly and more homogeneously with OX6, which were probably dendritic cells. In the superficial layers of the inflamed synovia, Ia-positive cells were found predominantly located in close contact with the cartilage. This tendency for the localization of Ia-positive cells was most pronounced in joints of rats with the aggravating form of arthritis (group II). In all experimental groups, a considerable number of Ia-positive cells was found in the joint space and along the articular cartilage. Some of these cells had morphological features of lymphocytes or monocytes, whereas others had macrophagelike characteristics.

The distribution pattern of cells positive for W3/25 was similar to that of Ia antigen in the inflamed synovium. A considerable accumulation of W3/25-positive cells perivascularly located was observed. Its amount was most pronounced in experimental groups I and II. Numerous W3/25-positive cells were presented in

Table 2 Location and Amount of the Different Cell Types in Normal and Arthritic Knee Joints. Grading from (−) Positive Cells Are Virtually Absent to (+ + +) an Accumulation of Positive Cells

	Control (untreated)			Arthritic knee joints on day 21								
				I group			II group			III group		
Knee joint	OX6	W3/25	OX8	OX6	W3/5	OX8	0X6	W3/25	OX8	OX6	W3/25	OX8
Synovium												
superficial layer	−	−	−	+	+	−	+ + +	+ +	−	+	+	−
dispersed	−	±	−	+	+	±	+ +	+ +	+	+	+	±
perivascular clusters	−	−	−	+ + +	+ + +	+	+ + +	+ + +	+	+ +	+	+
isolated small groups	−	−	−	−	−	−	±	−	−	−	−	−
Joint space	−	±	−	+ + +	+ +	+	+ + +	+ + +	+	+ +	+	+

the joint space, some of which were observed in close contact with the cartilage surface. Only very few cells positive for OX8 were found.

IV. CONCLUSIONS

Our data demonstrate that IND, like CY, exerts a dual effect on the AA development—aggravating or suppressive depending on the mode of its application; that is, before or after induction of AA with CFA. The augmentation of AA is manifested by hyperexpression of MHC class II antigens, whereas the suppression is accompanied by hypoexpression. A similar hyperexpression of these antigens was also observed in the aggravating form of AA after CY administration (2). Thus, IND and CY have an analogous effect on the expression of MHC class II antigens, although the mechanisms of their biological action are different (3). This suggests that the expression of MHC class II antigens probably acts as a key factor in the immunomodulatory efficiency of IND and CY.

In conclusion, the aberrant expression of MHC class II antigens may be used as a reliable test for the assessment of the immunomodulatory action of the pharmacological products.

REFERENCES

1. Marinova-Mutafchieva L, Altankov G, Neronov A, Goranov I. Relationship between the effect of cyclophosphamide on adjuvant arthritis severity, skin allograft rejection reaction and spleen cell cytotoxic activity in rats. Methods Find Exp Clin Pharmacol 1990; 12:103.
2. Marinova-Mutafchieva L, Goranova I, Goranov I. Positive effect of cyclophosphamide on the expression of MHC class II antigens. Methods Find Exp Clin Pharmacol 1990; 12:545.
3. Hadden JW, Renoux G, Chirigos M. The characterization of immunotherapeutic agents. In: Hadden JW, Szentivanyi A, eds. Immunology reviews. New York: Plenum Press, 1990:1.
4. Barclay AN, Mayrhofer G. Bone marrow origin of Ia-positive cells in the medulla of rat thymus. J Exp Med 1981; 153:1666.
5. Barclay AN. The localization of lymphocytes defined by monoclonal antibodies in rat lymphoid tissues. Immunology 1981; 42:593.

III

SYNTHETIC MURAMYL PEPTIDES

20

Protective Activity of MDP-Lys(L18) on Hantavirus Infection in Newborn Mice and Potentiation of Antigenicity by B30-MDP and MDP-Lys(L18) of Inactivated Hantavirus Strain B-1 Vaccine and Recombinant Hepatitis B Virus Surface Antigen

Ichiro Azuma, Yung Choon Yoo, Mizuho Tamura, Ryu Yoshida, Kumiko Yoshimatsu, and Jiro Arikawa
Hokkaido University, Sapporo, Japan

Koichi Yamanishi
Osaka University, Osaka, Japan

I. INTRODUCTION

Heat-killed mycobacterial cells suspended in mineral oil (Freund's complete adjuvant) were recognized to be the most potent immunoadjuvants for induction of cell-mediated immunity and humoral antibody formation (1). In 1971, it was shown that the cell wall skeleton fraction, especially the peptidoglycan portion of the wall, was the active immunoadjuvant component of the cells of mycobacteria and related bacteria (2,3).

In 1974, Ellouz et al. (4) reported that the minimal adjuvant-active subunit of bacterial cell walls was N-acetylmuramyl-L-alanyl-D-isoglutamine (muramyl dipeptide, MDP), and the biological activities of MDP were reported by several laboratories (5). A variety of MDP analogues and derivatives have been synthesized, and the relationship between their chemical structures and adjuvant activities has been elucidated. Although MDP and its analogues have potent adjuvant activities in vivo when they are administered as a water-in-oil emulsion together with antigen, MDP exhibits only limited adjuvant activity when administered as an aqueous solution owing to its rapid excretion into urine. Hydrophobic (acyl) derivatives of MDP analogues were synthesized and the stearoyl-MDP derivatives, MDL-Lys(L18) and B30-MDP (Fig. 1), were selected as adjuvants for the nonspecific stimulation of the host defense mechanism against bacterial and viral infections in patients with cancer (6).

(a)

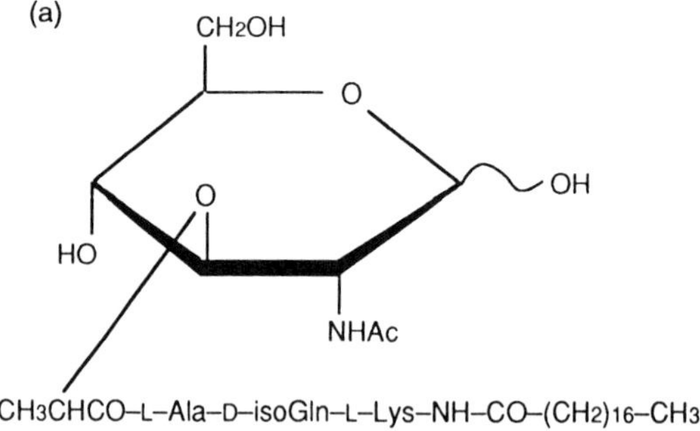

(b)

CH3(CH2)13
CHCO–OCH2
CH3(CH2)13
O
OH
O
HO
NHCOCH3
CH3C–CO–NH–CH–CO–NH–CH–CONH2
H CH3 (CH2)2
COOH

Figure 1 Chemical structure of (a) MDP-Lys(L18) and (b) B30-MDP.

This chapter describes the effects of both MDP derivatives for the stimulation of host defense mechanism against hantavirus and potentiation of antigenicity of vaccine of hantavirus strain B-1 virus and recombinant hepatitis B virus vaccine in mice.

II. MATERIALS AND METHODS

A. Cell Culture and Viruses

E6 clones of Vero cells (Vero-E6, ATCC c1008, CRL 1586) (7) and a subcloned line of CV-1 African green monkey cells designated CV-7 (8) were maintained in Eagle's minimal essential medium (EMEM, GIBCO, Grand Island, NY) supplemented with 5% fetal calf serum (FCS) and in Dulbecco's modified Eagle's medium (DMEM, GIBCO) with 10% FCS, respectively. The stock virus of strain Hantaan 76-118 (HTN), isolated from Korean field rodent *Apodemus agrarius coreae* (9), was prepared from the culture supernatants of Vero-E6 cells infected with HTN virus 4 days after infection and stored at –80°C as described previously (10). The titer of HTN virus was determined by the method described previously (8) using CV-7 cells and was shown to be 1.3×10^6 plaque-forming units (PFU)/ml. The preparation of inactivated B-1 virus vaccine of Seoul type hantavirus was described previously (11). The protein content of the stock solution of B-1 virus was 350 μg/ml by the Lowry method.

B. Animals

Specific pathogen–free (SPF) Balb/c female mice, 5 or 7 weeks old, and pregnant inbred Balb/c mice were purchased from the Shizuoka Laboratory Animal Center, Hamamatsu, Japan. Adult mice were maintained in the Laboratory of Animal Experiment at the Institute of Immunological Science, Hokkaido University, Sapporo, Japan, under laminar air-flow condition. Pregnant or newborn mice were maintained in plastic cages in vinyl film isolators. This study was carried out after permission from the Committee of Animal Experimentation, Institute of Immunological Science, Hokkaido University.

All animals were treated according to the Laboratory Animal Control Guidelines in our institute, which basically conform to those of the National Institutes of Health—American Association of Laboratory Animal Control.

C. Reagents

MDP-Lys(L18) and B30-MDP were obtained from Daiichi Pharmaceutical Co., Ltd., Tokyo, Japan. Two reagents were suspended in an appropriate amount of 0.1 M phosphate-buffered saline (PBS, pH 7.4). Yeast-derived recombinant hepatitis B surface antigen (rHBsAg) absorbed to alum (commercial name YHB-ST) was kindly supplied by Chemo-Sero Therapeutical Institute, Kumamoto, Japan. Human serum-derived HBsAg (hHBsAg) was obtained from Doto Chemical Co., Ltd., Obihiro, Japan.

D. Assay for HTN Virus Infection in Newborn Mice

A group of six to eight Balb/c newborn mice were inoculated subcutaneously with 4 LD_{50} (6.5×10^3 PFU/mouse) of HTN virus 1 day after birth. Analysis for the prophylactic effect of MDP-Lys(L18) was conducted by subcutaneous injection of 50 or 100 μg of MDP-Lys(L18) 1 day before viral infection. In analysis for therapeutic effect, mice received a single subcutaneous injection of 100 μg of MDP-Lys(L18) 4, 7, or 14 days after viral infection. The protective effect of MDP-Lys(L18) was determined by survival rates until 40 days after viral infection.

E. Immunization Schedules

The effect of both B30-MDP or MDP-Lys(L18) on antibody production against B-1 vaccine or rHBsAg was examined using 5- or 7-week old Balb/c mice, respectively. For the antibody production against B-1 vaccine, 6 mice received subcutaneous injection with 100 μl of 10-fold dilution of B-1 vaccine with or without B30-MDP. After 3 weeks, three mice of each group were boosted subcutaneously with the same injection with that of the first immunization. In the experiment using rHBsAg (YHB-ST), five mice were given intraperitoneal injection with 100 μl of fourfold dilution or stock suspension of rHBsAg with or without B30-MDP or MDP-Lys(L18), respectively. All mice were bled from the retro-orbital sinus every week and the titers of antibody were determined by indirect immunofluorescent antibody (IFA) test or enzyme-linked immunosorbent assay (ELISA).

F. Indirect IFA Test

The IFA test was carried out using serum specimens from mice immunized with B-1 vaccine with or without B30-MDP on acetone-fixed smears of strain HTN-infected Vero-E6 cells as antigen (12). Fluorescein isothiocyanate (FITC)–conjugated goat antimouse immunoglobulins (IgA + IgG + IgM) (Cappel Laboratories, USA) were used as the second antibody. The procedure was described previously in detail. IFA titers were expressed as the reciprocal of the highest dilution of antisera that resulted in specific immunofluorescence in the cytoplasm of infected cells.

G. ELISA

ELISA was performed by the method as described previously (13) with some modifications in flat-bottom microtiter plates (Falcon 3915, Becton Dickinson, Lincoln Park, NJ). The plates were coated with 50 μl of hHBsAg (protein content 1 μg/ml) in 50 mM bicarbonate buffer (pH 9.6) and incubated overnight at 4°C. After washing three times with washing buffer (PBS containing 0.05% Tween 20), serial twofold dilutions of sera in 3% skim milk in PBS were added to each well and then incubated for 2 h at 37°C. After washing, peroxidase-labeled goat antimouse IgG (Cappel Laboratories) in 3% skim milk solution was added to each well as the second antibody and incubated for 1 h at 37°C. Then the plates were washed and developed by the addition of 200 μl of substrate solution (22 mg of 2,2-azino-di-[3-ethylbenzo-thiazoline-6-sulfonic acid]diammonium salt [ABTS, Polysciences, USA] per milliliter in citrate buffer, pH 4.5, and 0.03% H_2O_2) per well. After 1 h at 37°C, absorbances were read at 405 nm using a Biochromatic Multiscan plate reader (Labsystems, USA). All samples were processed on the same day.

III. RESULTS

A. Protective Activity of MDP-Lys(L18) on HTN Virus Infection in Newborn Mice

To investigate the prophylactic effect of MDP-Lys(L18) on the HTN virus infection, newborn mice were injected with MDP-Lys(L18) 1 day before the infection of HTN virus at a lethal dose, 4 LD_{50}. As shown in Figure 2, the mice received 100 μg of MDP-Lys(L18) 1 day before the infection indicated a high survival rate (67%) until 40 days after the infection as compared with that of the control group (17%). Moreover, the protective activity of MDP-Lys(L18) also was shown at a dose of 50 μg (survival rate 50%). We next examined the effect of MDP-Lys(L18) on restoring the mice from the lethal infection with HTN virus. Figure 3 shows that the treatment of 100 μg of MDP-Lys(L18) 4 or 7 days after the infection caused potent protective activity (57 and 67%, respectively). However, the treatment of MDP-Lys(L18) 14 days after infection was nearly as effective (30%). All the mice of the control group died within 37 days after the infection. These results indicate that MDP-Lys(L18) confers protection from lethal HTN virus challenge in newborn mice.

B. Effect of B30-MDP on Antibody Production Against Inactivated B-1 Vaccine

The effect of B30-MDP on antibody production against B-1 vaccine was examined by comparing the antibody titers between mice immunized with B-1 vaccine alone

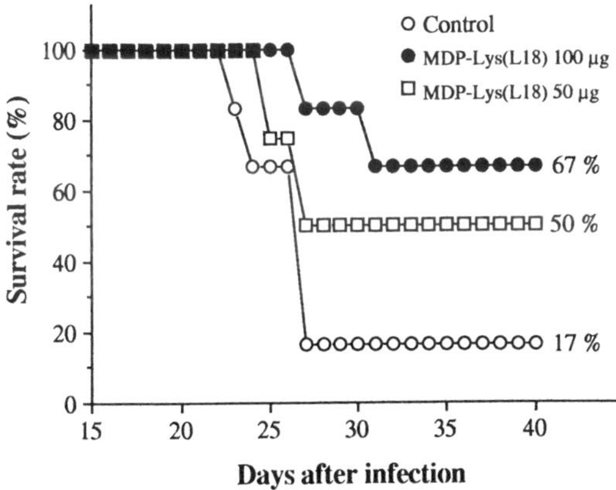

Figure 2 Prophylactic effect of MDP-Lys(L18) on HTN virus infection in newborn mice. Four or six Balb/c newborn mice were injected subcutaneously with 50 or 100 μg of MDP-Lys(L18) on <24 h of birth. All mice were inoculated subcutaneously with 4 LD_{50} (6.5×10^3 PFU/mouse) of HTN virus.

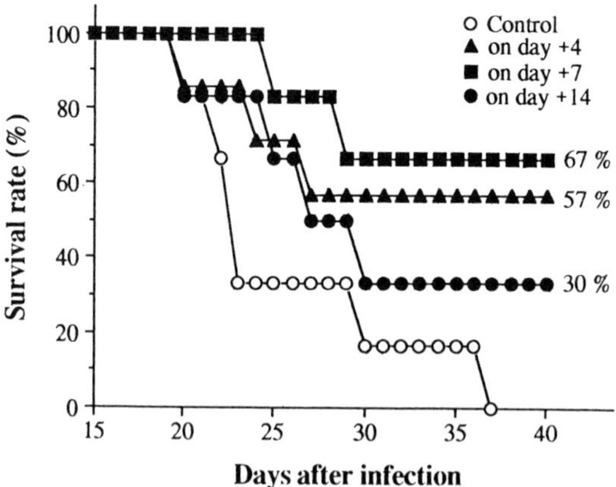

Figure 3 Therapeutic effect of MDP-Lys(L18) on HTN virus infection in newborn mice. Six or seven Balb/c newborn mice were inoculated subcutaneously with 4 LD_{50} (6.5 × 10^3 PFU/mouse) on HTN virus 1 day after birth. Then mice were injected subcutaneously with 100 μg of MDP-Lys(L18) on the indicated days.

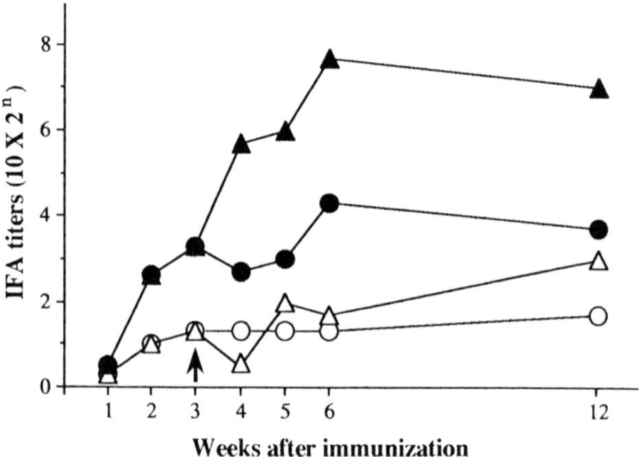

Figure 4 Effect of B30-MDP on the production of antibody against inactivated strain B-1 virus vaccine. Six Balb/c mice per group were immunized subcutaneously with 100 μl of 10-fold diluted B-1 virus vaccine with or without 100 μg of B30-MDP. Three mice of each group were boosted subcutaneously with the same injection 3 weeks after the first immunization as indicated by an arrow. The titers were measured by IFA on the fixed Vero-E6 cells infected with HTN virus on the indicated weeks. Open symbols (○, Δ) indicate titers for the group without B30-MDP, filled symbols (●, ▲) for with B30-MDP, circles (○,●) for single-immunized, and triangles (Δ, ▲) for boosted.

and mice with the mixture of B-1 vaccine and B30-MDP. As shown in Figure 4, B30-MDP remarkably accelerated the induction of antibody within 3 weeks after the single immunization. The antibody titer was about four times higher than that of the control group (immunized with B-1 vaccine alone). The maximal titer of the group with B30-MDP was much higher than that of the group without B30-MDP in both immunization schedules—single and booster immunization. Furthermore, it is noteworthy that the group immunized once with the mixture of B-1 vaccine and B30-MDP manifested higher titers than those of the group boosted with B-1 vaccine alone. The titer of the group with B30-MDP was still much higher level than that of the group without B30-MDP 12 weeks after immunization. These results indicate that B30-MDP augments the ability not only to induce antibodies against B-1 vaccine but also to maintain a high level of antibody titers for a long time.

C. Effects of B30-MDP or MDP-Lys(L18) on Antibody Production Against rHBsAg

Serum specimens from mice that received a single injection of rHBsAg with or without B30-MDP or MDP-Lys(L18) were used for examining the effect of both adjuvants on augmentation of antibody titers against rHBsAg. Figure 5 shows that B30-MDP enhances the production of antibodies against rHBsAg. The prominent differences of antibody titers between two groups with or without B30-MDP emerged from 5 weeks after immunization. MDP-Lys(L18) also enhanced the titers

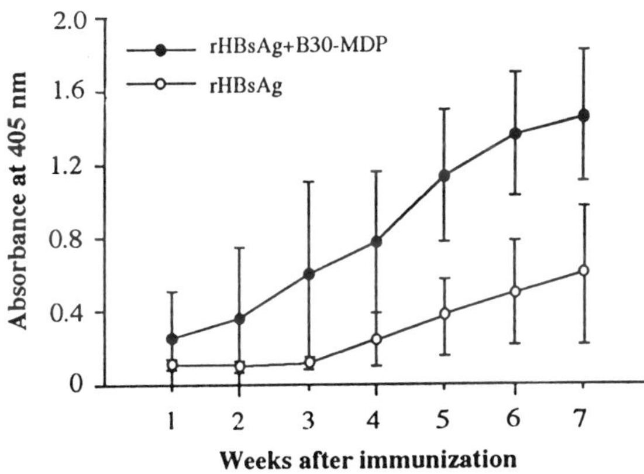

Figure 5 Effect of B30-MDP on the induction of antibody against rHBsAg. Five mice were injected intraperitoneally with 100 μl of fourfold diluted rHBsAg solution with or without 100 μg of B30-MDP. The titers were determined by ELISA. Absorbance value of 405 nm of serum specimens at 100-fold dilution were plotted. Bar indicates the standard deviation of each value.

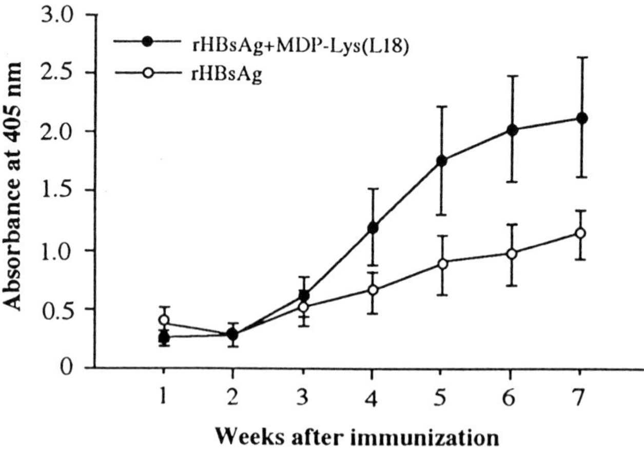

Figure 6 Effect of MDP-Lys(L18) on the production of antibody against rHBsAg. Five mice were immunized intraperitoneally with 100 μl of rHBsAg stock solution with or without 100 μg of MDP-Lys(L18). The titers were determined by ELISA. Absorbance value of 405 nm of serum specimens at 100-fold dilution were plotted. Bar indicates the standard deviation of each value.

of antibody against the recombinant protein (Fig. 6). The enhancing effect of MDP-Lys(L18) on induction of antibodies against rHBsAg was examined from 4 weeks after immunization, although the titers of two groups were almost the same within 3 weeks after immunization.

The results indicate that B30-MDP and MDP-Lys(L18) are possibly potent immunomodulators to enhance the ability to induce antibodies against recombinant proteins such as rHBsAg.

IV. DISCUSSION

It is well recognized that nonspecific host stimulation with immunostimulants is one of the most important factors in the treatment of patients with cancer and infectious diseases and a variety of immunostimulants have been developed (3). Since the discovery of MDP by Ellouz et al. (4), numerous biological activities have been reported (5). Chedid and coworkers (14) reported that MDP and its derivatives stimulated host resistance against *Klebsiella pneumoniae* infection in mice. Several MDP derivatives and related compounds such as murabutide, stearoyl-MDP derivatives (15–18), MTP-PE (19), FK-565 (20), FK-156 (20), and R.P. 40 639 (21) have been found to have host-stimulating activities against bacterial infections in experimental models. Matsumoto and coworkers (15) also examined 64 acyl-MDP derivatives as adjuvants for the stimulation of host resistance against nonspecific bacterial infection using a sepsis-type infection model with *Escherichia coli* in mice and selected 2 MDP derivatives, L18-MDP and MDP-Lys(L18). The stimulatory effects and their mechanisms of action of both stearoyl-MDP derivatives were examined in detail (22–31). Results of the protective experiments against Sendai virus and herpes simplex type 1 virus infections in mice suggest that MDP-Lys(L18) is effective not only for the prevention of the bacterial and fungal infections but also for the viral infection (25,26). It was also shown that MDP-Lys(L18) induced a variety of cytokines such as interleukin-1 (IL-1), IL-6, colony-stimulating factors (CSFs), tumor necrosis factor (TNF), interferon gamma (IFN-γ), and prostaglandin E_2 (PGE_2) in mice and humans (29,30), and it was suggested that these cytokines might play important roles for the stimulation of host resistance against infections. The above results clearly indicated that MDP-Lys(L18) is the potent cytokine inducer in vivo and a new immunotherapeutic agent for infectious diseases and cancer (31). In this study, we showed that MDP-Lys(L18) stimulated the host resistance against hantavirus infection in newborn mice. Leukocytosis was induced in newborn mice by the administration of MDP-Lys(L18) and maximal leukocyte number was observed on day 5 after treatment (data not shown). Recently, it was shown that MDP-Lys(L18) was effective in patients with cancer who were treated with radiation therapy and anticancer chemotherapies for the restoration of decreased neutrophils and platelets (24,25). During clinical trials, it was observed that the incidence of infectious diseases in the MDP-Lys(L18)–treated group was lower than that in the control group (32–34).

B30-MDP was used as an adjuvant for the potentiation of immunogenicity of viral and tumor cells in experimental models. Nerome et al. (35) reported the preparation of the liposome (MDP-virosome) consisting of B30-MDP, cholesterol, and influenzavirus antigen (purified hemagglutinin-neuraminidase). The B30-MDP–virosome of 100 nm diameter thus prepared enhanced both the circulating antibody response and the induction of cell-mediated immunity against the antigenic molecule located on its surface in an experimental mouse model. In the challenge experiment with influenza virus, significant survival rates were determined in the group of mice immunized with B30-MDP–virosome vaccine. Kaji et al. (36) have confirmed the safety of influenza-B30-MDP–virosome vaccine in a phase 1 clinical trial in humans. In this experiment, we have clearly shown that B30-MDP potentiates the antigenicity of the inactivated B-1 vaccine and rHBsAg. These results suggest that the liposome containing B30-MDP will be useful as an adjuvant-active vehicle for the inactivated vaccine or recombinant vaccine.

It also was shown that B30-MDP is effective for the potentiation of antigenicity of irradiated tumor cell vaccine. Kataoka et al. (37,38) have demonstrated that the immunization of strain 2 guinea pigs with x-ray–irradiated line 10 hepatoma or acute B-cell leukemia cells EN-L2C together with B30-MDP induced systemic and specific cellular tumor immunity in the hosts. We also have found that the immunization of Balb/c mice with x-ray–irradiated mouse lymphoma cells L5178Y-

ML25 combined with B30-MDP enhanced the activity of cytotoxic killer T cells to inhibit the metastasis of live L5178Y-ML25 cells (39).

In conclusion, immunoadjuvants play very important roles in the prevention of viral infections via potentiation of specific and nonspecific immunities in experimental models and humans.

V. CONCLUSIONS

The immunoadjuvant activity for the prevention of hantavirus infection and potentiation of antigenicity of viral vaccine of both muramyl dipeptide derivatives, MDP-Lys(L18) and B30-MDP, were examined in mouse experimental systems. MDP-Lys(L18) prevented hantavirus infection in newborn mice. B30-MDP and MDP-Lys(L18) were effective for the potentiation of antigenicity of Seoul type hantaviruses strain B-1 (inactivated) vaccine and recombinant hepatitis B virus surface antigen.

ACKNOWLEDGMENTS

This work was supported in part by Grants-in-Aid for Cancer Research from the Japanese Ministry of Health and Welfare for Comprehensive 10-Year Strategy for Cancer Control and for Scientific Research from the Japanese Ministry of Education, Science and Culture; by the Osaka Foundation for Promotion of Clinical Immunology; by the Special Grant-in-Aid for Promotion of Education and Science in Hokkaido University Provided by the Japanese Ministry of Education, Science and Culture; and also by the Uehara Memorial Foundation.

REFERENCES

1. Freund J. The mode of action of immunologic adjuvants. Adv Tuberc Res 1956; 7:130.
2. Azuma I, Kishimoto S, Yamamura Y, Petit J-F. Adjuvanticity of mycobacterial cell wall. Jpn J Microbiol 1971; 15:193.
3. Azuma I. Immunological and biochemical properties of bacterial fractions and related compounds with special reference to BCG cell wall skeleton and *N. rubra* cell wall skeleton. In: Yamamura Y, Azuma I, eds. Molecular and cellular networks for cancer therapy. Tokyo: Excerpta Medica, 1989:83.
4. Ellouz F, Adam A, Ciorubaru R, Lederer E. Minimal structural requirements for adjuvant activity of bacterial peptidoglycan subunits. Biochem Biophys Res Commun 1974; 59:1317.
5. Adam A, Lederer E. Muramyl peptides, immunomodulator, sleep factors, and vitamins. Med Res Rev 1984; 4:111.
6. Azuma I. An acyl-MDP derivative, MDP-Lys(L18) (romuritide). In: Stewart-Tull D,

ed. Adjuvants: Theory and practical applications. Oxford, England: Butterworth-Heinemann, 1993:in press.
7. Schmaljohn CS, Hasty SE, Harrison SA, Dalrymple JM. Characterization of Hantaan virus virions, the prototype virus of hemorrhagic fever with renal syndrome. J Infect Dis 1983; 148:1005.
8. Arikawa J, Schmaljohn A, Dalrymple JM, Schmaljohn CS. Characterization of Hantaan virus envelope glycoprotein antigenic determinants defined by monoclonal antibodies. J Gen Virol 1989; 70:615.
9. Lee HW, Lee PW, Johnson KM. Isolation of the etiologic agent of Korean hemorrhagic fever. J Infect Dis 1978; 137:298.
10. Yoo YC, Yoshimatsu K, Yoshida R, Tamura M, Azuma I, Arikawa J. Comparison of virulence between Seoul virus strain SR-11 and Hantaan virus strain 76-118 of hantaviruses in newborn mice. Microbiol Immunol 1993: 37:557.
11. Yamanishi K, Tanishita O, Tamura M, Asada H, Kondo K, Takagi M, Yoshida I, Konobe T, Fukai K. Development of inactivated vaccine against virus causing haemorrhagic fever with renal syndrome. Vaccine 1988; 6:278.
12. Arikawa J, Yao JS, Yoshimatsu K, Takashima I, Hashimoto N. Protective role of antigenic sites on the envelope protein of Hantaan virus defined by monoclonal antibodies. Arch Virol 1992; 126:271.
13. Byars NE, Nakano G, Welch M, Lehman D Allison AC. Improvement of hepatitis B vaccine by the use of a new adjuvant. Vaccine 1991; 9:309.
14. Chedid L, Parant M, Lefrancier P, Choay J, Lederer E. Enhancement of nonspecific immunity to *Klebsiella pneumoniae* infection by a synthetic immunoadjuvant (*N*-acetylmuramyl-L-alanyl-D-isoglutamine) and several analogs. Proc Natl Acad Sci USA 1977; 74:2089.
15. Matsumoto K, Ogawa H, Kusama T, Nagase O, Sawaki N, Inage M, Kusumoto S, Shiba T, Azuma I. Stimulation of nonspecific resistance to infection induced by 6-*O*-acyl muramyl dipeptide analogs in mice. Infect Immun 1981; 32:748.
16. Matsumoto K, Ogawa H, Nagase O, Kusame T, Azuma I. Stimulation of nonspecific resistance to infection induced by muramyl dipeptide. Microbiol Immun 1981; 25:1047.
17. Matsumoto K, Otani T, Une T, Osada Y, Ogawa H, Azuma I. Stimulation of nonspecific resistance to infection induced by muramyl dipeptide analogs substituted in the γ-carboxy group and evaluation of N^{α}-muramyl dipeptide N^{ε}-stearoyllysine. Infect Immun 1983; 39:1029.
18. Matsumoto K, Osada Y, Une T, Otani T, Ogawa H, Azuma I. Anti-infectious activity of the synthetic muramyl dipeptide analogue MDP-Lys(L18). In: Azume I, Jollès G, eds. Immunostimulants: Now and tomorrow. Tokyo: Japanese Scientific Society Press/Berlin: Springer-Verlag, 1987:79.
19. Schumann G. Biological activities of a lipophilic muramyl peptide (MTP-PE). In: Azuma I, Jollès G, eds. Immunostimulants: Now and tomorrow. Japanese Scientific Society Press/Berlin: Springer-Verlag, 1987:71.
20. Goto T, Aoki H. The immunomodulatory activities of acylpeptides. In: Azuma I, Jollès G, eds. Immunostimulants: Now and tomorrow. Tokyo: Japanese Scientific Society Press/Berlin: Springer-Verlag, 1987:99.

21. Floc'h F, Poirier J, Fizames C, Woehrle R. Pemelautide (R.P. 49 639). From experimental results to clinical trials: An illustration. In: Azuma I, Jollès G, eds. Immunostimulants: Now and tomorrow. Tokyo: Japanese Scientific Society Press/ Berlin: Springer-Verlag, 1987:183.
22. Otani T, Une T, Osada Y. Stimulation of nonspecific resistance to infection by muroctasin. Arzneim Forsch Drug Res 1988; 38(II):969.
23. Osada Y, Otani T, Sato M, Une T, Matsumoto K, Ogawa H. Polymorphonuclear leukocyte activation by a synthetic muramyl dipeptide analog. Infect Immun 1982; 38:848.
24. Shimoda K, Okamura S, Koawasaki C, Omori F, Matsuguchi T, Niho Y. Muroctasin [MDP-Lys(L18)] augments the production of granulocyte colony-stimulating factor (G-CSF) from human peripheral blood mononuclear cells in vitro. Int J Immunopharm 1990; 12:729.
25. Ishihara C, Hamada N, Yamamoto K, Iida J, Azuma I, Yamamura Y. Effects of muramyl dipeptide and its stearoly derivatives on resistance to Sendai virus infection in mice. Vaccine 1985; 3:370.
26. Ishihara C, Iida J, Mizukoshi N, Yamamoto N, Yamamoto K, Kato K, Azuma I. Effect of N^{α}-acetylmuramyl-L-alanyl-D-isoglutaminyl-N^{ε}-stearoyl-L-lysine on resistance to herpes simplex virus type 1 infection in cyclophosphamide-treated mice. Vaccine 1989; 7:309.
27. Nakajima R, Ishida Y, Akahane K, Sekiguchi M, Osada Y. Stimulatory effect of romuritide on hematopoiesis in monkeys. Arzneim Forsch Drug Res 1991; 41:60.
28. Nakajima R, Ishida Y, Yamaguchi F, Otani T, Ono Y, Nomura M, et al. Beneficial effect of muroctasin on experimental leukopenia induced by cyclophosphamide or irradiation in mice. Arzneim Forsch Drug Res 1988; 38:986.
29. Saiki I, Saito S, Fujita C, Ishida H, Iida J, Murata J, Hasegawa A, Azuma I. Induction of tumoricidal macrophages and production of cytokines by synthetic muramyl dipeptide analogs. Vaccine 1988; 6:238.
30. Yamaguchi F, Akasaki M, Tsukada W. Induction of colony-stimulating factor and stimulation of stem cell proliferation by injection of muroctasin. Arzneim Forsch Drug Res 1988; 38(II):980.
31. Azuma I. Development of cytokine-inducer romurtide; experimental study and clinical application. Trends Pharmacol Sci 1992; 13:425.
32. Furuse K, Sakuma A. Activation of the cytokine network by muroctasin as a remedy for leukopenia and thrombopenia. Arzneim Forsch Drug Res 1989; 39:915.
33. Tsubura E, Nomura T, Niitani H, Osamura T, Tanaka M, Ota K, et al. Restorative activity of muroctasin on leukopenia associated with anticancer treatment. Arzneim Forsch Drug Res 1988; 38:1070.
34. Sakamoto S, Okawa T, Ogawa N. Therapeutic effect of muroctasin on cancer patients with leukopenia during radiation therapy. Shin-yaku to Rinsho 1988; 38:1070 (in Japanese).
35. Neorme K, Yoshioka Y, Ishida M, Okuma K, Oka T, Kataoka T, et al. Development of a new type of influenza subunit vaccine made by muramyldipeptide-liposome: enhancement of humoral and cellular immune responses. Vaccine 1990; 8:508.
36. Kaji M, Kaji Y, Kaji M, Ohkuma K, Honda T, Oka T, et al. Phase 1 clinical tests of influenza MDP-virosome vaccine (KD-5382). Vaccine 1992; 10:663.

37. Kataoka T, Tokunaga T. A synthetic adjuvant effective in inducing antitumor immunity. Jpn J Cancer Res (Gann) 1988; 79:817.
38. Kataoka T, Kinomoto M, Takegawa M, Tokunaga T. Effect of a synthetic adjuvant for inducing anti-tumour immunity. Vaccine 1991; 9:300.
39. Yoo YC, Saiki I, Sato K, Azuma I. B30-MDP, a synthetic muramyl dipeptide derivative for tumour vaccination to enhance antitumor immunity and antimetastatic property in mice. Vaccine 1992; 10:792.

21

Immunotherapy of Infectious Postoperative Complications with Glucosaminylmuramyl Dipeptide

Rakhim M. Khaitov, Boris V. Pinegin, and Alexander A. Butakov
Ministry of Public Health of Russia, Moscow, Russia

Tatyana M. Andronova
Shemyakin-Ovehinnikav Institute of Bioorganic Chemistry, Russian Academy of Sciences, Moscow, Russia

I. INTRODUCTION

It is well known that peptidoglycans of the bacterial cell wall possess immunostimulating properties. The study of peptidoglycans as immunostimulating agents dates back to 1974 when it was established that the immunostimulating activity of Freund's adjuvant is predominantly connected with the muramyl peptide fraction of the *Mycobacterium tuberculosis* cell wall (1). These data provided the basis for obtaining the synthetic equivalent of the minimal cell wall structure, MDP (N-acetylmuramyl-L-alanyl-D-isoglutamine) (2), that shows equivalent activity to whole mycobacteria in complete Freund's adjuvant.

N-acetylglucosaminyl-N-acetylmuramyl-L-alanyl-D-isoglutamine (GMDP) was isolated during analysis of the antitumor drug blastolysine, which is a hydrolysate of *Lactobacillus bulgaricus* cell wall (3). As a further development of that project, Andronova and Ivanov (4) designed an original technique for the synthesis of GMDP that includes the condensation of dipeptide with disaccharide isolated from *Micrococcus lysodeikticus*.

There are many experimental studies showing that GMDP affects macrophage function and cellular immunity. It was found, for example, that GMDP induces Ia-antigen expression on macrophage cell membranes (5) and increases the production of superoxide radical, which is part of the bacteriostatic mechanism of macrophages (6). The administration of GMDP to C57Bl/6 mice carrying the EL-4 lymphoma stimulated lymphocytes specifically cytotoxic for the tumor (7).

The purpose of the present study was to evaluate both clinically and immunologically the efficacy of GMDP administration for the prophylaxis and treatment of postoperative septic complications after operations on the colon. Operations on the gastrointestinal tract carry a particularly high risk of septic complications.

II. MATERIALS AND METHODS

A. Clinical Protocol

GMDP therapy was carried out in patients with cancer of the colon and rectum undergoing surgery at Moscow Clinical Hospital No. 24. Patients were divided into three groups for the prophylaxis study: 43, 33, and 36 patients received 1, 2, or 3 mg of GMDP (or placebo) per os for 10 days prior to operation. No other drugs were administered to the patients during that period, although antibiotic treatment was commenced after operation: All patients received penicillin and gentamycin, and in patients developing septic complications additional antibiotic therapy was introduced as appropriate. Postoperative treatment with GMDP was carried out in three groups of patients (41, 35, 42 patients) with septic postoperative complications receiving 1, 2, or 3 mg of GMDP (or placebo) per os for 10 days. In each group, approximately half the patients received GMDP and half placebo on a double-blind basis. The efficacy of GMDP treatment was evaluated by mortality and the development of septic complications.

B. Laboratory Analyses

Clinical blood analysis (hemoglobin and blood formula), blood biochemistry (glucose, bilirubin, creatinine, urea, total protein, and albumin), as well as measurement of immunological parameters was carried out before GMDP administration and 3–4 days after termination of administration in the prophylaxis trial. The same analyses were performed before GMDP administration and 2–3 days after termination of administration during postoperative GMDP treatment.

Immunological analysis of B lymphocytes (3F3), NK cells (CD16), and T-lymphocyte subpopulations (CD5, CD4, CD8) was carried out by flow cytometry (Epics–Profile 2, Coulter). The proliferative activity of mononuclear cells after stimulation with phytohemagglutinin (PHA) was evaluated by a standard technique of [^{3}H]thymidine incorporation (8), and the functional activity of phagocytic cells was measured by chemiluminescence (LKB 1251 luminometer) (9) using cells attached to plastic (spontaneous chemiluminescence) and after stimulation by opsonized zymosan (induced chemiluminescence).

The change in each parameter (hematological, biochemical, immunological) after administration of placebo or GMDP was calculated as a percentage of pretreatment values. These changes were compared using Student's *t*-test (GMDP vs placebo).

III. RESULTS AND DISCUSSION

A. Clinical Endpoints

There was a decrease in the number of septic complications in patients with cancer of the colon undergoing prophylactic administration of GMDP (Table 1). The percentage of complications in patients receiving 1, 2, or 3 mg of GMDP per day was 42, 29, and 36%, whereas the percentage of complications in patients receiving the same doses of placebo was 84, 50, and 63%, respectively. The highest effect was observed with respect to the frequency of wound suppuration, pelvic suppurations, and pneumonia. It was demonstrated, using a confidence evaluation technique at a 5% significance level (10), that there was a statistically significant difference between the groups receiving GMDP (36%) and the total percentage of complications in the groups receiving placebo (66%). A total of two patients died in groups receiving GMDP in contrast to eight patients in the groups receiving placebo.

There also was a beneficial effect when using GMDP for treatment of septic complications postoperatively. The percentage of complications in patients receiving 1, 2, or 3 mg of GMDP each day was 60, 41, and 33%, whereas the percentage of complications in patients receiving similar doses of placebo was 80, 55, and 66%, respectively (Table 2), although the difference in the total percentage of complications in groups receiving GMDP (45%) and placebo (68%) was statistically nonsignificant. Four patients died in groups receiving GMDP, whereas 11 patients died in groups receiving placebo. Both in this case and in the prophylaxis trial, the difference in mortality was nonsignificant as compared with the placebo group.

It should be stressed that there were no adverse events or side effects attributable to the administration of GMDP.

Table 1 Mortality and Postoperative Septic Complications in Patients Given GMDP Prophylactically

Dose group (mg)	GMDP				Placebo			
	Subjects (no.)	Complications (no.)	Complications (%)	Deaths (%)	Subjects (no.)	Complications (no.)	Complications (%)	Deaths (%)
1	24	10	42	1	19	16	84	2
2	17	5	29	0	16	8	50	2
3	17	6	36	1	19	12	63	4
Total	58	21	36	2	54	36	67	8

Table 2 Mortality and Postoperative Septic Complications in Patients Given GMDP Therapeutically

Dose group (mg)	GMDP Subjects (no.)	GMDP Complications (no.)	GMDP Complications (%)	GMDP Deaths (%)	Placebo Subjects (no.)	Placebo Complications (no.)	Placebo Complications (%)	Placebo Deaths (%)
1	20	12	60	3	21	17	80	6
2	17	7	41	1	18	10	55	3
3	21	7	33	0	21	14	66	2
Total	58	26	45	4	60	41	68	11

B. Biochemical and Immunological Endpoints

It was found that all doses of GMDP (used both for prophylaxis and for treatment of septic complications) produced statistically significant increases in the number of leukocytes, principally neutrophils, with the effect being most evident at the dose of 3 mg/day (Fig. 1).

According to biochemical analysis, levels of hemoglobin, total protein, and albumin in patients receiving GMDP were higher than in patients receiving place-

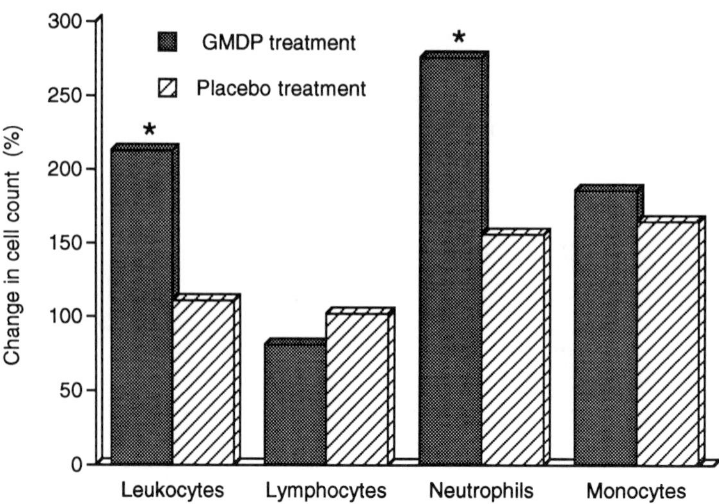

Figure 1 Changes in blood cell counts (total leukocytes and specific cell types) in patients treated prophylactically with GMDP at a dose of 3 mg daily. Here and in Figures 2 and 3, the post-treatment value is expressed as % of pre-treatment value, such that 100% represents no change. * indicates a significant difference ($P < .05$) in comparison to the placebo group.

bo. Bilirubin, creatinine, and urea levels were lower in patients receiving GMDP than in patients receiving placebo. Because of the high variance of individual parameters, differences were significant only with respect to the hemoglobin ratio during the administration of 2 and 3 mg GMDP and for bilirubin and creatinine after GMDP prophylaxis at a dose of 2 mg/day.

Immunological analysis showed that the greatest changes of immunological parameters were caused by prophylaxis with GMDP at a dose of 3 mg daily. In comparison with placebo, this dose (Fig. 2) caused an increase of the total T-lymphocyte population (CD5), T helper cells (CD4), and NK cells (CD16). A significant increase of T helper cells also was observed at the other GMDP doses (data not shown). The capacity of GMDP to influence CD4 lymphocytes may represent an important component of its immunostimulating action. A significant increase in B lymphocytes was observed only after postoperative treatment with GMDP, at which time immunoglobulin M (IgM) also increased significantly (data not shown). The remaining classes of immunoglobulins were unaffected by GMDP administration.

Results concerning the functional activity of lymphocytes and phagocytic cells are of great interest (Fig. 3). GMDP at all doses enhanced the mean proliferative response of lymphocytes to PHA, but due to large individual differences, this enhancement was not significant. The same was observed with chemiluminescence; this was enhanced by practically all doses of GMDP, but a significant increase took place only during prophylaxis at a dose of 2 and 3 mg daily (Fig. 3). It should be noted that there was a general decrease in chemiluminescence after operation in

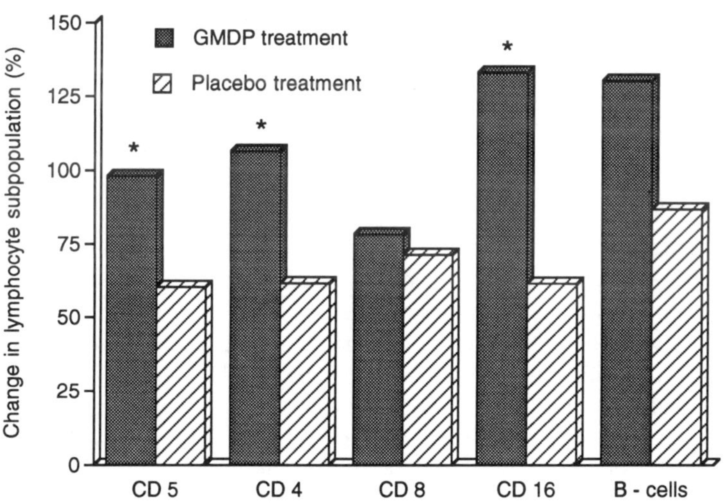

Figure 2 Changes in peripheral blood lymphocyte subpopulations in patients treated prophylactically with GMDP at a dose of 3 mg daily.

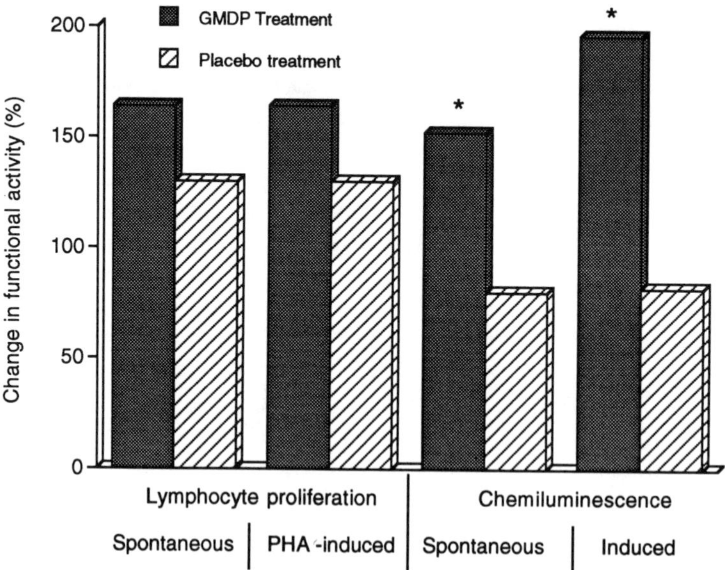

Figure 3 Changes in functional activity of lymphocytes and phagocytes in blood of patients treated prophylactically with GMDP at a dose of 3 mg daily. For each of the indices (proliferation and chemiluminescence), spontaneous and induced (PHA and zymosan, respectively) values are given.

patients receiving placebo, which was alleviated by GMDP treatment.

IV. CONCLUSIONS

The data provide evidence for the efficacy of GMDP in the control of septic complications arising from abdominal surgery when used prophylactically before surgery or after the operation if infections arise, although the results achieved significance only in the former case. GMDP-induced increases in lymphocyte and phagocyte numbers and function also were observed, which we believe could contribute to the therapeutic effect. Our immunological results are consistent with those of others (5,6,11) obtained in nonclinical models.

Postoperative septic complications are most commonly caused by opportunistic microorganisms such as *Staphylococcus, Streptococcus,* and *E. coli*. The major role in protection against these microbes belongs to phagocytes. The positive clinical effect of GMDP may be related to its ability to stimulate the functional activity of phagocytes and to increase the number of T helper cells, which could synthesize cytokines leading to further activation of phagocytes. However, we cannot exclude the possibility that the clinical effect of GMDP may be connected partly with its positive influence on liver activity as manifested in a decrease of creatinine and bilirubin levels.

Our results suggest that GMDP treatment may be a promising addition to other therapeutic modalities for the control of postoperative sepsis.

REFERENCES

1. Chedid L, Lederer E. Past, present and future of the synthetic immunoadjuvant MDP and its analogs. Biochem Pharmacol 1978; 27:2183.
2. Kusumoto S, Tarumi Y, Ikenaka K, Shiba T. Chemical synthesis of N-acetylmuramyl peptides with partial structures of bacterial cell wall and their analogs in relation to immunoadjuvant activities. Bull Chem Soc Jpn 1976; 49:533.
3. Bogdanov IG, Dalev PG, Gurevich AI, Kolosov MN, Mal'kova VP, Plemyannikova LA, Sorokina IB. Antitumor glycopeptides from *Lactobacillus bulgaricus* cell wall. FEBS Lett 1975; 57:259.
4. Andronova T, Ivanov V. The structure and immunomodulating function of glucosaminylmuramyl peptides. Sov Med Rev D Immunol 1991; 4:1.
5. Nesmeyanov VA, Khaidukov SV, Komaleva RL, Andronova TM, Ivanov VT. Muramylpeptides augment expression of Ia-antigens on mouse macrophages. Biomed Sci 1990; 1:151.
6. Balitsky KP, Umansky VY, Tarakhovsky AM, Andronova TM, Ivanov VT. Glucosaminylmuramyl dipeptide-induced changes in murine macrophage metabolism. Int J Immunopharmacol 1989; 11:429.
7. Pimenov AA, Rakhmilevich AL, Deev VV, Kirillova IV, Migdal TL, Andronova TM, Fuks BB. Activation of cellular immunity in mice under normal conditions and in tumor growth during treatment with glucosaminyl muramyl dipeptide. Vopr Med Khim 1990; 36:58.
8. Friemal H. Immunologische Arbeitsmethoden. Gena: VEB Gustav Fischer Verlag, 1976.
9. Zemskov VM, Barsukov AM, Beznosenko AA, et al. The study of functional state of human phagocytes (oxygen metabolism and cell mobility). In: Methodical recommendations. Moscow: 1988 (in Russian).
10. Hey JO. An introduction to Bayesian statistical inference for economists. New York: Martin Robertson, 1984:137.
11. Pimenov AA, Fuks BB, Andronova TM. Analysis of cellular mechanisms of antitumor immunity under the action of GMDP in combination with lipopolysaccharide, indomethacin and cyclophosphan. In: International Symposium on Structure, Biosynthesis and functions of Molecular Elements of Immune System. Pushchino 1987:39 (abstr) (in Russian).

22

Cellular and Molecular Mechanisms of Biological Activity of Muramyl Peptides

Vladimir Nesmeyanov, Tatyana Golovina, Tatyana Valyakina, Tatyana M. Andronova, and Vadim Ivanov
Shemyakin-Ovehinnikav Institute of Bioorganic Chemistry, Russian Academy of Sciences, Moscow, Russia

Svetlana Bykovskaya
Oncological Centre, Moscow, Russia

I. INTRODUCTION

Since muramyl dipeptide (MDP) was identified by Lederer et al. (1) as a minimal structure from the bacterial cell wall capable of replacing killed mycobacteria in complete Freund's adjuvant, large numbers of MDP analogues and derivatives, called muramyl peptides (MPs), have been synthesized and studied. Besides adjuvant activity, MPs demonstrate other biological activities that make them attractive for medical and veterinary fields. In particular, they have been shown to induce nonspecific resistance to bacterial and viral infections. Some of MPs (especially lipophylic derivatives) have been shown to cause growth inhibition and necrosis of solid tumors (1,2).

The practical use of MPs precedes studies of their mechanisms of biological activity. Nor–muramyl dipeptide (Nor-MDP) is undergoing clinical trials as a component of human antifertility vaccine (3), a steroyl derivative of MDP-lysine (Muroctasin) is being developed in Japan as an antitumor drug (4), and a phase I clinical trial of N-acetylglucosaminylmuramyl dipeptide (GMDP) for the prevention of septic complications in colorectal surgery has been completed in Russia.

In contrast, the cellular and molecular basis of the biological activity of MPs require further study. The purpose of this work was to study the effects of MPs on target cells and to characterize their molecular targets.

II. RESULTS AND DISCUSSION

A. Effect of Muramyl Peptides on Macrophages

Mononuclear phagocytes seem to be the primary targets of MPs. MPs increase the cytotoxicity of macrophages, inhibit their migration and induce secretion of certain cytokines, in particular, interleukin-1 (IL-1), IL-6, colony-stimulating factors (CSFs) (5).

Not long ago we showed that the immunomodulatory activity of MPs can be partially attributed to their effect on the expression of major histocompatibility (MHC) class II antigens on macrophages/monocytes. Upon study of N-acetylglucosaminyl–containing MPs (GMPs, Fig. 1), synthesized by us (6), we demonstrated that they caused dose-dependent enhancement of Ia-antigen expression by murine peritoneal macrophages in vitro and in vivo (7). The effect correlated with the adjuvant activity of GMPs: adjuvant nonactive compounds (e.g., L-GMDP,

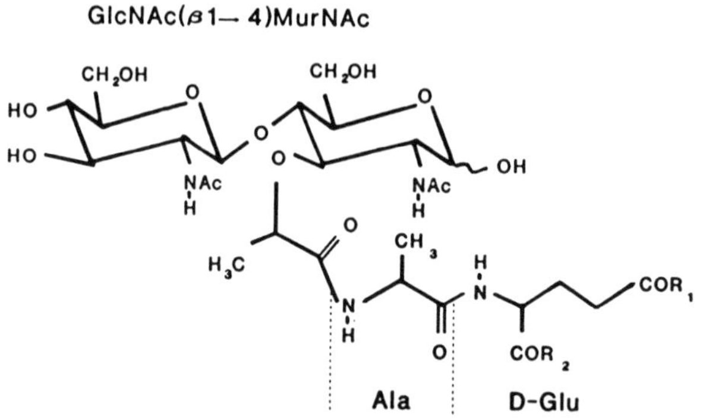

GlcNAc(β1→4)MurNAc

GMP	R_1	R_2
GMDP	OH	NH_2
GMDP-OBu	OBu	NH_2
GMDP-Lys	NH-CH(COOH)-$(CH_2)_4NH_2$	NH_2
GMDP-acid	OH	OH
GMDP-Lys-steroyl	NH-CH(COOH)-$(CH_2)_4$NH-CO-$C_{17}H_{35}$	NH_2

Figure 1 Structures of N-acetylglucosaminyl–containing muramyl peptides (GMPs).

Table 1 Correlation Between Ia-Inducing and Adjuvant Activities of Glucosaminylmuramyl Peptides

GMP	Increase in Ia-positive cells		Adjuvant active or nonactive
	1 μg/ml	10 μg/ml	
GMDP	18.01	25.18	+
L-GMDP	2.70	2.60	−
$(GMDP)_2$	14.32	19.61	+
GMDP(OBu)	13.39	7.29	+
GMDP-acid	23.90	28.50	+
GMDP-Lys	14.81	16.57	+
GMDP-Lys-Steroyl	25.49	25.70	+
GlcNAc-MurNAc	2.96	1.13	−

disaccharide GlcNAc-MurNAc) had no or very little effect on Ia expression (Table 1). Kinetic study has shown that after treatment of cells with GMDP, Ia-expression peaked at 6 and 18 h, which indicates the existence of two populations of macrophages—rapid and slow responding. This was proved by FACS separation of peritoneal cells into Ia-positive and Ia-negative followed by GMDP treatment: Only the former population was capable of rapid response. Enhanced expression of MHC class II antigens (HLA-DR) could also be induced by GMDP on human monocytes.

Thus, the anti-infectious and adjuvant activities of GMPs can be partially attributed to their effect on MHC class II antigen expression resulting in improved recognition of foreign antigens by T cells.

B. Effect of Muramyl Peptides on Tumor Cells

The improvement of antigen recognition by GMDP indicated to us that the same mechanism might be involved in GMPs' antitumor activity. It is well established that tumor cells have an altered array of surface antigens (8,9). Three groups of antigens are of major importance for immune recognition of tumor cells; namely, MHC class I antigens, adhesion molecules, and tumor-associated antigens (TAAs). We evaluated the effect of GMPs on the expression of MHC antigens and TAAs by three human tumor cell lines: lung adenocarcinoma A-549, melanoma BRO, and mammary gland carcinoma MaTu.

Treatment of these cell lines in vitro with GMDP for 18–48 h resulted in a dose-dependent increase in the number of cells expressing TAAs (Fig. 2). For various cell lines the effect peaked at various GMDP concentrations. The highest effect was observed for BRO and the lowest for MaTu cells. Kinetic studies have shown that

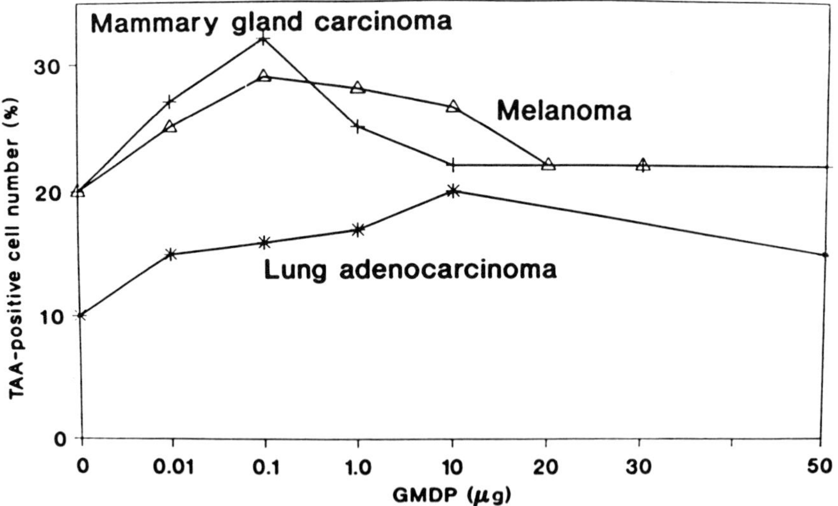

Figure 2 Dose-response plot of GMDP-induced expression of tumor-associated antigens by human tumor cell lines: melanoma BRO (–Δ–), lung adenocarinoma A-549 (–*–), and mammary gland carinoma MaTu (–+–).

optimal incubation time was 18–24 h. A biologically nonactive GMDP analogue with L-glutamine residue substituted for D-glutamine (L-GMDP), used as negative control, over a concentration range 0.001–50.0 μg/ml did not have any effect on TAA expression. In contrast, interferon gamma, like GMDP, augmented the expression of TAAs by all studied tumor cell lines, although the magnitude of the response was lower.

A higher increase in TAA expression was achieved if a second dose of GMDP was added at 24 or 48 h and even higher increase occurred if GMDP was added at 24 and 48 h: The number of TAA-bearing cells almost doubled for melanoma and increased three times for lung adenocarcinoma cells (Table 2). Thus, priming with GMDP sensitized tumor cells to subsequent GMDP treatment.

TAA-inducing activity was not restricted to GMDP but was characteristic for other GMDPs as well. For instance, GMDP-Lys had a higher effect than GMDP on the expression of mammary gland–specific antigen at a concentration an order of magnitude lower than GMDP. A lipophilic steroyl derivative of GMDP-Lys at a lower concentration had an even higher effect. As for GMDP, TAA induction increased after repeated GMDP-Lys–steroyl introduction (see Table 2). It is well documented that lipophilic MPs have high antitumor activity. Antitumor effect is characteristic for the MDP-Lys–steroyl (4), MDP-phosphatidylethanolamine (10). Thus, the higher TAA-inducing activity of GMDP-Lys–steroyl correlates with the

Table 2 Effect of Repeated Addition of Glucosaminylmuramyl Peptides on TAA Expression[a]

	BRO		A-549	
GMP addition at (h)	GMDP[b] (%)	GMDP-Lys-steroyl[c] (%)	GMDP[d] (%)	GMDP-Lys steroyl[c] (%)
No addition	20	20	10	10
0	28	28	15	15
0, 24	41	48	26	20
0, 24, 48	48	52	45	35

[a] TAA expression was evaluated at 72 h.
[b] 0.1 μg/ml.
[c] 0.001 μg/ml.
[d] 10 μg/ml.

antitumor effect of lipophilic MPs, which indicates that it might contribute to the antitumor effect of GMPs.

As has been shown by many investigators, the reduced expression of certain MHC class I antigens might result in an impaired antitumor response (8,11). We quantified MHC class I antigens on MaTu and BRO cells by using antibodies to β_2-microglobulin which is known to represent their invariant chain. Treatment of the above cell lines with GMDP in vitro did not alter β_2-microglobulin expression: Regardless of the dose of GMDP (0.1–10.0 μg/ml) and time of incubation (24–72 h), only 5% of MaTu and 8% of BRO cells were β_2-microglobulin positive before and after treatment.

C. Functional Implications of Altered TAA Expression

The data obtained showed that GMDP might enhance the antitumor response by two closely associated mechanisms; namely, by activating white blood cells, in particular, macrophages, and by facilitating recognition of tumor targets by augmenting the expression of cell membrane antigens, in particular, TAAs. To prove that the latter mechanism can contribute to the antitumor effect of GMPs, we compared the lytic activity of peripheral blood mononuclear cells (PBMCs) from normal donors with melanoma cells, nontreated or treated in vitro with GMDP. Upon optimal PBMC:target cell ratio lysability of melanoma cells after 24 h treatment with GMDP considerably increased for 8 of 13 donors (Table 3). In contrast, after 6 h of GMDP treatment, resulting in reduced expression of melanoma-associated antigen, a reduction in the lysability of melanoma cells was demonstrated in the majority of cases.

Table 3 Increase in Cytotoxicity (%) of Leukocytes from Normal Donors Against GMDP-Treated Cells

Donor	Effector/target ratio	6 h	24 h
1	50:1	−6.7	46.4
2	12:1	−24.0	34.8
3	12:1	−16.2	22.1
4	12:1	7.3	35.0
5	2:1	−14.9	nd
6	6:1	−9.0	nd
7	12:1	−16.2	−6.5
8	12:1	10.2	20.6
9	12:1	−7.9	−0.7
10	25:1	7.0	12.3
11	12:1	6.4	14.4
12	12:1	−10.7	4 3
13	12:1	0	49.0
14	12:1	−0.2	0.7
15	12:1	−0.9	5.6

nd, not determined.

The experimental data obtained are not sufficient to ascribe the increased lysis solely to enhanced TAA expression. MHC antigens do not seem to play the major role in this case, but it cannot be excluded that the expression of other cell membrane antigens, for example, adhesion molecules, which are essential for immune recognition of tumor cells, was altered.

D. Molecular Targets of Muramyl Peptides

The molecular basis of the interaction of MPs with target cells remains unclear, although receptor-mediated mechanisms seem to be the most probable. Despite a number of studies, the presence of specific surface receptors is still controversial. Silverman et al. (12) have found a low number of MP-binding sites on plasma membrane of murine macrophages. Tenu et al. (13), by using an affinity labeling technique, have demonstrated the presence of MP-binding protein on B cells.

Recently, by using flow cytometry and fluorescence polarization analysis, we have studied the binding of a fluorescein isothiocyanate (FITC)–labeled derivative of GMDP-lysine to a variety of cells and cell lines (14). Only cells permeabilized with beta-octylglucoside were specifically labeled. These results demonstrated that specific GMP-binding sites were located largely inside cells, namely, macrophages, transformed T lymphocytes, and neuroblastoma cells, but not inside myeloma cells. Surface receptors were not detected, although we cannot exclude their presence in low numbers (<1000/cell). Among normal cells besides macrophages, splenic T helper cells, but not B lymphocytes, were capable of specific binding of GMDP-Lys–FITC.

Two types of GMP-binding sites with dissociation constant (Kd) values of 21 and 540 nM were found inside macrophages. The binding was inhibited much more efficiently by GMDP than by fragments of this molecule (disaccharide GlcNAc-MurNAc, tripeptide Ala-D-iGln-Lys) or L-GMDP indicating the importance of intact structure and stereochemistry of the ligand (Fig. 3). Taking into account that these compounds were biologically nonactive, the aforementioned data showed correlation between ligand specificity of binding and ligand specificity of biological activity.

Affinity labeling of permeabilized with digitonin macrophages and WEHI-3 cells revealed that GMP-binding sites resided on proteins. The labeling was performed with GMDP-Lys–iodine-125–azidosalicilic acid. Cells were solubilized in sample buffer and subjected to sodium dodecylsulfate–polyacrylamide gel electrophoresis (SDS-PAGE). A number of iodine-125–labeled bands were observed on blots from the gels, but only few proteins (32, 38, 43 Kd) were labeled specifically: If the labeling was performed in the presence of cold GMDP, a reduction in intensity of corresponding bands was observed.

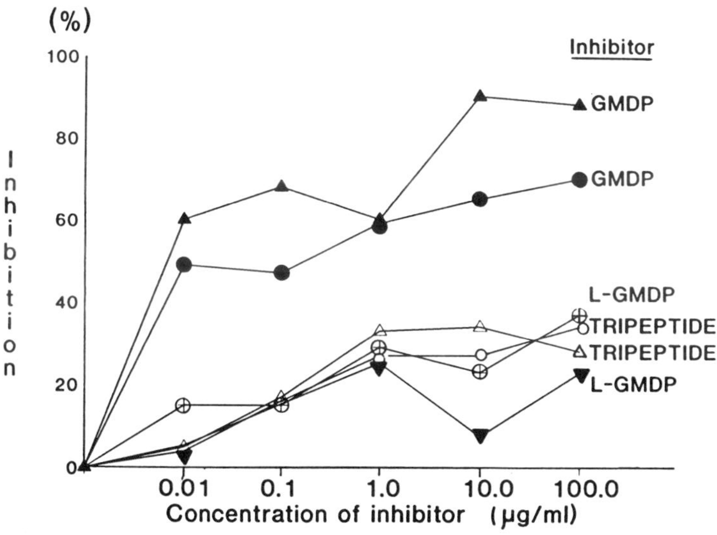

Figure 3 Inhibition of binding of GMDP-Lys-FITC to murine peritoneal macrophages (circles) and WEHI-3 cells (triangles) by L-GMDP and tripeptide Ala-D-Gln-Lys.

Another approach to identification of GMP-binding proteins included SDS-PAGE of WEHI-3 cell proteins followed by blotting of proteins onto nitrocellulose membrane and labeling them with GMDP-containing synthetic polymer. This polymer consisted of a polyacrylamide backbone with immobilized residues of GMDP and biotin. After incubating the blot with this polymer, MP-binding bands were visualized by staining with streptoavidin–horseradish peroxidase conjugate and 4-chloronaphtol. The membrane fraction of WEHI-3 cells was isolated and peripheral membrane proteins were extracted with 2.5 M KCl. SDS-PAGE revealed the presence of a number of protein bands, of which again 32- to 34-Kd protein was specifically labeled (Fig. 4a,b), as well as 18-Kd protein.

E. Anti-Idiotypic Antibodies to GMDP

To further characterize the receptor anti-idiotypic monoclonal antibodies (MAbs) to GMDP were produced. Initially we developed hybridomas that secreted MAb to GMDP. This antibody demonstrated specific binding to the GlcNAc-MurNAc-Ala fragment of GMDP. The configuration of glutamine residue was not crucial. Based on this MAb, we obtained a set of anti-idiotypic MAbs capable of competing with GMDP for the anti-GMDP MAb combining site as well as for GMDP-binding sites on target cells. Besides receptor-binding activity, C5/F3 and C5/F8 antibodies were able to enhance in Balb/c mice antibody response to ovalbumin; thus, resembling GMDP activity.

III. CONCLUSIONS

Our results show that an array of biological activities of GMPs can be attributed to their ability to augment expression of cell surface antigens, which are important for self-nonself discrimination.

There are specific GMP-binding sites inside responding cells, although we still cannot exclude the presence of a low number of cell surface receptors (<1000/cell). If surface receptors do not exist on some types of target cells (e.g., macrophages), two questions must be answered. First, does GMDP-binding to internal sites result in a biological signal? Second, if the answer to the first question is yes, how do GMPs manage to penetrate into target cells? We are continuing research along these lines.

ACKNOWLEDGMENTS

The authors thank Dr. N.V. Bovin for providing GMDP-biotin–containing polyacrylamide polymer and Prof. E. Revazova for tumor cell lines.

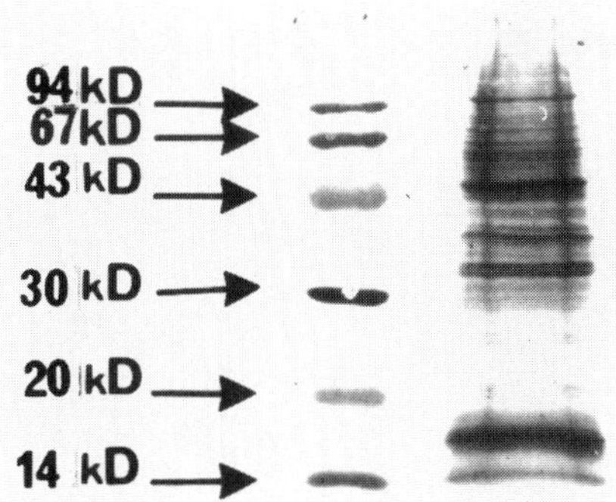

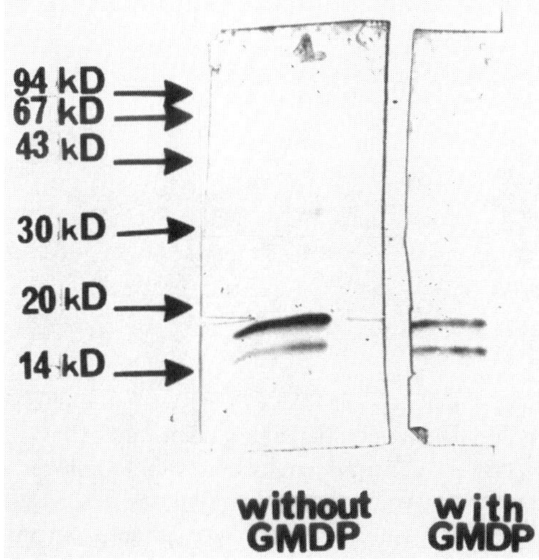

Figure 4 Blot from SDS-PAGE of proteins, eluted from WEHI-3 cell membranes by 2.5 M KCl. Blot was developed by using GMDP- and biotin-containing polymer followed by staining with streptoavidin-HRP conjugate and 4-chloronaphtol.

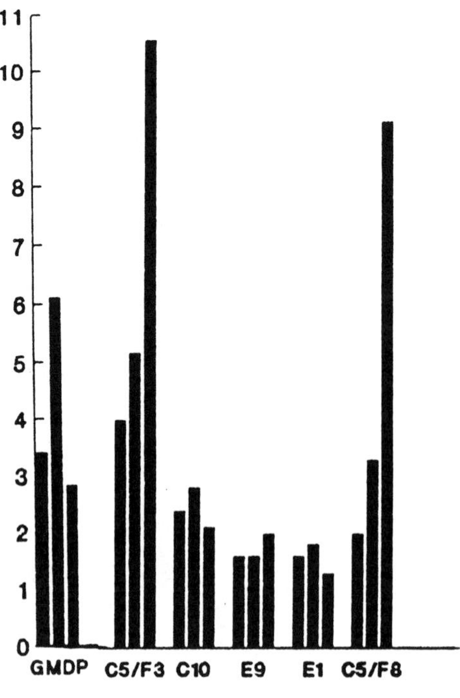

Figure 5 Adjuvant activity of anti-idiotypic antibodies to GMDP. Mice were immunized intraperitoneally with 25 μg of ovalbumin in saline-containing adjuvant (antibodies or GMDP). Two and 4 weeks later mice received 12.5 μg of ovalbumin without adjuvant. Antiovalbumin antibody titer was determined by solid-phase ELISA on day 7 after last immunization and compared with antibody titer from mice immunized without adjuvant.

REFERENCES

1. Adam A, Lederer E. Muramyl peptides: immunomodulators, sleep factors, and vitamins. Med Res Rev 1984; 4:111.
2. Andronova T, Ivanov V. The structure and immunomodulating function of glucosaminylmuramyl peptides. Sov Med Rev D Immunol 1991; 14:1.
3. Stevens VC, Cinader B, Powell JE, Lee AC, Kohn SW. Preparation and formulation of hCG anti-fertility vaccine: Selection adjuvant and vehicle. Am J Reprod Immunol 1981; 1:315.
4. Ichihara N, Kanazawa R, Sasaki S, Ono K, Otani T, Yamaguchi F, Une T. Phase I study and clinical pharmacological study of Muroctasin, Arzneimittel Forschung. Drug Res 1988; 38(II):1043.

5. Parant M, Chedid L. Macrophage activation by muramyl peptides. In: Biological properties of peptidoglycan. Berlin: Walter de Gruyter, 1986:353.
6. Ivanova VT, Andronova TM, Bezrukov MV, Rar VA, Makarov EA, Kozmin SA, Astapova MV, Barkova TI, Nesmeyanov VA. Bacterial cell wall glycopeptides: Synthesis, conformation and antitumor activity. Pure Appl Chem 1987; 59:317.
7. Nesmeyanov VA, Khaidukov SV, Komaleva RL, Andronova TM, Ivanov VT. Muramylpeptides augment expression of Ia-antigens on mouse macrophages. Biomed Sci 1990; 1:151.
8. Tanaka K, Yoshioka T, Bieberich C, Jay G. Role of the major histocompatibility complex class I antigens in tumor growth and metastasis. Ann Rev Immunol 1988; 6:359.
9. Schreiber H, Ward PL, Rowley DA, Stauss HJ. Unique tumor-specific antigens. Ann Rev Immunol 1988; 6:465.
10. Kleinerman ES, Murray JL, Snyder JS, Cunningham JE, Fidler IJ. Activation of tumoricidal properties in monocytes from cancer patients following intravenous administration of liposomes containing muramyl tripeptide phosphatidylethanolamine. Cancer Res 1989; 49:4665.
11. Hammerling GJ, Klar D, Pulm W, Momburg F, Moldenhauer G. The influence of major histocompatibility complex class I antigens on tumor growth and metastasis. Biochim Biophys Acta 1987; 907:245.
12. Silverman DHS, Krueger JM, Karnovsky ML. Specific binding sites for muramyl peptides on murine macrophages. J Immunol 1986; 136:2195.
13. Tenu JP, Adam A, Souvannovong V, Yapo A, Petit J-F, Douglas K. Photoaffinity labelling of macrophages and B-lymphocytes using 125-iodine-labelled aryl-azide derivatives of muramylpeptides. Int J Immunopharmacol 1989; 11:653.
14. Sumaroka MV, Litvinov IS, Khaidukov SV, Golovina TN, Kamraz MV, Komaleva RL, Andronova TM, Makarov EA, Nesmeyanov VA, Ivanov VT. Muramyl peptide binding sites are located inside target cells. FEBS Lett 1991; 295:48.

23

Effect of Liposome-Encapsulated Muramyl Tripeptide Phosphatidylethanolamine and Interferon-Gamma on *Klebsiella pneumoniae* Infection in T-Cell–Deficient Mice and In Vitro

Timo L. M. ten Hagen, Wim van Vianen, Elma A. T. Straathof, and Irma A. J. M. Bakker-Woudenberg
Erasmus University, Rotterdam, The Netherlands

I. INTRODUCTION

It is well known from clinical experience that antibiotic treatment of severe infections in immunocompromised patients is not always successful. Several factors may contribute to this lack of success. One important factor is the failure of the host defense mechanism to provide adequate support for antibiotic treatment at the site of infection. Administration of agents that stimulate the antimicrobial resistance might improve treatment of infections in the immunocompromised host (1–3). In this respect, it is of great importance to stimulate the nonspecific host defense, in particular the cells of the mononuclear phagocyte system (MPS). Immunomodulating agents, both synthetic compounds like muramyl peptide derivatives and natural compounds like cytokines, have been shown to be useful (4–6). Muramyl peptides have been found to be highly active in inducing tumoricidal (7–10), antiviral (11–14), and bactericidal (12,15,16) activities by macrophages. The precise mode of action by which muramyl peptides exert their protective effect is not completely understood. It has been shown that muramyl peptides activate macrophages in terms of increased oxygen and nitrogen intermediate production, enhanced major histocompatibility class (MHC) II expression, and enhanced monokine and enzyme production (17–20). It also was shown that administration of muramyl peptides resulted in an increased number of leukocytes in the blood of uninfected mice (16,21,22), which could compensate for the leukopenia in infected mice (12,16).

Interferon gamma (IFN-γ) is a cytokine mainly produced by activated T helper 1 cells (in mice) that has a broad, predominant proinflammatory spectrum of biological activities (23), such as antiviral activity, activation of macrophages, action of B-cell differentiation, and action on T lymphocytes and natural killer (NK) cells. The activation of macrophages consists of increased production of enzymes and toxic intermediates, enhanced MHC class II expression, enhanced bactericidal and tumoricidal activity (24).

Previous studies in our laboratory showed that treatment of mice with muramyl tripeptide phosphatidylethanolamine (MTPPE) and/or IFN-γ resulted in improvement of resistance to infection with *Listeria monocytogenes* and *Klebsiella pneumoniae* (15). Liposomal encapsulation of these agents not only reduced the toxicity of the drug by preventing exposure to the system (25) but also increased the efficacy by targeting of the agents to the MPS, which allow lower doses (15). In the treatment of infections induced by intravenous injection of *Klebsiella pneumoniae* with liposomal MTPPE (LE-MTPPE), a twofold lower dose was sufficient to obtain the same effect as the immunomodulator in the free form (15). Owing to encapsulation of IFN-γ (LE–IFN-γ) or MTTPE (LE-MTPPE) in liposomes, a 33- or 66-fold lower dose, respectively, was needed in the treatment of *L. monocytogenes* infection in mice to obtain the same results as with the agents in the free form (15). Coencapsulation of MTPPE and IFN-γ in the same liposome showed a synergistic effect of the two immunomodulators in the survival of mice from *L. monocytogenes* infection. Exposure of monolayers of peritoneal macrophages to the two immunomodulators simultaneously resulted in killing of intracellular *L. monocytogenes* (15), whereas exposure of the macrophages to MTPPE alone had no effect on the intracellular bacterial survival, and exposure to IFN-γ resulted only in inhibition of bacterial growth. The use of liposome-encapsulated immunomodulators, which target to cells of the MPS, seems straightforward when infection with intracellular bacteria like *L. monocytogenes* is involved. An important observation is that in a model of *K. pneumoniae* septicemia, we found that survival from infection with gram-negative extracellular bacteria could be increased by administration of liposome-encapsulated MTPPE (15). In the present study, it was investigated whether the observed increased survival of mice could be reproduced in T-cell–deficient mice. This model is highly clinically relevant, as immunocompromised patients often develop septicemia from a local infection. To extend the in vivo observations, and to evaluate possible immunomodulator formulations for therapeutic use, we investigated the effect in vitro of liposome-encapsulated MTPPE and IFN-γ on peritoneal macrophages in monolayers infected with *K. pneumoniae*.

II. MATERIALS AND METHODS

A. Animals

Specific pathogen–free, 11- to 13-week-old female C57Bl/Ka mice were used (ITRI-TNO, Rijswijk, The Netherlands).

B. Bacteria

K. pneumoniae, capsular serotype 2 (ATCC 43816) was used.

C. Reagents

Liposome-encapsulated muramyl tripeptide phosphatidylethanolamine (LE-MTPPE) and placebo liposomes (ELs) were kindly provided by Ciba Geigy Ltd. (Basel, Switzerland). Recombinant rat IFN-γ was kindly provided by Dr. P. van der Meide (ITRI-TNO). Liposomes containing IFN-γ (LE–IFN-γ) were prepared by shaking dry lipid lyophilisate with phosphate-buffered saline (PBS) containing IFN-γ, after which free IFN-γ was removed. The liposomes consisted of phosphatidylcholine and phosphatidylserine in a molar ratio of 7:3. At 7 days before inoculation with bacteria, treatment with LE-MTPPE was started and continued every 48 h, with the last dosage being administered at 24 h before bacterial inoculation. Monoclonal antibody (MAb) 17A2, directed against CD3, was kindly provided by Ann Vossen (Department of Immunology, Erasmus University, Rotterdam, The Netherlands). Cyclosporin A was obtained from Sandoz Pharma AG (Basel, Switzerland). Fluorescein isothiocyanate (FITC) was obtained from Sigma (St. Louis, MO).

D. Induction of T-Cell Deficiency

T-cell depletion was induced by intraperitoneal injection with MAb 17A2 at a single dose of 5 mg/kg at 8 days before bacterial inoculation. T-cell deficiency was induced by subcutaneous injection every other day of 100 mg/kg cyclosporin A in olive oil starting 8 days before bacterial inoculation of mice and continued for the period of the experiment.

E. Experimental Infection Caused by *K. pneumoniae*

Infections were induced by intraperitoneal inoculation of 10^3 colony-stimulating units (CFU) of *K. pneumoniae* into C57Bl/Ka mice. Survival was examined until 21 days after bacterial inoculation.

F. Monolayers of Peritoneal Macrophages Infected with *K. pneumoniae*

Monolayers of peritoneal macrophages from C57Bl/Ka mice were cultured at 37°C in chamberslides (Miles Laboratories, Inc., Naperville, IL) in culture medium containing Dulbecco's modified Eagle's medium and 15% fetal bovine serum. Cultures were exposed 24 h before inoculation with bacteria to LE-MTPPE, LE–IFN-γ or placebo liposomes, or to PBS. At zero time, bacteria were added in a ratio of 15 bacteria per macrophage. Uptake of *K. pneumoniae* by peritoneal macrophages was determined by counting the amount of intracellular fluorescein isothiocyanate (FITC)–labeled bacteria. The killing of *K. pneumoniae* by peritoneal macrophages was determined by preparing 10-fold serial dilutions of lysed macrophage cell suspensions.

G. Statistical Analysis

Statistical evaluation of differences in survival curves between different groups of mice was performed by using the log rank test. The Mann-Whitney test was used for evaluation of the differences in the number of intracellular bacteria between LE-MTPPE, LE–IFN-γ, and EL treated monolayers and for evaluation of killing of bacteria by macrophages.

III. RESULTS

The effect of treatment with LE-MTPPE on the survival of immunocompetent and immunodeficient mice from *K. pneumoniae* infection is shown in Figure 1. Repeated administration of LE-MTPPE at 3.1 μg increased the survival rate of immunocompetent mice from 0 to 67%. Repeated administration with placebo liposomes had no effect (data not shown). The protective effect of LE-MTPPE observed in the immunocompetent mice was investigated in T-cell–deficient mice. Mice were treated intraperitoneally with 100 μg of anti-CD3 MAb 17A2 to eliminate the T cells. Treatment with MAb 17A2 was shown to deplete the blood completely from T cells within 1 day and to significantly reduce T cells in the spleen. It also significantly delayed rejection of skin grafts on mice (data not shown). In these mice, which were immunodeficient because of T cell depletion, treatment with LE-MTPPE enhanced the survival significantly (Fig. 1), although significantly less compared with immunocompetent mice. In mice treated with 100 mg/kg cyclosporin A, the survival of mice was reduced to the level of the controls.

The effect of LE-MTPPE and LE–IFN-γ on the uptake and intracellular killing of *K. pneumoniae* by peritoneal macrophages in monolayers exposed to LE-MTPPE, LE–IFN-γ, EL, or PBS in vitro is shown in Table 1 and Figure 2, respectively. The uptake of *K. pneumoniae* was examined by counting intracellular FITC-labeled bacteria. As shown in Table 1, exposure of peritoneal macrophages to the immunomodulators did not have any effect on the uptake of bacteria significantly different from the controls. As shown in Figure 2, although a slight decrease in the amount of intracellular bacteria is seen due to exposure to the immunomodulators in the first 2 h after infection, eventually no significant differ-

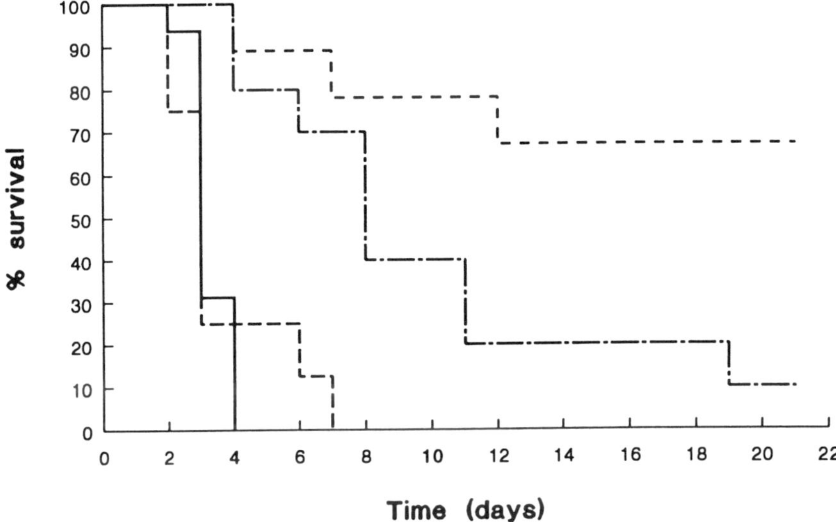

Figure 1 Effect of repeated prophylactic administration of 3.1 μg liposome-encapsulated MTPPE on the survival of immunocompetent and immunodeficient mice from *K. pneumoniae* infection. Mice were intraperitoneally inoculated with 10^3 viable bacteria at zero time. At 168, 120, 72, and 24 h before bacterial inoculation immunocompetent (- - - - -) and immunodeficient mice were intravenously treated with 3.1 μg LE-MTPPE. Immunodeficiency was induced by intraperitoneal injection of 5 mg/kg of antiCD3, 24 h before treatment with LE-MTPPE was started (- — - —) or by subcutaneous injections of 100 mg/kg of cyclosporin A at 48-h intervals for the period of the experiment starting 24 h before treatment with LE-MTPPE was started (- - - -). Immunocompetent control mice were treated with PBS instead of LE-MTPPE (——).

ences were observed compared with the PBS-treated controls. At 4 h after infection of the monolayers in all groups, the intracellular bacterial number increased dramatically.

IV. DISCUSSION

Treatment of severe infections in immunocompromised patients is often complicated and difficult. Owing to the impaired host immune system, failure of antibiotic treatment frequently occurs. Developments in the field of cancer therapy and transplantation have resulted in an increased number of patients with severe, often lethal, opportunistic infections. Therefore, it is important to develop therapies that enhance the host defense and support adequate treatment with antibiotics. Research done by our group with MTPPE and LE-MTPPE showed that treatment of mice with infections by *L. monocytogenes* or *K. pneumoniae* resulted in a dramatic decrease

Table 1 Effect of Liposome-Encapsulated MTPPE (LE-MTPPE) and Liposome-Encapsulated IFN-γ (LE-IFN-γ) on the Uptake of *Klebsiella pneumoniae* by Peritoneal Macrophages

Treatment[a]	No. of bacteria/cell	Percentage of cells containing bacteria at times after infection[b]	
		10 min	30 min
LE-MTPPE	1	93	85
	2	5	7
	2–5	2	4
	>5	1	4
LE-IFN-γ	1	95	90
	2	3	5
	2–5	1	3
	>5	1	2
EL	1	95	89
	2	3	6
	–5	2	3
	>5	1	2
PBS	1	91	78
	2	4	10
	2–5	3	7
	>5	1	5

[a] Monolayers of peritoneal macrophages were exposed to 0.25 μg LE-MTPPE/ml, 31 U LE–IFN-γ/ml, placebo liposomes (EL), or PBS for a period of 24 h before infection with *K. pneumoniae*.
[b] On each slide 400 cells were microscopically screened on the uptake of intracellular FITC-labeled bacteria.

of bacterial numbers in organs and blood. In the present study, we investigated whether activation of the cells of the MPS could increase the resistance of immunocompromised mice toward a naturally acquired septicemia due to *K. pneumoniae*.

An increased resistance in various models of *K. pneumoniae* infection due to treatment with MDP, LE-MTPPE, or other derivatives was observed by several investigators (15,26–32), although the precise mechanism is not entirely understood. In previous studies, we observed that depletion of tissue macrophages abrogated the effect on survival induced with LE-MTPPE, whereas cyclophosphamide-induced inhibition of recruitment of leukocytes did not seem as important (16). In the present study, it was observed that repeated treatment with LE-MTPPE enhanced the resistance of mice against a naturally acquired septicemia by *K. pneumoniae*. Mice

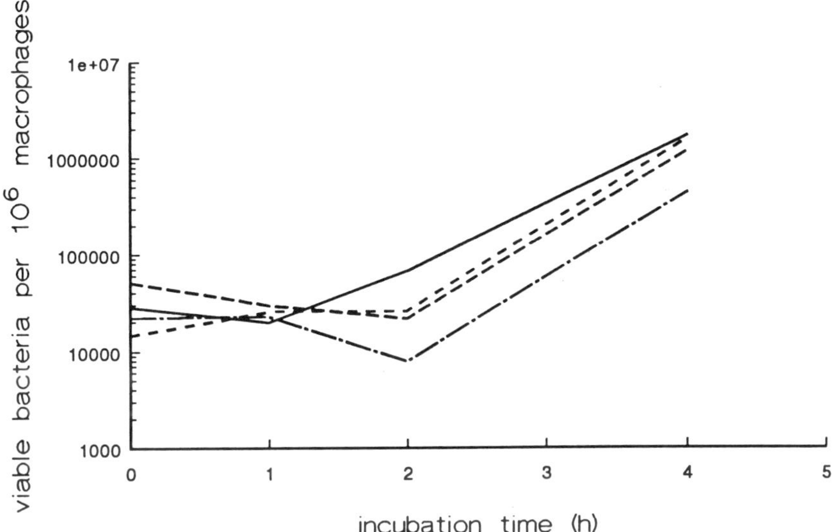

Figure 2 Effect of liposome-encapsulated MTPPE and liposome-encapsulated IFN-γ on killing of intracellular *K. pneumoniae* by peritoneal macrophages. Monolayers of macrophages were exposed for a period of 24 h before infection to 0.25 μg liposome-encapsulated MTPPE per milliliter (– – – –), 31 U liposome-encapsulated IFN-γ milliliter (– — – —), placebo liposomes (-------), or PBS (———).

depleted of their T cells also showed increased survival as a result of treatment with LE-MTPPE but not as pronounced as in immunocompetent mice. However, no increase in infection resistance after administration of LE-MTPPE was seen in mice rendered immunodeficient with cyclosporin A and consequently all mice died. These results clearly show that the activation of macrophages by LE-MTPPE is T-cell dependent. Observations made by Melissen et al. (15) indicated that only therapeutic administration with LE-MTPPE resulted in an enhanced survival of mice with a *K. pneumoniae* infection. It also was shown that the best results were achieved when treatment was started at least 24 h before inoculation. These results, and the observations in T-cell–deficient mice, suggest that communication between macrophages and T cells after treatment with LE-MTPPE is necessary to get optimal macrophage activation. The discrepancy between cyclosporin A– and MAb 17A2–treated mice is probably due to the broader spectrum of activity of cyclosporin A.

To get a better understanding of the results obtained in vivo, we performed in vitro experiments with monolayers of macrophages infected with *K. pneumoniae* after being exposed to the immunomodulators. The present study shows that exposure of macrophages to the immunomodulators did not influence bacterial uptake or killing. Melissen et al. (15) showed that exposure of macrophages to IFN-γ in the free form arrested growth of intracellular *L. monocytogenes,* whereas exposure of monolayers of macrophages to MTPPE and IFN-γ in the free form simultaneously resulted in killing of the intracellular bacteria (15). Also, Hockertz et al. (33) demonstrated a synergistic antiparasitic effect against *Leishmania donovani,* after incubation of macrophages with a combination of MTPPE and IFN-γ coencapsulated in liposomes. The discrepancy noticed between our results and the studies done by Melissen et al. (15) and Hockertz et al. (33) with *L. monocytogenes* and *Leishmania donovani* is most likely due to the nature of the bacteria. *L. monocytogenes* is an intracellular organism that is protected against killing by macrophages resulting in multiplication of intracellular bacteria. Therefore, it might be possible that when the macrophages become activated by exposure to immunomodulators, this bacterium is not equipped to survive the assault by the macrophages. Most of the research done on the effect of immunomodulators on killing by cultured macrophages is performed with intracellular parasites. *K. pneumoniae* organisms on the other hand are not intracellular surviving bacteria. These bacteria are poorly phagocytized in nonimmune serum by cells of the MPS. Only when opsonized with specific antibodies and complement, phagocytosis and killing of *K. pneumoniae* by the cells of the MPS occurs. However, our results in vivo show that mice can be protected by immunomodulators that mainly exert their effect on cells of the MPS. This indicates that activation of the cells of the MPS by LE-MTPPE can enhance uptake and killing of *K. pneumoniae* even in the nonimmune host. Whereas the cells of the MPS play an important role in the survival from an infection, both our in vivo results as well as our in vitro data indicate that T cells are necessary for LE-MTPPE–induced macrophage activation. Therefore, it may be concluded that the effect of LE-MTPPE on the cells of the MPS is not a direct activation of these cells.

REFERENCES

1. Drews J. The experimental and clinical use of immunomodulating drugs in the prophylaxis and treatment of infections. Infection 1984; 12:157–66.
2. O'Grady F. The second L.P. Garrod lecture. Strategies for potentiating chemotherapy in severe sepsis: Some experimental pointers. J Antimicrob Chemother 1984; 13:535.
3. Parant M. Host resistance against bacterial infections, immunomodulation by antimicrobials and synthetic immunoenhancers. In: Escobar MR, Litz JP, eds. The reticuloendothelial system, a comprehensive treatise. New York: Plenum Press, 1988:85.
4. O'Reilly T, Zak O. Enhancement of the effectiveness of antimicrobial therapy by muramyl peptide immunomodulators. Clin Infect Dis 1992; 14:1100–9.
5. Dukor P, Schumann G. Modulation of non-specific resistance by MTPPE. In: Majde JA, ed. Immunopharmacology of infect diseases. Vaccine adjuvants and modulators of non-specific resistance of infectious diseases. New York: Alan R. Liss, 1987:255.

6. Vogels MTE, Van der Meer JWM. Use of immune modulators in nonspecific therapy of bacterial infections. Antimicrob Agents Chemother 1992; 36(1):1–5.
7. Phillips NC, Riuox J, Tsao MS. Activation of murine Kuppfer cell tumoricidal activity by liposomes containing lipophilic muramyl peptides. Hepatology 1988; 8:1046–50.
8. Phillips NC, Tsao MS. Inhibition of experimental liver tumor growth in mice by liposomes containing a lipophylic muramyl dipeptide derivative. Cancer Res 1989; 49:936.
9. Reisser D, Jeannin JF, Lagadec P, Martin F. Comparative effect of muramyl dipeptide in vivo and in vitro on the tumoricidal activity of rat peritoneal macrophages. J Biol Respir Modif 1985; 4:460–3.
10. Sone S. Human monocyte activation to the tumoricidal state by liposome-encapsulated muramyl tripeptide and its therapeutic implications. In: Lopez-Berestein G, Fidler IJ, eds. Liposomes in the therapy of infectious disease and cancer. New York: Alan R. Liss, 1989:125.
11. Dietrich FM, Hochkeppel HK, Lukas B. Enhancement of host resistance against virus infections by MTP-PE, a synthetic lipophilic muramyl peptide-I. Increased survival in mice and guinea pigs after single drug administration prior to infection, and the effect of MTP-PE on interferon levels in sera and lungs. Int J Immunopharmacol 1986; 8:931–42.
12. Izbicki JR, Readler C, Anke A, et al. Beneficial effect of liposome-encapsulated muramyl-tripeptide in experimental septicemia in a porcine model. Infect Immun 1991; 59:126–30.
13. Masihi KN, Kröger H, Lange W, Chedid L. Muramyl peptides confer hepatoprotection against murine viral hepatitis. Int J Immunopharmacol 1989; 57:2495–501.
14. Masihi KN, Lange W, Rohde-Scholz B, Chedid L. Muramyl dipeptides inhibit replication of human immunodeficiency virus *in vitro*. AIDS Hum Retroviruses 1990; 6:393–9.
15. Melissen PMB, van Vianen W, Rijsbergen Y, Bakker-Woudenberg IAJM. Free versus liposome-encapsulated muramyl tripeptide phosphatidylethanolamine in treatment of experimental *Klebsiella pneumoniae* infection. Infect Immun 1992; 60:95–101.
16. Melissen PMB, van Vianen W, Bakker-Woudenberg IAJM. Roles of peripheral leukocytes and fixed tissue macrophages in the antibacterial resistance induced by free or liposome encapsulated phosphatidylethanolamide (MTPPE). Infect Immun 1992; 60:4891.
17. Imai K, Tanaka A. Effect of muramyl dipeptide, a synthetic bacterial adjuvant, on enzyme release from cultured mouse macrophages. Microbiol Immunol 1981; 25:51–62.
18. Broudy VC, Kaushansky K, Shoemaker SG, Aggerwal BB, Adamson JW. Muramyl dipeptide induces production of hemopoietic growth factors *in vivo* by a mechanism independent of tumor necrosis factor. J Immunol 1990; 144:3789–94.
19. LeClerc C, Chedid L. Macrophage activation by synthetic muramyl peptides. Lymphokines 1982; 7:1–21.
20. Mehta K, Juliano RL, Lopez-Berestein G. Stimulation of macrophage protease secretion via liposomal delivery of muramyl dipeptide derivatives to intracellular sites. Immunology 1984; 51:517–27.

21. Wachsmuth ED. Stimulation of cell proliferation in rabbits by MTP-PE, a lipophilic, muramyl dipeptide or lipopolysaccharide on superoxide anion-generating activities of macrophages. Infect Immun 1984; 45:82–6.
22. Yamaguchi F, Akasaki M, Tsukada W. Induction of colony-stimulating factor and stimulation of stem cell proliferation by injection of muroctasin. Arzneim Forsch Drug Res 1988; 38(II);nr. 7a:980–6.
23. Yzermans JNM, Marguet RL. Interferon-gamma: A review. Immunobiology 1989; 179:456–73.
24. Murray HW. Interferon-gamma, the activated macrophage, and host defence against microbial challenge. Am Intern Med 1989; 106:595–608.
25. Schumann GP, van Hoogevest P, Fankhauser P, et al. Comparison of free and liposomal MTPPE: Pharmacological, toxicological and pharmacokinetic aspects. In: Lopez-Berestein G, Fidler IJ, eds. Liposomes in the therapy of infectious disease and cancer. New York: Alan R. Liss, 1989:191.
26. Ausobsky JR, Scuitoo M, Trachtenberg LS, Polk HC, Jr. The role of muramyl dipeptide in the therapy of established experimental bacterial infection. Br J Exp Pathol 1984; 65:1–9.
27. Chedid LA, Parant MA, Audibert FM, et al. Biological activity of a new synthetic muramyl peptide adjuvant devoid of pyrogenicity. Infect Immun 1982; 35:417–24.
28. Coune A. Liposomes as drug delivery system in the treatment of infectious diseases. Potential applications and clinical experience. Infection 1988; 16:141–6.
29. Galland RB, Trachtenberg LS, Rynerson N, Polk HC, Jr. Nonspecific stimulation of host defences against bacterial challenge in immunosuppressed mice. Arch Surg 1983; 118:161–4.
30. Parant MA, Audibert FM, Chedid L, et al. Immunostimulant activities of a lipophilic muramyl dipeptide derivative and of desmuryl peptidolipid analogs. Infect Immun 1980; 27:826–31.
31. Polk HC, Jr, Galland RB, Ausobsky FR. Nonspecific enhancement of resistance to bacterial infection. Ann Surg 1982; 196:436–41.
32. Polk HC, Jr., Lamont PM, Galland RB. Containment as a mechanism of non-specific enhancement of defences against bacterial infection. Infect Immun 1990; 58:1807–11.
33. Hockertz S, Franke G, Kniep E, Lohmann-Matthes ML. Mouse interferon-γ in liposomes: Pharmacokinetics, organ-distribution, and activation of spleen and liver macrophages in vivo. J Interferon Res 1989; 11:177.

24

In Vitro Stimulatory Effects of Adamantylamide Dipeptide on Macrophage Enzyme Activities (Beta-Glucuronidase and Nitric Oxide Synthase)

Z. Zídek and K. Mašek
Czech Academy of Sciences, Prague, Czech Republic

I. INTRODUCTION

In the search for immunomodulatory compounds from a group of des-muramyl peptides, more than a decade ago we have synthesized a compound in which the essential moiety of muramyl dipeptide, L-alanine-D-isoglutamine, was linked to 1-aminoadamantane (1). This compound has been shown to possess a plethora of immune properties, including effects on both humoral and cellular immunity; namely, it increases the number of plaque-forming cells in sheep red blood cell (SRBC)–immunized animals, augments graft-versus-host disease (GvH) reaction, delayed-type hypersensitivity (DTH) response, and T-cell–dependent vascular permeability, and stimulates the production of tumor necrosis factor (TNF) alpha (TNF-α) and oxidative burst of macrophages but also possesses antiviral activity (2–8).

In this chapter, we describe some novel findings of the immunotropic activity of adamanthylamide dipeptide (AdDP); namely, its ability to stimulate the production of microbicidal/tumoricidal nitric oxide and macrophage beta-glucuronidase activity, an enzyme that is believed to be related to antigen processing.

II. MATERIALS AND METHODS

A. Animals and Cell Cultures

Specific pathogen–free female mice of the inbred strain CB7BL/1ONCrlBR (BL/10) or Balb/cAnNCrlBR (Balb/c) were purchased from the Charles River Wiga GmbH (Sulzfeld, Germany).

Resident peritoneal macrophages were obtained by rinsing the cavity with phosphate-buffered saline (PBS) containing 20 U/ml heparin. The cells were washed several times and then left to adhere for 2 h in the wells of Nunc (Roskilde, Denmark) 96-well tissue culture microplates.

Spleen mononuclear cells were obtained by a standard gradient centrifugation procedure using Lympho-paque (Nyegaard, Oslo, Norway).

The final concentration of the cells was always 2×10^6/ml. Culture medium contained gentamicin (0.005%), 2-mercaptoethanol (5×10^{-5} M) (Sigma, St. Louis, MO) and 10% FBS (Flow Labs, Irvine, CA). The cultures were run for 24 h in a Hereaus incubator (Hanau, Germany) (5% CO_2, 100% relative humidity).

B. Assay for Beta-Glucuronidase Activity

Cells were resuspended in a phenol-free RPMI 1640 medium (Whittaker, MD) containing 1mM p-nitrophenyl-beta-D-glucuronide (Sigma). Concentration of p-nitrophenyl, a metabolism product of beta-glucuronidase activity, was read at 405 nm using Uniskan II spectrophotometer (Labsystems, Helsinki Finland).

C. Determination of Nitrite Production

Cells were resuspended in RPMI 1640 medium (USOL, Prague, Czech Republic) containing 1.2 nM arginine. The content of nitrites, reflecting the nitric oxide (NO) generation, was assayed using Griess reagent. The reaction was read on a Uniskan II spectrophotometer at 540 nm.

D. Agents Tested

AdDP and muramyl dipeptide (MDP) were of our own production (Dr. M. Flegel). Interferon gamma (INF-γ) was purchased from Genzyme (Cambridge, MA), concanavalin A (ConA) from Aldrich (Steinheim, Germany) and lipopolysaccharide (LPS) (*Salmonella typhimurium*) from Sigma. They all were prepared in the complete RPMI 1640 medium. Their concentration in the cultures is to be found in Figures 1–4. LPS was always added 2 h following other agents used.

E. Statistical Analysis

Analysis of variance and Dunnett's *t*-test were employed for determination of statistically significant changes.

III. RESULTS

A. Beta-Glucuronidase Activity

Both AdDP and MDP were able to enhance the statistically highly significantly production of p-nitrophenol, a metabolism product of p-nitrophenyl-beta-D-glucuronide in both BL/10 and Balb/c peritoneal macrophages (Figs. 1 and 2). The effect

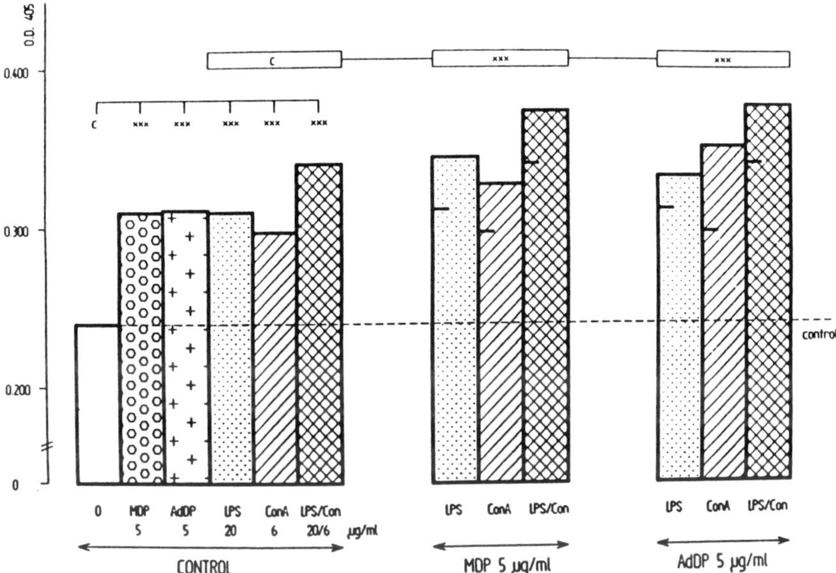

Figure 1 Beta-glucuronidase activity of resident peritoneal macrophages of Balb/c mice after 24-h cultivation.

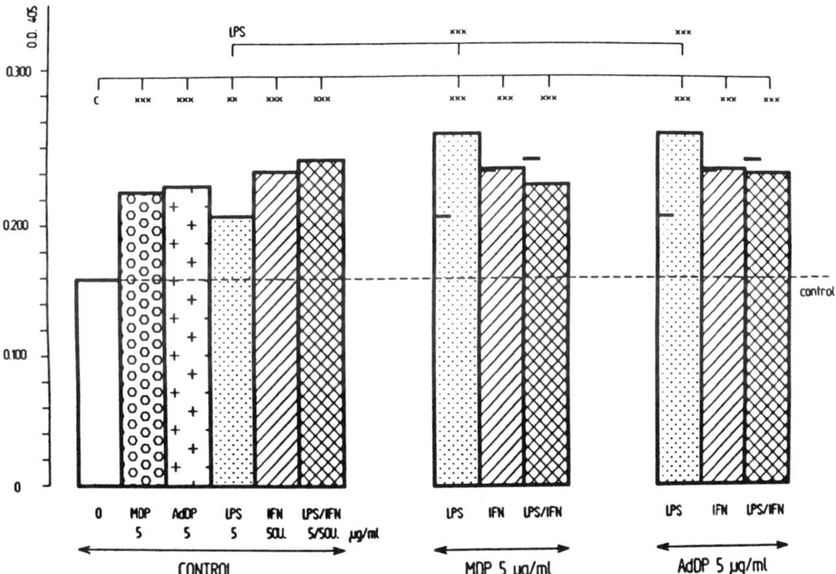

Figure 2 Beta-glucuronidase activity of resident peritoneal macrophages of C57BL/10 mice after 24-h cultivation.

was very similar to that produced by more complex stimulatory agents, LPS, ConA, or IFN-γ. It was interesting to observe that the effects of AdDP as well as of MDP were additive to the effects of LPS and ConA, even to their combination (Figs. 1 and 2), but not to the enhancing effect of IFN-γ (see Fig. 2).

B. Nitric Oxide Production

Concentration of nitrites in supernatants of macrophages cultured in the presence of AdDP or MDP was significantly enhanced but did not reach the values observed after LPS, IFN-γ, or LPS/IFN-γ combination. Neither of the immunomodulators was able to augment production of nitrites produced by IFN or IFN-LPS (Fig. 3). Production of nitrites by AdDP or MDP was much more pronounced in mononuclear splenocyte cultures (Fig. 4) than in peritoneal macrophages. The effect of both AdDP and MDP was even more prominent than the effect of LPS or ConA. After the combination of either of the immunomodulatory agents with LPS or ConA, the effect was comparable to the effect of MDP or AdDP themselves, with the effect of AdDP being even slightly higher than the effect of MDP (see Fig. 4).

IV. DISCUSSION

The results of experiments presented in this study have extended our knowledge on AdDP as a potent activator of immunocompetent cell functions. It has been shown that with respect to the enzymatic reactions investigated, AdDP is fully comparable with the potent molecule of MDP.

It is noteworthy that they both are equally potent as compared with LPS, ConA, and IFN-γ as regards the stimulation of macrophage beta-glucuronidase activity, but they are less effective in stimulating macrophage nitrite production. On the other hand, splenocyte mononuclear cells were more responsive to AdDP and MDP than to LPS or ConA. This difference could imply that activation of macrophages is most probably directly achieved by these agents in the case of beta-glucuronidase activity, but it probably needs a confluential participation of lymphocytes to reach a considerable production of reactive nitrogen radicals.

The clear-cut potential of AdDP to stimulate the activity of the enzymes beta-glucuronidase and NO synthase warrants further studies to ascertain whether this immunomodulatory agent does have tumoricidal and microbicidal effects under the conditions of in vivo host-mediated assays. Such experiments are currently underway.

V. CONCLUSIONS

Activity of beta-glucuronidase and NO synthase were determined under in vitro conditions in mouse resident peritoneal macrophages or splenocyte mononuclear cells. AdDP proved to be a potent stimulator of these immune-related enzymes. The

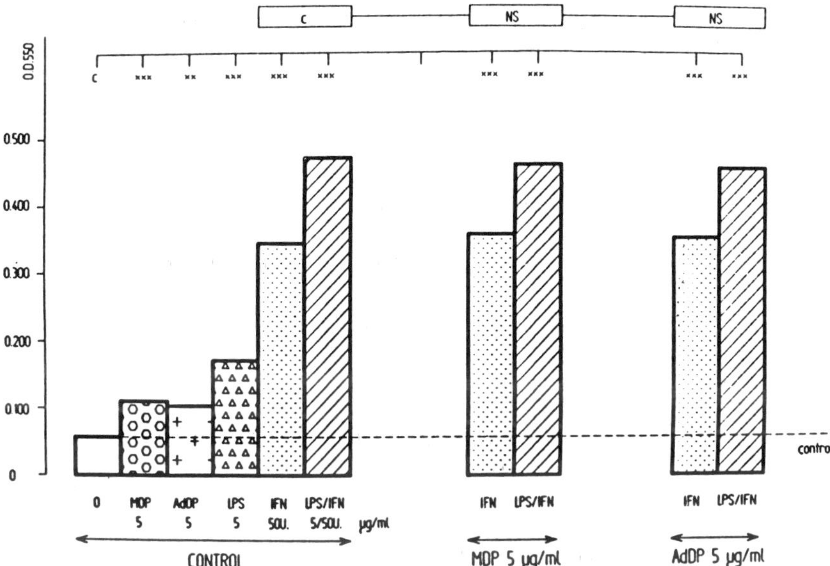

Figure 3 Concentration of nitrites in 24-h cultures of resident peritoneal macrophages (C57BL/10 mice).

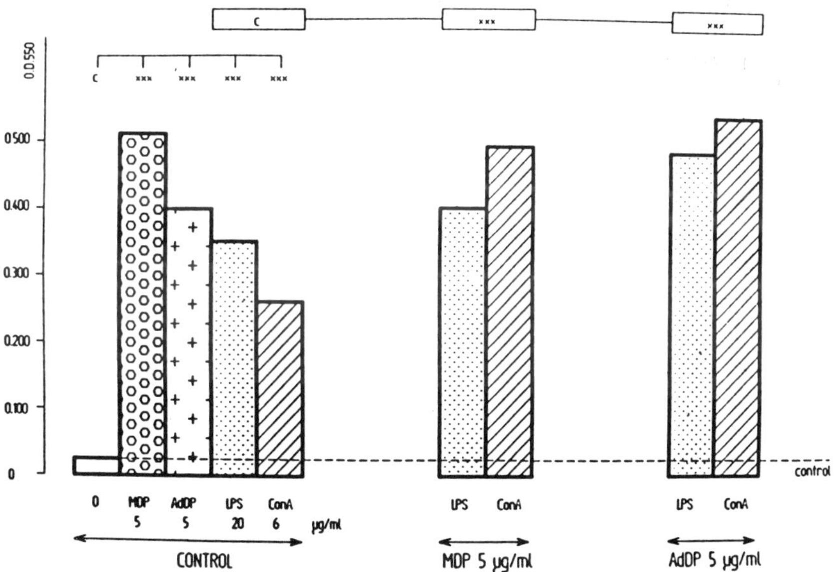

Figure 4 Concentration of nitrites in 24-h cultures of spleen mononuclear cells (C57BL/10 mice).

possible significance of these effects has been discussed from the point of view of bactericidal and/or tumoricidal potential of AdDP in host-mediated assays.

REFERENCES

1. Mašek K, Seifert J, Flegel M, Krojidlo M, Kolínský J. The immunomodulatory property of a novel synthetic compound adamantylamide dipeptide. Methods Find Exp Clin Pharmacol 1984; 6:667.
2. Mašek K. Immunopharmacological properties of synthetic peptides and their analogs. Adv Biosci 1988; 68:11.
3. Zídek Z, Mašek K, Svobodová M, Šůla K, Nouza K. Müller J. Some immunopharmacological properties of adamanthylamide dipeptide (AdDP). In: Masihi KN, Lange W, eds. Immunotherapeutic prospects of infectious diseases. Berlin: Springer-Verlag, 1990:363.
4. Šůla K, Zídek Z, Nouza K. Effects of adamantylamide dipeptide (AdDP) on cell-mediated immunity in vivo. In: Masihi KN, Lange W, eds. Immunotherapeutic prospects of infectious diseases. Berlin: Springer-Verlag, 1990:217.
5. Müller J, Nouza K, Macákova A, Šůla K, Zídek Z. Ex vivo and in vitro effects of adamantylamide dipeptide (AdDP) on macrophages and lymphocytes. In: Masihi KN, Lange W, eds. Immunotherapeutic prospects of infectious diseases. Berlin: Springer-Verlag, 1990:211.
6. Zídek Z, Loučková M, Šůla K, Nouza K, Mašek K. Effects of adamantylamide dipeptide (AdDP) on humoral immunity in mice. Int J Immunother 1991; 7:15.
7. Zídek Z, Mašek K, Franková D, Flegel M. T-cell–dependent immunobiological activity of a desmuramyl analog of muramyl dipeptide, adamantylamide dipeptide (AdDP) and D-isoglutamine. Int J Immunopharmacol 1993; 15: 631.
8. Masihi KN, Lange W, Rohde-Schultz B, Mašek K. Antiviral activity of immunomodulator adamantylamide dipeptide. Int J Immunother 1987; 3:89.

25

Biological Activity of Novel Adamantyl Tripeptides with the Emphasis on Their Immunorestorative Effect

Branka Vranešić, Jelka Tomašić, and Đurdica Ljevaković
Institute of Immunology, Zagreb, Croatia

Ivo Hršak
Ruđer Bošković Institute, Zagreb, Croatia

I. INTRODUCTION

Polymeric peptidoglycans are the ubiquitous constituents of bacterial cell walls that are built of large polysaccharide chains and short peptide units. Peptidoglycans have received increasing attention in the last 15 years because of their remarkable biological activities, particularly as potent immunomodulators. Versatile biological activities of peptidoglycans have been widely investigated and well documented (1). A large number of synthetic novel compounds representing smaller parts of the original peptidoglycan molecule have been synthesized and found to possess immunomodulating properties. It also was shown that a very specific peptide portion of peptidoglycans containing D-glutamic acid is essential for biological activity.

Peptidoglycan monomer (PGM) (produced by PLIVA, Pharmaceutical Works, Zagreb, Croatia) originating from *Brevibacterium divaricatum* is the disaccharide pentapeptide of the following composition: GlcNAc-MurNAc-L-Ala-D-isoglutaminyl-*meso*-diaminopimelyl(εNH_2)-D-Ala-D-Ala (2); it is water soluble, nonpyrogenic and nontoxic. Biological activities of PGM exhibited in various experimental models in mice include stimulation of humoral and cellular immune response, antitumor and antimetastatic activity, and inhibition of drug-metabolizing enzymes in liver. (3).

Muramyl dipeptide (MDP, for N-acetyl-muramyl-L-Ala-D-isoGln) has been identified as the smallest molecule isolated from mycobacteria that acts as biological response modifier (4). MDP (5) and several hundred of its chemically defined analogues and derivatives have been synthesized since in order to modulate, preferably improve, the properties of the parent molecule (6).

Structurally related compounds that have been derived from MDP can roughly be divided into three groups: (1) compounds that have a modified peptide moiety, (2) compounds that have a modified N-acetylmuramyl structure, and (3) desmuramyl peptides.

A very different, but not less interesting, molecule from both PGM and MDP is the rigid diamondlike structure of adamantane. Some compounds containing adamantyl residue exhibit antiviral activity; amantadine, for instance, is currently used in therapy of viral diseases (7). In general, introducing an adamantyl moiety to the substances with known biological activity improves their pharmacological properties and enhances their activity (8). Since the adamantyl moiety is very strange to the mammalian organism, the substances comprising adamantyl structure exhibit high metabolic stability; they are retained much longer in the body and are excreted from the organism almost unchanged. In addition, the very strong lipophilic character of the adamantane moiety enables easier penetration of the adamantane analogues through the phospholipid layer of the cell membrane facilitating the fast transport of the active substance into the cell. The compounds containing the adamantyl structure have been described, for example, as sedatives, antitumor agents, antibiotics, hypoglycemics, antidepressants, and drugs for Parkinson's disease.

In the view of a novel approach of stimulating the host immune response using immunomodulators to enhance resistance against viruses, relatively few efforts have been made, so far, to prepare novel compounds containing both peptidoglycan structures (muramyl dipeptide structure [MDP] in particular) and adamantyl residue. Only a few reports (9–12) concerning the synthesis of a novel adamantylamide dipeptide that consisted of the adamantyl moiety and dipeptide characteristic for MDP have been reported so far.

Our aim was to synthesize novel compounds containing adamantyl residue coupled to the peptide characteristic for peptidoglycan structures (des-muramyl peptide) through C-C bond at position 2 of the adamantane molecule. It was likely to expect that these compounds might exhibit antiviral properties in parallel with immunomodulating and antitumor activity.

The biological potential of des-muramyl peptides and similar compounds, as well as adamantyl-containing compounds, has been widely investigated and proved. But far less data on the immunorestorative activity of such compounds in immunocompromised hosts exist. Therefore, we considered it interesting to study the restorative activity of adamantyl tripeptides in mice immunosuppressed by x-ray irradiation and the chemotherapeutic drug cyclophosphamide.

II. SYNTHESIS OF ADAMANTYL TRIPEPTIDES

The dipeptide L-Ala-D-isoGln (the essential moiety of MDP) was linked to D,L-(2-adamantyl)-Gly yielding two diastereoisomers of (2-adamantyl)-Gly-L-Ala-D-

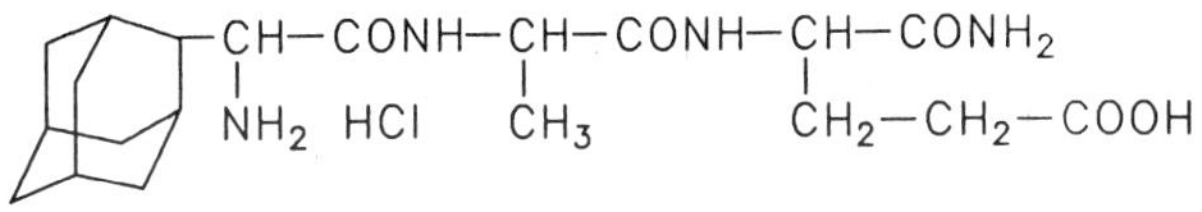

CH—CONH—CH—CONH—CH—$CONH_2$
NH_2 HCl CH_3 CH_2—CH_2—COOH

D-(2-adamantyl)-Gly-L-Ala-D-isoGln
(2-Ad-TP, Isomer 1)

L-(2-adamantyl)-Gly-L-Ala-D-isoGln
(2-Ad-TP, Isomer 2)

Figure 1 Structures of adamantyl peptides.

isoGln regarding adamantyl-Gly moiety. Since the enantiomeric mixture of D,L-(2-adamantyl)-Gly was used for the peptide synthesis, it appeared likely that the two components of the obtained peptide mixture were D-(2-adamantyl)-Gly-L-Ala-D-isoGln (isomer 1) and L-(2-adamantyl)-Gly-L-Ala-D-isoGln (isomer 2) (13,14). Diastereoisomers were easily separated on a Silica gel column and characterized, but only the use of stereospecific enzymes led to the assignation of the adamantyl-glycine moiety configuration in the respective diastereoisomers (Fig. 1) (15).

Both adamantyl tripeptide diastereoisomers are water-soluble, nonpyrogenic, and nontoxic compounds.

III. BIOLOGICAL ACTIVITY OF ADAMANTYL TRIPEPTIDES

A. Cytotoxic and Antiviral Activity

Cytotoxic effect and antiviral activity of novel adamantyl tripeptides, observed as the cell survival and the infectivity reduction of the influenza A virus H1N1 and H3N2 strains have been studied on Madin-Darby canine kidney (MDCK) cells in vitro.

Both isomers have been better tolerated by MDCK cells than amantadine. Four times higher concentrations of the novel compounds than of amantadine could be used without affecting either the morphology of the cells or the capability of the cell growth after one subsequent passage (13,14).

Isomers 1 and 2 reduced the infectivity of the tested influenza A virus strains. But each isomer showed an inhibitory effect at different minimal inhibitory concentration (MIC) and a different potency toward a different strain (13,14).

B. Immunoadjuvant Activity

The immunostimulating properties of isomers 1 and 2 were tested in vivo in rabbits inoculated with human gamma globulin (hGG) in incomplete Freund's adjuvant (IFA). Antibody formation against hGG (assayed immunoradiometrically) was significantly enhanced in animals treated with isomer 2 in comparison with the antibody levels in control group of animals immunized only with hGG in IFA (13,14).

C. The Influence on Macrophages and Leukocyte Functions In Vitro

In order to evaluate the in vitro effect of both isomers on immunocompetent cells, the influence on phagocytic activity of mouse and human phagocytes and human lymphocytes was measured.

The obtained results indicated that both adamantyl tripeptides could activate phagocytosis of mouse peritoneal macrophages and stimulate depressed basal phagocytosis or antibody-dependent cell-mediated cytotoxicity (ADCC) of human polymorphonuclear neutrophils (PMNs) (16).

The ability of tested compounds to stimulate lymphocyte proliferation and to modulate the mitogen-induced proliferation of lymphocytes also was observed (16).

D. Induction of Interferon Production in Human Mononuclear Cells

The influence of both isomers on in vitro interferon (IFN) production in human mononuclear cells was investigated and the results have shown that both compounds induce IFN-α synthesis. The inducing ability of either isomer alone did not differ; that is, antiviral titers in culture supernatants were the same for isomer 1 and isomer 2. But in combination with standard interferon inducers (concanavalin A [ConA], phorbol myristate acetate [PMA], or polyinosinic-polycitydylic acid [poly-I:C]) isomers 1 and 2, respectively, influenced the interferon production to a different extent. When mononuclear cells were stimulated with isomer 1 or 2 in combination with either ConA, PMA, or poly I:C, higher levels of interferon were induced than with each of the mentioned standard inducers alone (17).

E. Antimetastatic Activity

Both isomers were tested in the model of the MCa mammary carcinoma in CBA mice. Tested compounds, and particularly isomer 2 [L-(2-adamantyl)-Gly-L-Ala-D-isoGln] were shown to be more effective in reducing lung metastases than primary tumor growth and to be active in conditions in which a low number of metastases were formed (18). The effects of isomer 1 were much less impressive, which indicated that chirality might play a certain role in the antitumor and antimetastatic activity of adamantyl tripeptides.

The effectiveness of both isomers on lung metastases formation also was tested in C57BL/6 mice bearing B-16 melanoma. Following intravenous administration of isomer 1 as well as isomer 2, the dissemination of metastases to the lungs was significantly reduced to one third of the number detected in the untreated control groups (13,14).

F. Immunorestorative Activity

Following an intensive search for compounds that might exhibit an effect on the immune system, several different methods for evaluating potential new immunotherapeutics have been established. Animal models that are supposed to have at least some predictive value for the clinical outlook include among various methods a model with animals suppressed with chemical immunodepressants or with x-ray irradiation. But the data on immunorestorative activity of the low molecular weight peptidoglycans are rather scarce. Thus, the increase in protection against infections of *Candida albicans* or *Pseudomonas aeruginosa* in mice immunosuppressed by cyclophosphamide was observed after treatment with MDP and several of its derivatives (19). In another experimental model in rats immunosuppressed by sleep deprivation, the decrease in antibody production to sheep red blood cells (SRBCs) could be completely prevented with MDP treatment (20). Also, alveolar and peritoneal macrophages of mice treated with chemotherapeutic doxorubicin (Adriamycin) could be efficiently stimulated to tumoricidal function by liposome-entrapped muramyl tripeptide phosphatidylethanolamine (MTP-PE), a lipophilic derivative of MDP (21). A beneficial effect of romurtide (muroctasin, MDP-Lys [18]), a synthetic MDP derivative, on experimental leukopenia induced by cyclophosphamide or x-ray irradiation in mice was observed as well (22,23).

Our data support and extend these findings. The immunorestorative activity of adamantyl tripeptide diastereoisomers was followed in immunosuppressed mice and was compared with the similar restorative effect of PGM and its synthetic derivative Boc-Tyr-PGM (24).

The immunosuppression in mice was induced by cyclophosphamide (Table 1) and by x-ray total-body irradiation (Table 2).

Both the adamantyl tripeptide diastereoisomers, PGM, and Boc-Tyr-PGM expressed similar activity in enhancing the restoration of antibody production against sheep red blood cells (SRBCs) (assayed by Jerne's PFC test) (25) in experimental animals suppressed by either the cytostatic drug cyclophosphamide or x-ray total-body irradiation. It should be stressed that in irradiated mice with a very pronounced immunosuppression, the adamantyl tripeptides, PGM, and Boc-Tyr-PGM were able significantly to restore the tested immune function of recipient animals.

It has been well recognized that nonspecific host immunostimulation is one of the most important factors in the treatment of immunosuppressed hosts, such as patients with cancer. Consequently, the observed restorative activity might be taken into consideration when planning combined treatment of neoplastic diseases, where efficient antitumor activity is needed but unwanted side effects, that is, immunosuppression, should be avoided.

Table 1 Number of PFCs in the Spleens of Mice Immunosuppressed with Cyclophosphamide[a]

		Treatment		No. of PFCs ($\times 10^3$)		
Experiment	Group	Cyclophosphamide	2-Ad-TP, PGM Boc-Tyr-PGM	Mean	Range	%
A	1	−	−	101.7	87.2–112.8	100
	2	+	−	43.0	29.5–52.4[b]	42
	3	+	2-Ad-TP (Isomer 1)	65.6	51.6–86.8[c]	65
	4	+	2-Ad-TP (Isomer 2)	70.0	54.4–77.6[c]	69
	5	+	PGM	76.2	68.8–94.8[c]	75
B	1	−	−	198.1	132.4–268.4	100
	2	+	−	108.6	85.5–156.4[b]	55
	3	+	2-Ad-TP (Isomer 1)	219.0	144.7–237.5[c]	111
	4	+	2-Ad-TP (Isomer 2)	151.9	96.7–189.6	73
	5	+	PGM	161.7	131.7–208.9[c]	82
	6	+	Boc-Tyr-PGM	171.5	117.7–201.3	87

[a] Groups of five CBA mice were given cyclophosphamide (100 mg/kg) 7 days before immunization with SRBCs and were treated with 2-Ad-TP (isomer 1 and isomer 2), PGM, and Boc-Tyr-PGM (200 μg/mouse) 1 day before immunization.

[b] $P < .05$ in comparison to group 1.

[c] $P < .05$ in comparison to group 2.

Table 2 Number of PFCs in the Spleens of Mice Immunosuppressed by Total Body X-Ray Irradiation[a]

	Treatment		No. of PFCs ($\times 10^3$)		
Group	Irradiation	2-Ad-TP, PGM Boc-Tyr-PGM	Mean	Range	%
1	−	−	136.9	87.2–188.4	100
2	+	−	8.3	6.4–12.5[b]	6
3	+	2-Ad-TP (Isomer 1)	10.1	7.5–29.0[b,c]	11
4	+	2-Ad-TP (Isomer 2)	19.4	11.7–27.3[b,c]	14
5	+	PGM	18.2	11.2–28.3[b,c]	13
6	+	Boc-Tyr-PGM	23.6	8.3–40.7[b,c]	17

[a] Groups of five CBA mice were irradiated (3 Gy) 7 days before immunization with SRBCs and treated with 2-Ad-TP (isomer 1 and isomer 2), PGM, and Boc-Tyr-PGM (200 μg/mouse) 1 day before immunization.

[b] $P < .05$ in comparison to group 1.

[c] $P < .05$ in comparison to group 2.

IV. CONCLUSIONS

It may be said that novel adamantyl tripeptides exhibit both antiviral and immunostimulating properties. Immunostimulation and, even more important, the ability to enhance restoration of immune reactivity in the immunocompromised host is in its magnitude comparable with the effectiveness of low molecular weight peptidoglycans. But observed antiviral activity is less pronounced than the activity of amantadine on H1N1 and H3N2 strains of influenza A virus (13,14).

Therefore, being water soluble and nontoxic, i.e., suitable for parenteral application, with described activities these compounds seem to be appropriate material to be extensively studied for potential treatment of immunodeficient conditions and certain malignancies.

ACKNOWLEDGMENTS

This work was supported in part by a grant from the Commission of the European Communities for Science, Research and Development (CI1*CT90-0711) and the grants from the Ministry of Science, Informatics and Technology of the Republic of Croatia (1-07-044 and 1-08-151).

REFERENCES

1. Seidl PH, Schleifer KH, eds. Biological properties of peptidoglycan. Berlin: W. de Gruyter, 1986.
2. Keglević D, Ladešić B, Tomašić J, Valinger Z, Naumski R. Isolation procedure and properties of monomer unit from lysozyme digest of peptidoglycan complex excreted into the medium by penicillin-treated *Brevibacterium divaricatum* mutant. Biochim Biophys Acta 1979; 585:273.
3. Tomašić J, Hršak I. Peptidoglycan monomer originating from *Brevibacterium divaricatum*—Its metabolism and biological activities in the host. In: Schrinner E, Richmond MH, Seibert G, Schwarz U, eds. Surface structures of microorganisms and their interactions with the mammalian host. Weinheim: VCH, 1987:113.
4. Lederer E. Chemistry and biology of muramyl peptides. Proceedings of VIIIth International Symposium on Medicinal Chemistry. Vol. 1. Uppsala: 1984:13–26.
5. Merser C, Sinay P, Adam A. Total synthesis and adjuvant activity of bacterial peptidoglycan derivatives. Biochem Biophys Res Commun 1975; 66:1316.
6. Adam A, Lederer E. Muramylpeptides as immunomodulators. ISI Atlas of Science: Immunology 1988; 1:205.
7. Oxford JS. Anti-influenza virus activity of amantadine, rimantadine and analogues. In: De Clercq E, Walker RT, eds. Targets for the design of antiviral agents. Series A: Life Sciences, Vol. 73. New York: Plenum Press, 1984:159.
8. Gerzon K, Krumkalns EV, Brindle RL, Marshall FJ, Root, MA. The adamantyl group in medicinal agents. I. Hypoglycemic N-arylsulfonyl-N′-adamantylureas. J Med Chem 1963; 6:760.
9. Masek K, Seifert J, Flegel M, Krojidlo M, Kolinsky J. The immunomodulatory property of a novel synthetic compound adamantylamide dipeptide. Methods Find Exp Clin Pharmacol 1984; 6(11):667.

10. Masihi KN, Lange W, Rohde-Schulz B, Masek K. Antiviral activity of immunomodulator adamantylamide dipeptide. Int J Immunother 1987; III(2):89.
11. Masihi KN, Lange W, Schwenke S, Gast G, Huchshorn P, Palache A, Masek K. Effect of immunomodulator adamantylamide dipeptide on antibody response to influenza subunit vaccines and protection against aerosol influenza infection. Vaccine 1990; 8:159.
12. Walder P, Buchar E, Machkova Z, Vrba T, Flegel M, Janku I, Masek K. Pharmacokinetic profile of the immunomodulating compound adamantylamide dipeptide (AdDP), a muramyl dipeptide derivative in mice. Immunopharmacol Immunotoxicol 1991; 13(1&2):101.
13. Vranešić B, Tomašić J, Smerdel S, Kantoci D, Sava G, Hršak I. European patent application 90102600.5, Institute of Immunology, Zagreb, Croatia (Feb 09, 1990); US Patent 5,066,642 (Nov 19, 1991).
14. Vranešić B, Tomašić J, Kantoci D, Smerdel S, Sava G, Hršak I. Novel bioactive adamantylpeptides. In: Giralt E, Andreu D, eds. Peptides 1990. Leiden: Escom, 1991:885.
15. Vranešić B, Tomašić J, Smerdel S, Kantoci D, Benedetti F. Synthesis and antiviral activity of novel adamantylpeptides. Helv Chim Acta 1993; 76:1752.
16. Rabatić S. (unpublished results).
17. Mažuran R, Vranešić B, Ikić-Sutlić M, Šimrak D, Tomašić J. Isomers of D,L-(2-adamantyl)-glycyl-L-alanyl-D-isoglutamine induce interferon production in human mononuclear cells (unpublished results).
18. Sava G, Pacor S, Vranešić B, Tomašić J, Cocchietto M. Modification of the growth of MCa mammary carcinoma of CBA mouse by new adamantylpeptides. Int J Oncol 1993; 2:607.
19. Fraser-Smith EB, Waters RV, Matthews TR. Correlation between in vivo Anti-*Pseudomonas* and Anti-*Candida* activities and clearance of carbon by the reticuloendothelial system for various muramyl dipeptide analogs, using normal and immunosuppressed mice. Infect Immun 1982; 1:105.
20. Brown R, Price RJ, King MG, Husband AJ. Interleukin-1β and muramyl dipeptide can prevent decreased antibody response associated with sleep deprivation. Brain Behav Immun 1989; 3:320.
21. Hisano G, Fidler IJ. System activation of macrophages by liposome-entrapped muramyl tripeptide in mice pretreated with the chemotherapeutic agent adriamycin. Cancer Immunol Immunother 1982; 14:61.
22. Nakajima R, Ishida Y, Yamaguchi F, Otani T, Ono Y, Nomura M, Une T, Osada Y. Beneficial effect of muroctasin on experimental leukopenia induced by cyclophosphamide or irradiation in mice. Arzneim Forsch Drug Res 1988; 38(II):986.
23. Azuma I. Review: Inducer of cytokines *in vivo*: Overview of field and romurtide experience. Int J Immunopharmacol 1992; 3:487.
24. Vranešić B, Ljevaković Đ, Tomašić J, Ladešić B. A competitive radioimmunoassay for peptidoglycan monomer. Clin Chim Acta 1991; 202:23.
25. Jerne NK, Nordin AA, Henry C. The agar plaque technique for recognizing antibody-producing cells. In: Amos B, Koprowski H, eds. Cell bound antibodies. Philadelphia: Wistar Institute Press, 1963:109.

26

Preparation, Properties, and Biological Activity of *tert*-Butyloxycarbonyl-L-Tyrosyl Peptidoglycan Monomer

Ivo Hršak
Ruđer Bošković Institute, Zagreb, Croatia

Đurdica Ljevaković, Jelka Tomašić, and Branka Vranešić
Institute of Immunology, Zagreb, Croatia

I. INTRODUCTION

Peptidoglycans are common constituents of bacterial cell walls responsible for the physical integrity of bacteria, which also exhibit remarkably versatile biological activities (1). They are composed of glycan chains, which are built of β-1,4-linked N-acetylglucosamine and N-acetylmuramic acid residues and peptide units that consist of alternating L- and D-amino acids. The glycan chains are mostly cross-linked through relatively short-chain peptides.

Peptidoglycan monomer (PGM) (produced by PLIVA, Pharmaceutical Works, Zagreb, Croatia) is the basic repeating unit of peptidoglycan of *Brevibacterium divaricatum* obtained after lysozyme digestion of un-cross-linked peptidoglycan chains isolated from culture fluid of penicillin-treated bacteria (2). It has been shown that PGM exhibits strong immunomodulating (3,4), antitumor, and antimetastatic activities (5–7) and inhibits the function of drug-metabolizing enzymes in the liver (8). PGM is a chemically well-defined structure, obtained from natural sources, and it is water soluble, nonpyrogenic, and nontoxic. After intravenous and subcutaneous administration of PGM to mice, it is excreted rapidly in urine partly as unchanged PGM and partly as the corresponding pentapeptide and disaccharide owing to the action of N-acetylmuramyl-L-alanine amidase (9,10).

Boc-Tyr-PGM (Fig. 1) is the synthetic analogue of PGM and is composed of disaccharide-hexapeptide of the following structure: GlcNAc-MurNAc-L-Ala-D-*iso*Gln-*meso*-diaminopimelyl-[N^{ε}-(Boc-Tyr)]-D-Ala-D-Ala. There were two reasons

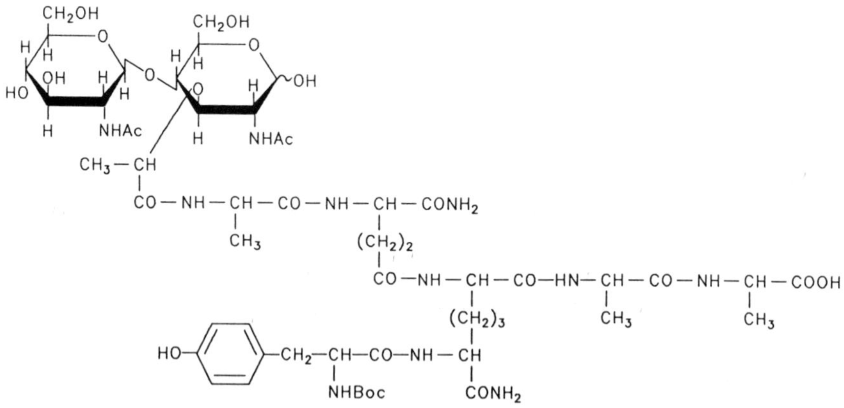

Figure 1 Structure of Boc-Tyr-PGM.

for its synthesis: (1) the introduction of tyrosine, a naturally occurring amino acid, could enhance the lipophilic character of the molecule and therefore its better penetration into the cell, which might result in the change of biological activity; and (2) the preparation of peptidoglycan monomer derivative that comprises a p-hydroxyl–substituted aromatic ring rendered it suitable for iodination and the use in a radioimmunoassay.

In this chapter, the biological activity of newly synthesized derivatives are described and compared with the peptidoglycan monomer.

II. SYNTHESIS AND PHYSICOCHEMICAL PROPERTIES OF Boc-Tyr-PGM

In an effort to enhance and/or influence the biological activity of PGM, its derivative Boc-Tyr-PGM was prepared (11,12) by chemical modification of the pentapeptide moiety of the molecule. Thus, the condensation of PGM and N-hydroxysuccinimide ester of Boc-Tyr in the presence of triethylamine, followed by chromatography on a Sephadex G-25, silica gel, and Bio Gel P-2 columns gave the above-mentioned compound. The structure of Boc-Tyr-PGM was confirmed by ^{1}H-NMR spectroscopy and amino acids analysis. It should be pointed out that synthesis of tyrosylated peptidoglycan monomer derivative was carried out with unprotected PGM molecule and tyrosine was coupled to the free amino group of diaminopimelic acid. Despite the complexity of the PGM molecule that has several reactive groups in both the disaccharide and the pentapeptide moiety, the nucleophilicity of free amino group of *meso*-diaminopimelic acid (A_2pm) was strong enough to form a stable peptide bond. Two reports concerning the synthesis of N-acyl derivatives of peptidoglycans, carried out with unprotected PGM (13) and the N-acetylglucosaminyl-β-(1 → 4)-N-acetylmuramyl tripeptide and tetrapeptides (14) have been published.

Boc-Tyr-PGM is a water-soluble molecule despite increased lipophilicity. It is nonpyrogenic in rabbits (0.3 mg/kg) and in vitro apparently nontoxic for macrophages and lymphocytes (S. Rabatić, unpublished results).

III. APPLICATION OF Boc-Tyr-PGM IN RADIOIMMUNOASSAY

In order to enable the study of distribution and the pharmacokinetics of PGM and its metabolites in body fluids, the competitive radioimmunoassay for its determination was developed (11). For this purpose, ^{125}I-labeled Boc-Tyr-PGM was prepared by radioiodination with Na ^{125}I using the well-known chloramine-T method. Specific activity of the prepared radioactive hapten was 3.21 MBq/μg (87 μCi/μg).

PGM was rendered immunogenic by coupling to bovine serum albumin (BSA) and subsequently the specific polyclonal anti-PGM antibodies were raised in rabbits by immunization of the animals with the prepared PGM-BSA conjugate.

It was proved that the introduction of the p-hydroxyl–substituted aromatic ring (tyrosine) into the PGM molecule and bulky atom of iodine did not damage the tracer immunoreactivity for anti-PGM antibodies.

Consequently, we were able to establish a reliable radioimmunoassay for PGM determination that has been shown to be sensitive over the range of 0.06–4.0 μg/ml. In order to inhibit N-acetylmuramyl-L-alanine amidase present in mammalian blood, all standards and samples contained either ethylenediaminetetraacetic acid (EDTA) or diethylenetriaminepentaacetic acid (DTPA) as inhibitor.

IV. SUSCEPTIBILITY OF Boc-Tyr-PGM TO N-ACETYLMURAMYL-L-ALANINE AMIDASE

Low molecular weight peptidoglycans were shown to be the substrates for the enzyme N-acetylmuramyl-L-alanine amidase from human and mammalian sera (9,10). Peptidoglycan monomer (PGM) originating from *Brevibacterium divaricatum* was hydrolyzed by the amidase yielding the disaccharide GlcNAc-MurNAc and the respective pentapeptide L-Ala-D-*iso*Gln-*meso*A_2pm(εNH_2)-D-Ala-D-Ala. In several experimental models, the pentapeptide exhibited the biological activity and tissue distribution comparable with the original PGM molecule (7,15–17).

In view of the findings described above, the susceptibility of Boc-Tyr-PGM to hydrolysis with the amidase was investigated. The treatment of Boc-Tyr-PGM with partially purified amidase under the conditions described for PGM (9) resulted in the regioselective hydrolysis of the lactylamide bond between the disaccharide and hexapeptide moieties. The products of the reactions were separated by ion exchange chromatography on SP-Sephadex C-25; the disaccharide formed by enzymic hydrolysis was eluted first followed by the hexapeptide. Further purification was carried out by chromatography on Sephadex G-25 and Bio Gel P-2. The structure of the hexapeptide was confirmed by amino acid analysis and ^{1}H-NMR spectroscopy (Đ. Ljevaković, to be published).

The hexapeptide L-Ala-D-*iso*Gln-*meso*-diaminopimelyl-[N^{ε}-(Boc-Tyr)]-D-Ala-D-Ala is a water-soluble, nonpyrogenic compound. Preliminary experiments carried out in mice immunized with sheep red blood cells (SRBCs) indicated that the hexapeptide stimulated the humoral immune response to this antigen and that the peptide portion alone retained the activity of the integral Boc-Tyr-PGM molecule in this model system (I. Hršak, to be published).

V. BIOLOGICAL ACTIVITY

A. Binding of ^{125}I-Boc-Tyr-PGM to Macrophages and Lymphocytes

Low molecular weight bacterial peptidoglycans influence functions of macrophages and lymphocytes, but the primary targets for their effects have not yet been clearly established. Since it has been shown that muramyl dipeptide (MDP, the smallest repeating peptidoglycan unit) binds to macrophages (18) and lymphocytes (19), we considered it interesting to investigate the binding ability of ^{125}I-Boc-Tyr-PGM to macrophages and lymphocytes as well.

Our preliminary data obtained by using PGM-coated sheep erythrocytes as indicators pointed to macrophages as the principal cells to which PGM binds (20). Now we have extended this investigation to the binding of ^{125}I-Boc-Tyr-PGM to purified populations of macrophages, T and B lymphocytes. Macrophages were collected from the peritoneal cavity of CBA mice. These adherent peritoneal cells were treated with antitheta serum in such a way that the final suspension contained less than 10% of lymphocytes. T and B cells were isolated from the spleen after removal of adherent cells followed by treatment with appropriate antisera. The final suspension contained more than 95% of B cells and more than 90% of T cells, respectively. Purified cell suspensions were divided into two portions. The first batch was preincubated with PGM (20 μg/ml/1 × 10^6 cells), whereas the other batch was not treated with PGM. Subsequently, the cells were washed and incubated with ^{125}I-Boc-Tyr-PGM for 10, 20, or 30 min. After washing, the radioactivity bound to the cells was determined. ^{125}I-Boc-Tyr-PGM binds to the T and B cells virtually to the same extent irrespective of preincubation with PGM (Fig. 2). On the other

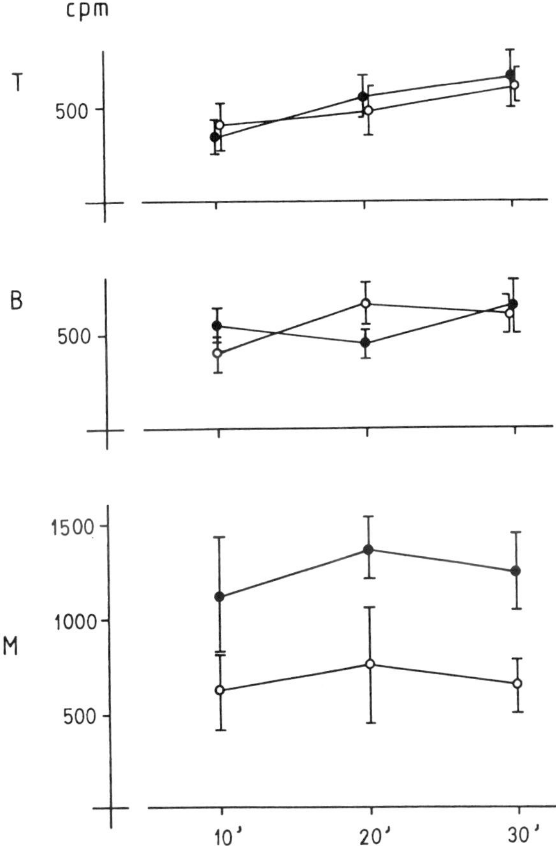

Figure 2 Radioactivity bound to 1×10^6 of T cells (T), B cells (B), and macrophages (M) after 10, 20, and 30 min of incubation with ^{125}I-Boc-Tyr-PGM. Mean values ±SEM of four experiments are shown. o–o cells preincubated with PGM; ●–● cells not preincubated with PGM.

hand, preincubation of macrophages with PGM evidently decreased subsequent binding of ^{25}I-Boc-Tyr-PGM. Thus, it seems that Boc-Tyr-PGM preferentially binds to macrophages and at a very low rate, if at all, to T or B lymphocytes. These data correlate with the findings of Silverman (18) describing the presence of specific binding sites for muramyl dipeptide (MDP) on murine macrophages.

B. Immunostimulating Activity

The immunostimulating activity of Boc-Tyr-PGM was tested in two strains of mice and the effect was correlated to the PGM activity (Table 1). Mice were immunized

Table 1 Effect of the Treatment with Boc-Tyr-PGM on the Production of PFCs in the Spleens of CBA and AKR Mice[a]

Strain of mice	Treatment	No. of PFCs Range	Mean	%
	Hanks' solution	95,000–238,000	188,906	100
CBA	Boc-Tyr-PGM	257,600–304,266[b]	295,866	157
	PGM	198,800–449,866[b]	313,413	166
	Hanks' solution	20,000–55,200	37,388	100
AKR	Boc-Tyr-PGM	31,730–68,670[b]	55,974	150
	PGM	27,200–61,070	48,668	130

[a] Groups of five animals received 200 μg of Boc-Tyr-PGM or PGM intravenously 1 day after the antigen (1×10^8 SRBCs); 3 days thereafter the number of antibody-forming cells was determined by Jerne's PFC assay.
[b] $P < .05$.

with SRBCs followed by the administration of tested compounds and thereafter the number of antibody-forming cells in their spleens was determined by Jerne's assay (21). Treatment with Boc-Tyr-PGM was at least as effective in stimulating plaque-forming cells (PFC) formation as was the treatment with the parent compound PGM. Evidently, coupling of tyrosine to the *meso*-diaminopimelic acid of PGM did not change its immunostimulating properties. This finding might be explained by the fact that the most important part of any existing peptidoglycan fragment for its biological activity is L-Ala-D-*iso*Gln (22), and this structure also remains unchanged in the PGM molecule after coupling of tyrosine.

C. Antimetastatic Activity

Antimetastatic activity of Boc-Tyr-PGM was tested in the mice bearing B-16 melanoma metastases in the lungs. The results are presented in Table 2.

Treatment with either Boc-Tyr-PGM or PGM reduced significantly the number of tumor metastases in recipient animals. There was no significant difference between the effects of single or double injection. PGM treatment seems to be slightly more effective, although the difference is not statistically significant. It might be concluded, therefore, that Boc-Tyr-PGM displayed antimetastatic activity similar to that of PGM.

VI. CONCLUSIONS

Boc-Tyr-PGM is a new synthetic derivative of PGM prepared by coupling of tyrosine to the peptide portion of the molecule. The introduction of the tyrosyl moi-

Table 2 Effect of Treatment with Boc-Tyr-PGM on the Number of B-16 Melanoma Metastases in the Lungs of C57/BL Mice[a]

		No. of metastatic nodules/mouse		
Treatment	Day	Mean ± S.E.	%	Inhibition %
Hanks' solution	–	12.3 ± 6.1	100	–
	3	7.6 ± 1.9[b]	61.3	38.7
Boc-Tyr-PGM	7	7.2 ± 4.0[b]	58.1	41.9
	3 + 7	7.0 ± 2.7[b]	56.4	43.6
	3	6.2 ± 4.4[b]	50.0	50.0
PGM	7	5.4 ± 1.7[b]	43.5	66.5
	3 + 7	5.3 ± 1.7[b]	42.7	67.3

[a] Groups of five mice received 1×10^5 B-16 melanoma cells intravenously on day 0. Boc-Tyr-PGM or PGM (1 mg/mouse) were injected intravenously on days 3, 7, or 3 and 7. Macroscopically detectable lung metastases were counted on day 23.
[b] $P < .05$.

ety enabled the labeling with radioactive iodine and subsequent application of ^{125}I-Boc-Tyr-PGM in the radioimmunoassay of PGM. Prepared labeled Boc-Tyr-PGM, besides the applications in radioimmunoassay, might be considered for use in pharmacokinetic studies of various peptidoglycan derivatives in the organism as well.

Boc-Tyr-PGM was shown to be a good substrate for N-acetylmuramyl-L-alanine amidase and was hydrolyzed to the respective disaccharide and hexapeptide.

In experimental models in mice, Boc-Tyr-PGM exhibited immunostimulating and antimetastatic activities. Although the structure of original PGM (a disaccharide-pentapeptide) molecule has been considerably changed by introduction of the sixth amino acid, the new synthetic derivative exhibited very similar biological activity to the activity of PGM.

ACKNOWLEDGMENT

This work was supported by the grants from the Ministry of Science, Informatics and Technology of Republic of Croatia (1-07-044 and 1-08-151).

REFERENCES

1. Seidl PH, Schleifer KH, eds. Biological properties of peptidoglycans. Berlin: W. de Gruyter, 1986.
2. Keglević D, Ladešić B, Tomašić J, Valinger Z, Naumski R. Isolation procedure and properties of monomer unit from lysozyme digest of peptidoglycan complex excreted into the medium by penicillin-treated *Brevibacterium divaricatum* mutant. Biochim Biophys Acta 1979:585:273.

3. Hršak I, Tomašić J, Pavelić K, Valinger K. Stimulation of humoral immunity by peptidoglycan monomer from *Brevibacterium divaricatum*. Z Immun Forsch 1979; 155:312.
4. Hršak I, Novak Đ, Tomašić J. Immunostimulating activity of peptidoglycan monomer (PGM) on *in vivo* primary response to sheep erythrocytes, *Salmonella typhimurium* and New Castle disease virus. Period Biol 1980; 82:147.
5. Hršak I, Tomašić J, Osmak M. Immunotherapy of B-16 melanoma with peptidoglycan monomer. Eur J Cancer Clin Oncol 1983; 19:681.
6. Sava G, Tomašić J, Hršak I. Antitumor and antimetastatic activity of the immunoadjuvant peptidoglycan monomer (PGM) in mice bearing MCa mammary carcinoma. Cancer Immunol Immunother 1984; 18:49.
7. Tomašić J, Hršak I. Peptidoglycan monomer originating from *Brevibacterium divaricatum*—Its metabolic and biological activities in the host. In: Schrinner E, Richmond MH, Seibert G, Schwarz U, eds. Surface structures of microorganisms and their interactions with the mammalian host. Weinheim: VCH, 1987:113.
8. Treščec A, Iskrić S, Ljevaković Đ, Hršak I, Tomašić J. The effects of immunomodulating peptidoglycan monomer and muramyl dipeptide on hepatic microsomal UDP-glucuronyltransferase and B-glucuronidase. Int J Immunopharmacol 1987; 9:371.
9. Valinger Z, Ladešić B, Tomašić J. Partial purification and characterization of *N*-acetylmuramyl-L-alanine amidase from human and mouse serum. Biochim Biophys Acta 1982; 701:63.
10. Vanderwinkel E, De Vlieghere M, De Pauw P, Cattalini N, Ledoux V, Gigot D, Ten Have J-P. Purification and characterization of *N*-acetylmuramyl-L-alanine amidase from human serum. Biochim Biophys Acta 1990; 1039:331.
11. Vranešić B, Ljevaković Đ, Tomašić J, Ladešić B. A competitive radioimmunoassay for peptidoglycan monomer. Clin Chim Acta 1991; 202:23.
12. Ljevaković Đ, Vranešić B, Tomašić J, Hršak I, Ladešić B. European patent application 91121871.7, Pliva, Zagreb, Croatia (Dec 19, 1991); US Patent 07/810,010 (Dec 20, 1991).
13. Šušković B, Vajtner Z, Naumski R. Synthesis and biological activities of some peptidoglycan monomer derivatives. Tetrahedron 1991; 47:8407.
14. Furuta R, Kawata S, Naruto S, Munami A, Kotani S. Synthesis and biological activities of *N*-acetylglucosaminyl-β-(1 → 4)-N-acetylmuramyl tri- and tetrapeptide derivatives. Agric Biol Chem 1986; 50:2561.
15. Hršak I, Tomašić J. Immunostimulatory and antimalignant activity of peptidoglycan monomer and its metabolites. Period Biol 1986; 88(Suppl.)1:22.
16. Treščec A, Iskrić S, Hršak I, Tomašić J. Effect of immunoadjuvant peptidoglycan monomer on liver cytochrome P-450. Biochem Pharmacol 1983; 32:2354.
17. Tomašić J, Valinger Z, Hršak I, Ladešić B. Metabolic fate of peptidoglycan monomer from *Brevibacterium divaricatum* and biological activity of its metabolites. In: Seidl PH, Schleifer KH, eds. Biological properties of peptidoglycans. Berlin: W. de Gruyter, 1986:203.
18. Silverman DHS, Krueger JM, Karnowsky ML. Specific binding sites for muramyl peptides on murine macrophages. J Immunol 1986; 136:2195.

19. Dziarsky R. Binding sites for peptidoglycan on mouse lymphocytes. Cell Immunol 1987; 109:231.
20. Hršak I, Kušić B. Binding of peptidoglycan monomer (PGM) on lymphocytes and macrophages. Period Biol 1990; 92:20.
21. Jerne NK, Nordin AA, Henry C. The agar plaque technique for recognizing antibody-producing cells. In: Amos B, Koprowski H, eds. Cell bound antibodies. Philadelphia: Wistar Institute Press, 1963:109.
22. Lederer E. Chemistry and biology of muramyl peptides. Proceedings of VIIIth International Symposium on Medicinal Chemistry, Vol. 1, Uppsala, 1984:13.

IV

MICROBIAL AND NATURAL IMMUNOMODULATORS

27

Application of Lentinan as Cytokine Inducer and Host Defense Potentiator in Immunotherapy of Infectious Diseases

Yukiko Y. Maeda and Hiromichi Yonekawa
The Tokyo Metropolitan Institute of Medical Sciences, Tokyo, Japan

Goro Chihara
Teikyo University, Kawasaki, Japan

I. INTRODUCTION

Lentinan, a fully purified (1→3)-β-D-glucan with (1→ 6)-β-D-glucopyranoside branches, obtained from *Lentinus edodes* (Berk.) Sing., an edible mushroom, has a marked antitumor activity in allogeneic, syngeneic, and autochthonous primary hosts (1–4), suppresses chemical and viral tumorigenesis (4–5), and prevents cancer recurrence or metastasis after surgery (6–8) in animal models. Lentinan consists of a triple helical structure, which is important for its biological activities.

Results of the clinical application of lentinan have proved prolongation of life span of the patients with advanced and recurrent stomach cancer (9–10) and breast cancer (11). Lentinan also showed an excellent therapeutic effect in patients with advanced esophageal (12) and squamous cell lung carcinoma (13) in combination with surgery and irradiation. Lentinan protected patients from the side effects of radiotherapy and improved various kinds of immunological parameters (10,12) with no toxic side effects in animal models and humans.

Lentinan induces various kinds of lymphocytokines and appears to represent host defense potentiators (HDPs), which might make the physiological constitution of the host highly cancer- and infection-resistant (14). This is a concept in Oriental medicine whose fundamental principle is to regulate homeostasis of the whole body and to bring the diseased person to normal state (14–15).

Lentinan also increases host resistance to various kinds of bacterial, viral, fungal, and parasitic infections (16–18). This chapter deals with the recent results of immunotherapy of lentinan in infectious diseases and the role of lentinan as a lymphocytokine inducer and a host defense potentiator.

II. EFFECT OF LENTINAN ON INFECTIOUS DISEASES

Lentinan increases host resistance against various kinds of infectious diseases as well as cancer, as is shown in Table 1. Lentinan shows therapeutic effect to bacterial, viral, fungal, and parasitic infections. Since many cancer and AIDS (acquired immune deficiency syndrome) patients die of various kinds of opportunistic infections, potentiation of host resistance against these infections may be very important.

A. Antibacterial Activity

Antibacterial activities of lentinan were examined using *Listeria monocytogenes* (19) and *Mycobacterium tuberculosis* (20). The effect of lentinan on *L. monocytogenes* infection in normal mice is shown in Figure 1. ICR mice were infected intravenously with 2.5×10^6 cells of *L. monocytogenes,* and lentinan was administered once from 2 to 14 days prior to the infection at a dosage of 1 or 10 mg/kg. A marked anti-infective activity was observed, and all the mice completely survived when lentinan was injected intravenously once from 4 to 10 days before the infection. The same activity was observed in CY-treated mice. CY was administered

Table 1 Effects of Lentinan Against Bacterial, Viral, and Parasitic Infections

Infections	Effects
Bacteria:	
Mycobacterium tuberculosis	Relapse prevention after termination of chemotherapy
Listeria monocytogenes	100% survival in lethal dosage of the bacteria
Viruses:	
Influenza A virus	Significant protection and prolongation in survival time
Adenovirus type 12	Suppression of oncogenesis and inhibition of tumor growth
VSV-encephalitis virus	Complete protection and cure in (Balb/c $\times$ C57BL/6)F_1 mice
Abelson virus	Reduction of tumor incidence and Abelson virus–induced tumor growth
HIV	Protection of toxicity of AZT and inhibited viral replication
Parasites:	
Schistosoma mansoni and *Schistosoma japonicum*	Potentiation of schistosome granuloma formation around eggs
Mesocestoides corti	Distinct reduction in the number of parasites

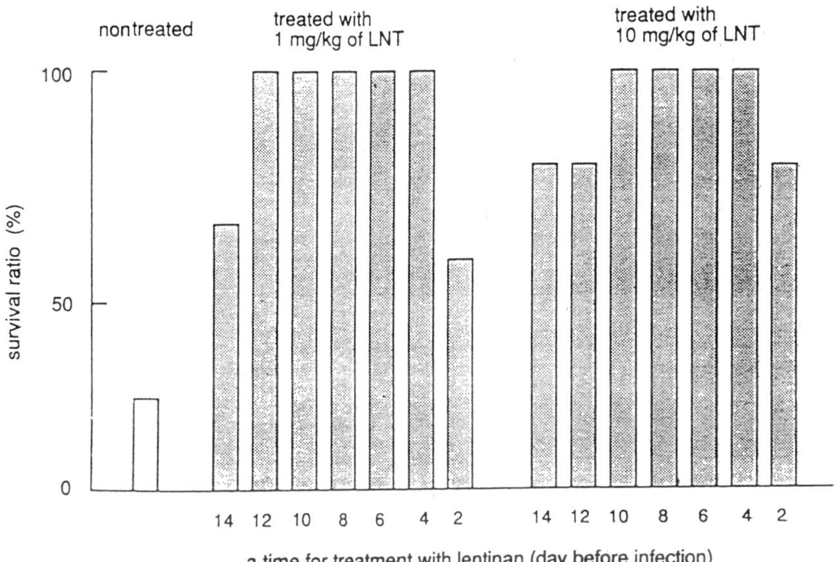

Figure 1 The anti-infective activity of lentinan on *Listeria monocytogenes*. CD-1 mice were injected intravenously with 2.5 × 10^6 of *L. monocytogenes* and lentinan injected intravenously once from day 2 to 14 prior to the infection at a dosage of 1 or 10 mg/kg. The survival mice were counted at day 14 after infection.

intraperitoneally at a dosage of 50 mg/kg on 1, 3, and 5 days prior to the infection. All CY-treated mice infected by *L. monocytogenes* completely survived by an injection of the dose of 3 mg/kg of lentinan 7 days prior to the infection, while almost all the CY-treated mice without lentinan treatment died.

The relapse-preventive effect of lentinan administered after the termination of chemotherapy with streptomycin (SM), isoniazid (INH), and refampsin (RFP) in experimentally infected tuberculosis in mice is shown in Figure 2. In the control group with chemotherapy alone, *M. tuberculosis* bacilli were detected in the lung of mice at 3 months after the termination of the chemotherapy, whereas no bacilli were detected except in the case of one mouse in the lentinan-treated group.

B. Antiviral Activity

Lentinan and its related polysaccharides have a marked antiviral activity and increase host resistance against various kinds of viral infections, such as adenovirus (5), vesicular stomatitis virus (VSV)–encephalitis virus and Abelson virus (21), human immunodeficiency virus (HIV) (22), and influenza virus (23).

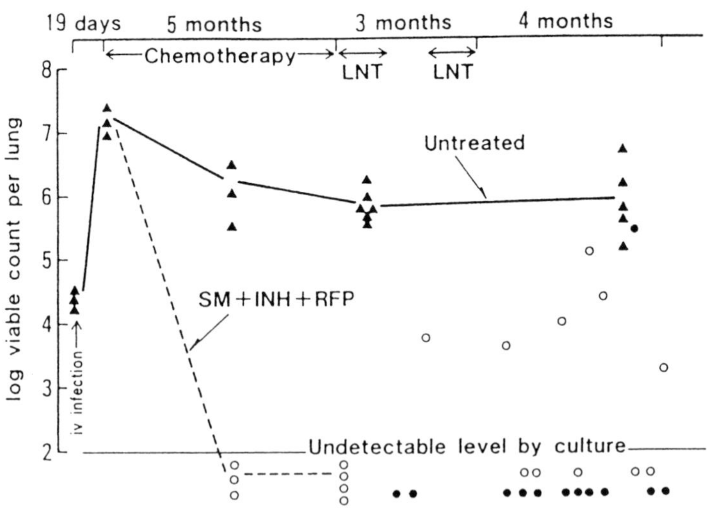

Figure 2 The relapse-preventing effect of lentinan given after termination of chemotherapy with SM + INH + RFP in experimentally infected tuberculous mice. SM, streptomycin, INH, isoniazid, RFP, rifampicin. ●, chemotherapy + lentinan, ○, chemotherapy alone, ▲, untreated control mice.

The effect of lentinan on influenza virus infection has been reported by Masihi (23). NMRI mice were pretreated with an aqueous solution of lentinan by an intranasal or intravenous route, and an aerosol of influenza virus was given 24 h later. A significant protection was conferred by 200 μg of lentinan given intranasally prior to the lethal influenza virus infection. These results were confirmed even at a lower dosage of 50 μg, as is shown in Figure 3.

Lentinan prevents adenovirus tumorigenesis as well as 3-methylcholanthrene–induced carcinogenesis (5). When newborn C3H/He mice were infected with 10^7 $TCID_{50}$/mouse of adenovirus type 12, tumors occurred in about 80% of the mice at day 80 after the infection, but when mice were injected intraperitoneally with 10 mg/kg of lentinan three times on days 14, 16, and 18 after the infection of adenovirus, the tumor occurrence ratio dropped to about 40%, as is shown in Figure 4.

C. Antiparasitic Activity

It was reported that host resistance against parasites such as malaria infection depends on cell-mediated immunity (24). A T-cell–oriented adjuvant like lentinan (25) might therefore have antiparasitic activity. Lentinan was effective against parasitic infections such as those caused by *Schistosoma mansoni, S. japonicum,* and *Mesocestoides corti,* a tapeworm. According to a report by Byrum (26), lentinan

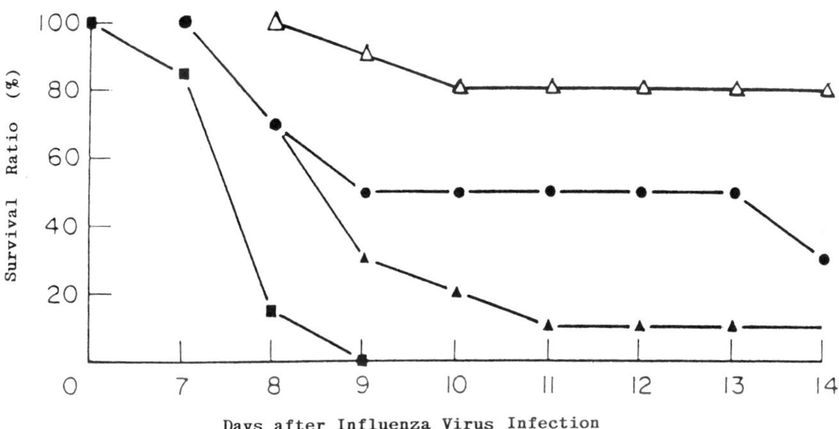

Figure 3 Effect of lentinan on aerosol influenza infection. Δ, lentinan 200 μg/ml, ●, lentinan 50 μg/ml, ▲, lentinan 10 μg/ml, ■, control mice.

potentiated granulomatous inflammatory responses against *S. mansoni* and *S. japonicum*. Lentinan potentiated lung granulomas in comparison with the control mice when CBA/J mice were sensitized with eggs of *S. mansoni*. Similar results were obtained in the case of *M. corti* (27). These responses are T-cell dependent, since they cannot be observed in nude mice.

III. MODE OF ACTION MECHANISM OF LENTINAN AS CYTOKINE INDUCER AND HOST DEFENSE POTENTIATOR

Lentinan appears to represent HDPs that can improve homeostasis of the host against cancer and infections (14). A leading principle of the functions of HDPs resides in that they can restore and/or augment the ability of responsiveness of host cells to lymphocytokines, hormones, or other intrinsic bioactive factors by stimulating maturation, differentiation, and proliferation of the cells important in host defense mechanisms. In other words, lentinan is able to improve the physiological constitution of the host against cancer and infections and to restore immune function of the patients.

A. Immunological Activities of Lentinan

Lentinan shows various kinds of immune activities, as listed in Table 2. Lentinan has no direct cytotoxicity against target cells and its effects are host mediated. In many cases, lentinan has bioactivities only in in vivo but not in in vitro experiments.

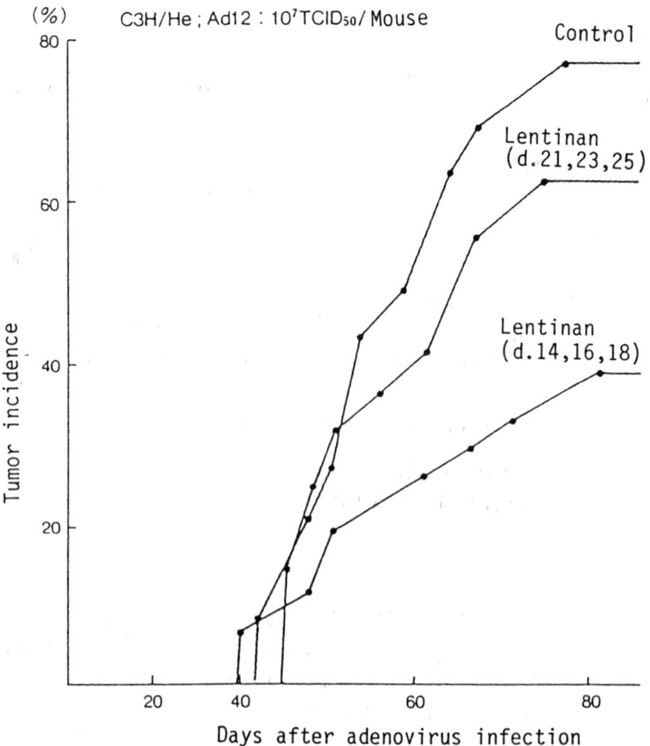

Figure 4 Inhibitory effect of lentinan on adenovirus type 12–induced oncogenesis. Newborn C3H/He mice were infected with 1×10^7 $TCID_{50}$ of adenovirus type 12 and treated intraperitoneally with 10 mg/kg of lentinan three times on days 14, 16, and 18 or 21, 23, and 25 postinfection. Control mice were infected but untreated. (Data are from Ref. 5.)

The antitumor activity of lentinan was abolished in neonatally thymectomized mice and decreased by administration of antilymphocyte serum (28–29). These results clearly support the concept that the antitumor activity of lentinan requires immunocompetent T-cell compartments. On the other hand, the effect of lentinan was inhibited by antimacrophage agents such as carrageenan or silica. Lentinan is a T-cell–oriented adjuvant in which macrophages play some role.

Nevertheless, various kinds of bioactivities of lentinan are separable between T-cell–mediated responses and T-cell–nonmediated responses (30). The characteristics of immunological properties of lentinan also can be separated into two main responses; that is, augmentation of various kinds of cytokine inducers and cytokine receptors in host defense mechanisms (31).

Table 2 Immunological Activities of Lentinan

T-cell participation	
Neonatal thymectomy	Abolished antitumor effect
Antilymphocyte serum	Decreased antitumor effect
Helper T cell in vitro	No observed effect
Helper T cell in vivo	Activation or restoration
Cytotoxic T cell in vitro	No observed effect
Cytotoxic T cell in vivo	Increased responsibility
Suppressor T cells	No induction
Delayed hypersensitivity	Potentiation or restoration
MIF-producing T cell	Activation
IL-2	Increased production
IL-3	Increased production
T cell–derived CSF	Increased production
NK cell participation	
NK cell in vitro	No effect
NK cell in vivo	Activation in C3H/He but not Balb/c mice
NK activity by poly I:C in vitro	More activation using lentinan-treated spleen cell
Macrophage participation	
Antimacrophage agents	Decreased antitumor activity
Macrophage:phagocytic in vivo	Very weak effect
Macrophage:cytotoxic in vitro	Not observed
Macrophage:cytotoxic in vivo	Activation
Macrophage:suppressive in vivo	Decreased prostaglandin E release
IL-1 and IL-6	Increased production
Antibody formation	
Antibody for SRBC	Increased production with T cell
Antibody-dependent Mϕ cytotoxity	Activation
Cellular reactions	
VDH reaction	Stimulation
Bradykinin-induced skin reaction	Stimulation
Local cellular reaction	Increase around tumor
Granuloma formation	Increase around *Schistosoma*
Complement participation	
Alternative pathway	Activation
Classic pathway	Activation
Total complement value	Increased production

B. Earliest Reactions After Lentinan Administration in Host

How lentinan affects the host at the earliest stage prior to the induction of many immune reactions has been studied. Various kinds of serum factors were induced soon after the administration of lentinan (32). They are acute-phase protein–inducing factor (APPIF) (33), vascular dilatation hemorrhage-inducing factor (VDHIF) (34), interleukin-1 (IL-1) (35), IL-3 (36), and IL-6 (37), colony-stimulating factor(s) (CSF) (36), and others. These serum factors are mainly produced by macrophages or T-lymphocytes and may act on lymphocytes, hepatocytes, vascular endothelial cells, and other cells. Many biological reactions such as vascular dilatation and hemorrhage reaction (VDH) or various kinds of immunological reactions are then activated in the host, as shown in Figure 5.

C. Genetical Consideration in the Activities of Lentinan

1. Acute-Phase Protein Responses and **ltn-1** *Gene*

Lentinan increased the amounts of serum protein components LA, LB, and LC in mice reaching their maximum level at day 6 after the intraperitoneal injection of 10 mg/kg of lentinan. They were identified as acute-phase proteins (APPs), as shown in Figure 6: LA was identified as haptoglobin-hemoglobin complex, LB as hemopexin, and LC as haptoglobin and ceruloplasmin (38). The amounts of serum amyroid A and P, complement component C3, and factor B also increased.

Lentinan-treated mice were divided into two distinct phenotypic groups: the one group comprises sensitive strains such as DBA/2, C57BL/6, SWR/J, or CD-1, which showed a marked increase in the levels of APPs, and the other group comprises resistant strains such as A/J, C3H/HeJ, Balb/c, MA/MyJ, or AKR/J mice, which remained at the same level of APPs as that of the nontreated control mice.

To characterize the genetic regulation of the APP induction, F_1 hybrids were generated from several combinations between APP-sensitive and APP-resistant strains, and an increase in the amount of hemopexin (LB) was examined to determine their sensitivity to lentinan. F_1 hybrids derived from parents with the same phenotype had the same phenotype as that of their parents. All F_1 hybrids between high responders such as DBA/2 or C57BL/10 and low responders such as A/J, C3H/HeJ, or AKR/J were low responders such as (DBA/2 × A/J)F_1, which indicated that APP production induced by lentinan was recessive in F_1 hybrids, as shown in Table 3. Besides, the backcross progeny, (DBA/2 × A/J)F_1 × DBA/2, between the F_1 and APPs-sensitive strain were segregated almost evenly into the high and low responders suggesting that a single gene of an autosome is mainly responsible for the induction of APPs. We thus designated this responsible gene as *ltn-1* (34) (Fig. 7).

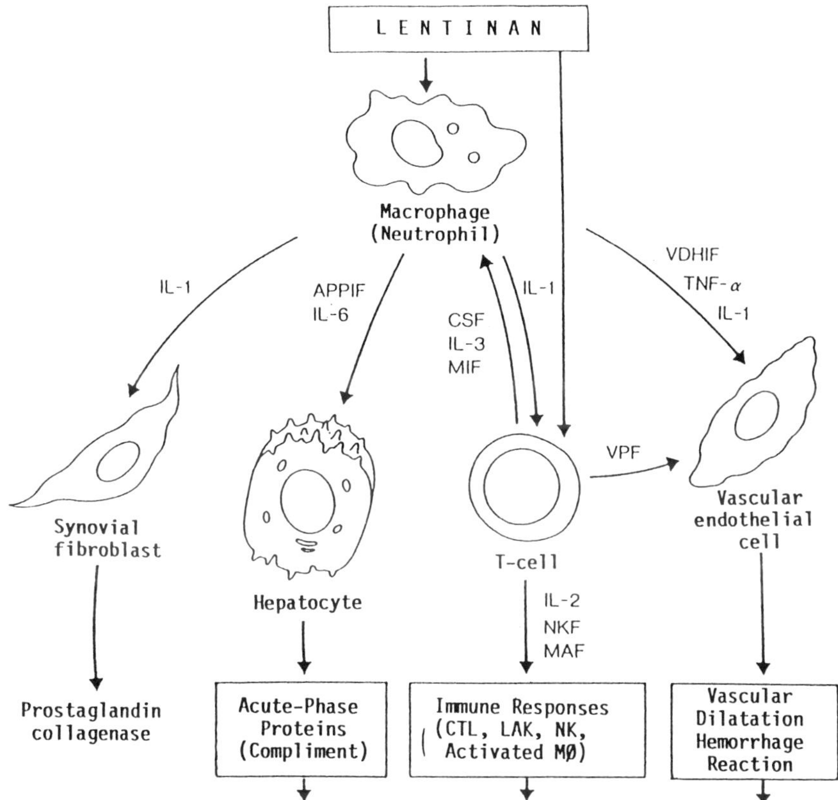

Figure 5 Early phase of the mechanism of action of lentinan and possible pathway for inflammatory and immune reactions. IL-1, IL-2, IL-3, and IL-6: interleukin-1, -2, -3, and -6; APPIF: acute-phase protein–inducing factor VDHIF: vascular dilatation and hemorrhage-inducing factor; CSF: colony-stimulating factor; MIF: migration inhibitor factor; TNF: tumor necrosis factor; VPF: vascular permeability factor; NKF: natural killer cell–activating factor; MAF: macrophage-activating factor; CTL: cytotoxic T lymphocyte; LAK: lymphokine-activated killer cell.

2. *Acute-Phase Protein-Inducing Factor (APPIF) and IL-6*

Since BSF-2/IL-6 has been reported to be a potent inducer for the production of APPs (39), induction of IL-6 mRNA was then examined in spleens of lentinan-treated mice. The profile of the increase in the amounts of IL-6 mRNA in SWR/J, a sensitive strain for the APP induction, showed two peaks at 4 h and at day 7 after the lentinan treatment, whereas only one peak at 4 h in MA/MyJ or AKR/J, APP-resistant strains, and in lentinan-resistant (MA/MyJ × SWR/J)F_1 mice as well

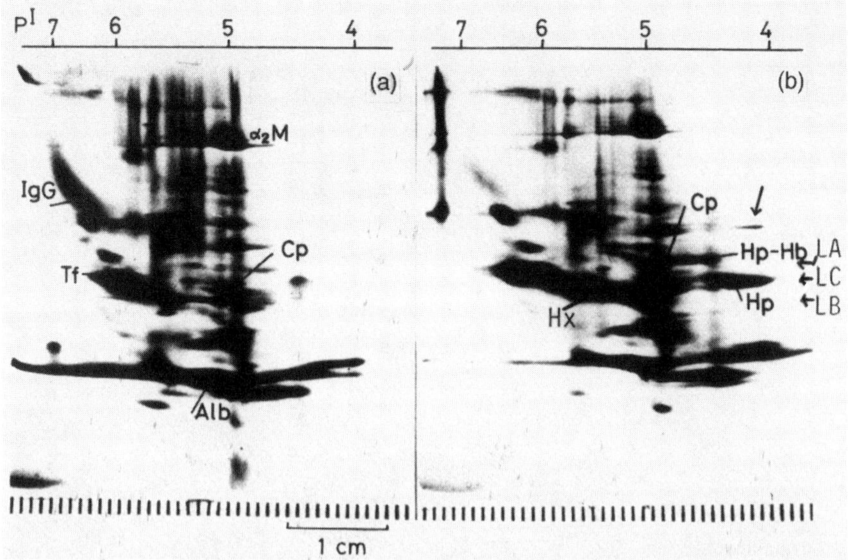

Figure 6 Identification of mouse serum proteins LA, LB, and LC induced by lentinan. Electrophoresis pattern of mouse serum samples by micro two-dimensional polyacrylamide gradient gel (2D-PAGE) was run 3 h. (a) control, (b) 7 days after lentinan injection. Hx; hemopexin; Hp: haptoglobin; Hp-Hb: haptoglobin-hemoglobin complex; Cp: ceruloplasmin; Tf: transferrin; α_2 M: α_2-macroglobulin; Alb: albumin; IgG: immunoglobulin.

(37). Therefore, the induction of IL-6 mRNA at day 7 is suggested to be essential for the induction of APPs. The *ltn-1* gene is responsible for APP induction, being involved in IL-6 mRNA expression. The *ltn-1* gene was different from the *lps-2* gene, because the strain distribution pattern of induction of IL-6 and APPs stimulated by LPS differed from that by lentinan.

We have found an inducing factor of acute-phase proteins (APPIF) in the serum obtained from mice at 6 h after an intraperitoneal injection of 10 mg/kg of lentinan (32). An intraperitoneal injection of 0.1 ml/mouse of the serum induces APPs after 4–7 days. Antimacrophage agents such as carrageenan or silica inhibit the production of APPIF and APPs. Although characterization of APPIF is still obscure, it might be a factor(s) before induction of IL-6 and APPs in the host.

3. *Vascular Dilatation and Hemorrhage (VDH) Induction and* ltn-2 *Gene*

Marked dilatation and subsequent hemorrhage of the venule is often observed in very localized areas such as ears, feet, and tail, peaked at day 4 after an intraperitoneal injection of lentinan, when antitumor-sensitive mice such as DBA/2 or A/J strain were used (Fig. 8).

Table 3 Phenotypes of F_1 Hybrids Treated with Lentinan in Hemopexin, an Acute-Phase Protein Production

Mice	Relative amount of hemopexin (η)	Phenotypes of F_1 hybrids
SWR/J	13.3 ± 2.9(7)	High
DBA/2J	12.4 ± 2.9(10)	High
(SWR/J × DBA/2)F_1	12.3 ± 2.8(5)	High
MA/MyJ	2.0 ± 0.6(4)	Low
AKR/J	1.4 ±0.3(6)	Low
(Ma/MyJ × AKR/J)F_1	1.7 ± 0.6(12)	Low
DBA/2J	12.4 ± 2.9(10)	High
A/J	2.2 ± 0.5(5)	Low
(DBA/2J × A/J)F_2	1.6 ± 0.2(5)	Low
(A/J × DBA/2J)F_1	1.8 ± 0.4(7)	Low
C57BL/10SnSle	8.7 ± 1.6(8)	High
C3H/HeJ	1.8 ± 0.4(8)	Low
(C57BL/10SnSle × C3H/HeJ)F_1	1.2 ± 0.4(15)	Low
AKR/J	1.4 ± 0.3(6)	Low
DBA/2J	12.4 ± 2.9(10)	High
(AKR/J × DBA/2J)F_1	1.3 ± 0.3(13)	Low
C57BL/10SnSle	8.7 ± 1.6(8)	High
A/J	2.2 ± 0.5(5)	Low
(C57BL/10SnSle × A/J)F_1	2.0 ± 0.2(7)	Low

Mice were injected intraperitoneally with 10 mg/kg of lentinan and 7 days later a relative amount of hemopexin in the serum of male and female mice was measured. (η), Number of mice used.

We also have observed the strain distribution pattern of VDH responsiveness to lentinan, which differed from that of responsiveness of APPs (Table 4). F_1 hybrids of mice between VDH-sensitive and VDH-resistant strains expressed VDH sensitive indicating that the sensitive phenotype was dominant. The backcross mice between the F_1 and a resistant strain were evenly divided into sensitive and resistant phenotypes in VDH induction indicating that a single gene was mainly responsible for VDH induction. We designated this gene as *ltn-2* (34,37).

The *ltn-2* gene may regulate the antitumor activity of lentinan because of the strain differences both in antitumor activity against sarcoma 180 and in preventive effect against chemical carcinogenesis are consistent with the strain variability in VDH induction (4), which is different from IL-6 or APP production. Further char-

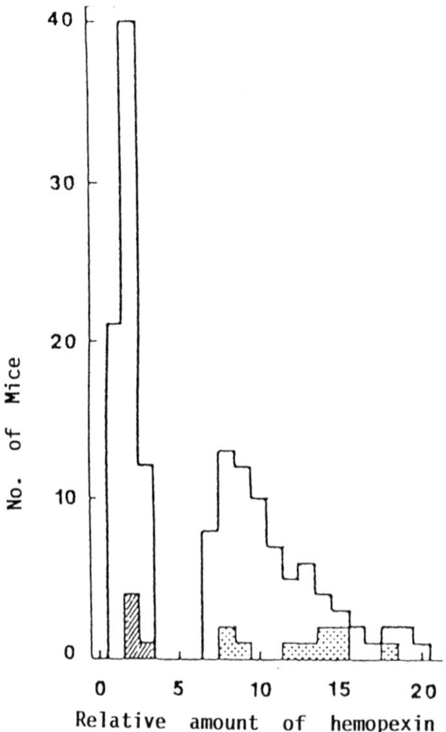

Figure 7 Segregation of (A/J × DBA/2J)F_1 × DBA/2J backcross progeny in response to lentinan in the induction of hemopexin. Mice were injected intraperitoneally with 10 mg/kg of lentinan and then the relative amount of hemopexin in the sera prepared after 3 days. Open column: (A/J × DBA/2J)F_1 × DBA/2J backcross mice; hatched column: A/J mice; stippled column: DBA/2J mice.

acterization on type *ltn-1* and *ltn-2* genes will make clear the unique reaction mechanisms of neutral antitumor polysaccharides.

D. Lentinan as Host Defense Potentiator

1. CSF, IL-1, IL-3, and IL-2

Lentinan triggers the increased production of CSFs and IL-3, which correlates with the IL-1-producing activity of macrophages (36). Increased production of IL-1 results in augmented maturation capable of inducing IL-2, natural killer (NK)-activating factor, and macrophage-activating factor (MAF). IL-1 also amplifies maturation of immature effector cells to mature cells and augments responsiveness to lymphocytokines such as IL-2, MAF, and others. Augmented production of cytokines induced by lentinan and its role in generating effector cells are summarized in Figure 9.

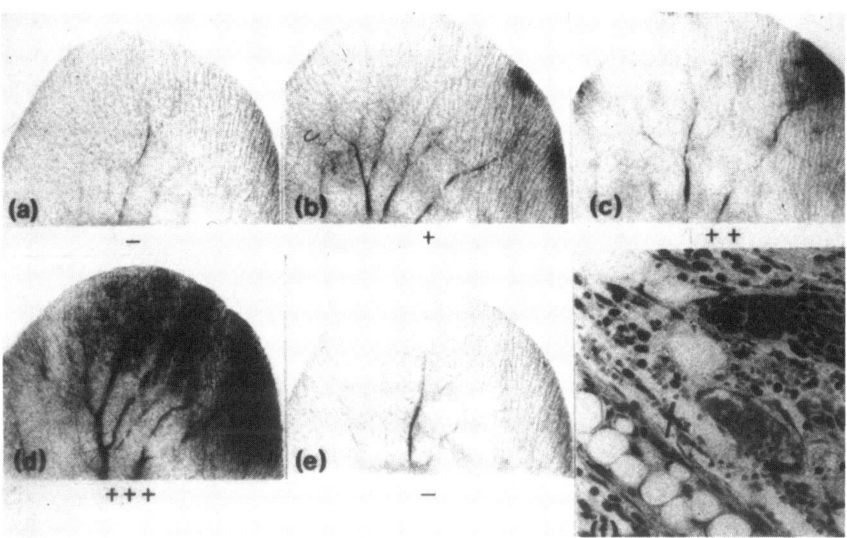

Figure 8 VDH reaction at the ear of mice after intraperitoneal injection of 10 mg/kg of lentinan. (a): untreated control(–); (b): day 1(+); (c): day 3(++); (d): day 4(+++); (e): day 7(–); (f): a micrograph of the ear with grade (+++) of VDH, which refers to red blood cells leaked out of blood vessels.

Table 4 VDH Induction by Lentinan in F_1 and Backcross Progeny Mice Derived from MA/MyJ and AKR/J Strains

Mice	VDH (No. of mice)				VDH-positive/ used mice (%)
	+++	++	++	−	
MA/MyJ	16	0	0	0	100
AKR/J	0	0	0	10	0
(MA/MyJ × AKR/J)F_1	20	14	9	0	100
(AKR/J × MA/MyJ)F_1	1	4	0	0	100
(MA/MyJ × AKR/J)F_1 × AKR/J	9	20	20	52	49

VDH expressed at the ears of mice was observed for 7 days after an intraperitoneal injection of lentinan at a dosage of 10 mg/kg.

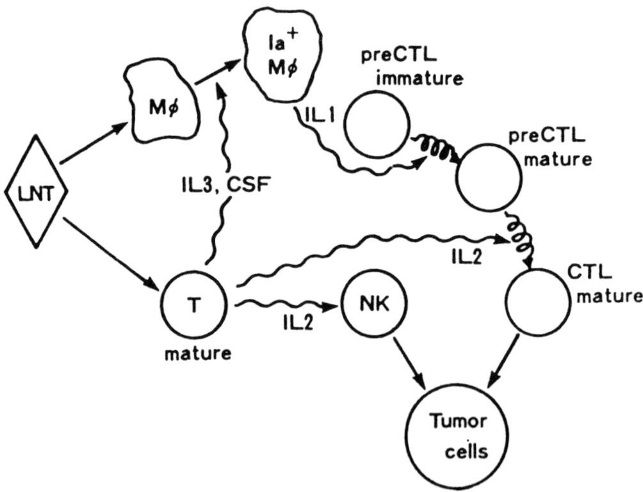

Figure 9 Cytokine production and induction of effector cells induced by lentinan. Straight lines show direct action, wavy lines show factors, and coiled lines show differentiation.

To evaluate the validity of the reaction scheme presented here, generation of allokiller cells from thymocytes was tested (40). Thymocytes, immature pre-T cells, harvested from Balb/c mice that had been treated with a minute amount of lentinan induced the augmented generation of allokiller cells against EL-4 lymphoma in the presence of IL-2. This fact clearly demonstrates that thymocytes of lentinan-treated mice are more reactive to IL-2 than untreated normal mice, as shown in Figure 10.

Lentinan cannot activate other effector NK cells when incubated in vitro, as a polynucleotide, poly I:C does. Nevertheless, the activity of NK cells cultured with poly I:C was augmented in an in vitro experiment when the spleen cells harvested from mice that had been treated with lentinan were used, as shown in Figure 11. It seems likely that in vivo application of lentinan results in augmentation of reactivity of NK cells to IL-2 (40).

2. *HDPs and Immunotherapy*

The mode of action mechanisms of lentinan demonstrates that it is a typical HDP. A simplified scheme of possible methods for immunotherapy by HDPs is shown in Figure 12, which demonstrates four directions: (1) Infusion of efferent cytokines or application of immunopotentiators capable of inducing cytokines. (2) Application of immunopotentiators capable of augmenting the reactivity of preeffectors to cytokines. (3) Combination of lymphokines and preeffector reactivity augmentors. (4) Application of immunopotentiators capable of augmenting the maturation of immature immune cells.

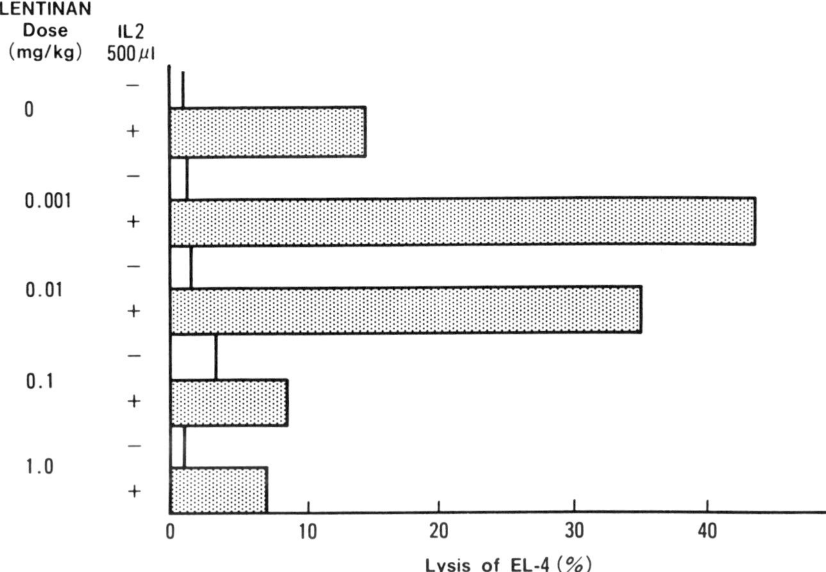

Figure 10 Lentinan augments sensitivity of thymocytes to IL-2. Augmented allokiller T-cell induction induced by a minute amount of lentinan from thymocytes (immature precursor T cells) in the presence of IL-2. Mixed lymphocyte culture, Balb/c ∞ C57BL/6, E:T = 3:1.

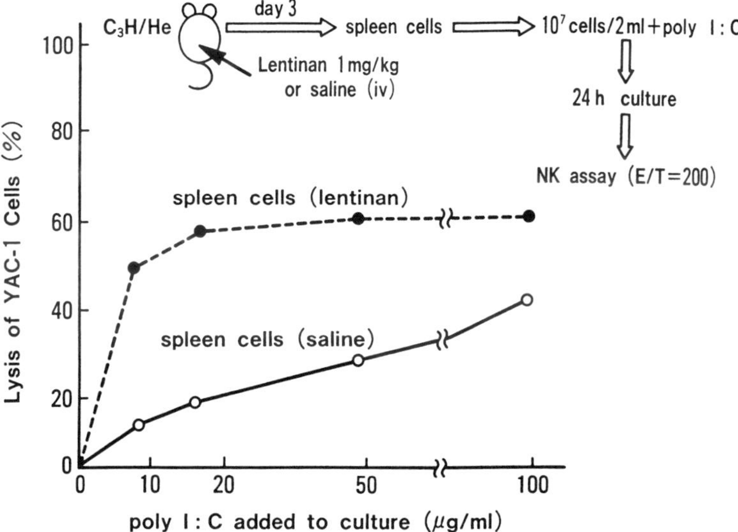

Figure 11 Augmentation of natural killer cell activation in vitro by lentinan with poly I:C using the spleen cells of lentinan-treated mice.

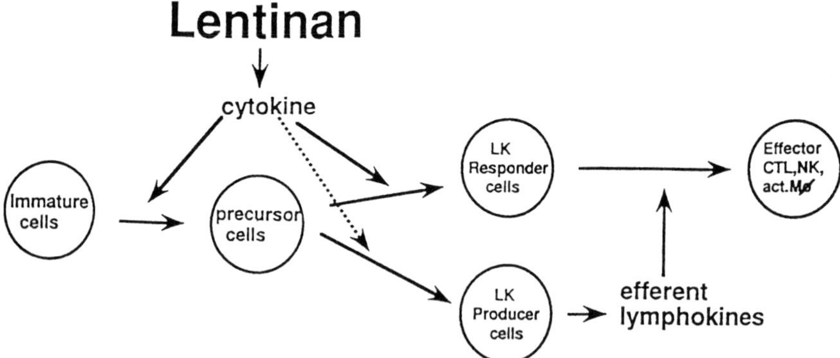

Figure 12 A simplified scheme for immunotherapy by HDPs like lentinan. LK: lymphokines; CTL: cytotoxic T lymphocytes; act.Mϕ : activated macrophages.

Considering the above-mentioned results, it is most probably that lentinan augments lymphokine responder cells and lymphokine-producing cells such as the production of IL-2. Lentinan augments the reactivity of precursor immature effector cells to lymphokine responder mature cells that have responsibility for various kinds of bioactive substances in the host. These are the most important characteristics of HDPs.

IV. CONCLUSIONS

Lentinan can restore or augment the ability of host cells to respond to lymphocytokines or other intrinsic bioactive factors and protect patients from infectious diseases or cancer metastases. Lentinan also can improve the physiological constitution of host defense mechanisms by restoring homeostasis and enhancing intrinsic resistance to diseases.

A fundamental principle in Oriental medicine is to regulate homeostasis of the whole body and to bring the diseased person to his or her normal state. Potentiating the physiological constitution in favor of host defense results in the activation of many vitally important cells for the maintenance of homeostasis. A wide variety of new HDPs with novel characteristics associated specifically with the nervous, endocrine, and immune systems should be further developed. These substances and their applications will undoubtedly play an important role in forming bridges among modern immunology and biology and traditional medicine.

ACKNOWLEDGMENT

We deeply thank Dr. Yasuko Takeda for her excellent help in the preparation of this manuscript.

REFERENCES

1. Chihara G, Maeda YY, Hamuro J, Sasaki T, Fukuoka F. Inhibition of mouse sarcoma 180 by polysaccharides from *Lentinus edodes* (Berk.) Sing. Nature 1969; 222:687.
2. Chihara G, Hamuro J, Maeda YY, Arai Y, Fukuoka F. Fractionation and purification of polysaccharides with marked antitumor activity, especially lentinan, from *Letinus edodes* (Berk.) Sing. Cancer Res 1970; 30:2776.
3. Zakány J, Chihara G, Fachet J. Effect of lentinan on tumor growth in murine allogeneic and syngeneic hosts. Int J Cancer 1980; 25:371.
4. Suga T, Shiio T, Maeda YY, Chihara G. Antitumor activity of lentinan in murine syngeneic and autochthonous hosts and its suppressive effect on 3-methylcholanthrene induced carcinogenesis. Cancer Res 1984; 44:5132.
5. Hamada C. Inhibitory effect of lentinan on tumorigenesis of adenovirus type 12 in mice. In Aoki T et al., eds. Manipulation of host defence mechanisms. Amsterdam: Excerpta Medica, 1981:76.
6. Rose WC, Reed FC, III, Siminoff P, Bradner WT. Immunotherapy of Madison 109 lung carcinoma and other murine cancer using lentinan. Cancer Res 1984; 44:1368.
7. Suga T, Yoshihama T, Tsuchiya Y, Shiio T, Maeda YY, Chihara G. Prevention of tumor metastasis and recurrence of DBA/2·MC·CS-1, DBA/2·MC·CS-T fibrosarcoma, MH-134 hepatoma and other murine tumor using lentinan. Int J Immunother 1989; 5:187.
8. Lapis K, Papay J, Paku S, Szende B. The effect of lentinan on the metastasis of Lewis lung carcinoma. Int J Immunother 1989; 5:195.
9. Taguchi T. Clinical efficacy of lentinan on patients with stomach cancer: End point results of a four year follow-up survey. Cancer Detect Prevent 1987; 1(Suppl.):333.
10. Ochiai T, Isono K, Suzuki T, Koide Y, Gunji Y, Nagata M, Ogawa N. Effect of immunotherapy with lentinan on patient survival and immunological parameters in patients with advanced gastric cancer: Results of a multi-centre randomized controlled study. Int J Immunother 1992; 8:161.
11. Yamashita A, Masuda E, Hattori Y, Kosaka A. Cellular and humoral factors in the antitumour action of lentinan on mammary tumors. Int J Immunother 1989; 5:177.
12. Sakamoto A, Isono K, et al. The effect of biological response modifiers using preoperative radiotherapy in advanced esophageal cancer patients. Biotherapy (Jpn.) 1990; 4:1524.
13. Kimura I, Ohnoshi T, Taguchi T, et al. A randomized trial of chest irradiation alone versus chest irradiation plus lentinan in squamous cell lung cancer in limited stage. Biotherapy (Jpn.) 1993; 4:601.
14. Chihara G, Maeda YY, Suga T, Hamuro J. Lentinan as a host defence potentiator (HDP). Int J Immunother 1989; 5:145.
15. Chihara G. Immunopharmacology of lentinan, a polysaccharide isolated from *Lentinus edodes*: Its application as a host defence potentiator. Int J Oriental Med 1992; 17:57.
16. Chihara G. Lentinan and its related polysaccharides as host defence potentiators: Their application to infectious diseases and cancer. In: Masihi KN, Lange W, eds. Immunotherapeutic prospects of infectious diseases. Heidelberg: Springer-Verlag, 1990:9.
17. Chihara G. Recent progress in immunopharmacology and therapeutic effects of polysaccharides. Dev Biol Standard 1992; 77:191.

18. Kaneko Y, Chihara G. Potentiation of host resistance against microbial infections by lentinan and its related polysaccharides. In: Friedman H, Klein TW, Yamaguchi H, eds. Microbial infections, role of biological response modifiers. New York: Plenum Press, 1992:201.
19. Kawamura T, Numazaki Y. Effect of lentinan to host resistance against *Listeria* infections. Proc Jpn Cancer Assoc 1980; 39:134.
20. Kanai K, Kondo E, Jacques PJ, Chihara G. Immunopotentiating effect of fungal glucans as revealed by frequency limitation of post chemotherapy relapse in experimental mouse tuberculosis. Jpn J Med Sci Biol 1980; 33:287.
21. Chang KSS. Lentinan-mediated resistance against VSV-encephalitis, Abelson virus–induced tumor, and trophoblastic tumor in mice. In: Aoki T et al., eds. Manipulation of host defence mechanisms. Amsterdam: Excerpta Medica, 1981:88.
22. Yoshida O, Nakagawa H, Yoshida T, Kaneko Y, Yamamoto I, Matsuzaki K, Uryu T, Yamamoto N. Sulfation of the immunomodulating polysaccharide lentinan: A novel strategy for antiviral to human immunodeficiency virus (HIV). Biochem Pharmacol 1988; 37:2887.
23. Irinoda K, Masihi KN, Chihara G, Kaneko Y, Katori T. Stimulation of microbicidal host defence mechanisms against aerosol influenza virus infection by lentinan. Int J Immunopharmacol 1992; 14:971.
24. Brown KN. Immunology of parasitic infections. Oxford, England: Blackwell Scientific, 1976:268.
25. Maeda YY, Chihara G. Lentinan, a new immunoaccelerator of cell-mediated responses. Nature 1992; 229:634.
26. Byrum JE, Sher A, DiPietro J, von Lichtenberg F. Potentiation of schistosome granuloma formation by lentinan, a T-cell adjuvant. Am J Pathol 1979; 94:201.
27. White TR, Thompson RCA, Penhale WJ, Chihara G. The effect of lentinan on resistance of mice to *Mesocestoides corti.* Parasitol Res 1988; 74:563.
28. Maeda YY, Hamuro J, Chihara G. The mechanisms of action of antitumor polysaccharides: The effect of anti-lymphocyte serum on the antitumor activity of lentinan. Int J Cancer 1971; 8:41.
29. Maeda YY, Chihara G. The effect of neonatal thymectomy on the antitumor activity of lentinan, carboxymethylpachymaran and zymosan, and their effects on various immune responses. Int J Cancer 1973; 11:153.
30. Maeda YY, Watanabe ST, Chihara C, Rokutanda M. Denaturation and renaturation of a β-1, 6; 1,3-glucan, lentinan, associated with expression of T-cell-mediated responses. Cancer Res 1988; 48:671.
31. Hamuro J, Chihara G. Lentinan, a T-cell oriented immunopotentiator: its experimental and clinical applications and possible mechanism of immune modulation. In: Fenichel RN, Chirigos MA, eds. Immune modulation agents and their mechanism. New York: Marcel Dekker, 1984:409.
32. Maeda YY, Chihara G, Ishimura K. Unique increase of serum proteins and action of antitumor polysaccharides. Nature 1974; 252:250.
33. Suga T, Maeda YY, Uchida H, Rokutanda M, Chihara G. Macrophage-mediated acute-phase transport protein production induced by lentinan. Int J Immunopharmacol 1986; 8:691.

34. Maeda YY, Sakaizumi M, Moriwaki K, Yonekawa H. Genetic control of the expression of two biological activities of an antitumor polysaccharide, lentinan. Int J Immunopharmacol 1991; 13:977.
35. Fruehauf JP, Ronnard GD, Herberman RB. The effect of lentinan on production of Interleukin-1 by human monocytes. Immunopharmacology 1982; 5:65.
36. Izawa M, Ohno K, Amikura K, Hamuro J. Lentinan augments the production of Interleukin-3 and colony stimulating factor(s) by T-cells. In: Aoki T et al., eds. Manipulation of host defence mechanisms. Amsterdam: Excerpta Medica, 1984:59.
37. Maeda YY, Sakaizumi M, Moriwaki K, Chihara G, Yonekawa H. Genetical control on lentinan-induced acute phase responses and vascular responses. Folia Histochem Cytobiol 1992; 30:207.
38. Manabe T, Takahashi Y, Okuyama T, Maeda YY, Chihara G. Identification of mouse serum proteins increased by the administration of antitumor polysaccharide lentinan, by micro two-dimensional electrophoresis. Electrophoresis 1983; 4:242.
39. Gauldie J, Richards C, Harnish D, Landsdorp P, Baumann H. Interferon β_2/B-cell stimulatory factor type 2 shares identity with monocyte derived hepatocyte stimulating factor and regulates the major acute phase protein response in liver cells. Proc Natl Acad Sci USA 1987; 84:7251.
40. Akiyama Y, Kashima S, Hayami T, Izawa M, Mitsugi K, Hamuro J. Immunological characteristics of antitumor polysaccharides, lentinan and its analogues, as immune adjuvants. In: Aoki T et al., eds. Manipulation of host defence mechanisms. Amsterdam: Excerpta Medica, 1981:227.

28

On the Chemistry and Biology of Bacterial Endotoxic Lipopolysaccharides

Otto Holst, Artur J. Ulmer, Helmut Brade, and Ernst T. Rietschel
Forschungsinstitut Borstel, Institut für Experimentelle Biologie und Medizin, Borstel, Germany

I. INTRODUCTION

Lipopolysaccharides (LPS) are characteristic constituents of the gram-negative bacterial cell wall, in which they are localized in the outer leaflet of the outer membrane (1). LPS represent the endotoxins of gram-negative bacteria. They exhibit a variety of biological activities in higher organisms, including humans; for example, pyrogenicity and lethal toxicity. On the other hand, LPS display essential physiological functions for the bacteria and are a prerequisite for bacterial survival.

LPS molecules possess binding sites for antibodies and other serum factors. Therefore, they are involved in specific recognition and elimination of bacteria by the host organism's defense system. However, they also may shield the bacterium from cellular host defense strategies, and thus play an important role in bacterial virulence. LPS are also potent immunostimulators in that they are capable of activating B lymphocytes, granulocytes, and mononuclear cells.

LPS of all gram-negative bacteria share a common architecture (Fig. 1), which consists of a lipid (lipid A) to which a saccharide portion is covalently linked. In LPS of various bacteria from different families, for example, *Enterobacteriaceae, Vibrionaceae,* and *Pseudomonadaceae,* the saccharide portion consists of an oligosaccharide unit with up to 15 monosaccharides (the core region) to which a polysaccharide of repeating units (the O-specific chain) is linked (smooth, or S-form, LPS). This type of LPS also is called wild-type LPS. A second type of LPS (rough, or R-type, LPS) that was first identified in enterobacteria (R mutants, as described

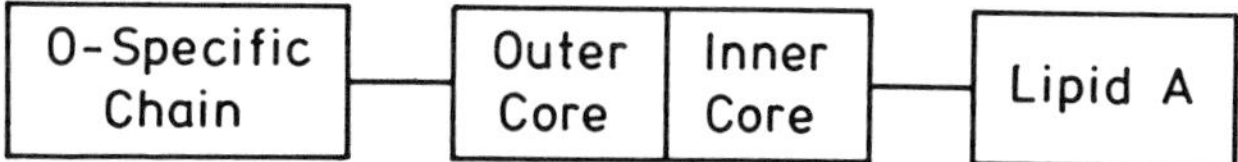

Figure 1 Scheme of the general structure of bacterial lipopolysaccharides.

for *Salmonella minnesota* or *Escherichia coli*) carries defects in the gene cluster responsible for the biosynthesis of the O chain (*rfb* locus) (2). However, it must be emphasized that other bacteria are of the wild type, for example, *Bordetella pertussis, Haemophilus influenzae, Neisseria meningitidis,* and *N. gonorrhoeae,* despite the fact that they possess no O-specific chain. Thus, the core oligosaccharide and the lipid A represent that structural principle of LPS present in all gram-negative bacteria investigated so far.

In this chapter, an overview is given on the present knowledge about the structural and biological properties of bacterial endotoxins.

II. STRUCTURE OF BACTERIAL ENDOTOXINS

A. O-Specific Chain

The O-specific chain represents, in most cases, a heteropolysaccharide consisting of repeating units made up of between one and eight sugar monomers (3–14). In enterobacterial LPS, up to 50 repeating units have been identified; in other bacteria, the number may be much smaller. The variety of monosaccharides seems to be unlimited. Neutral sugars (also branched), amino sugars with free or substituted amino groups, deoxy sugars, sugar acids (also in the amidated form), and sugar phosphates have been identified. Additionally, monosaccharides substituted by amino acids or which were at some positions acetylated or methylated have been found. By using this spectrum of possible monosaccharide variations, the bacteria are able to synthesize O-specific chains of broad structural diversity. This structural variety defines the variety of O antigenic epitopes (O antigenic determinants, O factors) that during infection lead to the production of highly specific antibodies. Several such O factors may be present in one repeating unit, and an O factor may be determined by only one monosaccharide unit or a part thereof.

The O factors of the O antigens determine the serological specificity of S-form bacteria. Many O factors have already been structurally characterized, for example, those of *Salmonella* (3,4), *Escherichia coli* (4–9), *Pseudomonas aeruginosa* (10), and *Serratia marcescens* (12–14), and thus have been determined as chemotypes. The chemical typing correlates very well with the serological one; this is true for LPS of *Salmonella* (Kaufmann-White scheme) and other bacteria, for example, *E. coli* or *Pseudomonas spp*. An example of the serology of *Salmonella* is presented in Figure 2, which shows the repeating unit of *S. typhi* in which O factors 12 and 9 are embedded (15,16). The latter factor is determined by the

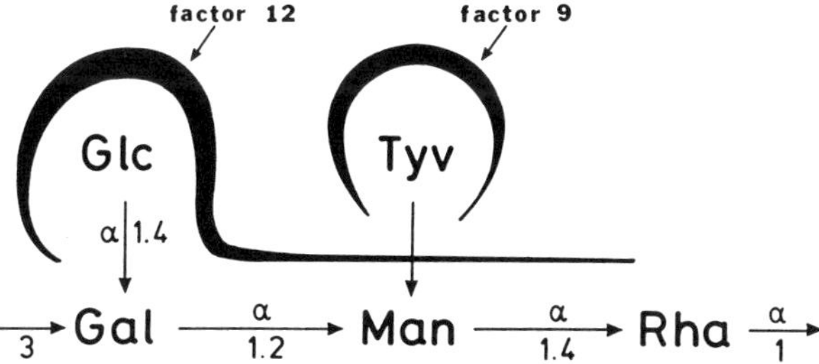

Figure 2 Chemical and antigenic structure of the repeating unit in the LPS of *Salmonella typhi*. Factor 9 is determined by the monosaccharide residue D-tyvelose (3,6-dideoxy-D-mannose), whereas factor 12 comprises the terminal glucose residue and the backbone trisaccharide.

monosaccharide tyvelose and is characteristic of *Salmonella* serogroup D_1. The terminal glucose residue together with the other three sugars, galactose, mannose, and rhamnose, determine the specificity of factor 12.

The biosynthesis of the O-specific chain (17) is performed independent of that of lipid A and the core oligosaccharide. In *E. coli* and *S. minnesota,* a repeating unit linked to bactoprenylpyrophosphate is first synthesized from activated monosaccharides, and only after the synthesis of the repeating unit is complete, is it transferred by a ligase to the core region of LPS. It should be noted that during polymerization of the O chain, a newly synthesized repeating unit is linked to the reducing end of the growing chain. The genes for O antigen biosynthesis may be located on or integrated into plasmids; therefore, it is possible to express O antigens in other bacteria (e.g., in *E. coli* K-12) (18–20).

The enzymes required for biosynthesis are membrane bound. Furthermore, they use sugar nucleotides as substrates and adenosine triphosphate (ATP). Consequently, the enzymes should face the cytoplasm, and the early steps in biosynthesis of LPS should take place at the inner surface of the cytoplasmic membrane. Evidence has been provided that in *S. typhimurium* the attachment of O antigen to the core takes place at the periplasmic face of the cytoplasmic membrane, that O antigen polymerase also faces the periplasm (21), and the transport of core oligosaccharides from the cytoplasmic to the periplasmic face of the cytoplasmic membrane requires energy in the form of proton motive force (22). However, it is unclear how LPS molecules are then exported to the outer leaflet of the outer membrane. Several alternatives have been proposed for this transport but none has so far been proven.

B. Core Region

A comparison of known structures of the core region (23) from LPS of different bacteria shows that despite the presence of structural differences chemical variation in this part of LPS is much more limited than in the structures of O-specific chains. In general, there are two carbohydrate components that are characteristic of the core region (Fig. 3): 3-deoxy-D-*manno*-2-octulosonic acid (or 2-keto-3-deoxy-D-*manno*-octonic acid, Kdo) (24) and heptopyranose (Hep*p*), the latter of which may be present in the L-*glycero*-D-*manno*-(L,D-) or in the D-*glycero*-D-*manno*-(D,D) configuration. Most of the investigated LPS contain only L,D-Hep*p*; for example, LPS of *S. minnesota* and *E. coli* (2). However, some LPS contain L,D- as well as D,D-Hep*p* (*Proteus mirabilis* [25], *Yersinia enterocolitica*), few LPS contain only D,D-Hep*p* (some phototrophic bacteria), and some LPS do not contain any heptose (*Bacteroides fragilis* [26]). D,D-Hep*p* has been described as the biosynthetic precursor of L,D-Hep*p* (27,28), and bacteria that contain in their LPS only D,D-Hep*p* should possess a defect in the gene locus responsible for the expression of ADP-D-*glycero*-D-*manno*-heptopyranose-6-epimerase.

Kdo has so far been identified in all investigated LPS, and owing to structural analyses, at least one Kdo residue (Kdo I in Fig. 4) is present in LPS that covalently links the core region to lipid A. Mutants devoid of Kdo (and thus of the complete core region) are not viable. Consequently, the smallest LPS consists of lipid A and one Kdo residue, and such an LPS has indeed been identified in a deep rough mutant of *Haemophilus influenzae* (29).

To date, Kdo I has been found to be replaced only in one LPS (from *Acinetobacter calcoaceticus* [30]) by a very similar compound; that is D-*glycero*-D-*talo*-2-octulopyranosonic acid (see Fig. 3) (U. Zähringer, personal communication, October 1990). The LPS from *A. calcoaceticus* also contains Kdo; however, the core region has not yet been structurally analyzed completely.

The general structure of the core region of LPS from various enterobacterial genera (*Escherichia, Salmonella, Shigella*) is shown in Figure 4. Formally, the core region may be subdivided (31) into two parts: the *inner* (linked to lipid A) and the *outer* core, which in S-form LPS is substituted by the O-specific chain. This sub-

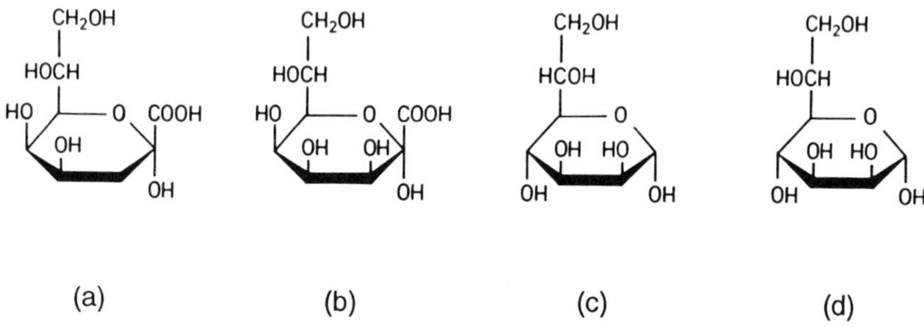

Figure 3 Formulas of (a), 3-deoxy-D-*manno*-octulopyranosonic acid (Kdo); (b), D-*glycero*-D-*talo*-octulospyranosonic acid (Ko); (c), L-*glycero*-D-*manno*-heptopyranose; (d), D-*glycero*-D-*manno*-heptopyranose.

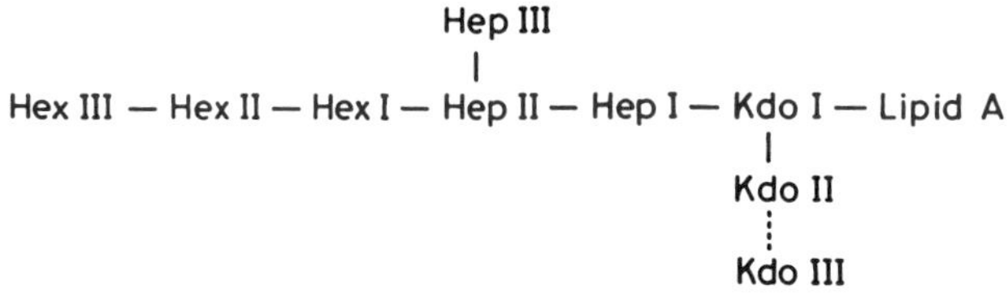

```
                                   Hep III
                                      |
Hex III — Hex II — Hex I — Hep II — Hep I — Kdo I — Lipid A
                                                |
                                              Kdo II
                                                :
                                              Kdo III
```

Figure 4 Scheme of frequently occurring sugar constituents of the core region of LPS. Dotted line: nonstoichiometric substitution.

division was originally devised according to the different composition in both regions; that is, hexoses and hexosamines in the outer core and heptose and Kdo in the inner core. Although hexose and hexosamine residues have been identified in the inner core region (e.g., D-Gal*p* in *E. coli* EH100 [32], L-Rha*p* in *E. coli* K-12 [33] and *Acinetobacter calcoaceticus* [34], D-Glc*p*N in *E. coli* R3 [35,36] and J-5 [37]), and heptose residues in the outer core region (e.g., L,D-Hep*p* in *E. coli* K-12 [38], D,D-Hep*p* in *Proteus mirabilis* [25]), the old subdivision nomenclature is still used for descriptive reasons.

The detailed structure of the Kdo region of the inner core had been unknown for a long time because of the difficulties in the analysis of Kdo-containing compounds. Kdo (24) (see Fig. 3) is a polyfunctional eight-carbon sugar possessing a carboxyl-group at C-1, a deoxy-group at C-3, and hydroxy groups at C-2, C-4, C-5, C-7, and C-8. Thus, Kdo is a ketose and its linkage therefore is acid labile. Treatment of Kdo with acid may furthermore lead to artifacts, like anhydro (39) and/or lactone derivatives. Since these artifacts cannot be determined quantitatively in photometry, it was necessary to find conditions that allow the quantification of Kdo in LPS, and it was only in the early 1980s that such hydrolysis conditions could be identified (40). Additionally, the methodology (methylation analysis) for the structural analysis of the Kdo region was developed (41), which allowed determination of the substitution pattern of Kdo I, Kdo II, and, if present, Kdo III and the arrangement of the Kdo residues (see Fig. 4).

A typical structural element of the core region of enterobacterial LPS is the tetrasaccharide α-D-Glc*p*-(1-3)-L-α-D-Hep*p*-(1–3)-L-α-D-Hep*p*-(1–5)-Kdo, which is also found in the LPS of some other bacteria. The smaller unit, L-α-D-Hep*p*-(1–3)-L-α-D-Hep*p*-(1–5)-Kdo, is present in LPS of various bacteria (23). These tri- and tetrasaccharides may be substituted at characteristic positions (23); that is at HO-4 of Kdo I (by Kdo, phosphate), at HO-4 of Hep I (phosphate, mono- or oligosaccharides), and at HO-7 of Hep II (Hep III, which in turn may be substituted at HO-7 by D-Glc*p*N, D-Gal*p*N, or D,D-Hep*p*). Furthermore, substituents at HO-7 (D-Gal*p*, 2-aminoethyl phosphate) and HO-8 (Kdo, L-Rha*p*, 4-amino-4-deoxy-L-arabinopyranose) of Kdo were identified, but to date it is unknown which of the Kdo residues is substituted.

The phosphate substitution of the core region is an important structural parameter; however, most of the structural features are unknown. There exist some reports on the linkage positions of phosphate, pyrophosphate, and phosphodiester groups, but in all cases the analysis has been performed by using chemical methods that may give rise to cleavage or migration of (pyro)phosphate groups and thus are not very reliable.

A relatively large number of genes participate in the biosynthesis of the core region, as was shown for the LPS of *E. coli* K-12 and *Salmonella typhimurium* (42–47). The genes are organized as a cluster in the *rfa* locus. Different mutations in this locus lead to different deep-rough phenotypes; for example, the Re, Rd, Rc, and Rd mutants of *S. minnesota* (2). In *S. typhymurium,* the *rfa* locus encodes for various monosaccharide transferases that mediate the biosynthesis of the core region: *rfaC* is responsible for linking Hep I to Kdo I, *rfaD* and *rfaE* encode for the biosynthesis of Hep II and Hep III, *rfaG* is responsible for the glucose-transferase of Glc I. Additionally, a *rfaH* region is present that encodes for a positive regulator responsible for the expression of the *rfa* genes; other genes of the *rfa* locus encode for the synthesis and transfer of the sugars distal of Glc I. In *E. coli* K-12, a similar organization of the *rfa* genes was found, and many genes show strong homology to those of *S. typhimurium.* The *rfa* gene cluster seems to consist of three regions: two conserved ones in the center of which two nonconserved genes are present (46). The organization of the *rfa* locus was suggested to resemble shingles on a roof, a row of transcriptional elements of which each possesses its own independent promoter but that additionally is regulated by a superposed regulatory element (43). Thus, a coordinated control of the genes as well as the diversity of the core region is possible.

C. Lipid A

The core oligosaccharide is covalently bound to the lipid A region. As was shown by the elementary work performed in the 1960s and 1970s (2) that led in the early 1980s to the elucidation of the lipid A structure from LPS of *E. coli* (54–57); and by comparative structural analyses of lipid A molecules from various LPS, most lipid A molecules consist of a disaccharide β-D-Glc*p*N-(1–6)-α-D-Glc*p*N that is phosphorylated (in most cases at positions C-1 and C-4′) and acylated. Characteristic acyl components are (R)-3-hydroxy fatty acids (52) that are partly esterified at their 3-hydroxy group by other fatty acids (3-acyloxyacyl groups). Figure 5(a) shows the structure of the lipid A from LPS of *E. coli.* Here, six fatty acids are nonsymmetrically distributed over the two GlcN residues: an ester-linked (R)-3-hydroxytetradecanoic acid (3OH-C14:0) that is esterified with tetradecanoic acid (C14:0) at HO-3′ of the nonreducing GlcN, which also carries an amide-linked 3OH-C14:0 esterified with dodecanoic acid (C12:0). The reducing GlcN carries unsubstituted 3OH-C14:0 in ester (HO-3) and amide linkage.

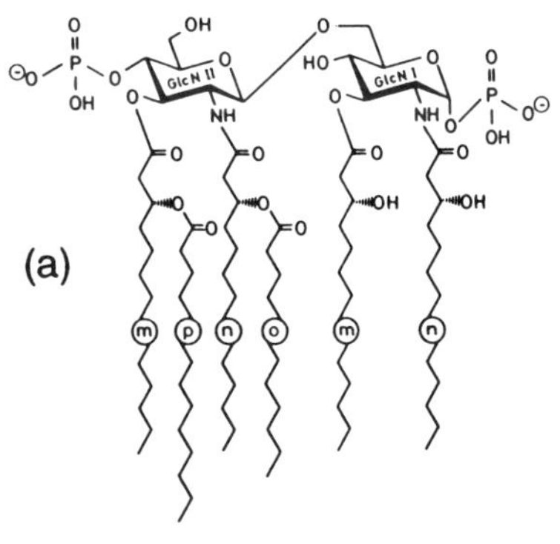

(a)

Lipid A	Number of C Atoms			
	m	n	o	p
Escherichia coli	14	14	12	14
Haemophilus influenzae	14	14	14	14
Campylobacter jejuni [a]	14	14	16	16

a) GlcN II is replaced by GlcN3N

(b)

Lipid A	Number of C Atoms		
	m	n	o
Rhodocyclus gelatinosus	10	10	12 (14)
Chromobacterium violaceum	10	12	12
Neisseria meningitidis	12	14	12

Figure 5 Structures of the hexaacyl lipid A forms with (a) asymmetric and (b) symmetric acylation pattern.

A second type of lipid A in which the fatty acids are symmetrically distributed was identified in the LPS of *Chromobacterium violaceum* (Fig. 5b) (53,54). This lipid A also contains a β-D-Glc*p*N-(1–6)-α-D-Glc*p*N disaccharide to both GlcN residues of which are linked an ester-linked (R)-3-hydroxydecanoic acid (3OH-C12:0) and an amide-linked 3-dodecanoyloxydodecanoic acid. At position C-2′ of the nonreducing GlcN, the amide-linked acyl group may also be 3-(S)-2-hydroxydodecanoyloxydodecanoic acid; thus two molecular species of lipid A are present.

It should be noted that lipid A from various LPS not only differ in the amount, quality, and distribution of fatty acids but also in phosphate substitution and, most interestingly, in the composition of amino sugars. Thus, lipid A is not one well-defined molecule but a family of lipids that are characteristic for LPS. In lipid A of several LPS, 2,3-diamino-2,3-dideoxy-D-glucopyranose (D-Glc*p*N3N) has been identified (55,56). This component may be present as monosaccharide (e.g., in the lipid A of *Rhodopseudomonas viridis* [55]) or as disaccharide (in the lipid A of *Campylobacter jejuni* [56]) or in combination with D-Glc*p*N (in the lipid A of *C. jejuni* [56]) (Fig. 5a). The lipid A from *R. viridis* is free of phosphate, and the monosaccharide D-Glc*p*N3N carries two amide-linked residues of 3OH-C14:0 but no ester-linked fatty acids. The lipid A preparation from LPS of *C. jejuni* was shown to possess pure β-(1–6)-linked disaccharides either of D-Glc*p*N or of D-Glc*p*N3N, and additionally a hybrid disaccharide of D-Glc*p*N (reducing sugar) and D-Glc*p*N3N (nonreducing sugar) that was the major molecular species.

Finally, it should be mentioned that in lipid A of various bacteria, for example, *Rhizobium* (57), *Legionella* (58,59), *Chlamydia* (60,61), some phototrophic and other bacteria (62), long-chain and long-chain (ω-1)–hydroxy fatty acids have been found.

The biosynthesis of lipid A has been investigated extensively in *E. coli* (17). One molecule UDP-D-Glc*p*NAc is in a first step acylated with 3OH-C14:0 at HO-3 by the (R)-3OH-C14:0-acyl-carrier–protein (ACP) complex and then de–N-acetylated and N-acylated with 3OH-C14:0 to yield UDP-2,3-diacyl-D-Glc*p*N. Cleavage of the glycosidic pyrophosphate linkage then results in 2,3-diacyl-D-Glc*p*N 1-phosphate (and UMP) that reacts with a second molecule of UDP-2,3-diacyl-D-Glc*p*N to one molecule tetraacyl-β-D-Glc*p*N-(1-6)-α-D-Glc*p*N 1-phosphate (and UDP). This disaccharide is then phosphorylated by ATP at position C-4′ to yield precursor Ia, which is substituted at HO-6′ via CMP-Kdo by two Kdo residues in α-(2-4)-linkage, and finally the molecule is fully acylated to the complete Re-LPS.

Around 1980, several research groups started to synthesize lipid A, and in 1985 the group of Shiba and Kusumoto succeeded in preparing completely synthetic *E. coli* lipid A (see Fig. 5a) (63), which then was proven to possess endotoxic properties identical to its natural counterpart (64–66). These findings finally falsified hypotheses that had stated that other bacterial compounds (proteins, polysaccharides) may be responsible for some of the endotoxic effects. Furthermore, many different lipid A analogues have been synthesized and biologically tested (63,67).

The topology of LPS and lipid A of *S. minnesota* and *E. coli* in the outer membrane has been investigated for the last 10 years, also by theoretical calculations and molecular modeling (68,69). Presently, the results indicate that the fatty acids, which are embedded in the outer leaflet of the outer membrane, are oriented parallel to one another and perpendicular to the plane of the membrane. The hydrophilic β-D-Glc*p*N-(1-6)-α-D-Glc*p*N 1,4′-bisphosphate backbone is located at a 45-degree angle relative to the membrane. The fatty acids are packed in hexagonal arrangement. Such well-arranged lipid A molecules result in a very rigid outer membrane structure.

III. ENDOTOXIC ACTIVITY OF LPS, LIPID A, AND LIPID A ANALOGUES

LPS represents the endotoxic principle of gram-negative bacteria inducing various pathophysiological effects like fever, hypotension, disseminated intravascular coagulation, local Shwartzman reaction, multiorgan failure, and shock, which in most severe situations may lead to death (70).

It is known today that the pathophysiological effects of endotoxin are not due to a direct action of LPS but are elicited by secondary mediators. Thus, fever is due to the action of so-called endogenous pyrogen (EP). EP was described as factors that are produced and released from nonlymphoid leukocytes. It is found in the serum in response to an exposure to LPS and induces fever (71). Formerly thought to be only one factor, it is now known that several different proteins can function as EP. Interleukin 1 (IL-1), interleukin 6 (IL-6), and tumor necrosis factor alpha (TNF-α) are three of such factors that are well defined, the genes of which have been cloned and sequenced (71). Described first as having distinct specific biological activities, it is now well established that these hormonelike polypeptides have ubiquitous activities on various cell types and physiological systems. Although termed monokines, they are not only produced by monocytes or macrophages but also by a variety of other cell types, including fibroblasts, endothelial cells, epithelial cells, Langerhans' cells, keratinocytes, and astrocytes (72–74).

The prominent role of LPS during infection has greatly stimulated approaches for a chemical-analytical elucidation of its structure and the recognition of its biologically active part. It was found that the hydrophobic lipid part, termed lipid A, represents the endotoxic principle of LPS; that is, the biological effects of LPS are reproduced in experimental animals by free lipid A (2,75). The successful chemical synthesis of lipid A and corresponding partial structures of LPS, such as *E. coli*–type lipid A (compound 506, LA-15-PP) and precursor Ia (compound 406, LA-14-PP, or lipid IVa) then provided the experimental basis to investigate the bioactive regions of LPS with regard to structure-activity relationships (75–78). Those compounds that enabled us to establish such relationships in our laboratory are summarized in Figure 6.

Based on the finding that the monokines IL-1, IL-6, and/or TNFα are involved in the manifestations of septicemia, experiments regarding structure-activity relationship of LPS, lipid A, and defined partial structures have been performed in vitro to investigate monokine production after stimulation of human peripheral blood mononuclear cells, monocytes, or murine macrophages as effector cells.

In our experiments, we were able to demonstrate a structure-dependent hierarchy of LPS and defined lipid A partial structures in their capacity to induce IL-1, IL-6, and TNFα (79–81). Figures 7 and 8 show typical results of experiments demonstrating this hierarchy. Similar results were always obtained independent of whether IL-1, IL-6, or TNFα production was determined. Through these in vitro experiments, we were able to confirm in vivo observations showing, for example, that lipid A constitutes the principle exerting all the activities of LPS. It also was found that full monokine-inducing activity is expressed by a molecule having two *gluco*-configured D-hexosamine (GlcN or GlcN3N) residues, two phosphoryl groups, and six fatty acids as present in *E. coli* lipid A (82). Lipid A partial structures deficient in one of these elements are less active or even nonactive in induc-

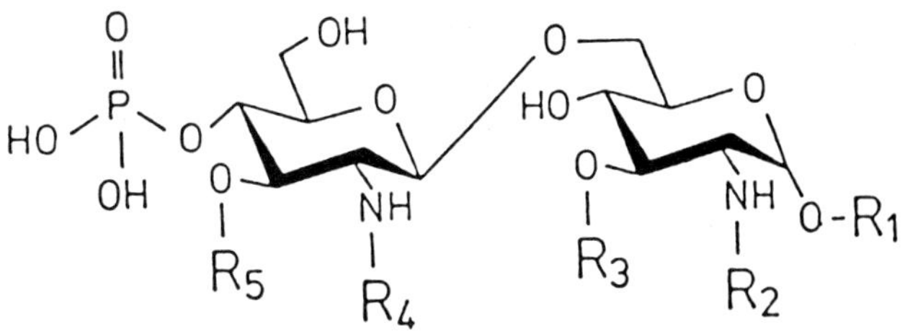

Compound	R_6	R_5	R_4	R_3	R_2	R_1	No. of acyl-residues
LA-14-PH (404)	P	C_{14}-OH	C_{14}-OH	C_{14}-OH	C_{14}-OH	α-O-P	4
LA-14-HP (405)	H	C_{14}-OH	C_{14}-OH	C_{14}-OH	C_{14}-OH	α-O-P	4
LA-14-PP (406)	P	C_{14}-OH	C_{14}-OH	C_{14}-OH	C_{14}-OH	α-O-P	4
LA-15-PH (504)	P	C_{14}-O-C_{14}	C_{14}-O-C_{12}	C_{14}-OH	C_{14}-OH	α-O-H	6
LA-15-HP (505)	H	C_{14}-O-C_{14}	C_{14}-O-C_{12}	C_{14}-OH	C_{14}-OH	α-O-P	6
LA-15-PP (506)	P	C_{14}-O-C_{14}	C_{14}-O-C_{12}	C_{14}-OH	C_{14}-OH	α-O-P	6
LA-16-PP (516)	P	C_{14}-O-C_{14}	C_{14}-O-C_{12}	C_{14}-OH	C_{14}-O-C_{16}	α-O-P	7
LA-19-PP (606)	P	H	C_{14}-OH	H	C_{14}-OH	α-O-P	2
LA-22-PP	P	C_{14}-OH	C_{14}-O-C_{14}	C_{14}-OH	C_{14}-O-C_{14}	α-O-P	6
LA-23-PP	P	C_{10}	C_{10}	C_{10}	C_{10}	α-O-P	4
LA-24-PP	P	C_{10}	C_{14}-OH	C_{10}	C_{14}-OH	α-O-P	4
PE-1	P	C_{14}-O-C_{14}	C_{14}-O-C_{12}	C_{14}-OH	C_{14}-OH	α-O-$(CH_2)_2$-O-P	6
PE-2	P	C_{14}-O-C_{14}	C_{14}-O-C_{12}	C_{14}	C_{14}	α-O-$(CH_2)_2$-O-P	6
PE-3	P	C_{14}-O-C_{14}	C_{14}-O-C_{12}	C_{14}	C_{14}	ß-O-$(CH_2)_2$-O-P	6
PE-4	P	C_{14}-OH	C_{14}-OH	C_{14}-OH	C_{14}-OH	α-O-$(CH_2)_2$-O-P	4

Figure 6 Chemical structure of lipid A and lipid A partial structures.

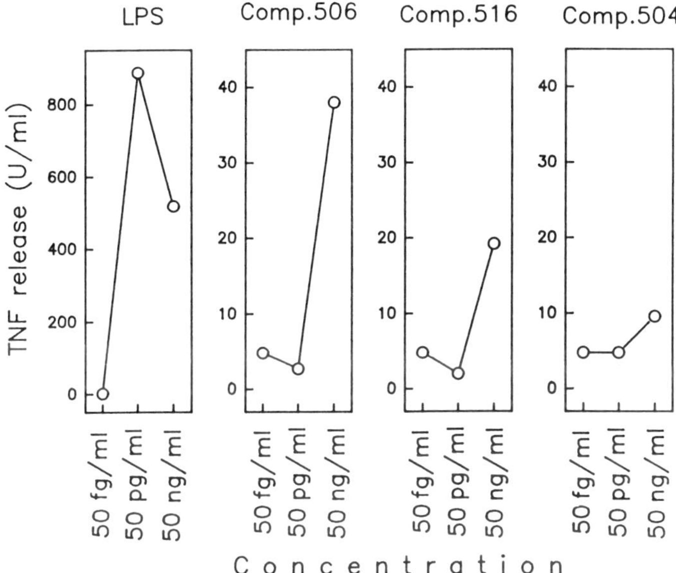

Figure 7 Induction of TNFα by LPS, lipid A, and lipid A partial structures in human PBMCs. Human PBMCs (4×10^6/ml) were stimulated with LPS, synthetic lipid A (compound 506), or synthetic lipid A partial structures (compound 516, 504) at concentrations ranging from 50 fg/ml to 50 ng/ml. After an incubation period of 12 h, the supernatants were harvested and tested for the presence of biologically active TNFα. Each value represents the mean of triplicate cultures.

ing monokines in human monocytes. For instance, the 1-dephospho (compound 504) and the 4′-dephospho (compound 505) synthetic partial structure of *E. coli* lipid A (compound 506) were less active than compound 506. The acyl-deficient tetraacylated lipid A precursor Ia (compound 406) is completely inactive in inducing IL-1, IL-6, and TNFα release in human monocytes (79–81,83). Also, heptaacyl lipid A (*S. minnesota* lipid A, compound 516) shows lower bioactivity than compound 506. In addition, the location of acyl residues is of importance as exemplified by the low bioactivity of compound LA-22-PP, which represents a hexaacylated lipid A of *Chromobacterium violaceum* type having, in contrast to compound 506, a symmetrical distribution of the fatty acids. The different biological activity of synthetic lipid A compounds seems to be regulated already at the level of mRNA expression of the monokines, since compound 406 is not able to induce IL-1 or TNFα mRNA production in human monocytes (95,84).

As mentioned, the phosphoryl groups are of great significance for the biological activity of lipid A. Recently, a new class of synthetic lipid A analogues, termed

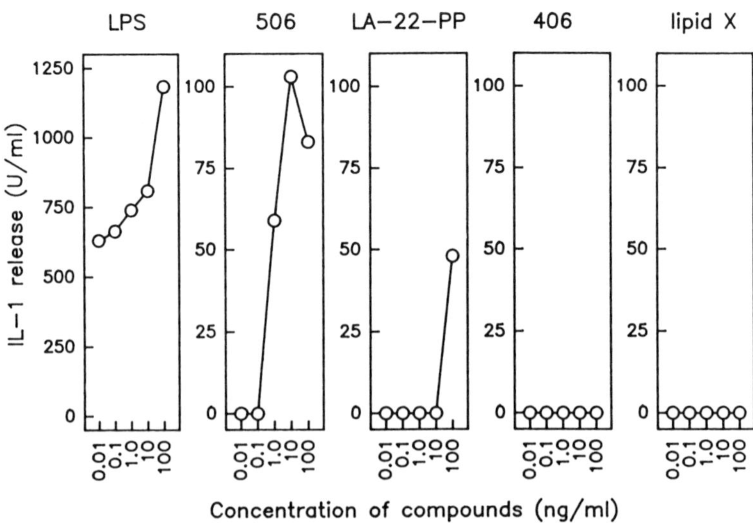

Figure 8 Induction of IL-1 by LPS, lipid A, and lipid A partial structures in human PBMCs. Human PBMCs (4×10^6/ml) were stimulated with LPS, synthetic lipid A (compound 506), or synthetic lipid A partial structures (LA-22-PP, compound 406, or lipid X) at concentrations ranging from 0.01 to 100 ng/ml. The results are given in IL-1 (U/ml). For further details see legend to Figure 7.

PE-1, PE-2, PE-3, and PE-4, have been prepared (85,86). These structures contain an oxyethyl-linked (-O-CH_2-CH_2-) phosphoryl group in position 1 of the reducing glucosaminyl residue (GlcN I) of lipid A. PE-1 is a hexaacylated analogue of *E. coli* lipid A (compound 506). PE-2 differs from PE-1 in having nonhydroxylated fatty acids at GlcN I. In PE-3, GlcN I is present in the β-anomeric form, and PE-4 is an analogue of tetraacyl precursor Ia (compound 406).

Investigations of the biological activity of these lipid A analogues revealed that the attachment of the glycosidic phosphoryl residue via an oxyethyl group had no effect on the biological activity of lipid A or related partial structures. Thus, PE-1 exhibited the same endotoxic properties as compound 506 in vivo and in vitro (87). Interestingly, when using such phosphonooxyethyl analogues of lipid A, namely, PE-2 and PE-3, it was found that the anomeric configuration of GlcN I is of great biological relevance as the α-anomer expressed high endotoxic activity, the β-anomer expressed low endotoxic activity. In conclusion, novel lipid A analogues are now available that can be more easily synthesized and purified and that are more stable than the former synthetic lipid A analogues. The stable α-phosphonooxyethyl glycoside makes the synthetic strategy more flexible and extends possibilities to create new compounds for laboratory and clinical applications.

The acyl residues at the lipid A backbone possess great significance regarding the biological activity for lipid A and LPS. This has, for example, been demonstrated previously by Loppnow et al. (88) who showed that LPS from *Rhodobacter capsulatus* exhibited only low monokine-inducing activities. LPS from *R. capsulatus* is composed of a lipid A backbone containing two amide-linked 3-oxo-myristic acids and two ester-linked 3OH-C10:0 fatty acids.

In all experiments performed so far, we were not able to stimulate IL-1, IL-6, or TNFα production in human monocytes by using the synthetic hexosamine monosaccharide precursor lipid X. In contrast, other investigators reported the induction of IL-1 and TNFα by synthetic lipid X (89,90). However, the stimulatory activity of synthetic lipid X was demonstrated to be due to bioactive contaminants (91). Interestingly, chemically synthesized monosaccharide analogues corresponding to the nonreducing part of lipid A (GLA27 and GLA60) have been shown to induce TNFα or IL-1, although lacking pyrogenic activity or Shwartzman reactivity (92,93). Furthermore, in comparison to *E. coli*–type lipid A (compound 506) the effective concentration of GLA27 is about 30 times lower, which supports the conclusion that only structures having a disaccharide (with two hexosamine residues) express full bioactivity in vitro and without exception in vivo.

It should be noted that *E. coli*–type lipid A (compound 506) is not as active as LPS in inducing monokines. First, the minimal concentration of lipid A necessary for the stimulation of monokine production by human mononuclear cells (MNCs) is about 10-fold higher than the minimal bioactive concentration of LPS. Second, the amount of monokines (IL-1, IL-6, or TNFα) produced after stimulation with lipid A is lower than after stimulation with LPS (see Fig. 7 for TNF production). Therefore, the hydrophilic oligosaccharides derived from the core region of LPS also were analyzed and it was found that in fact Kdo-containing oligosaccharides express monokine-inducing capacity (94,95). These oligosaccharides, however, are only active in high amounts. Other studies show that synthetic Kdo-containing structures fail to induce IL-1 (96). Nevertheless, the significance of the core oligosaccharides bound to lipid A is obvious: free lipid A is significantly less active than LPS and, therefore, the core oligosaccharide bound to lipid A is of apparent importance for biological activity of LPS. However, the number of glycosyl residues appears to be of minor importance for bioactivity, as LPS from the *H. influenzae* mutant I69, having only one Kdo bound to the lipid A backbone, expresses qualitatively and quantitatively similar activity than *S. abortus-equi* (S-form) LPS or *S. minnesota* Re LPS (R-form) (A. J. Ulmer, unpublished observations). Whether only Kdo may contribute to the enhanced bioactivity of LPS or whether Kdo can be substituted by other saccharides has to be investigated. It has been discussed that the oligosaccharides bound to lipid A may interact with a second type of receptor for LPS and thus result in an optimal response of the cells (97). Indeed, there is growing evidence for more than one specific LPS-binding site on macrophages that may recognize different substructures of LPS or lipid A, respectively (97–100). On the other hand, the oligosaccharide residues may change the physicochemical properties of LPS and contribute to conformational changes within the LPS molecule. Thus, LPS and lipid A may have different binding receptor properties.

It is of great interest that human monocytes have different structure requirements of lipid A for inducing monokine production than murine macrophages. Thus, compound 406, which is not active on human monocytes, shows endotoxic activities in vitro and in vivo in mice (101,102). A comparative study concerning the activity of compound 406 in vitro, using whole blood from different species, indicates that compound 406 is inactive only in primates and humans (A. J. Ulmer, unpublished observations). The reason for the different structural requirements of lipid A in different animal species and humans are still unknown.

In summary, we conclude from our experiments that the biological activity of lipid A and lipid A partial structures depends on the phosphorylation and acylation pattern of the hexosamine disaccharide. Maximal monokine-inducing activity is displayed by the bisphosphorylated lipid A possessing six acyl residues, which structurally corresponds to *E. coli*–type lipid A (compound 506). Partial structures lacking one of these components, structures containing different constituents, or structures with a different pattern of constituents are either less or not active in inducing monokines. Although the α-glycosyl phosphate group is an important constituent for the expression of biological activity of lipid A, its spacing from the backbone by introduction of an oxyethyl group has no considerable effect on activity. In addition to the activity relationship of the structure of lipid A, it is obvious that core region oligosaccharides contribute to the bioactivity of lipid A. In this respect only one Kdo is sufficient.

IV. INHIBITORY ACTION OF LIPID A ANALOGUES

In recent studies, we have analyzed the combined effect of LPS and their partial structures on monokine-induction. These investigations were performed in the hope that nonactive partial structures may compete with endotoxin-mediated reactions. Such a competitive action should differ from the so-called endotoxin-induced LPS resistance or LPS tolerance, which constitutes an actively induced refractory state of cells (in vitro) or animals (in vivo) rather than a competitive inhibitory effect (103–105). Up to now a variety of different lipid A analogues have been tested for their inhibitory effect on LPS-induced reactions in vitro and in vivo.

In fact, it was found that lipid A partial structures inhibit lipid A or endotoxin-induced activity. Low inhibitory activity has been reported for lipid X, the monosaccharide lipid A precursor in different animal or cellular models (81,106–108). More efficient antagonists of LPS-mediated endotoxic effects were found within the group of disaccharide lipid A analogues. Of special interest was the action of lipid A precursor Ia (compound 406) on endotoxin-induced IL-1, IL-6, and TNFα production by human monocytes. We (Fig. 9) and later others found that the synthetic precursor Ia is able to effectively inhibit monokine production induced by LPS (81,84,95,106,109). Precursor Ia exerted its inhibitory effect not by an inhibition of the release of monokines but at an earlier stage of monokine production: Already the LPS-induced specific mRNA production for monokines (IL-1 and TNFα) was found to be blocked by precursor Ia, as shown by Northern blot analysis (see Fig. 9). Similar results were obtained by Kovach et al. (83), using the structurally identical lipid IV_A isolated from *S. typhimurium,* and by Takayama et al. (110), using pentaacyl diphosphoryl lipid A of *Rhodobacter sphaeroides*. Furthermore, the nonactive LPS of *R. capsulatus* also shows inhibitory activity when investigated on LPS-induced monokine production in a specific manner (88).

Investigations on the structure-activity relationship of the inhibitory effects of disaccharide lipid A analogues proved that compound 406 and its phosphonooxyethyl analogue PE-4 are the most potent inhibitory compounds known so far (87). A lipid A partial structure with only two acyl groups (e.g., compound 606) (Fig. 10) is

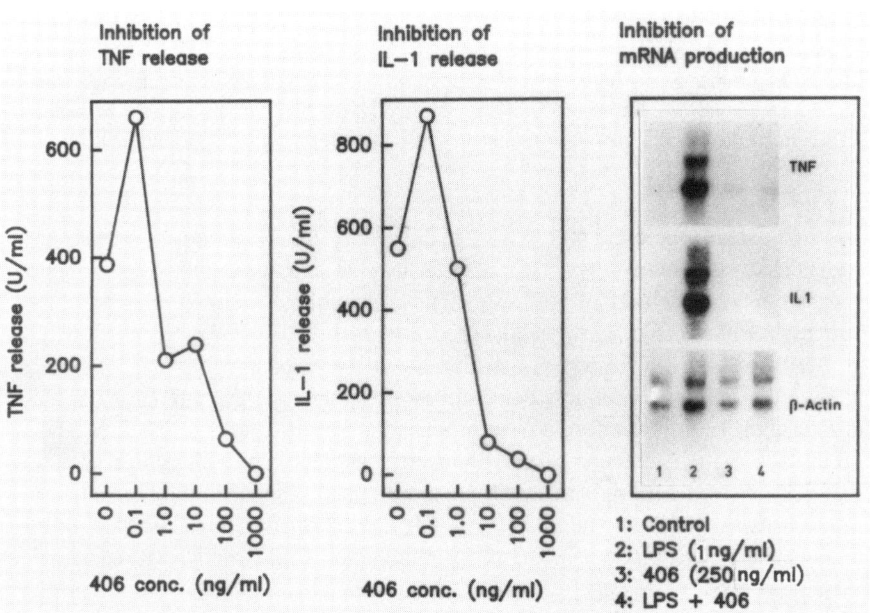

Figure 9 Inhibition of LPS-induced TNFα and IL-1 production by compound 406. PBMCs (4×10^6/ml) were preincubated in the presence of compound 406 at concentrations as indicated. After 1 h, LPS were added at a concentration of 1 ng/ml. TNFα mRNA, IL-1 mRNA, and β-actin mRNA was measured by Northern blot analysis after 4 h of incubation, the release of TNFα and IL-1 in the culture supernatant after 12 h of incubation.

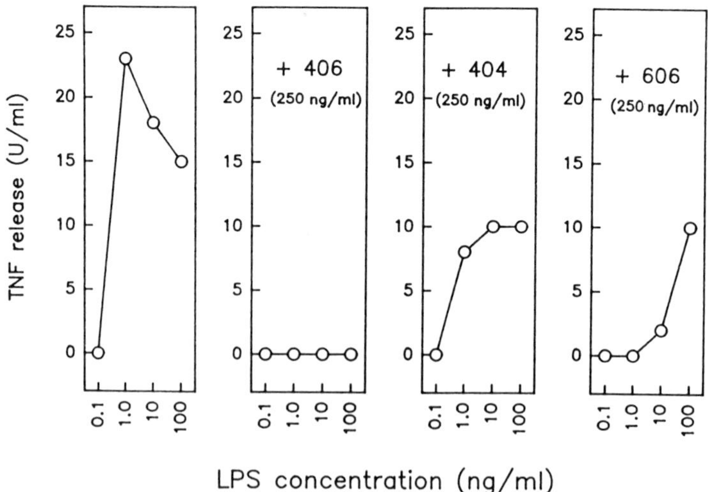

Figure 10 Inhibition of LPS-induced TNFα production by compound 406, 404, or 606. PBMCs (4×10^6/ml) were preincubated for 1 h in the presence of 406, 404, or 606 at 250 ng/ml, respectively. After 12 h of stimulation with LPS at different concentrations, the supernatants were harvested and assayed for TNFα activity in bioassays. The results are expressed as TNFα units per milliliter. Each value represents the mean of three independent cultures.

less active than compound 406. Also, the phosphorylation pattern is of importance for antagonistic activity (102). This indicates that the structure-activity requirements for the inhibitory activity of lipid A analogues also follows defined rules as do the structure requirements for the stimulatory activity.

All these findings raised the question about the mechanism of the inhibitory action. We presently assume a competitive mechanism for the inhibition by precursor Ia on LPS-induced monokine production. This assumption was first based on the observations that compound 406 exerted its inhibitory activity in a specific way and only on LPS induction. Monokine production induced by phorbolester, *S. epidermidis,* Calmette-Guérin bacillus, *Staphylococcus aureus* COWAN I, or by a lipopeptide (N-palmitoyl-(S)-[2,3-bis(palmitoyl-oxy)-(2RS)-propyl]-(R)-cysteinyl-seryl-lysyl-lysyl-lysyl-lysine was not affected (81,84,95). Second, it was found that compound 406 and PE-4 inhibited the binding of LPS to human and murine monocytes/macrophages (see below).

The results show that low acylated lipid A analogues, especially compound 406 and PE-4, represent potent antagonists of LPS-induced activities in human systems in vitro. It is possible that they may be able to prevent the LPS-mediated production of shock-inducing monokines also in vivo and thus represent a new approach in preventing and treating systemic endotoxemia.

V. BINDING OF LPS, LIPID A, OR LIPID A ANALOGUES TO MONOCYTES/MACROPHAGES

Interaction of LPS with a target cell (e.g., monocytes/macrophages) of an infected host is a prerequisite for the induction of endotoxic effects. The first event that can be observed is the binding to the surface membrane. Different cellular binding structures have been described: CD18 recognizes *E. coli* via LPS and is present on all leukocytes (111). It is noncovalently linked to either CD11a (LFA-1), CD11b (iC3b receptor), or CD11c (112). However, this LPS-recognizing antigen does not seem to be involved in activation of cytokine secretion by monocytes/macrophages by LPS (113). Other LPS-binding proteins that were detected on monocytes/macrophages are a 73-kd protein described by Morrison et al. (114–116) and a 40-kd protein detected by Kirikae (117). Although there is evidence that such proteins may serve as functional receptors for LPS (118), their participation in the activation of monocytes/macrophages certainly requires further investigation. On the other hand, it has unequivocally been proven that CD14 constitutes the prominent LPS-binding structure on monocytes/macrophages. Thus, interaction of LPS with CD14 is necessary for the specific binding and activation of human monocytes or murine macrophages (119–122). This finding does not exclude the presence of other functional LPS-binding structures on the cells. The affinity of LPS to CD14 seems to be rather low but can be augmented by an LPS-binding protein (LBP). The latter serum constituent forms a complex with LPS and enhances considerably the binding affinity to CD14. CD14 is bound to the membrane by a phosphoinositol anchor, and according to present knowledge, this anchor is not capable of transmitting transmembrane signals (123). Therefore, a further, up to now undefined membrane structure is likely to be involved in the LPS-mediated activation of monocytes.

Knowledge of the LPS receptors and/or binding structures leads to the question whether the different biological activities of lipid A partial structures is based on their different binding affinities to cells. Therefore, studies were performed to determine the binding of human monocytes or the murine macrophagelike cell line J774.1 (102,121). Because radiolabeled different lipid A partial structures were not available, the experiments were designed as inhibition experiments to determine the inhibition of binding of ^{125}I-LPS by unlabeled structures. Assuming that inhibition of ^{125}I-LPS binding by lipid A partial structures is of competitive nature, there should be a correlation between inhibition of binding and affinity to a given binding structure (or receptor). In these studies, the structure-activity relationship again shows that inhibition of binding of ^{125}I-LPS by lipid A analogues is a function of the degree of phosphorylation and the acylation pattern (Fig. 11) (102). The influence of the phosphorylation pattern was demonstrated by experiments that show that compound 404 (lacking the 1-phosphate group) as well as 405 (lacking the 4′-phosphate group) are less inhibitory for the binding of ^{125}I-LPS than compound 406 (102). In addi-

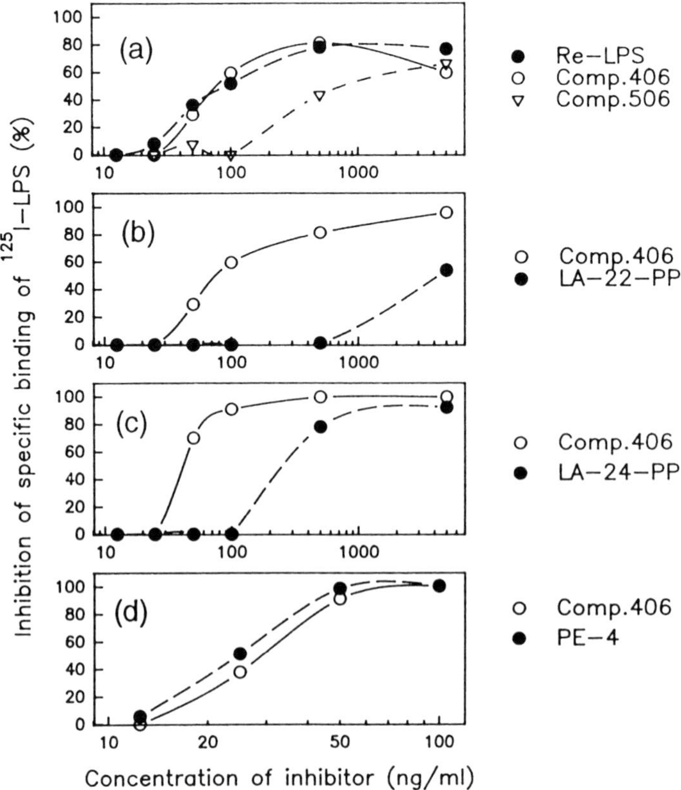

Figure 11 Inhibition of specific binding of ^{125}I-LPS to human monocytes by synthetic lipid A and related partial structures. Human monocytes (7×10^5/well) were incubated with ^{125}I-LPS in the presence or absence of unlabeled LPS, lipid A, or related partial structures (compound 506, 406, LA-22-PP, LA-24-PP, or PE-4) at different concentrations as indicated. The results are expressed as percent of inhibition of specific binding of ^{125}I-LPS to the cells. Each value represents the mean of three independent wells.

tion, the anomeric configuration of GlcN I was found to be of great biological relevance, as the α-anomer expressed high and the β-anomer low binding inhibitor capacity. For example, the tetraacylated compound 406 shows a more potent inhibitory activity than hexaacyl compound 506 or the bisacyl compound 606 (102,121). On the other hand, compound 506 is a stronger competitor than hexaacyl compound LA-22-PP, the latter having, in contrast to compound 506, a symmetrical distribution of the fatty acids. Also the nature of the acyl residues is of importance for the binding of lipid A analogues. Compound LA-23-PP, which contains four residues of nonhydroxylated decanoic acid (C10:0), as well as compound LA-24-PP, which carries two decanoic acids in addition to two hydroxylated acyl residues, have a pronounced reduced binding capacity compared with compound 406 (102,121).

Plotting the data according to Lineweaver-Burk, we can show that unlabeled LPS as well as PE-4 inhibit the binding of fluorescein isothiocyanate (FITC)–LPS in a competitive manner (Fig. 12). This indicates that the binding of LPS as well as the inhibition of binding of LPS by the investigated compounds is a specific, that is, true, antagonistic phenomenon.

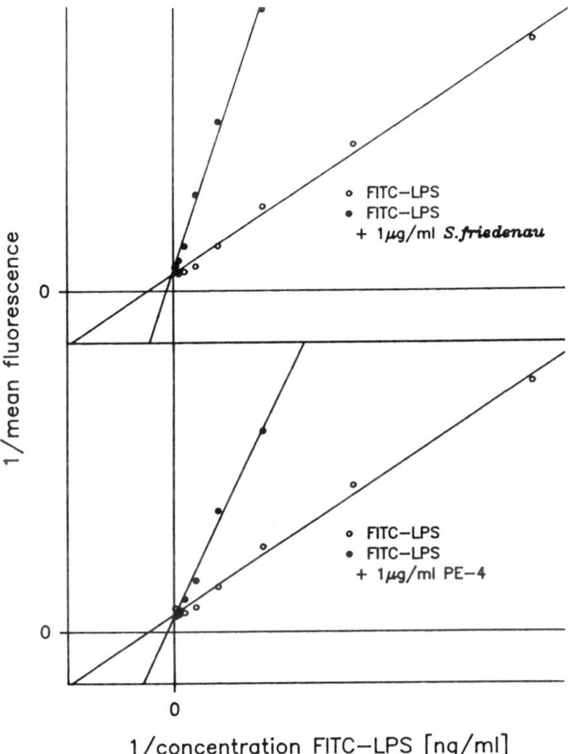

Figure 12 Lineweaver-Burk plot of binding inhibition of FITC-LPS by unlabeled LPS or PE-4. Human MNCs were labeled with FITC-LPS (0.01–20 μg/ml) in the presence of PE-4 (1 μg/ml) or unlabeled LPS (1 μg/ml). The fluorescence of labeled monocytes was determined by flow cytometry in an Epics Profile II. The results were presented in a Lineweaver-Burk plot. The regression coefficients were >0.99, respectively.

Our results show that the structure-activity relationship for the biological activity as well as for the binding capacity depends on the phosphorylation and acylation pattern. Between the structures expressing the optimal biological activity (monokine induction) and optimal binding affinity marked differences are recognized. Whereas these structures optimally expressing endotoxicity and binding contain a bisphosphorylated hydrophilic β–GlcN-(1-6)–αGlcN disaccharide backbone, optimal biological activity is seen with compound 506 having six acyl residues, and maximal binding affinity with compound 406 possessing four acyl residues. These results prove that optimal binding is not sufficient for induction of optimal biological activity. Therefore, one has to assume that in addition to binding a second event or feature (e.g., internalization or a second receptor) is necessary for maximal activation of cells.

Taken together, our results provide strong evidence for the concept that inhibition of LPS-cell interaction by compound 406 and other lipid A partial structures is based on a competitive inhibition of a specific LPS binding protein of the responder cell. In this way, these compounds deny access of LPS and thus may be able to prevent the fatal endotoxin-induced reactions of an infected host. Therefore, synthetic lipid A analogues, in particular compounds 406 and PE-4, could represent promising candidates for the development of new immunomodulators that may be used to prevent gram-negative septic shock and related disorders.

REFERENCES

1. Rietschel ET, Brade H. Bacterial endotoxins. Sci Am 1992; 267:54.
2. Galanos C, Lüderitz O, Rietschel ET, Westphal O. Newer aspects of the chemistry and biology of bacterial lipopolysaccharides, with special reference to their lipid A component. In: Goodwin TW, ed. International review of biochemistry, biochemistry of lipids II, Vol. 14. Baltimore: University Park Press, 1977:239.
3. Lüderitz O, Freudenberg MA, Galanos C, Lehmann V, Rietschel ET, Shaw DH. Lipopolysaccharides of gram-negative bacteria. In: Razin S, Rottem S, eds. Membrane lipids of procaryotes. Current topics in membranes and transport, Vol. 17. New York: Academic Press, 1982:79.
4. Jann K, Jann B. Structure and biosynthesis of O-antigens. In: Rietschel ET, ed. Handbook of endotoxin, Vol. 1: Chemistry of endotoxin. Amsterdam: Elsevier Science Publishers, 1984:186.
5. Jann B, Shashkov AS, Gupta DS, Jann K. The O18 antigens (lipopolysaccharides) of *Escherichia coli*. Structural characterization of the O18A, O18B and O18B1-specific polysaccharides. Eur J Biochem 1992; 210:241.
6. Gamian A, Romanowska E, Ulrich J, Defaye J. The structure of the sialic acid-containing *Escherichia coli* O104 O-specific polysaccharide and its linkage to the core region in lipopolysaccharide. Carbohydr Res 1992; 236:195.
7. Kogan G, Jann B, Jann K. Structure of the O24 antigen of *Escherichia coli*, a neuraminic acid-containing polysaccharide. Carbohydr Res 1993; 238:335.

8. Kogan G, Shashkov AS, Jann B, Jann K. Structure of the 056 antigen of *Escherichia coli,* a polysaccharide containing 7-substituted α-N-acetylneuramic acid. Carbohydr Res 1993; 238:261.
9. Weintraub A, Leontein K, Widmalm G, Vial PA, Levine MM, Lindberg AA. Structural studies of the O-antigenic polysaccharide of an enteroaggregative *Escherichia coli* strain. Eur J Biochem 1993; 213:859.
10. Knirel YA. Polysaccharide antigens of *Pseudomonas aeruginosa.* CRC Crit Rev Microbiol 1990; 17:273.
11. Gamian A, Romanowska E, Dabrowski U, Dabrowski J. Structure of the O-specific, sialic acid containing polysaccharide chain and its linkage to the core region in lipopolysaccharide from *Hafnia alvei* strain 2 as elucidated by chemical methods, gas-liquid chromatography/mass spectrometry, and ^{1}H nmr spectroscopy. Biochemistry 1991; 30:5032.
12. Oxley D, Wilkinson SG. Structure of the 021 antigen from *Serratia marcescens.* Carbohydr Res 1991; 212:187.
13. Oxley D, Wilkinson SG. Structure of a mannan isolated from the lipopolysaccharide of the reference strain (S3255) for a new serogroup of *Serratia marcescens.* Carbohydr Res 1991; 212:213.
14. Oxley D, Wilkinson SG. Structure of a neutral glycan from the lipopolysaccharide of reference strains for *Serratia marcescens* serogroups O2 and O3. FEMS Microbiol Lett 1992; 99:209.
15. Lüderitz O, Galanos C, Risse HJ, Ruschmann E, Schlecht S, Schmidt G, Schulte-Holthusen H, Wheat R, Westphal O, Schlosshardt J. Structural relationships of *Salmonella* O and R antigens. Ann NY Acad Sci 1966; 133:349.
16. Lüderitz O, Staub AM, Westphal O. Immunochemistry of O and R antigens. Bacteriol Rev 1966; 30:192.
17. Raetz CRH. Biochemistry of endotoxins. Annu Rev Biochem 1990; 59:129.
18. Morona R, Brown MH, Yeadon J, Heutzenroeder MW, Manning PA. Effect of lipopolysaccharide core synthesis on the production of *Vibrio cholerae* O-antigen in *Escherichia coli* K-12. FEMS Microbiol Lett 1991; 82:279.
19. Cheah K-C, Beger DW, Manning PA. Molecular cloning and genetic analysis of the rfb region from *Shigella flexneri* type 6 in *Escherichia coli* K-12. FEMS Microbiol Lett 1991; 83:213.
20. Yao Z, Liu H, Valvano MA. Acetylation of O-specific lipopolysaccharides from *Shigella flexneri* 3a and 2a occurs in *Escherichia coli* K-12 carrying cloned *S. flexneri* 3a and 2a *rbf* genes. J Bacteriol 1992; 174:7500.
21. McGrath BC, Osborn MJ. Localization of the terminal steps of O-antigen biosynthesis in *Salmonella typhimurium.* J Bacteriol 1991; 173:649.
22. McGrath BC, Osborn MJ. Evidence for energy-dependent transposition of core lipopolysaccharide across the inner membrane of *Salmonella typhimurium.* J Bacteriol 1991; 173:3134.
23. Holst O, Brade H. Chemical structure of the core region of bacterial lipopolysaccharides. In: Morrison DC, Ryan JL, eds. Bacterial endotoxic lipopolysaccharides, Vol. 1. Boca Raton, FL: CRC Press, 1992:135.
24. Unger FM. The chemistry and biological significance of 3-deoxy-D-*manno*-2-octulosonic acid (Kdo). Adv Carbohydr Chem Biochem 1981; 38:323.

25. Radziejewska-Lebrecht J, Mayer H. The core region of *Proteus mirabilis* R110/1959 lipopolysaccharide. Eur J Biochem 1989; 183:573.
26. Weintraub A, Zähringer U, Lindberg AA. Structural studies on the polysaccharide part of the cell wall lipopolysaccharide from *Bacteroides fragilis* NCTC 9343. Eur J Biochem 1985; 151:657.
27. Coleman WG Jr. The *rfaD* gene codes for ADP-L-*glycero*-D-*manno*-heptose-6-epimerase: An enzyme required for lipopolysaccharide core biosynthesis. J Biol Chem 1983; 258:1985.
28. Kocsis B, Kontrohr T. Isolation of adenosine 5′-diphosphate-L-*glycero*-D-*manno*-heptose, the assumed substrate of heptose transferase(s), from *Salmonella minnesota* R595 and *Shigella sonnei* Re mutants. J Biol Chem 1984; 259:11858.
29. Helander IM, Lindner B, Brade H, Altmann K, Lindberg AA, Rietschel ET, Zähringer U. Chemical structure of the lipopolysaccharide of *Haemophilus influenzae* strain I-69 Rd^-/b^+: Description of a novel deep-rough chemotype. Eur J Biochem 1988; 177:483.
30. Kawahara K, Brade H, Rietschel ET, Zähringer U. Studies on the chemical structure of the core-lipid A region of the lipopolysaccharide of *Acinetobacter calcoaceticus* NCTC 10305: Detection of a new 2-octulosonic acid interlinking the core oligosaccharide and lipid A. Eur J Biochem 1987; 163:489.
31. Rietschel ET, Brade H, Brade L, Kaca W, Kawahara K, Lindner B, Lüderitz T, Tomita T, Schade U, Seydel U, Zähringer U. Newer aspects of the chemical structure and biological activity of bacterial endotoxins. In: Cate JW, Buller HR, Sturke A, Levin J, eds. Bacterial endotoxins: Structure, biomedical significance, and detection with the *Limulus amebocyte lysate* test. New York: Alan R. Liss, 1985:31.
32. Holst O, Röhrscheidt-Andrzejewski E, Cordes H-P, Brade H. Isolation and identification of 3-deoxy-7-O-α-D-galactopyranosyl-D-*manno*-2-octulopyranosonate from the inner core region of the lipopolysaccharide of *Escherichia coli* EH100. Carbohydr Res 1989; 188:212.
33. Holst O, Brade H. Isolation and identification of 3-deoxy-5-O-α-L-rhamnopyranosyl-D-*manno*-2-octulopyranosonate from the inner core region of the lipopolysaccharide of *Escherichia coli* K-12. Carbohydr Res 1990; 207:327.
34. Brade H, Tacken A, Christian R. Isolation and identification of a rhamnosyl-rhamnosyl-3-deoxy-D-*manno*-octulosonic acid trisaccharide from the lipopolysaccharide of *Acinetobacter calcoaceticus* (10303 NTCC London). Carbohydr Res 1987; 167:295.
35. Haishima Y, Holst O, Brade H. Structural investigation on the lipopolysaccharide of *Escherichia coli* rough mutant F653 representing the R3 core type. Eur J Biochem 1992; 203:217.
36. Haishima Y, Holst O, Brade H. Structural investigation on the lipopolysaccharide of *Escherichia coli* rough mutant F653 representing the R3 core type: Correction. Eur J Biochem 1992; 207:1129.
37. Kaca W, de Jongh-Leuvenink J, Zähringer U, Rietschel ET, Brade H, Verhoef J, Sinnwell V. Isolation and chemical analysis of 7-O-(2-amino-2-deoxy-α-D-glucopyranosyl)-L-*glycero*-D-*manno*-heptose as a constituent of the lipopolysaccharide of the UDP-galactose epimerase-less mutant J-5 of *Escherichia coli* and *Vibrio cholerae*. Carbohydr Res 1988; 179:289.

38. Holst O, Zähringer U, Brade H, Zamojski A. Structural analysis of the heptose/hexose region of the lipopolysaccharide from *Escherichia coli* K-12 strain W3100. Carbohydr Res 1991; 215:323.
39. Volk WA, Salomonsky NL, Hunt D. *Xanthomonas sinensis* cell wall lipopolysaccharide. I. Isolation of 4,7-anhydro- and 4,8-anhydro-3-deoxy-octulosonic acid following acid hydrolysis of *Xanthomonas sinensis* lipopolysaccharide. J Biol Chem 1992; 247:3881.
40. Brade H, Galanos C, Lüderitz O. Differential determination of the 3-deoxy-D-*manno*octulosonic acid residues in lipopolysaccharides of *Salmonella minnesota* rough mutants. Eur J Biochem 1983; 131:195.
41. Tacken A, Rietschel ET, Brade H. Methylation analysis of the heptose/3-deoxy-D-*manno*-2-octulosonic acid region (inner core) of the lipopolysaccharide from *Salmonella minnesota* rough mutants. Carbohydr Res 1986; 149:279.
42. Mäkelä PH, Stocker BAD. Genetics of lipopolysaccharide. In: Rietschel ET, ed. Handbook of endotoxin, Vol. 1: Chemistry of endotoxin. Amsterdam: Elsevier Science, 1984:59.
43. Austin EA, Graves JP, Hite LA, Parker CT, Schnaitman CA. Genetic analysis of lipopolysaccharide core biosynthesis by *Escherichia coli* K-12: Insertion mutagenesis of the *rfa* locus. J Bacteriol 1990; 172:5312.
44. Pradel E, Schnaitman CA. Effect of *rfaH* (*sfrB*) and temperature on expression of *rfa* genes of *Escherichia coli* K-12. J Bacteriol 1991; 173:6428.
45. Pradel E, Parker CT, Schnaitman CA. Structures of the *rfaB, rfaI, rfaJ,* and *rfaS* genes of *Escherichia coli* K-12 and their roles in assembly of the lipopolysaccharide core. J Bacteriol 1992; 174:4736.
46. Klena JD, Pradel E, Schnaitman CA. Comparison of lipopolysaccharide biosynthesis genes *rfaK, rfaL, rfaY,* and *rfaZ* of *Escherichia coli* K-12 and *Salmonella typhimurium.* J Bacteriol 1992; 174:4746.
47. Sirisena DM, Brozek KA, MacLachlan PR, Sanderson KE, Raetz CRH. The *rfaC* gene of *Salmonella typhimurium.* Cloning, sequenzing, and enzymatic function in heptose transfer to lipopolysaccharide. J Biol Chem 1992; 267:18874.
48. Rietschel ET, Wollenweber H-W, Brade H, Zähringer U, Lindner B, Seydel U, Bradaczek H, Barnickel G, Labischinski H, Griesbrecht P. Structure and conformation of the lipid A component of lipopolysaccharides. In: Rietschel ET, ed. Handbook of endotoxin, Vol. 1: Chemistry of endotoxin. Amsterdam: Elsevier Science, 1984:187.
49. Rosner MR, Khorana HG, Satterthwait AC. The structure of lipopolysaccharide from a heptose-less mutant of *Escherichia coli* K-12. II. The application of ^{31}P NMR spectroscopy. J Biol Chem 1979; 244:5918.
50. Zähringer U, Lindner B, Seydel U, Rietschel ET, Naoki H, Unger FM, Imoto M, Kusumoto S, Shiba T. Structure of de-O-acylated lipopolysaccharides from the *Escherichia coli* Re mutant strain F515. Tetrahedron Lett 1985; 26:6321.
51. Seydel U, Lindner B, Wollenweber H-W, Rietschel ET. Structural studies on the lipid A component of enterobacterial lipopolysaccharides by laser desorption mass spectrometry. Location of acyl groups at the lipid A backbone. Eur J Biochem 1985; 145:505.
52. Wollenweber H-W, Rietschel ET. Analysis of lipopolysaccharide (lipid A) fatty acids. J Microbiol Methods 1990; 11:195.

53. Hase S, Rietschel ET. The chemical structure of the lipid A component of lipopolysaccharides from *Chromobacterium violaceum* NCTC 9694. Eur J Biochem 1977; 75:23.
54. Wollenweber H-W, Seydel U, Lindner B, Lüderitz O, Rietschel ET. Nature and location of amide-bound (R)-3-acyloxyacyl groups in lipid A of lipopolysaccharides from various gram-negative bacteria. Eur J Biochem 1984; 145:265.
55. Mayer H, Krauss JH, Yokota A, Weckesser J. Natural variants of lipid A. In: Friedman H, Klein TW, Nakano M, Nowotny A, eds. Endotoxin. New York: Plenum, 1990:45.
56. Moran AP, Zähringer U, Seydel U, Scholz D, Stütz P, Rietschel ET. Structural analysis of the lipid A component of *Campylobacter jejuni* CCUG 10936 (serotype O:2) lipopolysaccharide: Description of a lipid A containing a hybrid backbone of 2-amino-2-deoxy-D-glucose and 2,3-diamino-2,3-dideoxy-D-glucose. Eur J Biochem 1991; 198:459.
57. Bhat UR, Mayer H, Yokota A, Hollingsworth RI, Carlson RW. Occurrence of lipid A variants with 27-hydroxyoctacosanoic acid in lipopolysaccharides from members of the family *Rhizobiaceae*. J Bacteriol 1991; 173:2155.
58. Sonesson A, Jantzen E, Bryn K, Larsson L, Eng J. Chemical composition of a lipopolysaccharide from *Legionella pneumophila*. Arch Microbiol 1989; 153:72.
59. Moll H, Sonesson A, Jantzen E, Marre R, Zähringer U. Identification of 27-oxo-octacosanoic acid and heptacosane-1,27-dioic acid in *Legionella pneumophilia*. FEMS Microbiol Lett 1992; 97:1.
60. Nurminen M, Rietschel ET, Brade H. Chemical characterization of *Chlamydia trachomatis* lipopolysaccharide. Infect Immun 1985; 48:573.
61. Brade L, Schramek S, Schade U, Brade H. Chemical, biological, and immunochemical properties of the *Chlamydia psittaci* lipopolysaccharide. Infect Immun 1986; 54:568.
62. Bhat UR, Carlson RW, Busch M, Mayer H. Distribution and phylogenetic significance of 27-hydroxyoctacosanoic acid in lipopolysaccharides from bacteria belonging to the alpha-2 subgroup of *Proteobacteria*. Int J Syst Bacteriol 1991; 41:213.
63. Kusumoto S. Chemical synthesis of lipid A. In: Morrison DC, Ryan JL, eds. Bacterial endotoxic lipopolysaccharides, Vol. 1. Boca Raton, FL: CRC Press, 1992:81.
64. Kotani S, Takada H, Tsujimoto M, Ogawa T, Takahashi I, Ikeda T, Otsuka K, Shimauchi H, Kasai N, Mashimo J, Nagao S, Tanaka A, Tanaka S, Harada K, Nagaki K, Kitamura H, Shiba T, Kusumoto S, Imoto M, Yoshimura H. Synthetic lipid A with endotoxic and related biological activities comparable to those of a natural lipid A from an *Escherichia coli* Re-mutant. Infect Immun 1985; 49:225.
65. Galanos C, Lüderitz O, Rietschel ET, Westphal O, Brade H, Brade L, Freudenberg M, Schade U, Imoto M, Yoshimura H, Kusumoto S, Shiba T. Synthetic and natural *Escherichia coli* free lipid A express identical endotoxic activities. Eur J Biochem 1985; 148:1.
66. Homma JY, Matsuura M, Kanegasaki S, Kawakubo Y, Kojima Y, Shibukawa N, Kumazawa Y, Yamamoto A, Tanamoto K, Yasuda T, Imoto M, Yoshimura H, Kusumoto S, Shiba T. Structural requirements of lipid A responsible for the functions: A study with chemically synthesized lipid A and its analogues. J Biochem Tokyo 1985; 98:395.

67. Takada H, Kotani S. Structure-function relationships of lipid A. In: Morrison DC, Ryan JL, eds. Bacterial endotoxic lipopolysaccharides, Vol. 1. Boca Raton, FL: CRC Press, 1992:107.
68. Seydel U, Brandenburg K. Supramolecular structure of lipopolysaccharides and lipid A. In: Morrison DC, Ryan JL, eds. Bacterial endotoxic lipopolysaccharides, Vol. 1. Boca Raton, FL: CRC Press, 1992:225.
69. Kastowsky M, Gutberlet T, Bradaczyk H. Molecular modelling of the three-dimensional structure and conformational flexibility of bacterial lipopolysaccharide. J Bacteriol 1992; 174:4798.
70. Young LS, Stevens P, Kaijser B. Gram-negative pathogens in septicaemicy infections. Scand J Infect Dis 1982; 31(Suppl.):78.
71. Dianarello CA. Was the original endogenous pyrogen interleukin-1? In: Bomford R, Henderson B, eds. Interleukin-1, inflammation and disease. Amsterdam: Elsevier, 1989:17.
72. Schindler R, Dinarello A. Interleukin 1. In: Habenicht A, ed. Growth factors, differentiation factors, and cytokines. Berlin: Springer-Verlag, 1990:85.
73. Van Snick J, Nordan RP. Interleukin 6. In: Habenicht A, ed. Growth factors, differentiation factors, and cytokines. Berlin: Springer-Verlag, 1990:163.
74. Bendtzen K. Interleukin 1, interleukin 6 and tumor necrosis factor in infection, inflammation and immunity. Immunol Lett 1988; 19:183.
75. Galanos C, Lüderitz O, Rietschel ET, Westphal O, Brade H, Brade L, Freudenberg MA, Schade U, Imoto M, Yoshimura S, Kusumoto S, Shiba T. Synthetic and natural *Escherichia coli* free lipid A express identical endotoxic activities. Eur J Biochem 1985; 148:1.
76. Imoto M, Yoshimura H, Shimamoto T, Sakaguchi N, Kusumoto S, Shiba T. Total synthesis of *Escherichia coli* lipid A, the endotoxically active principle of cell-surface lipopolysaccharide. Bull Chem Soc Jpn 1987; 60:2205.
77. Kusumoto S, Yamamoto M, Shiba T. Chemical synthesis of lipid X and lipid Y, acylglucosamine-1-phosphates isolated from *Escherichia coli* mutants. Tetrahedr Lett 1984; 25:3727.
78. Rietschel ET, Brade L, Schade U, Seydel U, Zähringer U, Kusumoto S, Brade H. Bacterial endotoxins: Properties and structure of biologically active domains. In: Schrinner E, Richmond MH, Seibert G, Schwarz U, eds. Surface of microorganisms and their interactions with the mammalian host. Weinheim: Verlag Chemie, 1988:1.
79. Loppnow H, Brade L, Brade H, Rietschel ET, Kusumoto S, Shiba T, Flad H-D. Induction of human interleukin 1 by bacterial and synthetic lipid A. Eur J Immunol 1986; 16:1263.
80. Feist W, Ulmer AJ, Musehold J, Brade H, Kusumoto S, Flad H-D. Induction of tumor necrosis factor-alpha release by lipopolysaccharide and defined lipopolysaccharide partial structures. Immunobiology 1989; 179:293.
81. Wang M-H, Flad H-D, Feist W, Brade H, Kusumoto S, Rietschel ET, Ulmer AJ. Inhibition of endotoxin-induced interleukin 6 production by synthetic lipid A partial structures in human peripheral blood mononuclear cells. Infect Immun 1991; 59:4655.
82. Rietschel ET, Kirikae T, Feist W, Loppnow H, Zabel P, Brade L, Ulmer AJ, Brade H, Seydel U, Zähringer U, Schlaak M, Flad H-D, Schade U. Molecular aspects of

the chemistry and biology of endotoxin. In: Sies H, Flohé L, Zimmer G, eds. Molecular aspects of inflammation (42. Colloquium Mosbach, 1991). Berlin: Springer-Verlag, 1991:207.

83. Kovach NL, Yee E, Munford RS, Raetz CRH, Harlan JM. Lipid IVa inhibits synthesis and release of tumor necrosis factor induced by lipopolysaccharide in human whole blood ex vivo. J Exp Med 1990; 172:77.
84. Feist W, Ulmer AJ, Wang M-H, Musehold J, Schlüter C, Gerdes J, Herzbeck H, Brade H, Kusumoto S, Diamantstein T, Rietschel ET, Flad H-D. Modulation of lipopolysaccharide-induced production of tumor necrosis factor, interleukin 1, and interleukin 6 by synthetic precursor Ia of lipid A. FEMS Microbiol Immunol 1992; 89:73.
85. Kusama T, Soga T, Shioya E, Nakayama K, Nakajima H, Osada Y, Ono Y, Kusumoto S, Shiba T. Synthesis and antitumor activity of lipid A analogs having a phosphonooxyethyl group with α- or β-configuration at position 1. Chem Pharm Bull 1990; 38:3366.
86. Kusama T, Soag T, Ono Y, Kumazawa E, Shioya E, Osada Y, Kusumoto S, Shiba T. Synthesis and biological activities of analogs of a lipid A biosynthetic precursor: 1-*O*-phosphonooxyethyl-4′-*O*-phosphono-disaccharides with (*R*)-3-hydroxytetradecanoyl or tetradecanoyl groups at positions 2, 3, 2′ and 3′. Chem Pharm Bull 1991; 39:1994.
87. Ulmer AJ, Heine H, Feist W, Kusumoto S, Kusama T, Brade H, Rietschel ET, Flad H-D. Biological effects of synthetic phosphooxyethyl analogues of lipid A and lipid A partial structures. Infect Immun 1992; 60:3309.
88. Loppnow H, Libby P, Freudenberg MA, Kraus JH, Weckesser J, Mayer H. Cytokine induction by lipopolysaccharide (LPS) corresponds to the lethal toxicity and is inhibited by nontoxic *Rhodobacter capsulatus* LPS. Infect Immun 1990; 58:3743.
89. Lasfargues A, Chaby R. Endotoxin-induced tumor necrosis factor (TNF): selective triggering of TNF and interleukin-1 production by distinct glucosamine-derived lipids. Cell Immunol 1988; 115:165.
90. Sayers TJ, Macher I, Chung J, Kugler E. The production of tumor necrosis factor by mouse bone marrow-derived macrophages in response to bacterial lipopolysaccharide and a chemically synthesized monosaccharide precursor. J Immunol 1987; 138:2935.
91. Aschauer H, Grob A, Hildebrand J, Schuetze E, Stuetz P. Highly purified lipid X is devoid of immunostimulatory activity. Isolation and characterization of immunostimulating contaminants in a batch of synthetic lipid X. J Biol Chem 1990; 265:9159.
92. Homma JH, Matsuura M, Kumazawa Y. Structure activity relationship of chemically synthesized nonreducing parts of lipid A analogs. Adv Exp Med Biol 1990; 256:101.
93. Saiki I, Maeda H, Murata J, Takahashi T, Sekiguchi S, Kiso M, Hasegawa A, Azuma I. Production of interleukin 1 from human monocytes stimulated by synthetic lipid A subunit analogues. Int J Immunopharmacol 1990; 12:297.
94. Lebbar S, Cavaillon J-M, Caroff M, Ledur A, Brade H, Sarfati R, Haeffner-Cavaillon N. Molecular requirement for interleukin 1 induction by lipopolysaccharide-stimulated human monocytes: Involvement of the heptosyl-2-keto 3-deoxyoctulosonate region. Eur J Immunol 1986; 16:87.

95. Loppnow H, Brade H, Duerrbaum I, Dinarello CA, Kusumoto S, Rietschel ET, Flad H-D. IL-1 induction capacity of defined lipopolysaccharide and partial structures. J Immunol 1989; 142:3229.
96. Lasfargues A, Ledur A, Charon D, Szabo L, Chaby R. Induction by lipopolysaccharide of intracellular and extracellular interleukin 1 production: Analysis with synthetic models. J Immunol 1987; 139:429.
97. Haeffner-Cavaillon N, Cavaillon J-M. Involvement of the LPS receptor in the induction of interleukin-1 in human monocytes stimulated with endotoxins. Ann Inst Pasteur Immunol 1987; 138:473.
98. Morrison DC. The case for specific lipopolysaccharide receptors expressed on mammalian cells. Microbiol Pathogenesis 1989; 7:389.
99. Golenbock DT, Hampton RY, Raetz CRH, Wright SD. Human phagocytes have multiple lipid A-binding sites. Infect Immun 1990; 58:4069.
100. Tahri-Jouti M-A, Mondange M, Le Dur A, Auzanneau F-I, Charon D, Girard D, Chaby R. Specific binding of lipopolysaccharides to mouse macrophages. II. Involvement of distinct lipid A substructures. Mol Immunol 1990; 27:763.
101. Galanos C, Lehmann V, Lüderitz O, Rietschel ET, Westphal O, Brade H, Brade L, Freudenberg MA, Hansen-Hagge T, Lüderitz T, McKenzie G, Schade U, Strittmatter W, Tanamoto K, Zähringer U, Imoto M, Yoshimura H, Yamamoto M, Shimamoto T, Kusumoto S, Shiba T. Endotoxic properties of chemically synthesized lipid A part structures: Comparison of synthetic lipid A precursor and synthetic analogues with biosynthetic lipid A precursor and free lipid A. Eur J Biochem 1984; 140:221.
102. Kirikae T, Schade FU, Zähringer U, Kirikae F, Brade H, Kusumoto S, Kusama T, Rietschel ET. The influence of the hydrophilic backbone and the hydrophobic fatty acid regions of lipid A on macrophage binding and cytokine induction. FEMS Immunol Med Microbiol 1994; 8:13.
103. Beeson PB. Tolerance to bacterial pyrogens. I. Factors influencing its development. J Exp Med 1947; 86:29.
104. Freudenberg MA, Galanos C. Induction of tolerance to lipopolysaccharide (LPS)-D-galactosamine lethality by pretreatment with LPS is mediated by macrophages. Infect Immun 1988; 56:1352.
105. Mathison JC, Virca GD, Wolfson E, Tobias PS, Glaser K, Ulevitch RJ. Adaptation to bacterial lipopolysaccharide controls lipopolysaccharide-induced tumor necrosis factor production in rabbit macrophages. J Clin Invest 1990; 85:1108.
106. Wang M-H, Flad H-D, Feist W, Musehold J, Kusumoto S, Brade H, Gerdes J, Rietschel ET, Ulmer AJ. Inhibition of endotoxin or lipid A-induced tumor necrosis factor production by synthetic lipid A partial structures in human peripheral blood mononuclear cells. Lymphokine Cytokine Res 1992; 11:23.
107. Golenbock DT, Will JA, Raetz CRH, Proctor RA. Lipid X ameliorates pulmonary hypertension and protects sheep from death due to endotoxin. Infect Immun 1987; 55:2471.
108. Lam C, Hildebrandt J, Schütze E, Rosenwirth B, Proczor RA, Liehl E, Stütz P. Immunostimulatory, but antiendotoxin, activity of lipid X is due to small amounts of contaminating N,O-acylated disaccharide-1-phosphate: In vitro and in vivo reevaluation of the biological activity of synthetic lipid X. Infect Immun 1991; 59:2351.
109. Wang M-H, Feist W, Herzbeck H, Brade H, Kusumoto S, Rietschel ET, Flad H-

D, Ulmer AJ. Suppressive effect of lipid A partial structures on lipopolysaccharide or lipid A-induced release of interleukin 1 by human monocytes. FEMS Microbiol Immunol 1990; 64:179.

110. Takayama K, Quereshi N, Beutler B, Kirkland TN. Diphosphoryl lipid A from *Rhodopseudomonas sphaeroides* ATCC 17023 blocks induction of cachectin in macrophages to lipopolysaccharide. Infect Immun 1980; 57:1336.
111. Wright SD, Jong MTC. Adhesion-promoting receptors on human macrophages recognize *E. coli* by binding to lipopolysaccharide. J Exp Med 1986; 164:1876.
112. Detmers PA, Wright SD. Adhesion-promoting receptors on leukocytes. Curr Opin Immunol 1988; 1:10.
113. Wright SD, Detmers PA, Aida Y, Adamowski R, Anderson DC, Chad Z, Kabbash LG, Pabst MJ. CD18-deficient cells respond to lipopolysaccharide in vitro. J Immunol 1990; 144:2566.
114. Lei MG, Morrison DC. Specific endotoxic lipopolysaccharide-binding proteins on murine splenocytes. I. Detection of lipopolysaccharide-binding sites on splenocytes and splenocyte subpopulations. J Immunol 1988; 141:996.
115. Lei MG, Morrison DC. Specific endotoxic lipopolysaccharide-binding proteins on murine splenocytes. II. Membrane localization and binding characteristics. J Immunol 1988; 141:1006.
116. Lei MG, Stimpson SA, Morrison DC. Specific endotoxic lipopolysaccharide-binding proteins on murine splenocytes. III. Binding specificity and characterization. J Immunol 1991; 147:1925.
117. Kirikae T, Kirikae F, Schade UF, Yoshida M, Kondo S, Hisatsune K, Nishikawa S-I, Rietschel ET. Detection of lipopolysaccharide-binding proteins on membrane of murine lymphocyte and macrophage-like cell lines. FEMS Microbiol Immunol 1991; 76:327.
118. Morrison DC, Lei M-G, Kirikae T, Chen T-Y. Endotoxin receptors on mammalian cells. Immunobiology 1993; 187:212.
119. Wright SD, Ramos RA, Tobias PS, Ulevitch RJ, Mathison JC. CD14, a receptor for complexes of lipopolysaccharide (LPS) and LPS binding protein. Science 1990; 249:1431.
120. Schütt C, Ringel B, Nausch M, Bazil V, Horejsi V, Neels P, Waizel H, Jonas L, Siegl E, Friemel H, Plantikow A. Human monocyte activation induced by an anti-CD14 monoclonal antibody. Immunol Lett 1988; 19:321.
121. Ulmer AJ, Feist W, Heine H, Kirikae T, Kirikae F, Kusumoto S, Kusama T, Brade H, Schade U, Rietschel ET, Flad H-D. Modulation of endotoxin-induced monokine release in human monocytes by lipid A partial structures inhibiting the binding of ^{125}I-LPS. Infect Immun 1992; 60:5145.
122. Tobias PS, Ulevitch RJ. Lipopolysaccharide binding protein and CD14 in LPS dependent macrophage activation. Immunobiology 1993; 187:227.
123. Haziot A, Chen S, Ferrero E, Low MG, Silber R, Goyert SM. The monocyte differentiation antigen, CD14, is anchored to the cell membrane by a phosphatidylinositol linkage. J Immunol 1988; 141:547.

29

Immunomodulating Activities of Streptococcal Lipoteichoic Acids

Haruhiko Takada
Kagoshima University Dental School, Sakuragaoka, Kagoshima, Japan

Shozo Kotani
Osaka College of Medical Technology, Higashitenma, Kita-ku, Osaka, Japan

I. INTRODUCTION

Lipoteichoic acids (LTA) were initially referred to as intracellular teichoic acids, since they were identified as glycerophosphate-containing macromolecules in the particulate ribosome-membrane fraction of many gram-positive bacteria (1). Subsequent studies revealed that LTA is covalently linked to a glycolipid moiety of the cytoplasmic membrane in contrast to wall teichoic acids that are covalently linked to cell wall peptidoglycans and are either glycerol phosphate–containing or ribitol phosphate–containing compounds (2) (Fig. 1). Amphipathic LTA are distributed in *Streptococcus* spp. with some exceptions, *Staphylococcus* spp., *Lactobacillus* spp., *Bacillus* spp., *Leuconostoc mesenteroides,* and others. LTA are generally macromolecules in which a hydrophilic poly(1,3)glycerophosphate (PGP) backbone frequently substituted by alanyl esters or glycosidic groups covalently linked to a hydrophobic phosphatidylglycolipid, although there seem to be minor modifications among species.

The physicochemical properties and location of LTA in gram-positive bacterial cells have similarity to endotoxic lipopolysaccharide (LPS) of gram-negative bacteria, so several investigators anticipated that LTA has LPS-like bioactivities. Among them, Wicken and Knox made significant contributions to the early studies, and published a few review papers on LPS-like bioactivities of LTA (2,5–7), although these included only brief and limited descriptions on most of the bioactivities. They reported that LTA and LPS shared many bioactivities such as mitogenicity, stim-

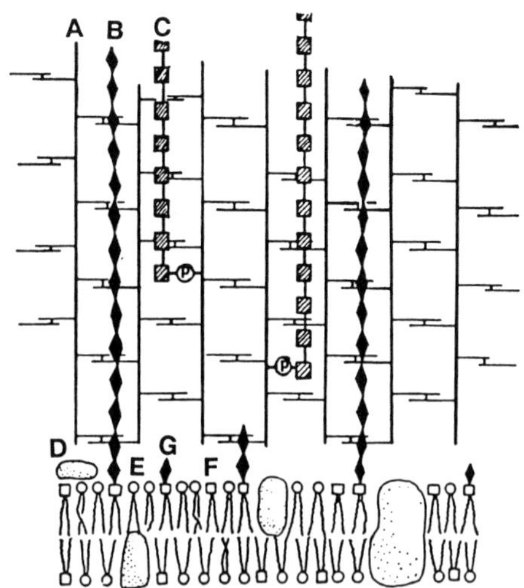

Figure 1 Proposed model of the cell wall–membrane complex of gram-positive bacteria. A, peptidoglycan; B, lipoteichoic acid (LTA); C, wall teichoic acid; D, protein; E, phospholipid; F, glycolipid; G, phosphoglycolipid. (From Refs. 3 and 4.)

ulation of bone resorption, anticomplementary activity, Shwartzman reactivity, stimulation of the reticuloendothelial system, stimulation of nonspecific immunity, macrophage lysosomal enzyme release, and *Limulus* activity. LTA, however, did not exhibit pyrogenicity and lethal toxicity in contrast to endotoxic LPS.

Based on these findings, one of us (S.K.) started a project study on possible antitumor and related activities of nontoxic LTA as a useful biological response modifier (BRM). In the first several years, LTA from *Streptococcus pyogenes* ATCC 21059 (Sv) was mainly used as the starting material, because this is the parent strain of *S. pyogenes* ATCC 21060 (Su). Cultures of the latter strain (Su) treated with penicillin G and hydrogen peroxide were designated as OK-432 and have been widely used in patients with cancer as an effective antitumor agent (8–10). An LTA preparation from *S. pyogenes* ATCC 21059 exhibits antitumor activities and tumor necrosis factor (TNF)–inducing activity in mice as described below. Incidentally, Hamada et al. (11) showed that the LTA, but not deacylated LTA, stimulated murine peritoneal macrophage cultures to exert significant cytotoxicity against TU5 tumor cells (derived from murine kidney). However, the progress of studies on immunobiological activities of *S. pyogenes* LTA has been seriously hampered by the poor reproducibility of preparations that elicit serum TNF in primed mice. This difficulty has been overcome by Tsutsui et al. (12), who studied two fractions of LTA from *Enterococcus hirae* (*Streptococcus faecium*) ATCC 9790 (13) that are different in chemical composition. They demonstrated significant differences in the bioactivities between these two fractions, LTA-1 and LTA-2, namely higher activities of the latter.

This chapter, after touching on the antitumor and TNF-inducing activities of *S. pyogenes* LTA, deals with recent developments in the study of *E. hirae* LTA-2. We aimed to identify the chemical entity responsible for the antitumor and related immunobiological activities of *E. hirae* LTA-2 and to synthesize a bioactive LTA in cooperation with Kusumoto et al. at Osaka University, although we have not yet accomplished our purpose.

II. ANTITUMOR AND TNF-INDUCING ACTIVITIES OF *S. PYOGENES* LTA

A test LTA specimen was extracted from *S. pyogenes* ATCC 21059 using phenol-water at room temperature followed by Sepharose 6B column chromatography (14). The ratios of glycerol, fatty acids, and alanine were consistent with those reported by Ofek et al. (15) for *S. pyogenes* LTA. The LTA preparation exhibited distinct antitumor effects against both ascites-type and solid-type Meth A fibrosarcoma in Balb/c mice (16). As shown in Figure 2, the LTA specimen, 10 μg (per dose) of

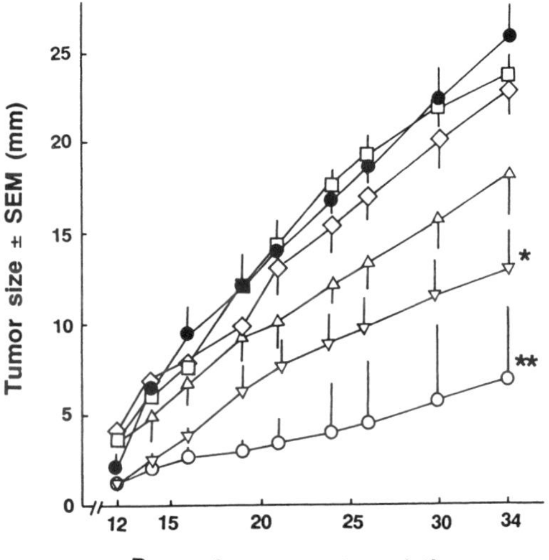

Figure 2 Suppressive effects of *S. pyogenes* cellular components on the growth of a solid Meth A fibrosarcoma in Balb/c mice. Test material was given intraperitoneally to groups of Balb/c mice (eight per group) for 4 successive days beginning 24 h after tumor inoculation. ∇ Whole organisms, 100 μg; ○ LTA, 10 μg; □ peptidoglycan, 10 μg; ◇ C-carbohydrate, 10 μg; Δ M protein, 10 μg; ● nontreated control. Significantly different from the control on day 34: $^{**}P \leq .01$, $^{*}P \leq .05$. (From Ref. 16.)

which was intraperitoneally injected on 4 successive days 24 h after tumor inoculation, markedly suppressed the growth of solid-type Meth A tumors. This effect was more powerful than that of 100 μg of whole cells. Cell wall peptidoglycans, group-specific C-carbohydrate, and M-protein (10 μg each per dose) did not exhibit significant activity in this system. Similar differential results were also obtained using the ascites-type Meth A system; namely, among test cellular components, only the LTA fraction had antitumor activity comparable with whole cells in terms of increased survival rate (17). To elucidate the possible mechanism, LTA and other streptococcal preparations were examined for TNF induction by administration into *Propionibacterium acnes*–primed CD-1 mice (14). Table 1 shows that under the assay conditions where whole *S. pyogenes* cells significantly induced serum TNF, the cell envelope fraction essentially consisting of cell walls and cytoplasmic membrane induced significantly more TNF than whole cells. The cell wall and the protoplast membrane fraction was only marginally active in inducing serum TNF. Among cellular components, the LTA fraction exerted considerable TNF-inducing activity, but the other cellular components, including peptidoglycans, were almost inactive.

The toxicity of intravenously injected LTA was examined in terms of lethality in normal and primed mice in comparison with that of LPS (summarized in Table 2). In normal ICR mice, LTA was devoid of lethal toxicity up to 20 mg/mouse in

Table 1 **TNF-Inducing Ability of Cellular Fractions or Components of *S. pyogenes* ATCC 21060**[a]

Preparation	Dose (μg/mouse)	Cytotoxicity (unit/ml)
Whole cells	1000	13,000
Cellular fractions		
Cell envelope	1000	62,500
	500	5,000
Cell wall	300	200
Protoplast membrane[b]	300	20
Cellular components		
LTA	100	19,000
M protein	100	20
Peptidoglycan	100	20
C-carbohydrate	100	20
Nucleic acids	100	20

[a] Groups of four *P. acnes*–primed CD-1 mice were injected intravenously with test materials. Nine days later, serum specimens were collected, heated at 56°C for 30 min, and assayed for cytocidal activity against L-929 cells. TNF activity was expressed as units per milliliter by referring to the reciprocal of the final serum dilution that caused 50% of the target cells to die.
[b] Prepared using a phage-associated lysin.
From Ref. 17; with permission.

Table 2 Comparison of the Lethal Toxicity of *S. pyogenes* LTA and *S. enteritidis* LPS in Normal and Primed Mice[a]

Test mouse	LD50 (μg/mouse)	
	LTA	LPS
ICR mouse	>20,000	250
P. acnes–primed[b] ICR mouse	>1,000	0.8
Galactosamine-sensitized[c] C57BL/6 mouse	ca 500	0.01
Galactosamine-sensitized C3H/HeN mouse	ca 150	0.0001

[a] Serially 10-fold decreased doses of LTA of *S. pyogenes* ATCC 21059 (Sv) or *S. enteritidis* LPS-*W* (DIFCO) in 0.2 ml of pyogene-free saline was injected intravenously into normal, *P. acnes*–primed or galactosamine-sensitized mice (four to six mice for each test point). The number of mice (dead/total) was recorded for 3 days after the challenge injection, and the LD_{50} was calculated.
[b] Mice were primed 11 days previously with an intraperitoneal injection containing 1.5 mg of formalin-treated *P. acnes* cells.
[c] Mice were injected intravenously with 16 mg of galactosamine hydrochloride and then they immediately received challenge injection of test materials.
Based on Refs. 14 and 17.

contrast to LPS from *Salmonella enteritidis* that killed 100 and 50% of test mice at doses of 500 and 250 μg/mouse, respectively (14). In *P. acnes*–primed mice, even at a dose as low as 3.13 and 0.8 μg/mouse, *S. enteritidis* LPS killed 100 and 50% of the mice, respectively. LTA did not kill any mice up to a dose of 1 mg/mouse (14). In galactosamine-sensitized mice, the LD_{50} of LTA and LPS was about 500 and 0.01 μg per C57BL/6 mouse and about 150 and 0.0001 μg per C3H/HeN mouse, respectively (see Table 2).

TNF was assumed to be responsible for manifesting the lethal toxicity of LPS in galactosamine-sensitized mice (18). Therefore, the toxicity in galactosamine-loaded mice of serum TNF induced by LTA was compared with that induced by LPS (19). To rule out the possible influence of LPS remaining in test serum specimens to the assay results, LPS-nonresponsive C3H/HeJ mice were included as test animals in addition to highly sensitive C3H/HeN mice and moderately sensitive C57BL/6 mice. As anticipated, the LTA-induced TNF did not kill galactosamine-loaded C3H/HeJ mice as well as C3H/HeN mice and C57BL/6 mice under the experimental conditions where the LPS-induced TNF that exhibited the cytocidal activity on L-929 cells comparable with that of LTA-induced TNF killed all of the mice. Incidentally, the cytocidal activity against L-929 cells of the LTA-induced TNF was completely inhibited by anti–TNF-α murine serum in a similar way to that of the LPS-induced TNF. This suggests that the LTA- and LPS-induced TNF share an antigenic epitope.

As mentioned in the introduction, bioactivities of *S. pyogenes* LTA preparations were poorly reproducible, and several preparations were scarcely able to induce serum TNF in primed mice. To screen for a suitable bacterial species from which to prepare constantly bioactive LTA, whole cells of a gram-positive bacterial species known to be LTA positive were examined for TNF-inducing ability in *P. acnes*–primed CD-1 mice. The results presented in Table 3 revealed that the TNF-

Table 3 **Induction of TNF by Whole Cells from Various Gram-Positive or Acid-Fast Bacteria in *P. acnes* Primed Mice**[a]

Test species (strain)	Cytotoxicity on L-929 cells (unit/ml)	Amphiphiles	
		LTA[b]	Others[c]
S. aureus			
(Cowan I)	1,200	+	
(Cowan)	20	+	
(209P)	720	+	
(209P-JC)	2,900	+	
(Smith)	32	+	
S. epidermidis	2,500	+	
S. pyogenes			
(Su, ATCC 21060)	13,000	+	
(Sv, ATCC 21059)	62,500	+	
(ATCC 19618)	12,500	+	
S. mutans	62,500	+	
S. faecalis (478)	62,500	+	
S. sanguis, biotype a	62,500	+	
S. salivarius (HHT)	2,500	+	
L. plantarum (ATCC 8014)	62,500	+	
L. casei (ATCC 27092)	13,000	+	
B. subtilis (AHU 1037)	4,400	+	
(W23)	25,000	+	
B. megaterium (AHU 1373)	6,500	+	
B. coagulans (AHU 1634)	62,500	+	
B. cereus	3,400	+	
B. licheniformis (AHU 1371)	4,000	+	
L. monocytogenes (EGD)	62,500	+	+
C. diphtheriae (PW8;RIMD 0343056)	150	−	+
P. acnes (ATCC 11808)	20	−	+
M. bovis (BCG)	62,500	−	+
M. smegmatis	62,500	−	+
M. phlei	62,500	−	+

[a] Groups of *P. acnes* primed CD-1 mice (four per group) were injected intravenously with whole bacterial cells (1 mg/mouse). The serum TNF levels were determined as described in the footnote to Table 1.

[b] Refer to Ref. 4.

[c] Refer to Ref. 20.

inducing ability of the whole cells varied widely among species. In consideration of various conditions and TNF-inducing potency, we selected *S. faecalis* (*E. hirae*) as the target bacterial species from which to prepare bioactive LTA.

III. BIOACTIVITIES OF *E. HIRAE* LTA AND RELATED SYNTHETIC COMPOUNDS

A crude fraction was obtained from *E. hirae* ATCC 9790 cells by hot aqueous-phenol extraction. After nuclease digestion, the fraction was applied to a Sepharose CL-6B column followed by hydrophobic interaction chromatography of Octyl-Sepharose CL-4B to separate LTA-1 and LTA-2 essentially as described by Fischer et al. (21). The chemical structures shown in Figure 3 were proposed for LTA-1 and LTA-2 and their glycolipid moiety by Fischer (4), although none of these fractions had yet been isolated in a sufficiently homogeneous state.

We prepared the glycolipid moiety of the LTA-1 and LTA-2 fractions by two methods: HGL-1 and HGL-2 were obtained as a chloroform layer by shaking acid-

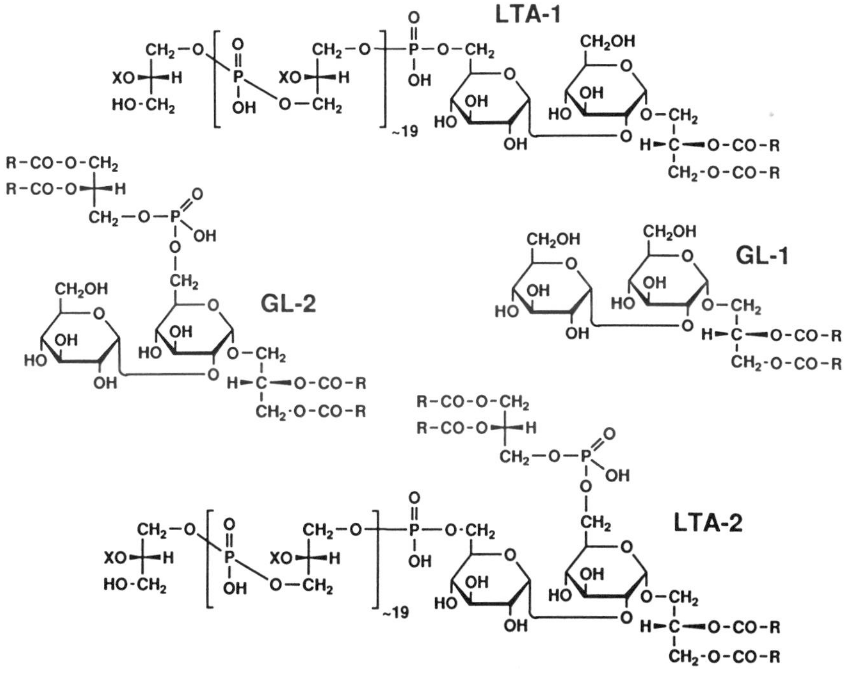

Figure 3 Chemical structures of *E. hirae* LTA-1 and LTA-2 and their glycolipid moieties proposed by Fischer (4). X; mono-, di-, tri-, and tetra-α-glucosyl residue: R; acyl group.

hydrolyzed LTA-1 and LTA-2 (with 1N HCl at 95°C for 20 min) with chloroform-methanol-water (1:1:1, v/v/v) as described by Tsutsui et al. (12); NGL-1 and NGL-2 were prepared directly from *S. pyogenes* ATCC 21059 and *E. hirae* ATCC 9790, respectively, by extraction with organic solvent according to Ishizuka and Yamakawa (22) with modifications. The NGL preparations obtained without prior acid hydrolysis were more intact and more homogeneous than the HGL preparations.

A. Cytokine-Inducing Activities

Intravenous injection of LTA-2 induced serum TNF-α and interleukin-6 (IL-6) activity into C3H/HeN mice primed by an intravenous injection of muramyl dipeptide (MDP) (100 μg) 4 h earlier (Table 4). Both LTA-1 or GL preparations irre-

Table 4 Induction of Serum TNF-α and IL-6 by Intravenous Injection of Whole Cells, LTA and Glycolipids Derived from *E. hirae* ATCC 9790 in MDP-Primed C3H/HeN Mice[a]

Test material	Dose (μg/mouse)	TNF-α[b] ng/ml	IL-6[c] U/ml
Whole cells	1,000	233 ± 14	886 ± 244
	500	158 ± 40	406 ± 72
LTA-1	100	<0.1	27.6 ± 13.4
LTA-2	100	297 ± 33	969 ± 48
	20	300 ± 27	
	4	113 ± 46	
	0.8	30 ± 12	
HGL-1	100	13 ± 3	9.8 ± 2.6
HGL-2	100	<0.1	17 ± 9
NGL-1	100	<0.1	0.8 ± 0.6
NGL-2	100	<0.1	0.5 ± 0.1
LPS of *S. abortus-equi*	0.8	225 ± 37	2,230 ± 325

[a] Groups of 3 C3N/HeN mice were primed with an intravenous injection of MDP (100 μg). Four hours later, the mice received intravenous injections of test materials for the induction of TNF-α and IL-6. The cytokine levels were determined in serum specimens collected 90 min after elicitation, and the data are expressed as the mean ± SEM.

[b] TNF-α activity was determined by measuring cytotoxic activity against actinomycin D–treated L-929 cells and was calculated as nanograms per milliliter based on the ratio of the 50% cytotoxic dose of a test specimen to that of standard recombinant human TNF-α (rHuTNF-α: a gift from the Dainippon Pharmaceutical Co. Ltd., Osaka, Japan). The TNF-α levels determined in the present study were of murine and not human TNF. Thus, strictly speaking, the nanograms per milliliter values described here mean the values equivalent to rHuTNF-α. The TNF activity induced by LTA-2 and LPS was confirmed to be due to TNF-α by its complete neutralization with an antimurine TNF-α rabbit antibody (a gift from Suntory Inc., Osaka, Japan).

[c] The IL-6 activity was determined by measuring the capacity to support the in vitro growth of the IL-6-independent cell line MH60.BSF2 (23).

spective of the source were devoid of TNF-inducing activity and exhibited only weak IL-6–inducing activity. LTA-2, but not LTA-1, also induced serum interferon gamma (IFN-γ) in *P. acnes*–primed C3H/HeN mice (data not shown). Incidentally, Tsutsui et al. (12) revealed that the deacylation of LTA-2 by alkaline hydrolysis diminished its bioactivities, including the induction of serum TNF. Thus the intact glycolipid moiety of LTA is probably essential but insufficient to induce cytokines. There is a discrepancy between our negative results and those reported by Tsutsui et al. (12) regarding the TNF-inducing and other bioactivities of HGL-2. This discrepancy may derive from the degree of HGL purity such as the extent of contamination with unhydrolyzed LTA-2 in Tsutsui's specimens. Anyway, the results obtained using NGL fractions that are more homogeneous may be more reliable than those with HGL fractions.

Along with the above studies of natural products, Fukase et al. (24,25) synthesized SGL-1, SGL-2, SLTA(4)-1, and SLTA(4)-2 (Fig. 4) whose structures essentially correspond to the proposed structures of GLA-1, GL-2, LTA-1, and LTA-2, respectively, except some differences from natural products (see Sec. V). None of these compounds induced significant levels of serum TNF-α or IL-6 in MDP-primed C3H/HeN mice.

With regard to stimulatory effects on murine peritoneal macrophage cultures, the bioactivity and product relationships were practically the same as those seen by means of the above in vivo assay (Table 5). However, some of the specimens lacking distinct activities in assays in vivo weakly activated murine peritoneal macrophage cultures to induce IL-6 and cell-associated thymocyte activating factor (TAF, probably IL-1).

So far, SLTA(4)-2, a synthetic "analogue" of LTA-2, was totally inactive in all of the in vivo and in vitro assays. This synthetic compound has a structure in which tetra(glycerophosphate) is linked to a phosphatidylglycolipid (SGL-2) and was synthesized as the first step in an attempt to mimic the structure of LTA-2 proposed by Fischer (4).

B. Antitumor Activity

Consecutive intravenous injection of MDP (100 μg) and LTA-2 (200 μg) in Meth A fibrosarcoma–bearing Balb/c mice caused strong hemorrhagic necrosis of the tumor and resulted in marked regression or complete cure of the fibrosarcoma (Fig. 5). LTA-2 did not weaken or kill the treated mice under conditions where three of six mice that received an intravenous injection of *S. enteritidis* LPS (1 μg) died within a few days and the three mice surviving LPS elicitation showed a complete cure of the tumors. By contrast, LTA-1 and four synthetic compounds caused only weak hemorrhagic tumor necrosis, but none of them caused any significant regression of the tumors (see Fig. 5). The bacterial glycolipids, NGL-1 and NGL-2 (100 μg), were virtually ineffective in terms of both induction of hemorrhagic necrosis and tumor regression. This study could not confirm the anti–Meth A fibrosarcoma effects of LTA-1 in MDP-primed mice that Tsutsui et al. (12) reported (see Sec. VI).

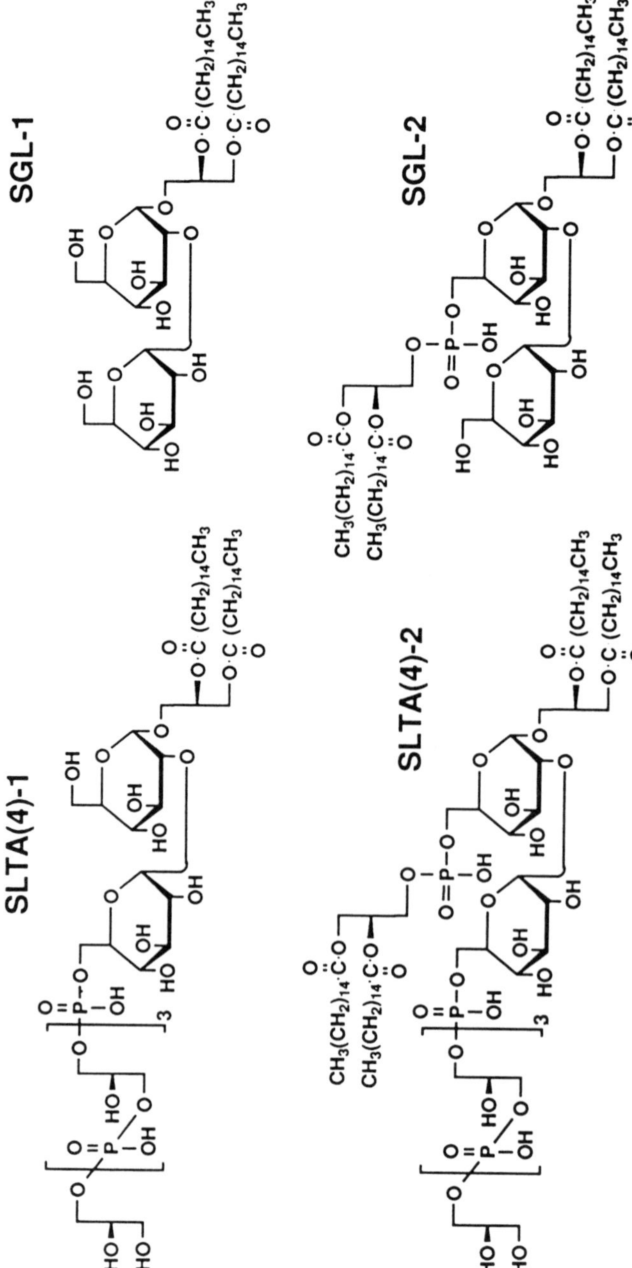

Figure 4 Chemical structures of synthetic "analogs" of LTA-1 and LTA-2 and their glycolipids.

Table 5 Induction of Cytokines by LTA and Glycolipids, Bacterial and Synthetic, in Murine Peritoneal Macrophage Cultures[a]

Test material	Dose (μg/ml)	TNF-α ng/ml	IL-6 U/ml	CA-TAF SI; mean ± SEM[b]
LTA-1	100	0.2	12.5	6.0 ± 0.9[c]
	10	<0.05	15.0	9.6 ± 1.6[c]
	1	<0.05	11.5	5.7 ± 0.3[c]
	0.1	<0.05	8.0	4.2 ± 0.9[d]
LTA-2	100	1.9	15.5	4.5 ± 0.2[c]
	10	0.2	188.0	15.1 ± 0.7[c]
	1	<0.05	95.0	12.1 ± 2.4[d]
	0.1	<0.05	22.5	4.7 ± 0.5[c]
HGL-1	100	0.7	12.5	11.6 ± 0.5[c]
	10	<0.05	11.5	7.9 ± 2.1[d]
	1	<0.05	9.5	4.0 ± 0.5[d]
	0.1	<0.05	7.0	3.1 ± 0.5
HGL-2	100	<0.05	15.0	6.7 ± 1.2[d]
	10	<0.05	17.0	3.5 ± 0.6[d]
	1	<0.05	27.5	3.2 ± 0.5[c]
	0.1	<0.05	6.5	2.9 ± 0.3[d]
SGL-1	100	<0.05	7.5	4.7 ± 0.7[c]
	10	<0.05	6.0	3.3 ± 0.2[d]
	1	<0.05	4.6	3.0 ± 0.2
SGL-2	100	<0.05	4.5	4.9 ± 0.8[d]
	10	<0.05	4.8	3.2 ± 0.7
	1	<0.05	4.6	2.4 ± 0.6
SLTA(4)-1	100	<0.05	2.8	0.9 ± 0.1
	10	<0.05	2.3	1.7 ± 0.3
	1	<0.05	3.3	2.1 ± 0.1
SLTA(4)-2	100	<0.05	<2.0	1.0 ± 0.1
	10	<0.05	<2.0	6.4 ± 0.8[c]
	1	<0.05	2.2	6.3 ± 1.1[d]
LPS	10	4.8	43.0	10.7 ± 0.8[c]
	1	4.3	44.0	10.8 ± 0.9[c]
	0.1	1.3	45.0	13.6 ± 1.4[c]
None	0	<0.05	<2.0	2.6 ± 0.4

[a] Monolayer cultures of thioglycolate-induced C3H/HeN peritoneal macrophages were incubated in triplicate with test or reference materials for 24 h. The pooled culture supernatants were assayed for TNF-α and IL-6, and macrophage cultures disrupted by alternate freezing and thawing were tested for cell-associated thymocyte activating factor (CA-TAF). TNF-α and IL-6 activities were determined as described in the footnotes to Table 4. TAF activity was determined by a conventional comitogenic assay for IL-1 measurement using thymocytes from C3H/HeJ mice and phytohemagglutinin (1 μg/ml) as the target cells and mitogen, respectively. The results were expressed as the stimulation index (SI), where SI = the radiolabeled deoxyrudine uptake in test culture/that in control (PHA alone) culture.

[b] The CA-TAF activity in test cultures was significantly higher than in control cultures according to Student's *t*-test.

[c] $P \leqslant .01$.

[d] $P \leqslant .05$.

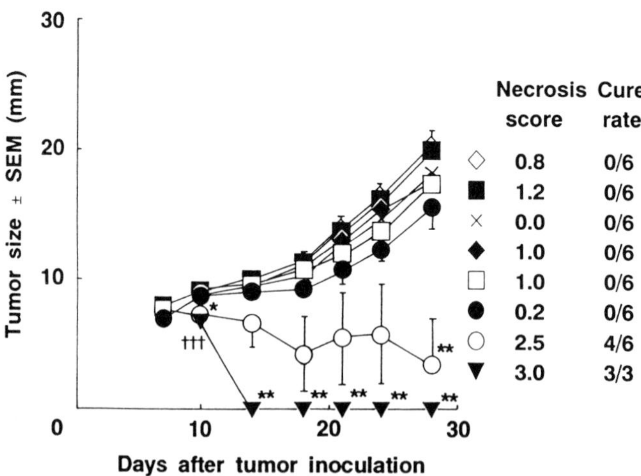

Figure 5 Regressive and curative effects of bacterial LTA and synthetic analogues of both LTA and glycolipids on Meth A fibrosarcoma established in Balb/c mice. Groups of 6 Balb/c mice bearing Meth A tumors (7–8 mm average diameter) were given intravenous injections of MDP (100 μg/mouse). Four hours later, the animals received intravenous test material (200 μg/mouse except for reference LPS, which was 1 μg/mouse): LTA-2 (○), LTA-1 (●), SLTA(4)-2 (◇), SLTA(4)-1 (◆), SGL-2 (□), SGL-1 (■), and reference LPS of *S. enteritidis* (DIFCO) (▲). A group that received MDP alone served as the control (x). Three of six mice that received LPS died within 3 days of receiving the test LPS, as shown by †. The necrosis score is the mean score for hemorrhagic necrosis 3 days after the intravenous injection of test and reference specimens (the maximum score being 3.0). The cure rate is expressed as the number of test animals in which Meth A tumors were completely cured against the number of those surviving at the end of the experimental period of 32 days. Statistically significant difference from the control (animals that received MDP alone) as determined by Student's *t*-test; $P \leq .05$ (*), $P \leq .01$(**).

C. Pyrogenicity and *Limulus* Activity

We then examined the pyrogenicity in rabbits and the *Limulus* activity of LTA-1 and LTA-2 specimens as representative bioactivities characteristic of endotoxic LPS. Eight of 12 rabbits injected with LTA-2 (100 μg/kg, i.v.) exhibited a slightly positive febrile response; namely, an increase in rectal temperature of slightly higher than 0.6°C over a 5-h observation period. By contrast, none of the four rabbits that received an intravenous injection of LTA-1 (100 μg/kg) had febrile responses. Thus, the pyrogenicity of the LTA-2 was at least 1000-fold less than a reference *S. abortus-equi* LPS.

The LTA-2 specimen was practically inactive in the *Limulus* test, being 1.31×10^{-5} equivalent to reference *E. coli* O111:B4 LPS by Endospecy (Seikagaku Kogyo

Co., Tokyo, Japan) (26). Thus, the weak but significantly positive *Limulus* potency reported by Fine et al. (27) was not confirmed.

IV. REACTIVITY OF *E. HIRAE* LTA AND PGP WITH ANTI–*S. PYOGENES* (OK-432) MONOCLONAL ANTIBODY (TS-2) AND NEUTRALIZATION OF THE CYTOKINE-INDUCING ACTIVITY OF *E. HIRAE* LTA-2 BY TS-2

Recently, Okamoto et al. (28) prepared a murine immunoglobulin M (IgM)–type monoclonal antibody (Mab, designated TS-2) to OK-432, an antitumor streptococcal product. They reported that TS-2 effectively neutralized the ability of OK-432 to induce IFN-γ in human peripheral blood mononuclear cell cultures. Here, we demonstrated by an enzyme-linked immunosorbent assay (ELISA) that TS-2 reacted with LTA and PGP preparations from *E. hirae* and PGP from *S. pyogenes* (Fig. 6A) and various bacterial LTA preparations (data not shown). Considering both the effective dose and the maximum extent of the reaction, among the tested *E. hirae* products, LTA-2 was the most reactive, followed by LTA-1 and PGP. There was a big difference in the reactivity between LTA-2 and PGP. The reactivity of *E. hirae* LTA-1 was similar to that of the reference PGP specimen from *S. pyogenes* (ATCC 21060) that was the immunizing antigen for the preparation of TS-2 in an effective dose but was lower than *S. pyogenes* PGP in the maximum extent reaction. NGL-1 and NGL-2 preparations that correspond to the glycolipid portion of *E. hirae* LTA-1 and LTA-2, respectively, were totally nonreactive with TS-2. Among synthetic compounds (Fig. 6B), both SLTA(4)-1 and SLTA(4)-2 were moderately reactive with TS-2. Although there were some differences in the reactivity between the two, it was somewhat lesser than that to the reference *S. pyogenes* (ATCC 21060) PGP. Neither SGL-1 nor SGL-2 was reactive.

The literature indicates that the immunodeterminants of LTA are mainly located in PGP itself, and its glycosyl (5) and alanine (30) substituents. The results of study using MAb raised against penicillin and hydrogen peroxide treated whole *S. pyogenes* (ATCC 21060) cells (OK-432) seem to be essentially consistent with those of the above reports. Nevertheless, although the glycolipid moiety, natural or synthetic, was totally nonreactive, *E. hirae* LTA, especially LTA-2, STA(4)-1, and STA(4)-2, consisting of PGP and glycolipid portions, exhibited significantly higher reactivity with TS-2 than *E. hirae* PGP. This suggests that the glycolipid moiety exerts indirect, but significant, effects on the reactivity of the test LTA with TS-2, and that the substituents on the PGP moiety are not exclusively immunodeterminant, at least in the reaction of *E. hirae* LTA with the MAb TS-2.

We then studied possible neutralizing activity of TS-2 on the cytokine-inducing activities of LTA-2. Table 6 shows that TS-2 can almost completely neutralize the

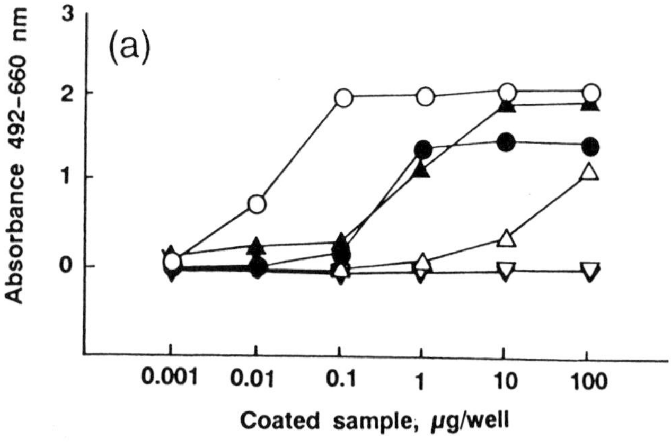

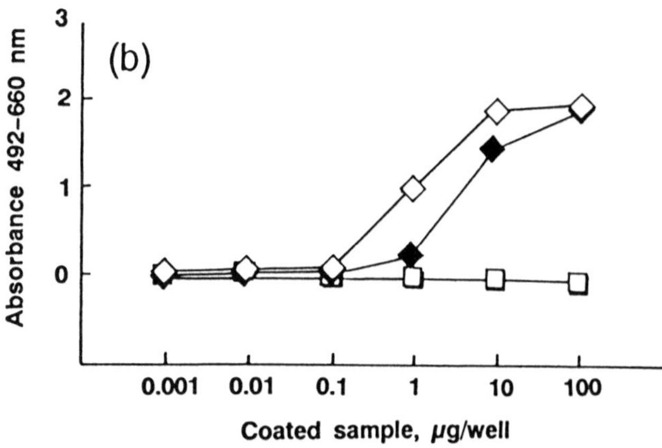

Figure 6 The serological reactivity of bacterial and synthetic LTA as well as related compounds in an ELISA using anti–OK-432 MAbs (TS-2 and TS-1). Microtiter plates were coated with test material (100 μl/well), which was serially 10-fold diluted from 1 to 1×10^{-5} mg/ml, and blocked with Block Ace (Dainippon Pharmaceutical Co.). After an incubation at room temperature for 1 h, 50 μl of TS-2 and TS-1 (5 μg protein/ml) was added to the wells, which were incubated overnight at 4°C. The wells were washed three times with 0.05 M phosphate-buffered saline, pH 7.4 (PBS) containing 0.05% Tween 20 (PBST), then 200-fold diluted peroxidase conjugated to a goat antimouse immunoglobulin (Cappel, West Chester, PA.) was added, and the plates were incubated at room temperature for 1 h. *O*-phenylenediamine (OPD) solution (25 mg of OPD, 5 ml of methanol, 25 μl of 31% peroxide, and 45 ml of 0.2 M citrate buffer, pH 5.5) was added to the wells. Five minutes later, the reaction was stopped with 2N sulfuric acid, and the absorbance at 492 nm was measured, with reference to the absorbance at 655 nm. Symbols: ○ LTA-2, ● LTA-1, Δ PGP, ∇ NGL-2, ▼ NGL-1, and ▲ reference PGP of *S. pyogenes* in panel (a); ◇ SLTA(4)-2, ◆ SLTA(4)-1, □ SGL-2, and ■ SGL-1 in panel (b). The reference murine IgM–type MAb to OK-432, TS-1 (29), did not react with any test materials, except SGL-1, which showed moderate reactivity.

Table 6 The Inhibitory Effects of Anti-OK-432 Monoclonal Antibody (TS-2) on the Cytokine-Inducing Activities of LTA-2 in Murine Macrophage Cultures[a]

Concentration (μg/ml) Preincubation		Culture	TNF-α (ng/ml)	IL-6 (U/ml)
LTA-2	TS-2	LTA-2	(%) control	
80	450	10	ND[b] (0)	5.0 (7)
80	150	10	0.25 (20)	8.4 (12)
80	15	10	1.30 (103)	16.9 (24)
80	0	10	1.26 (100)	71.9 (100)
None	None	0	ND	0.2

[a] Aqueous LTA-2 (80 μg/ml) was incubated with an equal volume of TS-2 (0 to 450 μg/ml) at 37°C for 2 h. The mixture was diluted with culture medium to a final concentration of 10 μg LTA/ml and then added to the monolayer of peritoneal macrophages from C3H/HeN mice. After 24 h, TNF and IL-6 activity in the culture supernatant was measured as described in Table 4.
[b] Not detected.

TNF-α- and IL-6-inducing activity of LTA-2 in murine peritoneal macrophage cultures, although the former needs more antibody. Similar neutralizing effects of TS-2 were noted also with the induction of serum TNF-α and IL-6 by intravenous injection of LTA-2 into MDP-primed mice. The neutralizing effects on IL-6 induction seemed to be partial (Table 7).

The results obtained from the ELISA and the neutralization studies suggest that the structure recognized by the MAb TS-2 is not exclusively located in PGP and may extend to the glycolipid moiety and that the glycolipid moiety itself was not reactive with TS-2, but this structure or its linkage with PGP may play a role in the induction of TNF-α and IL-6 by *E. hirae* LTA-2. Incidentally, the reaction of

Table 7 Inhibitory Effects of Anti-OK-432 Monoclonal Antibody (TS-2) on the Serum Cytokine-Inducing Activities of LTA-2 in MDP-Primed C3H/HeN Mice[a]

	Serum TNF		Serum IL-6	
Test material	ng/ml	% inhibition	U/ml	% inhibition
LTA-2	8.75 ± 1.15		859 ± 109	
LTA-2 + TS-2	0.85 ± 0.05	90.3	244 ± 25	71.6

[a] LTA-2 (4 μg) was preincubated with or without TS-2 (160 μg of protein) in 500 μl of physiological saline at 37°C for 2 h. The mixture was injected intravenously into C3H/HeN mice primed by an intravenous injection of MDP (100 μg) 4 h before. Cytokine levels were determined in serum specimens collected 90 min after elicitation with test materials.

biologically inactive, synthetic LTA "analogues" with TS-2, which can neutralize the cytokine-inducing bioactivity of *E. hirae* LTA-2, indicated that a structure shared by both LTA-2 and SLTA(4)-1 or SLTA(4)-2 is necessary but is insufficient in itself to exhibit bioactivity. The possibility remains that the antibody linked with the antigenic epitope may interfere with the manifestation of activities of the bioactive center located in parts of LTA remote from the antigenic epitopes.

V. DISCUSSION ON SYNTHETIC LTA "ANALOGUES"

The synthetic analogue, SLTA(4)-2, that was prepared by partially mimicking *E. hirae* LTA-2 differed from the bacterial product with respect to the following: (1) parts of the four fatty acid residues on the lipid anchor of bacterial LTA-2 are unsaturated, whereas all of four acyl groups on phosphatidylglycolipids in the synthetic products are saturated; (2) the degree of polymerization of the PGP backbone is assumed to be $9 \sim 40$ in the bacterial product, whereas that of the synthetic analogue was 4; and (3) there were no substituents in the tetra(glycerophosphate) moiety of the synthetic compound in contrast to the unusually high glycosyl substitution consisting of mono-, di-, tri-, and tetra-glycosyl residues in *E. hirae* LTA (31,32). Among these structural differences, we examined the possible role of unsaturated fatty acids in the biological activities inherent to natural LTA-2. The LTA-2 specimen was hydrogenated to obtain a derivative of LTA-2 in which the fatty acid residues in the lipid anchor portion were saturated. As shown in Table 8, this markedly reduced the cytokine-inducing capacity of the bacterial LTA-2 in both in vivo (A) and in vitro (B) assays. This finding, however, does not necessarily exclude the possible involvement of other structural differences as described above. Furthermore, an unknown constituent present in the LTA-2 fraction, but different from Fischer's compound for which the structure shown in Figure 3 is proposed, might be responsible for the immunobiological activities described with the bacterial LTA-2 specimen used in our study. However, the possibility that extraneous contaminants such as LPS are involved can be excluded by the negligible *Limulus* activity of bacterial LTA-2.

VI. CONCLUSIONS

There have been a number of studies on the biological activities (in both bacterial system and mammalian organisms) of LTA prepared from various gram-positive bacteria, as extensively reviewed in an article by Fisher (4). However, there is a wide range of differences in chemical structure and physicochemical properties among LTA specimens prepared from various species. This is true even with LTA preparations from the same bacterial species and from the same strain. The (micro)heterogeneity noted with LTA preparations from the same strain may

Table 8 Effects of Hydrogenating the LTA-2 Fraction on Its Cytokine-Inducing Activity

A. Administration into MDP-primed C3H/HeN mice[a]

Test material	Dose (μg/mouse)	Serum TNF-α ng/ml	Serum IL-6 U/ml
LTA-2	20	139 ± 4.4	859 ± 189
Hydrogenated LTA-2	20	0.1 ± 0.1	77 ± 13

B. Stimulation of murine macrophage cultures[b]

Test material	Dose (μg/ml)	TNF-α ng/ml	IL-6 U/ml
LTA-2	100	<0.05	7.2
	10	32.2	187.5
	1	6.4	87.5
	0.1	3.2	23.8
Hydrogenated LTA-2	100	<0.05	0.3
	10	<0.05	8.1
	1	<0.05	2.4

[a] Assay procedures are the same as those described in Table 4.
[b] Assay procedures are the same as those described in Table 5.

be derived from either minute differences in cultural, according the physiological conditions of the bacterial cells as the starting material or uncontrollable fluctuations in preparation procedures. Different molecular species may incur different biological activities among LTA specimens resulting in fluctuations in the bioactivities of various LTA prepared from even one strain, since a seemingly minor modification of chemical or physicochemical properties of an LTA may significantly influence its bioactivities. This may explain the discrepancies among the reported biological activities of LTA, and may complicate identifying the principle responsible for the biological activities of a variety of LTA. In fact, there is practically no exact information about the chemical entity of LTA carrying each of the reported biological properties.

We have often experienced marked fluctuations in the TNF-α–inducing and antitumor activities among *S. pyogenes* LTA prepared at different times. LTA from *E. hirae* is also subject to similar uncertainties, although to a far lesser extent than that of *S. pyogenes* LTA. In view of the fact that the fully bioactive LTA-2 fraction obtained from *E. hirae* LTA is by no means homogeneous, further studies are under progress in Kusumoto's laboratory, Faculty of Science, Osaka University.

The objectives are to isolate the active principle of *E. hirae* LTA responsible for the cytokine-inducing and antitumor activities in a satisfactorily homogeneous state to elucidate its entire chemical structure and to prepare a corresponding compound by chemical synthesis. The synthesis of a compound mimicking the principle responsible for the TNF-α–inducing and anti–Meth A activities in MDP-primed mice and further chemical modification of the active compound may culminate in the creation of novel compounds whose biological activities shift in favor of the host even compared with LTA, which has a highly favorable balance between beneficial and harmful effects than endotoxic bacterial LPS.

ACKNOWLEDGMENTS

The authors are grateful to Dr. Susumu Kokeguchi and Dr. Keijiro Kato (Department of Oral Microbiology, Okayama University Dental School) for supplying bacterial LTA and related preparations; to Dr. Shoichi Kusumoto, Dr. Koichi Fukase, Dr. Takuya Yoshimura, and Dr. Yasuo Suda (Faculty of Science, Osaka University) for supplying synthetic LTA analogues; to Dr. Tetsuo Komuro (National Institute of Hygienic Sciences, Osaka Branch) for pyrogenicity test; to Dr. Shigenori Tanaka, Dr. Hiroshi Tamura, and Dr. June Aketagawa (Tokyo Institute, Seikagaku Co.) for Endospecy test; and to Dr. Mitsunobu Sato (Second Department of Oral and Maxillofacial Surgery, Tokushima University School of Dentistry) for supplying anti–OK-432 MAbs, TS-1, and TS-2. We also wish to express our thanks to Dr. Yoshihiro Kawabata and Dr. Rieko Arakaki (Department of Microbiology and Immunology, Kagoshima University Dental School) and Dr. Takeshi Yoshida, Dr. Motoo Saito, Dr. Akihiro Yamamoto, Dr. Hiroko Usami, and Dr. Naoko Tanaka (Tokyo Institute for Immunopharmacology, Inc.) for excellent collaboration in this study and their generous permission to quote data from the papers in press or under submission and those unpublished.

REFERENCES

1. Critchley P, Archibald AR, Baddiley J. The intracellular teichoic acid from *Lactobacillus arabinosus* 17-5. Biochem J 1962; 85:420.
2. Knox KW, Wicken AJ. Immunological properties of teichoic acids. Bacteriol Rev 1973; 37:215.
3. van Driel D, Wicken AJ, Dickson MR, Knox KW. Cellular location of the lipoteichoic acids of *Lactobacillus fermenti* NCTC 6991 and *Lactobacillus casei* NCTC 6375. J Ultrastruct Res 1973; 43:483.
4. Fischer W. Bacterial phosphoglycolipids and lipoteichoic acids. Handbook Lipid Res 1990; 6:123.
5. Wicken AJ, Knox KW. Lipoteichoic acids: A new class of bacterial antigen. Science 1975; 187:1161.
6. Wicken AJ, Knox KW. Biological properties of lipoteichoic acids. In: Schlessinger D,

ed. Microbiology 1977. Washington DC: American Society for Microbiology, 1977:360.

7. Wicken AJ, Knox KW. Bacterial cell surface amphiphiles. Biochim Biophys Acta 1980; 604:1.
8. Kimura I, Ohnoshi T, Yasuhara S, Sugiyama M, Urabe Y, Fujii M, Machida K. Immunochemotherapy in human lung cancer using the streptococcal agent OK-432. Cancer 1976; 37:2201.
9. Sakurai Y, Tsukagoshi S, Satoh H, Akiba T, Suzuki S, Takagaki Y. Tumor-inhibitory effect of a streptococcal preparation (NSC-B116209). Cancer Chemother Rep 1972; 56:9.
10. Uchida A, Hoshino T. Clinical studies on cell-mediated immunity in patients with malignant disease. I. Effect of immunotherapy with OK-432 on lymphocyte subpopulation and phytomitogen responsiveness in vitro. Cancer 1980; 45:476.
11. Hamada S, Yamamoto T, Koga T, McGhee JR, Michalek SM, Yamamoto S. Chemical properties and immunobiological activities of streptococcal lipoteichoic acids. Zbl Bakt Hyg A 1985; 259:228.
12. Tsutsui O, Kokeguchi S, Matsumura T, Kato K. Relationship of the chemical structure and immunobiological activities of lipoteichoic acid from *Streptococcus faecalis* (*Enterococcus hirae*) ATCC 9790. FEMS Microbiol Immunol 1991; 76:211.
13. Farrow AE, Collins MD. *Enterococcus hirae,* a new species that includes amino acid assay strain NCDO 1258 and strains causing growth depression in young chickens. Int J Syst Bacteriol 1985; 35:73.
14. Yamamoto A, Usami H, Nagamuta M, Sugawara Y, Hamada S, Yamamoto T, Kato K, Kokeguchi S, Kotani S. The use of lipoteichoic acid (LTA) from *Streptococcus pyogenes* to induce a serum factor causing tumour necrosis. Br J Cancer 1985; 51:739.
15. Ofek I, Beachey EH, Jefferson W, Campbell GL. Cell membrane–binding properties of group A streptococcal lipoteichoic acid. J Exp Med 1975; 141:990.
16. Usami H, Yamamoto A, Yamashita W, Sugawara Y, Hamada S, Yamamoto T, Kato K, Kokeguchi S, Ohokuni H, Kotani S. Antitumour effects of streptococcal lipoteichoic acids on Meth A fibrosarcoma. Br J Cancer 1988; 57:70.
17. Usami H, Yamamoto A, Kato K, Kokeguchi S, Hamada S, Yoshida T, Kotani S. Antitumor and TNF-inducing activities of streptococcal lipoteichoic acids. A Report of the 7th Annual Meeting, Japanese Strepto-Club, 1988:183–99.
18. Lehmann V, Freudenberg MA, Galanos C. Lethal toxicity of lipopolysaccharide and tumor necrosis factor in normal and D-galactosamine–treated mice. J Exp Med 1987; 165:657.
19. Usami H, Yamamoto A, Sugawara Y, Hamada S, Yamamoto T, Kato K, Kokeguchi S, Takada H, Kotani S. A nontoxic tumour necrosis factor induced by streptococcal lipoteichoic acids. Br J Cancer 1987; 56:797.
20. Kotani S, Nagao A, Tamura T, Okamura H, Nagata A, Aoyama K, Kusumoto S, Kokeguchi S, Kato K, Fujii N, Usami H, Yoshida T, Akagawa K, Tanaka S, Komuro T, Ikeda-Fujita T, Kato Y, Utsunomiya J. Purification and endotoxin-like bioactivities of a novel amphiphile from *Mycobacterium bovis* BCG. In: Masihi KN, Lange W, eds. Immunotherapeutic prospects of infectious diseases. Berlin: Springer-Verlag, 1990:19.

21. Fischer W, Koch HU, Haas R. Improved preparation of lipoteichoic acids. Eur J Biochem 1983; 133:523.
22. Ishizuka I, Yamakawa T. Glycosyl glycerides from *Streptococcus hemolyticus* strain D-58. J Biochem 1968; 64:13.
23. Matsuda T, Hirano T, Kishimoto T. Establishment of an interleukin 6 (IL 6)/B cell stimulatory factor 2-dependent cell line and preparation of anti-IL 6 monoclonal antibodies. Eur J Immunol 1988; 18:951.
24. Fukase K, Matsumoto T, Ito N, Yoshimura T, Kotani S, Kusumoto S. Synthetic study of lipoteichoic acid of gram positive bacteria. I. Synthesis of proposed fundamental structure of *Streptococcus pyogenes* lipoteichoic acid. Bull Chem Soc Jpn 1992; 65:2643.
25. Fukase K, Yoshimura T, Kotani S, Kusumoto S. Synthetic study on lipoteichoic acid of gram positive bacteria. II. Synthesis of the proposed fundamental structure of *Enterococcus hirae* lipoteichoic acid. Bull Chem Soc Jpn 1994; 67: in press.
26. Obayashi T, Tamura H, Tanaka S, Ohki M, Takahashi S, Arai M, Masuda M, Kawai T. A new chromogenic endotoxin-specific assay using combined limulus coagulation enzymes and its clinical applications. Clin Chim Acta 1985; 149:55.
27. Fine DH, Kessler RE, Tabak LA, Shockman GD. Limulus lysate activity of lipoteichoic acids. J Dent Res 1977; 56:1500.
28. Okamoto M, Kaji R, Kasetani H, Yoshida H, Moriya Y, Saito M, Sato M. Purification and characterization of interferon-γ–inducing molecule of OK-432, a penicillin-killed streptococcal preparation, by monoclonal antibody neutralizing interferon-γ–inducing activity of OK-432. J Immunother 1993; 13:232.
29. Sato M, Kaji R, Urata M, Yoshida H, Yanagawa T, Miyamoto K, Azuma M, Furumoto N, Saito M. Monoclonal antibody to the streptococcal preparation OK-432: Tissue OK-432 localization and analysis of interaction between OK-432 and macrophages or NK cells in human salivary adenocarcinoma-bearing nude mice given OK-432. J Biol Response Modifiers 1988; 7:212.
30. McCarty M. The role of D-alanine in the serological specificity of group A streptococcal glycerol teichoic acid. Proc Natl Acad Sci USA 1964; 52:259.
31. Cabacungan E, Pieringer RA. Evidence for a tetraglucoside substituent of the lipoteichoic acid of *Streptococcus faecium* ATCC 9790. FEMS Microbiol Lett 1985; 26:49.
32. Leopold K, Fischer W. Separation of the poly(glycerophosphate) lipoteichoic acids of *Enterococcus faecalis* Kiel 27738, *Enterococcus hirae* ATCC 9790 and *Leuconostoc mesenteroides* DSM 20343 into molecular species by affinity chromatography on concanavalin A. Eur J Biochem 1991; 196:475.

30

Modulation of the Immune System by Bacterial Products: Hapten-Specific Humoral Immune Responses Induced by Lipopeptides Conjugated to T Helper Cell Epitopes

Wolfgang G. Bessler
Institut für Immunbiologie der Universität, Freiburg, Germany

Werner Beck, Karl-Heinz Wiesmüller, and Günther Jung
Institut für Organische Chemie der Universität, Tübingen, Germany

I. INTRODUCTION

Cell wall components of Gram-negative bacteria such as lipoprotein (1,2), lipopolysaccharide (LPS) (3), and protein I (4) constitute highly active immunological components functionally activating murine and human B lymphocytes and monocytes/macrophages. Biologically active synthetic analogues of the bacterial cell wall constituents have been prepared by various laboratories. The lipid A part of lipopolysaccharide (LPS) from Gram-negative bacteria, which is responsible for most of its biological activity, has been synthesized (5). Synthetic muramyl dipeptide (MDP) derived from bacterial peptidoglycan constitutes an effective immunoadjuvant (6). Synthetic segments of the porin OmpF from *Escherichia coli* B act as polyclonal B-lymphocyte activators (7). We have synthesized the N-terminal lipopeptide of the bacterial lipoprotein (Fig. 1), which acts as an immunoadjuvant and as a leukocyte activator (7).

II. METHODS

A. Chemical Synthesis of Lipopeptide Adjuvants

Synthesis of Pam_3Cys conjugates and T helper cell epitopes was performed as follows: Peptides were synthesized on a peptide synthesizer (Applied Biosystems 430A)

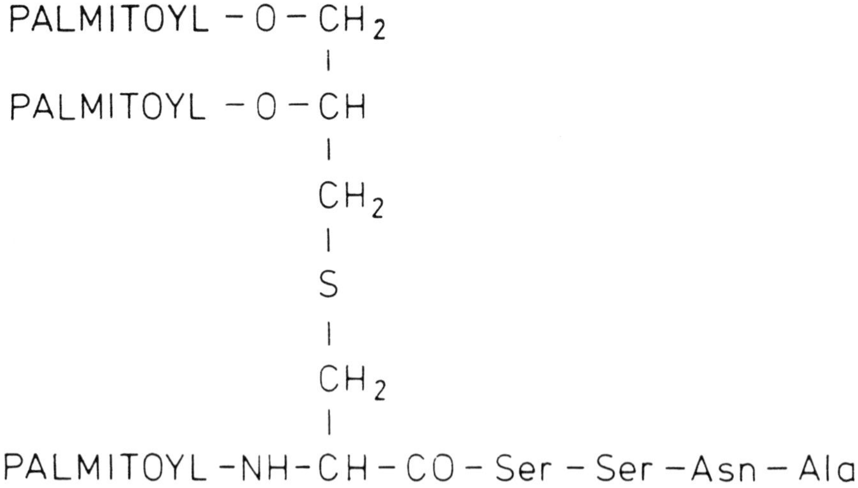

PALMITOYL – O – CH_2
|
PALMITOYL – O – CH
|
CH_2
|
S
|
CH_2
|
PALMITOYL – NH – CH – CO – Ser – Ser – Asn – Ala

Figure 1 Structural formula of the N-terminal lipopeptide Pam_3Cys-Ser-Ser-Asn-Ala from the lipoprotein of *E. coli.*

using our own programs developed for Fmoc/tBu strategy. Fmoc-protected amino acids and p-alkoxybenzyl alcohol resin (Novabiochem AG, Läufelfingen, Switzerland) were used. The coupling reactions were performed using a 2.5-molar excess of Fmoc-amino acid, HOBt and TBTU (2-[1H-benzotriazol-1-yl]-1,1,3,3 tetramethyluroniumtetrafluoroborate) and a 5-molar excess of diisopropylethylamine. The peptides were removed from the resin and deprotected by treatment with 95% trifluoroacetic acid for 2.5 h in the presence of thioanisole and thiocresol. Complete removal of the Mtr-protecting group from the arginine residues was obtained after an additional treatment with 95% trifluoroacetic acid (TFA), 2% thioanisole, 1.5% phenol, and 1.5% ethanedithiol for 1 h at 50°C. The peptides were purified by repeated precipitation from acetic acid with diethyl ether followed by semipreparative HPLC (acetonitrile, water, 0.1% TFA) and lyophilization. The peptides were analyzed by ion-spray mass spectrometry and showed the correct amino acid analyses. The racemization was determined by gas chromatography on Chirasil-Val and found to be below 2% for each of the amino acids. Pam_3Cys and Pam_3Cys-Ser were synthesized as described (8).

B. Preparation of Lipopeptide-Antigen-Conjugates

1. Antigen Adjuvant Mixtures

Lipopeptides can be obtained commercially (Fa. Rapp, Polymere, Eugenstr. 38/1, Tübingen, Germany) or can be prepared by chemical synthesis (8a). For the intraperitoneal, intracutaneous or subcutaneous application in combination with antigens, the water-soluble lipopeptide Pam_3Cys-Ser-$(Lys)_4$ is suspended in aqueous solution with antigen by gentle shaking at a 1:1 (w/w) ratio. Water-insoluble lipopeptides like Pam_3Cys-Ser-Ser-Asn-Ala are suspended together with antigen at a 1:1 (w/w) ratio in aqueous solution or in saline by sonification.

2. *Antigen Adjuvant Conjugates*

For covalent coupling of antigens to lipopeptide adjuvants, the following procedures are recommended:

1. Conjugation of peptides or proteins soluble in dimethylformamide (DMF): Two μmol of peptide or protein binding determinants is dissolved in 0.5–1.0 ml of DMF, and 8 mmol (9.2 mg) of solid Pam_3Cys-Ser-Gly-OSu (8) is added. A homogeneous solution is obtained by gentle heating and sonication, and 4 mmol of organic base (N-ethylmorpholine) is added. After stirring for 12 h, 1–2 ml of chloroform:methanol (1:1) is added, and the mixture is cooled in an ice bath for 2 h. The sediment is washed with 1 ml of cold chloroform:methanol (1:1), dissolved in tert-butylalcohol:water (3:1), sonicated if necessary, and freeze dried.
2. Peptides and proteins soluble in water: Two micromoles of peptide/protein are dissolved in 0.8 ml of water, and 4 mmol (6.5 mg) Pam_3Cys-Ser-$(Lys)_4$ is added. The mixture is thoroughly sonicated and a pH of 5.0–5.5 is set up. After 5 mg of EDC ([1,3-dimethyl-aminopropyl]-3-ethylcarbodiimide hydrochloride) dissolved in 100 ml of water has been added, the mixture is stirred at room temperature for 18 h and then dialyzed twice against 1 L of distilled water. The content of the dialysis tubing is freeze dried.

III. LIPOPEPTIDES AS LEUKOCYTE ACTIVATORS

The synthetic lipopeptide N-palmitoyl-S-(2,3-bis[palmitoyloxy]-[2RS]-propyl-[R]-cysteinyl-[S]-seryl-[S]-seryl-[S]-asparaginyl-[S]-alanine (Pam_3Cys-Ser-Ser-Asn-Ala) is identical to the lipoprotein N-terminus with respect to its peptide composition but contains solely palmitic acid residues. This and other lipopeptide analogues are biologically most active substances. They constitute B-lymphocyte and macrophage activators for a wide variety of murine inbred strains; no nonresponder strain could be detected so far (9–11). Lipopeptides induce B-cell differentiation into immunoglobulin (Ig)–secreting plasma cells, as determined by a hemolytic Jerne plaque assay. On days 3, 4, and 5 after the addition of lipopeptide to splenic cell cultures, predominantly IgM, $IgG_{2(a+b)}$, and IgG_3 were induced, whereas no significant IgA responses could be detected. The amount of Ig produced was comparable to LPS, with the isotype pattern being slightly different (11). Lipopeptides also induce lymphokine production, phagocytosis, induction of superoxide dismutase, or tumor cytotoxicity in human and murine monocytes/macrophages (12–15). Furthermore, lipopeptides constitute potent activators of human neutrophil granulocytes (16).

IV. LIPOPEPTIDES AS ADJUVANTS

A. Lipopeptides as Replacement for Freund's Adjuvant

Synthetically prepared lipopeptides constitute potent immunoadjuvants when administered in mixture with antigens (17,18). The newly synthesized lipopeptide analogues Pam_3Cys-Ser-$(Lys)_4$, (Pam_3Cys-Ser-$(Lys)_4$), and Pam_3Cys-Ser-polyoxyethylene 3000 (B. Kleine et al., submitted) are well soluble in aqueous media. Pam_3Cer-Ser-$(Lys)_4$ was markedly able to increase the specific in vivo Ig response against DNP-BSA: 14 days after the injection of a mixture of antigen and lipopeptide adjuvant into mice, a pronounced rise in anti–DNP-IgG antibodies was found; antibody titers in animals given only the hapten or adjuvant were marginal. Boostering with a mixture of DNP–bovine serum albumin (BSA) and Pam_3Cys-Ser-$(Lys)_4$ resulted in very high DNP-specific IgG responses (120–180 μg/ml) comparable with the response obtained by immunization with DNP-BSA in combination with complete Freund's adjuvant (18).

B. Lipopeptides Antigen Conjugates

Lipopeptides also act as potent immunoadjuvants when coupled covalently to low molecular weight antigens. After being covalently linked to haptens like digoxin, or DNP, for example, the lipopeptides rendered these molecules immunogenic, as shown by in vivo immunization of mice (18,19). Furthermore, the toxic effects of digoxin or amanitin were no longer observed after coupling the toxins to lipopeptide (G. Jung et al., in preparation). In a similar way, various peptides coupled to the lipopeptide anchor could induce pronounced peptide-specific immune responses in vivo and in vitro. Thus, a partial sequence from the epidermal growth factor receptor (EGF-R) coupled to Pam_3Cys-Ser induced an epitope-specific anti–EGF-R response (20). Because the EGF-R tetradecapeptide constitutes most probably a murine T helper cell epitope, the corresponding lipopeptide gave rise to boost effects and memory T helper cells. Lipopeptide-antigen conjugates also were used to prepare human immunodeficiency virus (HIV)–specific antibodies. We prepared oligopeptide segments of the HIV-1 and HIV-2 proteins p17, gp41, and gp120 by chemical synthesis. After covalent attachment of the segments to lipopeptides, antigen-specific antibodies were induced in vivo in mice and in vitro in murine and human cell culture systems. The kinetics of antibody induction and the isotypes were determined.

Finally, lipopeptide antigen conjugates can be effectively used for the preparation of murine or human monoclonal antibodies. Thus, we immunized in vitro human peripheral blood lymphocytes with HIV-peptide-lipopeptide conjugates or toxin-lipopeptide conjugates. The immunized cell cultures were fused to cells of the heteromyeloma line CB-F7 (provided by Dr. v. Baehr, Charité Berlin), and we obtained stable hybridomas producing antigen-specific IgM and IgG (21).

C. Induction of Immunological Memory by the Addition of T Helper Cell Epitopes

We could show that the response against synthetically prepared melittin or fragments thereof can be further enhanced by the additional introduction of a T helper cell epitope into the lipopeptide-hapten conjugate. The helper cell epitope applied, which is specifically presented by the Balb/c H-2^d haplotype, consists of a 16 amino acid oligopeptide (FISEAIIHVLHSRHPG) derived from sperm wale myoglobin. Antibodies obtained after three immunizations with the conjugates recognized the synthetic and the native mellitin molecule. The immune-enhancing effect was most pronounced for the peptide fragments Mel1-16 and Mel17-26 conjugated to lipopeptide-Th-peptide conjugates. Our results show that it is possible markedly to enhance a weak hapten-specific immune response by coupling the haptens to lipopeptide conjugated to a haplotype specific T helper cell epitope. These findings could be of importance for the optimization of immunization procedures and for the development of novel synthetic vaccines (22).

V. LIPOPEPTIDES AND VACCINES

A. Lipopeptides as Additives to Bacterial Vaccines

When added to Enterobacteriaceae vaccines, lipopeptides were found to provide protection for mice against lethal *Salmonella* infections. Thus, lipopeptides could substitute for 90% of a vaccine consisting of acetone-killed *S. typhimurium.* For Pam_3Cys-Ser-Ser-Asn-Ala a pronounced effect was found at an adjuvant concentration of 175 μg; the water-soluble lipopeptide adjuvant Pam_3Cys-Ser-$(Lys)_4$ exhibited an optimum at around 90 μg/mouse. These adjuvant properties were mediated by an enhancement of the humoral immune response without involving the cellular immune response (23).

B. Lipopeptide-Based Vaccines Give Protection Against Foot-and-Mouth Disease

Lipopeptide-based vaccines against foot-and-mouth disease were able to protect guinea pigs against lethal viral infection (24–26). We were able to design and synthesize an effective vaccine against foot-and-mouth disease by coupling the lipopeptide adjuvants to suitable VP1-segments of the viral protein. The resulting conjugates constitute the smallest totally synthetic vaccine known so far that gives full protection. We could show that a single injection of Pam_3Cys-Ser-Ser-(VP1 135-

154) was sufficient to protect seven out of seven guinea pigs against challenge with the homologous virus type present in the vaccine. Remarkably, in comparison with a mixture of Pam_3Cys diastereomers, only half the amount of pure diastereomer (R,R)-lipopeptide adjuvant-antigen conjugate is needed to induce full protection (25).

C. Induction of a Cytotoxic T-Lymphocyte Response In Vivo Against Viral Infections

Conjugates of lipopeptides with viral oligopeptides were able to induce peptide-specific cytotoxic T lymphocytes in vivo. Especially for the construction of vaccines, it is important that the lipopeptide immunoadjuvants can be used not only to improve humoral immune reactions but also, after coupling to viral peptide segments, to induce cytotoxic T lymphocytes that kill virus-infected cells. A cytotoxic response against influenza virus has been induced in mice (27), and corresponding experiments using lipopeptide-adjuvant-HIV peptide conjugates to induce cytotoxic T cells recognizing HIV-infected targets are in progress (G. Haas et al., unpublished data).

VI. MECHANISM OF LIPOPEPTIDE INDUCED ADJUVANTICITY

As outlined above, lipopeptides are potent stimulants for B lymphocytes and for cells of the Endoplasmatic Reticulum (monocytes, macrophages). After their addition to cell cultures, lipopeptides bind to defined membrane proteins, including proteins of the MHC. Binding proteins of 34 kd could be isolated by lipopeptide-affinity chromatography. Separated by two-dimensional electrophoresis, the binding proteins consisted of three slightly different polypeptides. It is not yet known whether the 34 kd protein possesses functional activity in the signal transduction process during lipopeptide activation (28–30). The analysis of already known second-messenger pathways during B-cell stimulation by lipopeptides showed that neither cyclic adenosine monophosphate (cAMP), cyclic guanosine monophosphate (cGMP), nor phosphatidylinositol metabolism are changed after applying the stimuli (31). Remarkably, 2 min after their addition to the cells, lipopeptides are found in different compartments within the cell, including the nucleus (32–34).

We also investigated how alterations of the lipopeptide molecule influenced its biological properties. Studies using mostly leukocyte-stimulation assays indicated that various replacements in both the lipid and peptide moieties of the lipopeptide are possible without significant loss of activity provided that particular structural features of the molecule are maintained (9,10). For example, protease resistance was enhanced by the incorporation of α-methylserine residues into the peptide chain, or the sulfur atom was replaced by a methylene group (35,36). As a rule, (R,R)-configurated Pam_3Cys derivatives are immunologically more active than those containing the R,S-configurated stereoisomers (25). The lipopeptide Pam_3Cys-Ser-Ser-Asn-Ala with an amino acid sequence identical to the N-terminus of native lipoprotein and two other lipopeptides, Pam_3Cys-Ser-$(Lys)_4$ and Pam_3Cys-Ser-$(Glu)_4$, were compared for their stimulatory activity toward Balb/c splenocytes. All three compounds exhibited comparable activities (18); however, Pam_3Cys-Ser-$(Lys)_4$ and Pam_3Cys-Ser-$(Glu)_4$, because of their increased solubility, were active at lower concentrations than the native Pam_3Cys-Ser-Ser-Asn-Ala. We concluded therefore, that the peptide composition, once the minimal dipeptide chain length has been reached (17), is not the major factor determining mitogenicity. The MHC–linked antigen presentation of lipopeptide antigen conjugates eliciting humoral or cellular immune responses is presently under investigation.

VII. LIPOPEPTIDES AS NOVEL ADJUVANTS

In summary, synthetic lipopeptide analogues derived from bacterial lipoprotein constitute potent immunoadjuvants in vitro and in vivo when administered in mixture with or covalently conjugated to antigen (37). Lipopeptide antigen conjugates display a powerful built-in adjuvanticity in the form of a membrane anchor compound derived from a natural immunostimulating bacterial protein that is present in large amounts in the Enterobacteriaceae species of the intestine. The enzymatic degradation in plasma and tissue is diminished by the rapid uptake of the immunogen into cell membranes or by the formation of micellar deposits. As a result, a long-term delivery of the antigen to the immune system occurs. Finally, however, lipopeptides are fully biodegradable into harmless amino acids and fatty acids, and S-glyceryl-cysteine is a metabolite from *E. coli* lipoprotein found in urine and feces. Remarkably, the administration of the lipopeptide requires no further adjuvant or additive, in particular the harmful Freund's adjuvant can be avoided (26,37). No toxic side effects or tissue damage have been observed so far in seven animal species for many years (38).

VIII. CONCLUSIONS

Lipopeptides derived from the bacterial lipoprotein constitute potent immunoadjuvants in vitro and in vivo. Lipopeptides were admixed to a great variety of protein antigens thus avoiding the use of Freund's adjuvant. When added to Enterobacteriaceae vaccines, lipopeptides markedly enhance the vaccine effect. After the coupling of lipopeptides to haptens or low molecular weight antigens, which are not immunogenic per se, a markedly specific antibody response is induced often after only one application of the conjugate. The humoral response can be further enhanced by introducing haplotype-specific T helper cell epitopes into the conjugate, as shown by us in the mellitin system. Lipopeptide antigen conjugates also can be applied as synthetic vaccines; for example, they give protection against foot-and-mouth disease. Last, by the coupling of lipopeptides to epitopes presented by major histocompatibility class (MHC) I molecules, the in vivo priming of cytotoxic T cells can be achieved, as it could be demonstrated by us for influenza virus CTL epitopes. The novel synthetic lipopeptides and lipopeptide antigen conjugates are nontoxic and nonpyrogenic; they can easily be prepared by chemical synthesis in gram amounts with high purity and reproducibility.

REFERENCES

1. Melchers F, Braun V, Galanos C. The lipoprotein of the outer membrane of *E. coli*: A B-lymphocyte mitogen. J Exp Med 1975; 142:473–82.
2. Bessler WG, Ottenbreit BP. Studies on the mitogenic principle of the lipoprotein from the outer membrane of *Escherichia coli*. Biochem Biophys Res Commun 1977; 76:239–46.
3. Andersson J, Sjöberg O, Möller G. Induction of immunoglobulin and antibody synthesis *in vitro* by lipopolysaccharides. Eur J Immunol 1972; 2:349–53.
4. Bessler WG, Henning U. Protein I and Protein II from the outer membrane of *Escherichia coli* are mouse B-lymphocyte mitogens. Z Immun Forsch 1979; 155:387–98.
5. Galanos C, Lüderitz O, Rietschel ET, Westphal O, Brade H, Brade L, Freudenberg M, Schade U, Immotot M, Yoshimura H, Kashimoto S, Shiba T. Synthetic and natural *Escherichia coli* free lipid A express identical endotoxic activities. Eur J Biochem 1985; 148:1–5.
6. Riveau G, Chedid L. Comparison of the immuno- and neuropharmacological activities of MDP and murabutide. Prog Leukoc Biol 1987; 6:213–22.
7. Vordermeier H-M, Hoffmann P, Gombert FO, Jung G, Bessler WG. Synthetic peptide segments from the Escherichia coli porin OmpF constitute leukocyte activators. Infect Immun 1990; 58:2719–24.
8. Metzger J, Wiesmüller K-H, Schaude R, Bessler WG, Jung G. Synthesis of novel immunologically active tripalmitol-S-glycerylcysteinyl lipopeptides as useful intermediates for immunogen preparations. Int J Pept Protein Res 1981; 37:46–57.

8a. Bessler WG, Jung G. Lipopeptides as novel adjuvants. Prg. Immunol 1992; 588–593.

9. Bessler WG, Cox M, Lex A, Suhr B, Wiesmüller KH, Jung G. Synthetic lipopeptide analogues of bacterial lipoprotein are potent polyclonal activators for murine B-lymphocytes. J Immunol 1985; 135:1900–5.
10. Prass W, Ringsdorf H, Bessler W, Wiesmüller K-H, Jung G. Lipopeptides of the N-terminus of *E. coli* lipoprotein: synthesis, mitogenicity and properties in monolayer experiments. Biochim Biophys Acta 1987; 900:116–28.
11. Kleine B, Sprenger R, Martinez-Alonso C, Coutinho A, Bessler WG. Analysis of the polyclonal B cell activation by a synthetic analogue of bacterial lipoprotein—qualitative and quantitative approach and comparison with bacterial lipopolysaccharide (LPS). Immunology 1987; 61:29–34.
12. Bessler WG, Flad H-D, Schade U, Rietschel E. Lymphokine induction in murine and human leukocytes by bacterial lipoprotein and a synthetic analogue and by bacterial protein I. Lymphokine Res 1984; 3:234.

13. Hauschildt S, Hoffmann P, Beuscher HU, Dufhues G, Heinrich P, Wiesmüller KH, Jung G, Bessler WG. Activation of bone marrow-derived macrophages stimulated with bacterial lipopeptide. Cytokine production, phagocytosis and Ia expression. Eur J Immunol 1990; 20:63–8.
14. Hoffmann P, Heinle S, Schade UF, Loppnow L, Ulmer AJ, Flad HD, Bessler WG. Stimulation of human and murine adherent cells by bacterial lipoprotein and synthetic lipopeptide analogues. Immunobiology 1988; 177:158–70.
15. Hoffmann P, Weismüller K-H, Metzger J, Jung G, Bessler WG. Induction of tumor cytotoxicty in murine bone marrow-derived macrophages by two synthetic lipopeptide analogues. Hoppe Seylers Z Physiol Chem 1989; 370:575–82.
16. Seifert R, Schultz G, Richter-Freund M, Metzger J, Wiesmüller K-H, Jung G, Bessler WG, Hauschildt S. Activation of superoxide formation and lysozyme release in human neutrophils by the synthetic lipopeptide Pam_3Cys-Ser-$(Lys)_4$: Involvement of G-proteins and syngerism with chemotactic peptides. Biochemistry J 1990; 267:795–802.
17. Lex A, Wiesmüller KH, Jung G, Bessler WG. A synthetic analogue of *Escherichia coli* lipoprotein, tripalmitoyl pentapeptide, constitutes a potent immune adjuvant. J Immunol 1986; 137:2676–81.
18. Reitermann A, Metzger J, Wiesmüller K-H, Jung G, Bessler WG. Lipopeptide derivatives of bacterial lipoprotein constitute potent immune adjuvants combined with or covalently coupled to antigen or hapten. Biochem Hoppe Seylers 1989; 370:343–52.
19. Böltz T, Jung G, Wiesmüller K-H, Metzger J, Bessler WG. Enhancement of the immune response using a synthetic B-lymphocyte mitogen covalently linked to antigens. In: Mani J, Dornand J, eds. Lymphocyte activation and differentiation. Berlin: Walter de Gruyter, 1988:379–82.
20. Jung G, Wiesmüller K-H, Becker G, Bühring HJ, Bessler WG. Verstärkte Produktion spezifischer Antikörper durch Präsentation der antigenen Determinanten mit kovalent verknüpften Lipopeptid-Mitogenen. Angew Chem 97:883–5; Angew Chem Int Ed Engl 1985; 24:872–3.
21. Hoffmann P, Jiminez-Diaz M, Loleit M, Tröger W, Wiesmüller K-H, Metzger J, Jung G, Kaiser I, Stöcklin I, Lenzner S, Peters JH, Grimm R, Schäfer E, Bessler WG. Preparation of human and murine monoclonal antibodies: antigens combined with or conjugated to lipopeptides constitutes potent immunogens for in vitro and in vivo immunizations. Hum Antibod Hybridomas 1990; 1:137–44.
22. Wiesmüller K-H, Deres K, Shi L, Wei D, Wernet P, Jung G. Lipopeptide-helper T cell epitope-CTL epitope conjugate induces monoclonal antibodies against the CTL epitope. In: Eptom R, ed. Innovation and perspectives in solid phase synthesis and related technologies: peptides, polypeptides and oligonucleotides. Andover: Intercept, 1992:499–502.
23. Schlecht S, Wiesmüller K-H, Jung G, Bessler WG. Lipopeptide als natürliche Adjuvantien für Impfstoffe aus Gram-negative Bakterien. Naturwissenschaften 1993; 80:9–17.
24. Wiesmüller K-H, Jung G, Hess G. Novel low-molecular-weight synthetic vaccine against foot-and-mouth-disease containing a potent B-cell and macrophage activator. Vaccine 1989; 7:29–34.
25. Wiesmüller K-H, Hess G, Bessler WG, Jung G. Diastereomers of tripalmitoyl-S-glyceryl-L-cysteinyl carrier adjuvant systems induce different mitogenic and protective

immune response. In: Holmstedt B, Frank H, Testa B, eds. Chirality and biological activity. New York: Alan R. Liss, 1990:267–272.
26. Wiesmüller K-H, Bessler WG, Jung G. Solid phase peptide synthesis of lipopeptide vaccines eliciting epitope-specific B-, T-helper and T-killer cell response. Int J Pept Protein Res 1992; 40:255–60.
27. Deres K, Schild H, Wiesmüller K-H, Jung G, Rammensee HG. In vivo priming of virus-specific cytotoxic T-lymphocytes with synthetic lipopeptide vaccine. Nature 1989; 342:561–4.
28. Scheuer WV, Biesert L, Bessler WG. Binding a synthetic analogue of mitogenic bacterial lipoprotein to murine major histocompatibility complex (MHC) gene products. Biol Chem Hoppe Seylers 1986; 367:1085–94.
29. Biesert L, Scheuer W, Bessler WG. Interaction of mitogenic bacterial lipoprotein and a synthetic analogue with mouse lymphocytes. Isolation and characterization of binding proteins. Eur J Biochem 1987; 162:651–7.
30. Bessler WG, Klein B, Biesert L, Schlecht S, Schaude R, Wiesmüller K-H, Metzger J, Jung G. Bacterial surface components as immunomodulators. In: Masihi KN, Lange W, eds. Immunotherapeutic prospects of infectious diseases. Berlin: Springer Verlag, 1990:37–48.
31. Hauschildt S, Lückhoff A, Langhorne J, Wiesmüller K-H, Jung G, Bessler W, Cambier JC. Increase in the intracellular free calcium concentration is not an obligatory early event in lipopeptide-induced B-cell activation. Immunology 1991; 73:366–8.
32. Wolf B, Uhl B, Hauschildt S, Metzger J, Jung G, Bessler WG. Interaction of the lymphoid cell line BCL_1 with a lipopeptide analogue of bacterial lipoprotein: distribution of the activator within the cells detected by electron energy loss spectroscopy (EELS). Immunobiology 1989; 180:93–100.
33. Wolf B, Hauschildt S, Uhl B, Metzger J, Jung G, Bessler WG. Localization of the cell activator lipopeptide in bone marrow-derived macrophages by electron energy loss spectroscopy (EELS). Immunol Lett 1989; 20:121–6.
34. Uhl B, Wolf B, Schwinde A, Jung G, Bessler WG, Hauschildt S. Intracellular localization of a lipopeptide macrophage activator: Immunocytochemical investigations and EELS analysis on ultrathin cryosections of bone marrow-derived macrophages. J Leuko Biol 1991; 50:10–18.
35. Metzger J, Olma A, Leplawy MT, Jung G. Synthesis of a B-lymphocyte activating α methylserine containing lipopeptide. Z Naturforsch 1987; 42B:1195–1201.
36. Metzger J, Bessler WG, Hoffmann P, Strecker M, Lieberknecht A, Schmidt U. Lipopeptides containing 2-(Palmitoylamino)-6,7(Palmitoyloxy)heptanoic acid: synthesis, stereospecific stimulation of B-lymphocytes and macrophages and adjuvanticity in vitro and in vivo. J Med Chem 1991; 34:1969–74.
37. Bessler WG. Synthetic lipopeptide immunomodulators derived from bacterial lipoprotein: tools for the standardization of in vitro assays. Develop Biol Stand 1992; 77:49–56.
38. Wiedemann F, Link R, Pumpe K, Jacobshagen U, Schaefer HE, Wiesmüller K-H, Hummel RP, Jung G, Bessler WG, Böltz T. Histopathological studies on the local reactions induced by complete Freund's adjuvant (CFA), bacterial lipopolysaccharide (LPS), and synthetic lipopeptide (P_3C) conjugates. J Pathol 1991; 164:265–71.

31

Stimulating Effect of an Oral Preparation from *Lactobacillus bulgaricus* (strain LB51) on Host Resistance to Experimental Bacterial Infections

Poly Popova, Guenka Guencheva, Edoardo Pacelli, and Ginka Opalchenova
National Drug Institute, Sofia, Bulgaria

Assen Bogdanov
DEODAN Laboratories, Sofia, Bulgaria

I. INTRODUCTION

It has been found that the nonspecific host resistance to bacterial infections can be enhanced by several immunoadjuvants such as mycobacteria, muramyl peptides, and chitin derivatives (1–3).

At the beginning of the century, the Bulgarian medical doctor Stamen Grigorov discovered a special kind of lactobacilli in samples of Bulgarian yogurt and called them *Lactobacillus bulgaricus*. Later on the well-known Russian scientist Illya Metchnikoff discovered the beneficial qualities of *Lactobacillus bulgaricus*. According to him, the beneficial effects of Bulgarian "sour milk" (yogurt) was due to the lactic acid produced by the lactobacilli. In his classic paper, Metchnikoff linked the longevity of the Bulgarian people with the consumption of yogurt. He stressed the fact that *L. bulgaricus* was the only bacterium that had never damaged the human organism. Metchnikoff was the first scientist who created a healing preparation based on *L. bulgaricus* and called it Lactobacilline.

In 1951, Dr. Ivan Bogdanov isolated from Bulgarian yogurt a special strain of *L. bulgaricus*—I. Bogdanov strain tumoronecroticance 51—ATCC 21815 or LB51.

DEODAN is an oral preparation obtained from lyzozyme lysates of the *L. bulgaricus* strain LB51. It has been demonstrated to possess antitumor effects in mice and humans (4). The structure of the active principle of this preparation was shown to be a glycopeptide with a molecular weight of 10,000 d (dalton), containing amino sugars, amino acids, glucose, galactose, glycerophosphate-teichoic acids, and muramyl peptides (5).

In the present study, we investigated the protective activity of DEODAN against bacterial challenge. Two models were used, one with the extracellular microorganism *Klebsiella pneumoniae* and the other with the intracellular microorganism *Listeria monocytogenes*.

II. MATERIALS AND METHODS

A. Animals

Male ICR mice 6–8 weeks old were used. They were bred under standard conditions with free access to water and food. Rabbits weighing 2.0–2.4 kg were used for pyrogenicity tests. During the 5 days before the experiments were performed, the rabbits were adapted to settle down in experimental chambers, in which the neck of the animal was loosely fixed, for at least 4 h a day.

B. Drug

DEODAN (DeoDan Laboratories) was dissolved in pyrogen-free saline and was applied orally in mice using an intubation needle for 3, 6, and 10 consecutive days at a dose of 150 mg/kg daily. The substance did not contain endotoxin as checked by the Limulus amebocyte lysate kit (M. A. Bioproducts, Walkersville, MD).

C. Bacterial Challenge

1. *Listeria monocytogenes* serovar 4B (Bulgarian Microbial Collection) was administered intraperitoneally to mice at a volume of 0.5 ml from suspension with concentration 1×10^8 microorganisms/ml.
2. *Klebsiella pneumoniae* strain 52145 (Institute Pasteur Collection) at a volume of 0.1 ml/10 g body weight was injected intraperitoneally to mice. The bacterial suspension contained 10^2 microorganisms/ml.

The mortality was registered up to 12 days following the challenge.

D. Assay for Spreading Ability and Phagocytic Activity of Mouse Peritoneal Macrophages

Twenty-four hours after the last administration of DEODAN, the peritoneal cavities of mice were washed twice with 5 ml RPMI 1640 (Sigma, St. Louis, MO) containing 5 U/ml heparin. The cells were washed, resuspended in RPMI 1640, and their viability was assessed by trypan blue dye exclusion test.

Macrophages (1×10^6/ml) were incubated in 24-well tissue culture plates for 24 h and the number of spreading cells was counted by microscopic examination. The spreading ability was calculated as follows.

$$\frac{\text{number of spreading cells}}{\text{total cells}} \times 100 = \%$$

Escherichia coli C 600 labeled with 5 μCi/ml [^{3}H]thymidine ([^{3}H]TdR) was added to the macrophage monolayers at a concentration of 1 × 10^8 bacterial cells/ml. After incubation for 1 h at 37°C and washing, 2 ml of 0.2% Triton x100 was added to each dish. Each group was assayed in triplicate and the incorporated radioactivity was measured in a liquid scintillation counter (LKB-Wallac 1215 Rackbeta).

E. Estimation of Interleukin-1 Production by Mouse Peritoneal Macrophages

Twenty-four hours after the last application of DEODAN, the macrophages from DEODAN-treated and control mice were incubated in 24-well tissue culture plate (2.5 × 10^5/well) for 72 h at 37°C, 5% CO_2 in RPMI 1640 medium, supplemented with 5% heat-inactivated fetal calf serum (FCS, Boehringer-Manheim, Germany), 100 U/ml penicillin, and 100 μg/ml streptomycin. Culture supernatants were filtered through 0.22-μm Milipore filters and interleukin-1 (IL-1) activity was measured by the thymocyte proliferation response (6). The cultures were incubated for 72 h at 37°C, 5% CO_2, with submitogenic doses of concanavalin A (ConA) (1 μg/ml). Twelve hours before the end of the incubation period, the cultures were labeled with [^{3}H]TdR (0.5 μCi/well), the cells were harvested, and their radioactivity was measured in a liquid scintillation counter.

F. Measurement of Macrophage Cytokine Production by the Rabbit Pyrogen Test

Mice were treated orally with 150 mg/kg DEODAN for 10 days and the controls with 0.9% NaCl. Twenty-four hours after the last application, the peritoneal cavities of mice were washed twice with 5 ml RPMI 1640 containing 5 U/ml heparin. The cells were washed and incubated in six-well tissue culture plates for 24 h at 37°C, 5% CO_2 in RPMI 1640 medium, supplemented with 5% heat-inactivated (FCS, Flow), 100 U/ml penicillin, and 100 μg/ml streptomycin. The filtered supernatant from 1 × 10^6 cultured cells was injected intravenously at a volume of 3 ml per rabbit and the changes in the rectal temperatures of rabbits were recorded.

G. Measurement of Tumor Necrosis Factor (TNF-α) Production in Mice

DEODAN or saline was administered orally for 3, 6, and 10 consecutive days. Two experimental designs were employed. In the first, sera were drawn 6 h after the last application. In the second, 4 h after the last dosing, mice were triggered intravenously with 5 μg/mouse Lipopolysaccharide (LPS) from *E. coli* (Sigma). Sera were collected 2 h after triggering.

H. Statistics

The results were statistically evaluated by Student's *t*-test. A value of $P < .05$ was considered significant.

III. RESULTS

A. Effect of 3 and 6 Days of Pretreatment with DEODAN on the Course of *L. monocytogenes* in Mice

The administration of 150 mg/kg DEODAN prior to bacterial challenge induced a protection against this infection (Fig. 1). In the control saline-treated group, the mortality was 80% on day 12. In the DEODAN-pretreated group, a decrease in the

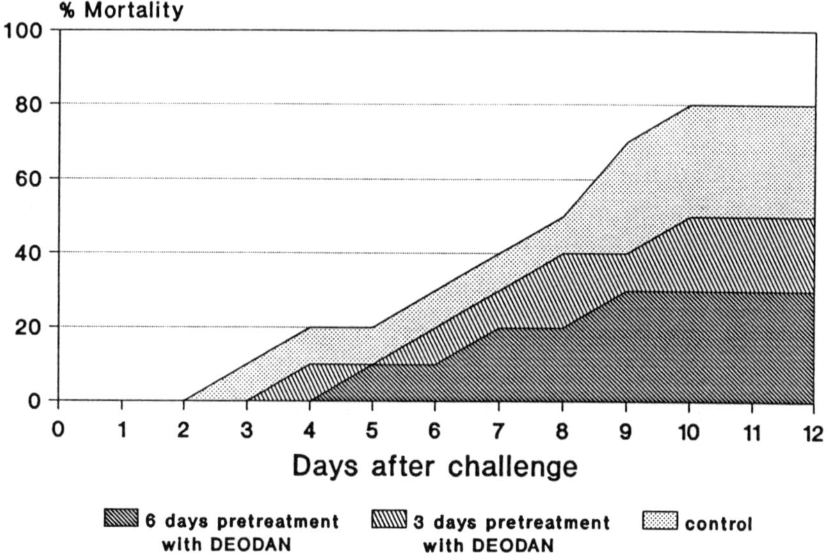

Figure 1 Effect of DEODAN on the mortality rate of mice challenged with *L. monocytogenes*. ICR mice (10/group) were treated orally for 10 consecutive days with DEODAN (150 mg/kg) or with saline (control). On day 3 or on day 6 of treatment, all mice were injected intraperitoneally with *L. monocytogenes* (0.5 ml/mouse) from bacterial suspension containing 1×10^8 microorganisms/ml. The mortality was registered for up to 12 days after challenge.

mortality rate was observed, with the protective effect being more pronounced when the drug was applied for 6 days before challenge than for 3 days before challenge.

B. Effect of 3 and 6 Days of Pretreatment with DEODAN on the Course of *K. pneumoniae* in Mice

A significant protection was observed in mice receiving DEODAN orally for 6 consecutive days prior to challenge with *K. pneumoniae* (Fig. 2). The level of protection was lower when the mice were pretreated with the preparation for 3 days only.

C. Effect of DEODAN on the Spreading Ability and Phagocytic Activity of Peritoneal Macrophages

The peritoneal macrophages from DEODAN-treated mice exhibited an increased spreading ability and phagocytic activity (Fig. 3). Compared with controls, these

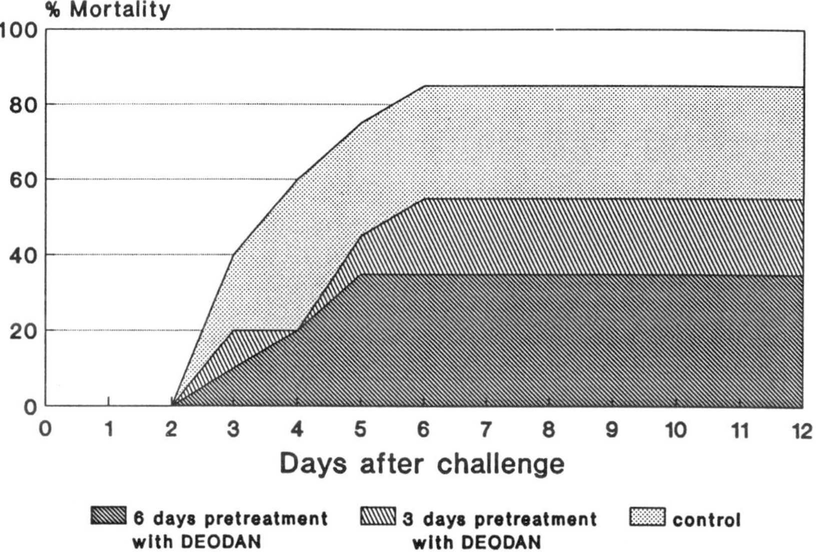

Figure 2 Effect of DEODAN on the mortality rate of mice challenged with *K. pneumoniae.* ICR mice (10/group) were treated orally for 10 consecutive days with DEODAN (150 mg/kg) or with saline (control). On day 3 or on day 6 of treatment all mice were injected intraperitoneally with *K. pneumoniae* (0.1 ml/10 g body weight) from bacterial suspension containing 10^2 microorganisms/ml. The mortality was registered for up to 12 days after challenge.

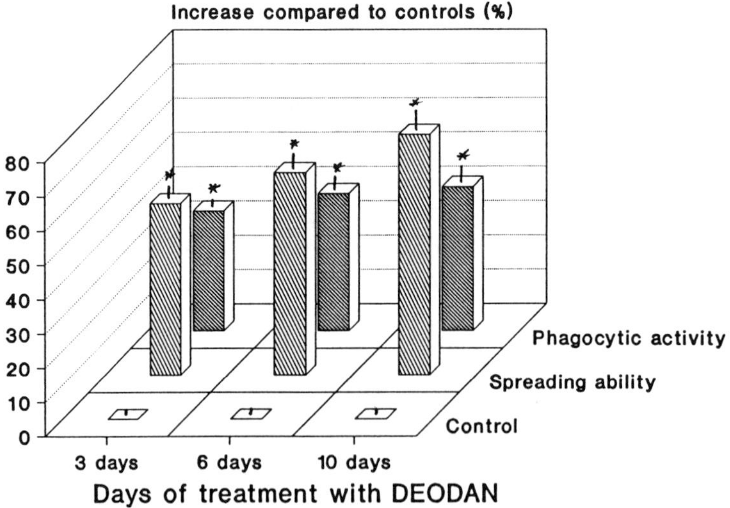

Figure 3 Effect of DEODAN on the spreading ability and on phagocytic activity of mouse peritoneal macrophages. ICR mice (5/group) were treated orally with DEODAN (150 mg/kg) or with saline (control). On days 3, 6, and 10 of treatment, the peritoneal macrophages were collected and their spreading ability and phagocytic activity was assessed. Data are expressed as the mean of three estimations ±SD.* Comparison with control: $P < .01$.

parameters were found to be correspondingly 50 and 35% higher on day 3 of DEODAN administration and reaching 70 and 42% on day 10.

D. Effect of DEODAN on the Macrophage Production of Cytokines Assessed in the Rabbit Pyrogen Test

DEODAN was applied orally to mice at a dose of 150 mg/kg for 10 days and their peritoneal macrophages were cultured for 24 h. The control animals received nonpyrogenic saline only. The pyrogenicity of the culture supernatant was assayed after intravenous injection into rabbits. Figure 4 shows the elevation of rectal temperature of rabbits after the intravenous injection of macrophage culture supernatant from DEODAN-treated mice and control animals. The macrophage culture supernatant from DEODAN-treated mice caused a febrile response with a monophasic elevation of the temperature peaking at 1 h 15 min. The temperature curve of the control rabbits injected with supernatant from mice treated orally for 10 days with nonpyrogenic saline showed no significant elevation (Fig. 4).

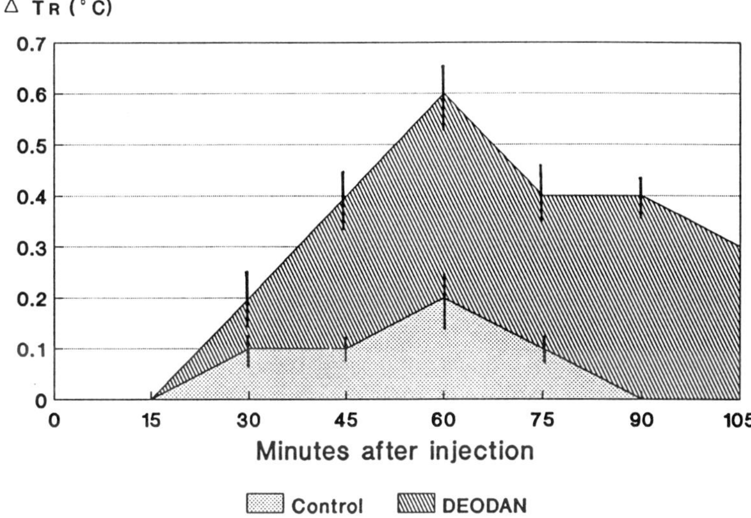

Figure 4 Rectal temperature curves in rabbits after intravenous injections of culture supernatant of peritoneal macrophages harvested from ICR mice treated orally with DEODAN (150 mg/kg for 10 days) and cultured for 24 h. The control rabbits were injected with supernatants from mice treated orally with saline. Each point represents means of ΔT^0R of three rabbits ± SD.

E. Effect of DEODAN on Macrophage IL-1 Production Measured by the Thymocyte Proliferation Assay

As shown in Figure 5, a slight enhancement of IL-1 production was observed in the culture supernatant of mouse peritoneal macrophages after 6 and 10 days of oral treatment with DEODAN.

F. Effect of DEODAN on TNF-α Levels

We investigated whether the oral administration of DEODAN (150 mg/kg) induced endogenous TNF-α in mice (Fig. 6). When applied alone, the drug caused an increase of serum TNF-α, with the highest levels being observed on day 3 of treatment. The augmenting effect of DEODAN on TNF-α synthesis was observed when DEODAN-treated mice were also triggered intravenously with LPS.

IV. DISCUSSION

DEODAN has shown antitumor effects in clinical trials in humans and in mice with transplanted tumors (4). In the present study, we examined the in vivo efficacy of

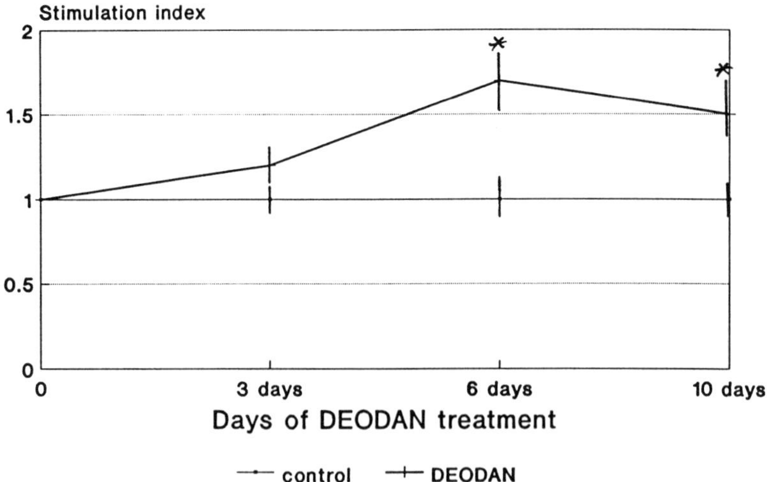

Figure 5 Effect of DEODAN on IL-1 production by mouse peritoneal macrophages. ICR mice were treated orally with DEODAN (150 mg/kg) or with saline (control). On days 3, 6, and 10 of treatment, peritoneal macrophages were collected and cultured for 72 h. IL-1 activity of culture supernatant was determined by the thymocyte proliferation assay (1 μg/ml ConA). *Comparison with control: $P < .01$.

DEODAN in two models of experimental bacterial infections. We used the intracellular and the extracellular pathogens *L. monocytogenes* and *K. pneumoniae.*

We have shown that both infections can be influenced by 6 days of pretreatment of mice with the preparation. Decreasing the pretreatment to only 3 days reduced the prophylactic effect of DEODAN.

The macrophages play the main role in affording protection in *L. monocytogenes* infection. Our previous studies demonstrated that the peritoneal macrophages from DEODAN-treated mice showed morphological signs of cell activation (7). We showed that following 3, 6, and 10 days of DEODAN treatment, the spreading ability and phagocytic activity of macrophages were enhanced. Therefore, the increased resistance to *L. monocytogenes* infection can be attributed to the macrophage activation by DEODAN.

It has been reported that macrophage-derived cytokines IL-1, TNF-α, and IL-6 are involved in this infection in mice (8–11). Since it is believed that these cytokines are among the main mediators of pyrogenicity, we investigated the febrile response of rabbits to the macrophage culture supernatant from DEODAN-treated mice. Following intravenous injection of macrophage supernatants, rabbits developed a monophasic fever that reaches a peak elevation after 45–60 min. Such rapid onset fevers have been demonstrated after intravenous injection of recombinant cytokines in rabbits (12,13). This suggests the presence of pyrogenic cytokines

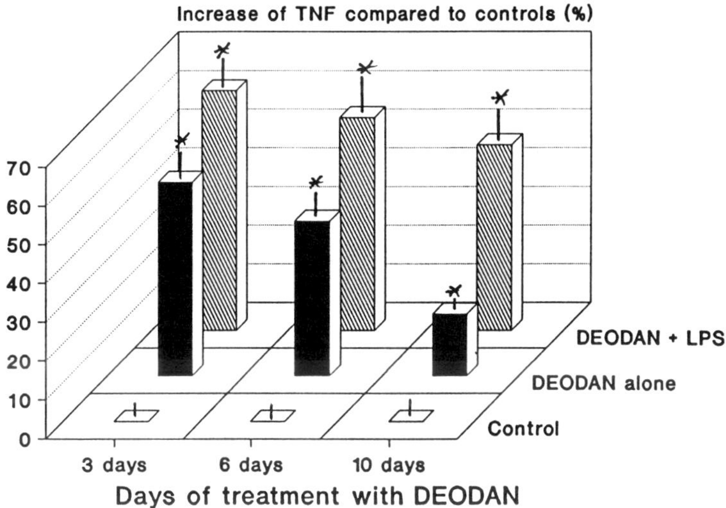

Figure 6 Effect of DEODAN on endogenous TNF-α production in mice. Mice (5/group) were treated orally with DEODAN (150 mg/kg) or with saline (control) for 3, 6, or 10 consecutive days. Six hours after the last treatment, the serum levels of TNF-α were assayed using an ELISA kit (black filled bars). Four hours after the last administration of DEODAN, mice were triggered intravenously with LPS (5 μg/mouse), sera were drawn 2 h afterwards and TNF-α levels were assayed using an ELISA kit (hatched bars). The results were expressed as percentage increase of TNF compared with controls *($P < .01$). Each bar represents the mean ± SD of triplicates. SD was <5% for all experimental values.

in the supernatants from macrophages of DEODAN-treated mice. In our previous neutralization experiments, we have shown that DEODAN induces the production of IL-1, TNF-α, and IL-6 (14). In the present study, we also demonstrated the induction of IL-1 and TNF-α after DEODAN treatment. On the basis of these results, we assume that the effect of the drug on the course of *L. monocytogenes* infection is due to macrophage activation in terms of increased phagocytosis and cytokine production.

We have also shown that DEODAN pretreatment decreases the mortality rate in mice infected with *K. pneumoniae*. Because the neutrophils are mainly involved in protection in this extracellular infection, it seems that DEODAN activates their functions too. We found increased TNF-α levels in the sera of DEODAN-treated mice. This elevation was higher when LPS was used as a trigger following the administration of DEODAN. In view of these data, we suppose that DEODAN acts as a primer and the gram-negative *K. pneumoniae* as a trigger of TNF secretion. It is known that TNF can activate granulocyte functions and enhance the respiratory burst induced by exogenous substances such as bacterial products (15,16).

Therefore, the increased resistance of mice orally treated with DEODAN to infection with *K. pneumoniae* may partly be due to the neutrophil-activating effect of TNF induced by the preparation.

Considering the major role of other cytokines such as interferons and colony-stimulating factors during both infections, future research has to be directed to evaluation of these cytokines after oral administration of DEODAN.

REFERENCE

1. Iida J, Une T, Ishihara C, Nishimura K, Tokura S, Mizukoshi N, Azuma I. Stimulation of non-specific host resistance against Sendai virus and *Escherichia coli* infections by chitin derivatives in mice. Vaccine 1987; 5:270.
2. Matsumoto K, Osada Y, Une T, Otani T, Ogawa H, Azuma I. Anti-infectious activity of the synthetic muramyl dipeptide analogue MDP-Lys (18). In: Azuma I, Jolles G, eds. Immunostimulants: now and tomorrow. Berlin: Japan Scientific Society Press, Springer-Verlag, 1987:79.
3. Ishihara C, Hamada N, Yamamoto K, Azuma I. Effect of stearoly-N-acetylmuramyl-L-alanyl-D-isoglutamine on host resistance to *Corynebacterium kutscheri* infection in cortisone-treated mice. Vaccine 1984; 2:261.
4. Bogdanov I, ed. Observations on the therapeutic effect of the anti-cancer preparation from *Lactobacillus bulgaricus* "LB51" tested on 100 oncologic patients. Sofia: Sofia Press, 127.
5. Bogdanov I, Dalev PG, Gurevich AI, Koslov VP, Malkova VP, Plemyannikova LA, Sorokina IB. Antitumor glycopeptides from *Lactobacillus bulgaricus* cell wall. FEBS Lett 1975; 57:259.
6. Mizel SB, Oppenheim JJ, Rosenstreich DL. Characterization of lymphocyte-activating factor (LAF) produced by the macrophage cell line, P338D. I. Enhancement of LAF production by activated L lymphocytes. J Immun 1978; 120:1497.
7. Popova P, Guencheva G, Davidkova G, Bogdanov A, Pacelli E, Opalchenova G, Kutzarova T, Koychev C. Stimulating effect of DEODAN (an oral preparation from *Lactobacillus bulgaricus* "LB51" on monocytes/macrophages and host resistance to experimental infections. Int J Immunopharmacol 1993; 15:25.
8. Havell EA. Evidence that tumor necrosis factor has an important role in antibacterial resistance. J Immun 1989; 143:2894.
9. Havell EA, Sehgal PB. Tumor necrosis factor-independent IL-6 production during murine listeriosis. J Immun 1991; 146:756.
10. Kurz-Jones EA, Virgin HW, IV, Unanue ER. In vivo and in vitro expression of macrophage membrane interleukin-1 in response to soluble and particulate stimuli. J Immun 1986; 137:10.
11. Nakane A, Numata A, Minagawa T. Endogenous tumor necrosis factor, interleukin-6 and gamma interferon levels during *Listeria monocytogenes* infection in mice. Infect Immun 1992; 60:523.
12. Dinarello CA, Cannon JG, Wolff SM, Bernheim HA, Beutler B, Cerami A, Figari IS, Palladino A, O'Connor JV. Tumor necrosis factor (cachectin) is an endogenous pyrogen and induces production of interleukin 1. J Exp Med 1986; 163:1433.

13. Murakami N, Sakata Y, Watanabe T. Central action sites of interleukin 1 for inducing fever in rabbits. J Physiol 1990; 428:299.
14. Guencheva G, Popova P, Davidkova G, Mincheva V, Mihailova S, Bogdanov A, Pacelli E, Auteri A. Determination of cytokine release after in vivo and in vitro administration of DEODAN (a preparation from *Lactobacillus bulgaricus* "LB51") by the rabbit pyrogen test. Int J Immunopharmacol 1992; 14:1429.
15. Klebanoff SJ, Vadas MA, Harlan JM, Sparks LH, Gamble JR, Agosti JM, Waltersdorph AM. Stimulation of neutrophils by tumor necrosis factor. J Immun 1986; 136:4220.
16. Trautinger F, Hammerle AF, Poeschl G, Micksche M. Respiratory burst capability of polymorphonuclear neutrophils and TNF alpha serum levels in relationship to the development of septic syndrome in critically ill patients. Leukoc Biol 1991; 49:449.

32

Can Orally Applied Immunomodulators Improve the Local Defense?

H. Wolf, C. Ruedl, and G. Wick
University of Innsbruck Medical School, Innsbruck, Austria

I. INTRODUCTION

The concept of mucosa-associated lymphoid tissue (MALT), represented by the gastrointestinal tract, genitourinary tract, and respiratory tract, inner ear, oral cavity, and others is based on lymphocyte recirculation between distinct and also distant mucosal lymphoid tissues. In general, this circulation is regulated by specific recognition elements on high endothelial venules (HEVs) and their corresponding adhesion molecules that are expressed as homing receptors on lymphocyte membranes (1).

The communication between different mucosal sites, leading to the induction of antigen-specific secretory immunoglobulin A (sIgA) secretion, forms the generally applicable basis of oral immunization strategies.

In previous studies (2), we followed lymphocyte kinetics and organ distribution after oral administration of the immunomodulating bacterial lysate (LUIVAC). We could demonstrate an enhanced localization of intestinal lamina propria lymphocytes (LPLs) and Peyer's patch lymphocytes (PPLs) in the respiratory tract of syngeneic recipients that had received these cells intravenously.

In the present work, we investigated the rate of Ig synthesis (IgG, IgM, and IgA) in the gut (LPs, PPs), lung, and spleen of mice vaccinated orally with the bacterial lysate. For this purpose, we developed a new and very sensitive assay based on time-resolved fluoroimmunoassay (TR-FIA) (3).

II. MATERIALS AND METHODS

A. Animals and Immunization Procedure

Inbred 5- to 7-week old female Balb/c mice were obtained from the Research Institute for Laboratory Animal Breeding, Himbreg, Austria, and maintained in the Central Laboratory Animal Facilities of the Medical School, University of Innsbruck, Austria.

The applied bacterial lysate (LUIVAC, kindly provided by M. Ellenrieder, Luitpold-Pharma GmbH, Munich, Germany) contains at least 10^{10}/g of each of the following heat-inactivated, mechanically disintegrated, and lyophilized bacteria: *Staphylococcus aureus, Streptococcus pyogenes, S. mitis, S. pneumoniae, Haemophilus influenzae, Branhamella catarrhalis,* and *Klebsiella pneumoniae*. The bacterial lysate was mixed in a ratio of 3:100 with mannit dissolved in phosphate-buffered saline (PBS, pH 7.2).

The mice were immunized orally by gavage for 5 consecutive days with 15 mg bacterial lysate/kg body weight in a total volume of 200 μl of PBS. After a 12-day interval, the immunization was repeated for another 5 days. One week later, the animals were killed and the organs removed for cell isolation.

B. Isolation and Characterization of Cells

Splenocytes, LPLs, and PPLs were isolated as described previously (1). Intraparenchymal lung lymphoid cells were isolated by a modification of the method described by Abraham et al. (4). Methods of cell staining and FACS analysis and sorting were performed as described (2).

C. Examination of Immunoglobulin Production Rate (IgG, IgM, IgA) of Lymphocytes

After oral stimulation with the immunomodulator (LUIVAC), Ig synthesis of intestinal LPLs, PPLs, lung lymphocytes, and, for comparison, splenocytes were detected by means of a modified ELISPOT-test connected with TR-FIA. This test was developed by Ruedl and coworkers in our laboratory (3).

D. Detection of Total and Specific Serum Antibodies

Patterns of immunoglobulins (IgG, IgA, IgM) in serum of immunized and unimmunized animals were evaluated by TR-FIA technique mentioned above and described by Ruedl et al. (2).

E. Statistical Evaluation

Data obtained are expressed as means ±SEM. For comparison of experimental groups, the unpaired Student's *t*-test was applied.

III. RESULTS AND DISCUSSION

In the present work, we investigated whether orally applied immunomodulators can improve the local defense. By means of a new and very sensitive TR-FIA, we examined the effect of a pneumotropic bacterial lysate (LUIVAC) on the Ig response in various lymphoid organs such as the spleen, intestinal LPs, PPs, and lung. The nonspecific and antigen-specific Ig secretion of lymphocytes from these organs was measured to find out whether the orally applied immunomodulator is able to induce local protection on distant mucosal effector sites where infections frequently occur.

After phenotypic characterization by FACS analysis (data not shown here), we found the following Ig isotype pattern that was characteristic for each isolated cell type.

As shown in Figure 1, the bacterial lysate did not provoke any significant antibody response in PPs, LPLs, and spleen but was effective in inducing a remarkable IgA increase in the lungs. In fact, 12 days after the last immunization (two vaccination periods), an enhanced secretion rate of IgA was observed compared with untreated animals. The antigen-specific IgA response (180%) as well as the non-

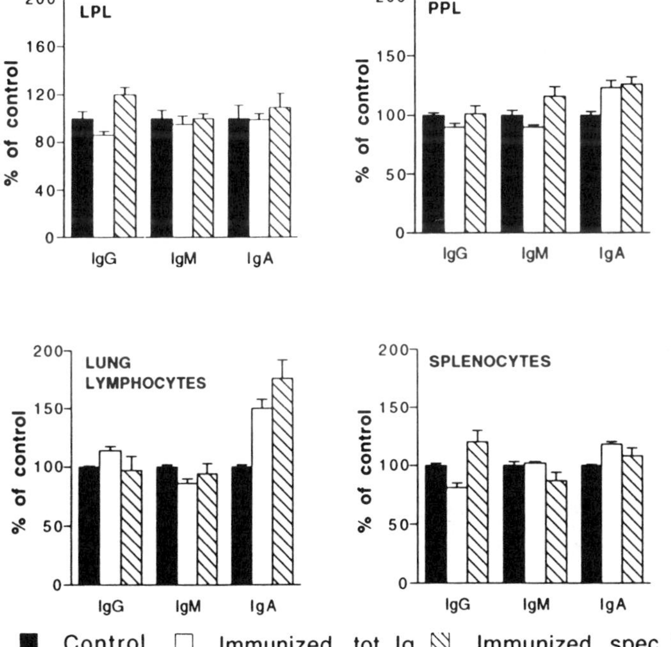

Figure 1 Immunoglobulin synthesis after oral administration of immunostimulator.

specific IgA signal (150%) were highly significantly increased, whereas the IgG and IgM synthesis remained unchanged.

To assess whether the immunomodulator also could enhance the systemic immune response, we measured the total and the antigen-specific antibodies (IgG, IgA, IgM) in peripheral blood of orally vaccinated mice. By means of TR-FIA technique, we found a small but significant (30%, $P < .03$) increase of antigen-specific IgA after two immunization cycles (Fig. 2). No significant differences in specific IgG and IgM could be observed. Total immunoglobulin levels revealed no marked differences between treated animals and untreated controls (Fig. 2).

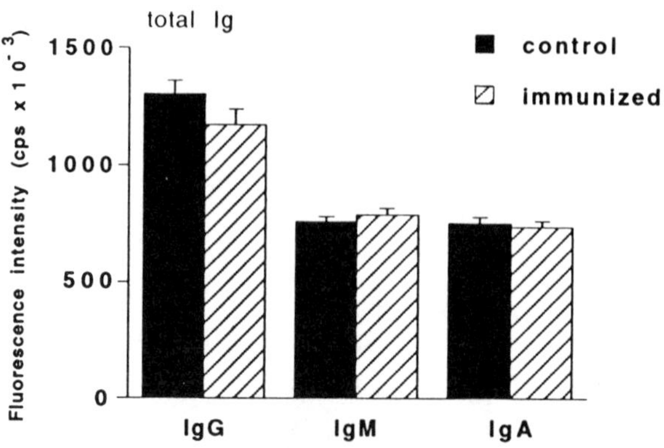

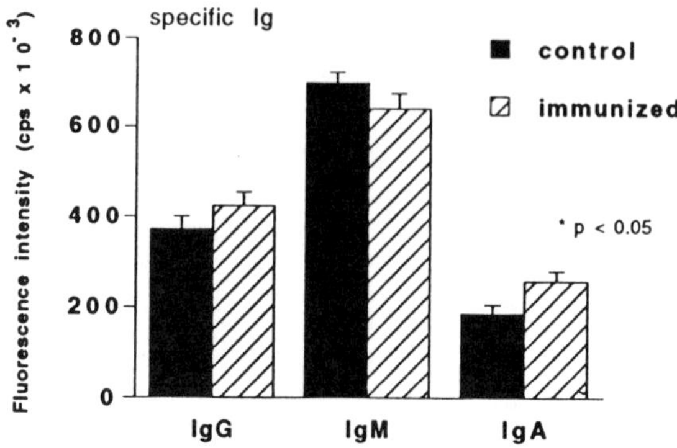

Figure 2 Serum immunoglobulin levels.

We conclude that priming of the gut-associated lymphoid tissue (GALT) with bacterial immunomodulators (LUIVAC) results in enhanced IgA production in the lungs that may raise protection from respiratory infections. These findings correlate well with our previous observations on the migratory behavior of immunocompetent intestinal lymphocytes that indicated oral immunization leads to an increased appearance of LPLs and PPLs in the lungs after intravenous injection of these gut cells into syngeneic recipients (2).

ACKNOWLEDGMENT

This work was supported by the Kamillo Eisner-Foundation, Hergiswil, Switzerland.

REFERENCES

1. McGhee JR, Mestecky J, Dertzbaugh MT, Eldridge JH, Hirasawa M, Kiyono H. The mucosal immune system: from fundamental concepts to vaccine development. Vaccine 1992; 10:75.
2. Ruedl C, Albini B, Böck G, Wick G, Wolf H. Oral administration of a bacterial immunomodulator enhances murine intestinal lamina propria and Peyer's patch lymphocyte traffic to the lung: possible implications for infectious disease prophylaxis and therapy. Int Immunol 1993; 5:29.
3. Ruedl C, Wick G, Wolf H. A novel and sensitive method for detection of antibody-secreting cells using time-resolved fluorescence. Immunol Methods 1994; 168:61.
4. Abraham E, Freitas AA, Coutinho AA. Purification and characterization of intraparenchymal lung lymphocytes. J Immunol 1990; 144:2117.

33

Immunoglobulins in Serum and Saliva of Children with Recurrent Infections of the Respiratory Tract During Treatment with an Oral Immunomodulator

A. van Aubel, C. Hofbauer, U. Elsasser, and A. Kämmereit
Luitpold Pharma, Munich, Germany

R. J. Riedl-Seifert
Pediatrician and Allergologist, Kassel, Germany

I. INTRODUCTION

The stimulation of the mucosa-associated lymphoid tissue (MALT) by oral application of different antigens is well established (1). Antigens presented to the MALT induce the maturing of B lymphocytes that are mainly disposed to immunoglobulin A (IgA) synthesis. In spite of the induction of specific IgA by oral application of different antigens, only a few oral vaccines and immunomodulators have been developed for clinical application.

LW 50020 is an oral immunomodulator that consists of the lysate of seven potentially pathogenic bacteria for the respiratory tract. The lysate contains at least 10^9 bacteria of each of the following strains: *Streptococcus pneumoniae, S. pyogenes, S. mitis, Staphylococcus aureus, Haemophilus influenzae, Klebsiella pneumoniae,* and *Branhamella catarrhalis.*

LW 50020 is indicated for the treatment of recurrent infections of the respiratory tract. The application scheme includes a 28-day priming phase followed by a 28-day therapy-free interval and a 28-day booster phase. The daily dosage is one 3-mg tablet.

In animal models, LW 50020 exhibited a protection against infection of mice with *S. pneumoniae* (2), *K. pneumoniae* (3), and influenza virus (4). LW 50020 induced an increase of interferon gamma (INF-γ) and a decrease of elastase in the bronchoalveolar lavage fluid of mice infected with *S. pneumoniae* (5). In an influenza virus model, alveolar macrophages of mice treated with LW 50020 exhibited a higher activity measured by cyclic adenosine monophosphate (AMP) (6). LW 50020 increased the oxidative burst of polymorphonuclear leukocytes of rabbits triggered by opsonized zymosan (7).

The Peyer's patches of mice immunized with a high dosage of LW 50020 showed an increase of IgA-positive plasma cells. Specific IgA in serum rose after immunization and booster (G. Wick, personal communication). In the lung homogenate, a higher activity of secretory IgA (sIgA) could be measured (G. Wick, personal communication).

In placebo-controlled double-blind studies in children and adults, the safety and efficacy of LW 50020 have been demonstrated (8–10).

II. PATIENTS AND METHODS OF THE sIgA STUDY

In a prospective placebo-controlled double-blind pilot study, 65 children (aged 4–10 years) with recurrent infections of the respiratory tract were randomized in six pediatric centers. The 4- to 6-year-old children should have suffered from at least 10 infections in the last year, and the 7- to 10-year-old ones from at least eight infections of the respiratory tract. Patients also were recruited if they had suffered from at least four severe infections with more than a 2-week duration during the last year. Patients with severe immunodeficiency, severe diseases, and immunosuppressive or immunomodulating therapy were excluded from the study.

Homogeneity of both treatment groups was confirmed for age, height, weight, sex, and passive smoking. Concerning the number of patients with IgG subclass deficiency and the number of infections of the lower respiratory tract, the data were in favor of the placebo-treated group (not statistically significant).

One week before starting the study and then from the beginning of the study (E0) every 2 weeks until the end at the 12th week (E6), samples of nonstimulated morning saliva were taken. At the same time, actual clinical findings and clinical history for the last 2 weeks were recorded. Blood samples were drawn the week before starting the study (ER), after the first immunization period (E2), and at the end of the study (E6) (Table 1).

Table 1 Flow chart of the sIgA Study

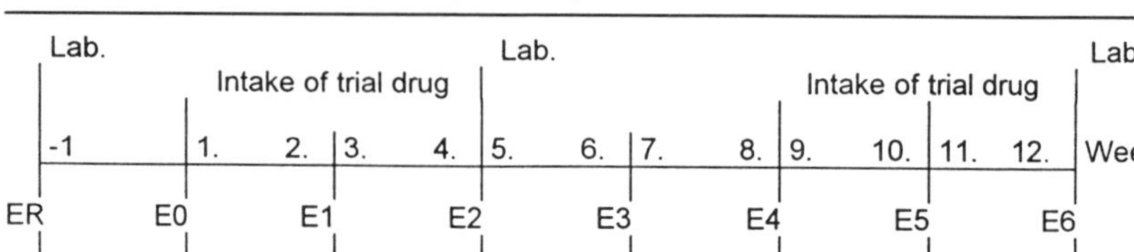

E = examination; R = recruiting visit.

A. Laboratory Investigations

Blood samples were sent to a central laboratory, where safety parameters (differential blood picture, clinical chemistry, immunoglobulins, IgG subclasses) were determined. An aliquot of the saliva was investigated at the immunological laboratory of Luitpold Pharma for total sIgA and specific antibodies against LW 50020. The analytical staff was blind as to the assignment of the samples.

B. Materials and Methods

Saliva samples were stored at –70°C before performing the assay. After thawing in a water bath (25°C), the samples were centrifuged for 10 min at 10,000*g* to remove cells and mucoid substance. The supernatant was taken for the measurements.

1. ELISA Procedure for Measuring Total sIgA

Volumes of 100 μl were used throughout the test. The microtiter plates were washed with phosphate-buffered saline (PBS) between each incubation step, supplemented with 0.05% Tween 20, three times. Incubation temperature was held constant at 37°C. Microtiter plates (Dyantech, M129A) were coated with antihuman IgA (Sigma I-2261) at a concentration of 12 μl/ml in PBS for 1 h. A second coating layer of 1% bovine serum albumin (BSA) in PBS was added for 30 min. Duplicates of saliva samples were applied in an appropriate dilution (1:512 to 1:1024). sIgA from colostrum (Behring) was used as standard. Incubation was carried out for 1 h. Rabbit immunoglobulin-peroxidase to human secretory component (Dako P166, 1:1000 in PBS/0.05% Tween) was added. Incubation was carried out for 30 min. The substrate (ABTS) was dissolved in phosphate citrate sodium perborate buffer (pH 5.0) and added. After 30 min, the absorbance was measured at 405 nm (reference 630 nm) with an ELISA reader (Dynatech).

In a semilogarithmic plot the correlation between absorbance and sIgA quantity could be approximated by a linear function in a range from 200–1600 μg/l. Concentrations of secretory IgA were determined by extrapolation from the standard curve.

2. ELISA Procedure for Measuring Specific IgA

Volumes of 100 μl were used throughout the test. The plates were washed as described above. Microtiter plates were coated for 1 h with the different bacterial monolysates at a concentration of 10 μg/ml in bicarbonate buffer pH 9.6. One percent BSA/PBS was added for 30 min. Saliva samples were diluted (1:16 to 1:32); each sample was done in duplicate. After 1 h of incubation, biotinylated antihuman IgA (Amersham RPN 1184) was added (1:500 in 1% BSA/PBS). After another hour, streptavidin-biotinylated peroxidase preformed complexes (Amersham RPN 1051) 1:400 in PBS/Tween 0.05% were added. Incubation was carried out for 30 min. As substrate ABTS was added, the assay was measured as described above.

Table 2 Difference to Baseline Values of Immunoglobulins in the Serum (difference not statistically significant between both groups)

	LW 50020			Placebo		
	n	mean	s	n	mean	s
IgA E2-E0	33	18.07	37.97	31	13.74	40.75
IgE E2-E0	32	−3.76	84.17	31	17.00	99.91
IgM E2-E0	33	18.48	51.78	31	22.74	47.70
IgG E2-E0	30	0.34	1.22	31	0.28	0.51

III. RESULTS

A. Immunoglobulins in Serum

The mean difference to baseline was not statistically significant for IgG, IgA, IgM, and IgE between groups (*t*-test; Table 2). IgE decreased in both groups but was more pronounced in the group treated with LW 50020. At the start of the study, twice as many children as in the placebo-treated group had IgG subclass deficiency in the LW 50020–treated group (Table 3). Those children reflected a history with more severe infections of the respiratory tract.

B. Immunoglobulins in the Saliva

Statistical analysis revealed that sIgA of ER and E0 could be pooled. Baseline values of sIgA showed a great intraindividual and interindividual variation but the difference was not statistically significant (LW 50020 28.3 ± 42.96 mg/dl; placebo 20.8 ± 24.77 mg/dl; Fig. 1). The mean sIgA during the study period was always lower in the placebo-treated patients. Patients with IgG subclass deficiency showed lower sIgA in both the LW 50020-treated and placebo-treated groups (Fig. 2). The lowest levels were seen in placebo-treated patients with IgG subclass deficiency.

Table 3 Patients with IgG Subclass Deficiency at Start of the sIgA Study

	LW 50020		Placebo	
	n	%	n	%
IgG deficiency				
no	22	64.7	25	80.6
yes	12	35.3	6	19.4

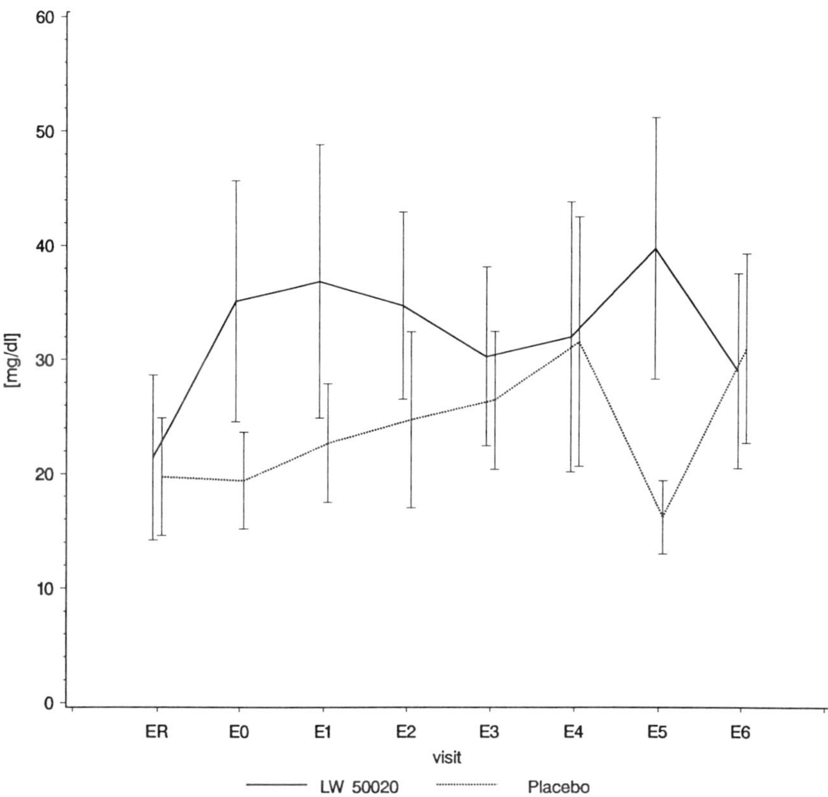

Figure 1 Total secretory IgA in the saliva of children with recurrent infections (mean ± standard error of mean). Treatment groups: 34 patients LW 50020, 31 patients placebo.

Patients younger than 6 years had lower sIgA than older patients (Fig. 3). Passive smoking had a decreasing effect on secretory antibodies in saliva: Lower levels were measured in patients with smokers in the family with a slight increase in the group of LW 50020–treated patients (Fig. 4). Owing to the limited number of patients and the high variation, the differences of this explorative investigation were not statistically significant between the groups.

C. Specific Antibodies Against Bacterial Antigens in Saliva

When possible, measurements of specific antibodies against the bacterial lysate mixture and each of the seven bacterial lysates were performed. Differences between

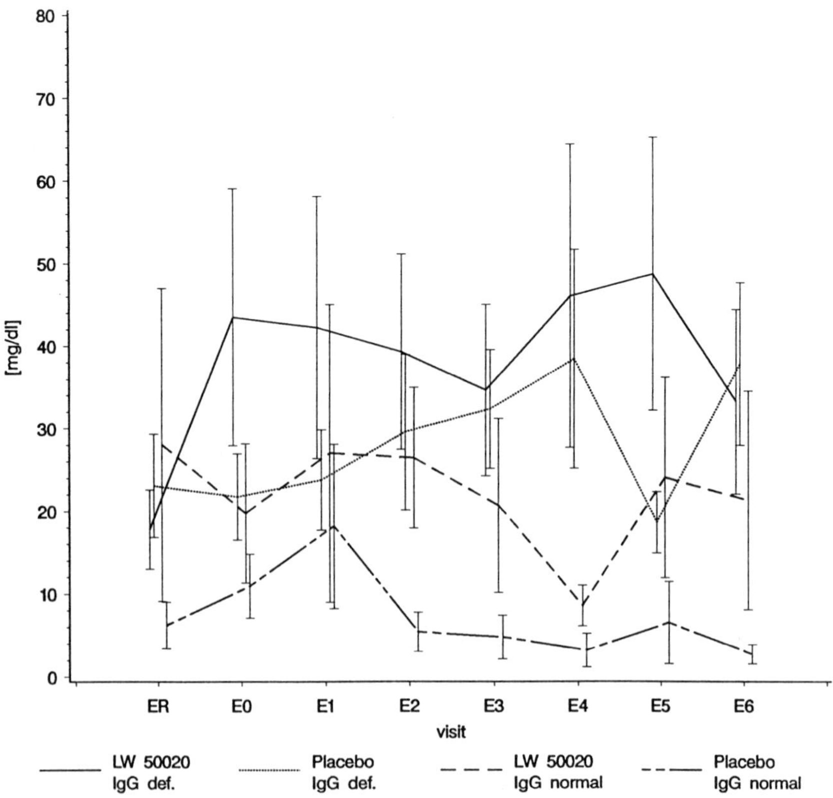

Figure 2 Secretory IgA in the saliva of patients with normal IgG subclasses and IgG subclass deficiency at start of the study (mean ± standard error of mean). Treatment groups: 34 patients LW 50020, 31 patients placebo.

the LW 50020-treated and placebo-treated group were not statistically significant, but the values of the first group were always higher than values of the placebo-treated group. Higher values were measured in the ELISA against LW 50020 (Fig. 5) and *S. pneumoniae* (Fig. 6). The assays with the other lysates reflected a lower amount of specific antibodies in this test system. After assessing single patient–related graphs of specific antibodies the LW 50020–treated group seemed to be divided into a nonresponder and responder cluster: after immunization with LW 50020, patients with higher baseline values showed a higher increase of antibody titers (Fig. 7 and 8). On the other hand, some patients with IgG subclass

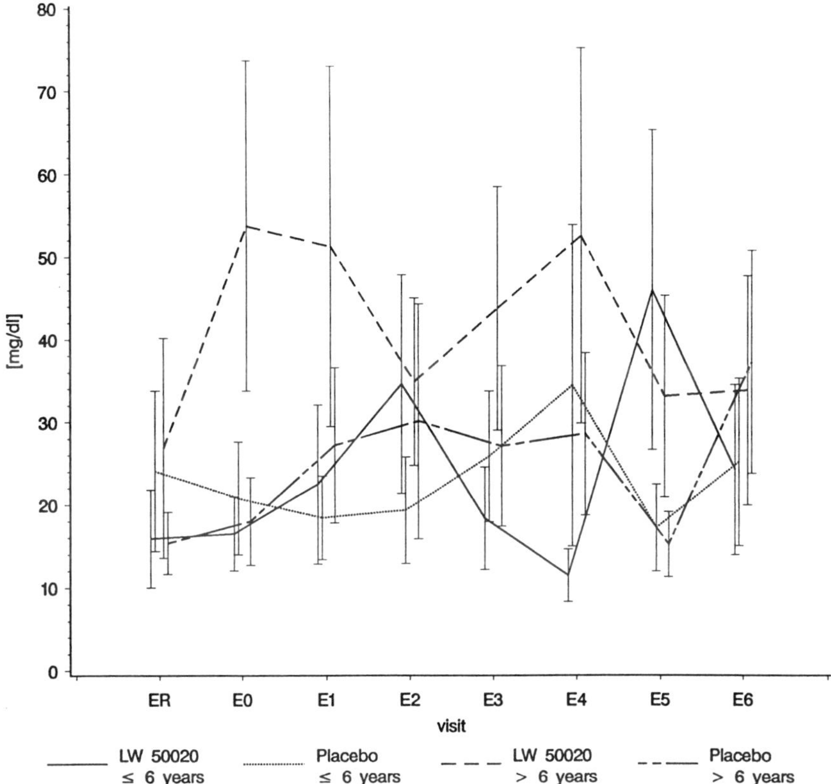

Figure 3 Secretory IgA in the saliva stratified by age (mean ± standard error of mean). Treatment groups: 34 patients LW 50020, 31 patients placebo.

deficiency and low sIgA showed a very low amount of specific antibodies in the saliva.

IV. DISCUSSION

Sampling of saliva for the measurement of antibodies, especially sIgA, is difficult: Until now, no standardization for collection and determination of antibodies has been established. The values depend on the test system and the standard (no international standard existing) used. In contrast to the literature, the amount of individual IgA

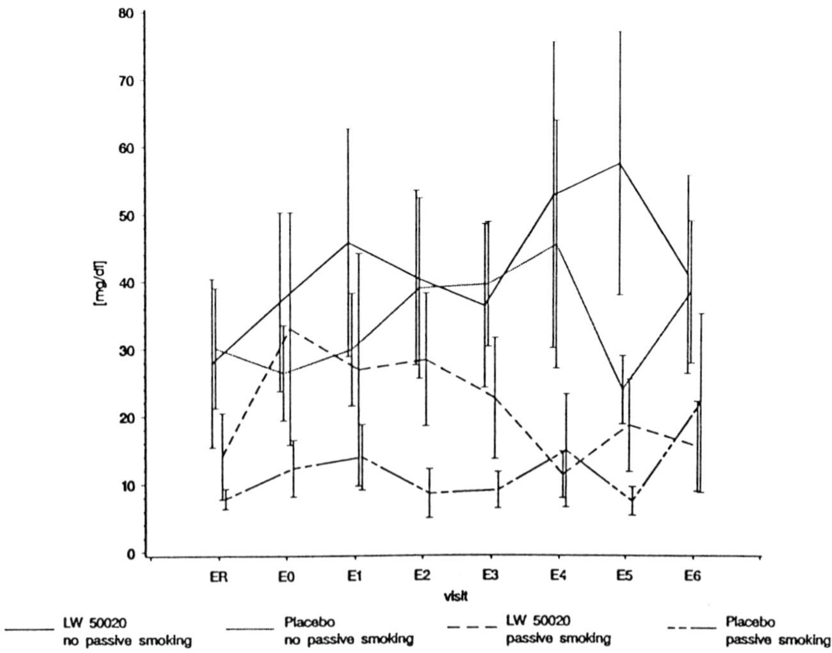

Figure 4 Secretory IgA in the saliva. Passive smoking and not passive smoking patients (mean ± standard error of mean). Treatment groups: 34 patients LW 50020, 31 patients placebo.

varied much more than expected. This may depend on the highly sensitive assay or correlate to the investigated infection-prone patients. In cases of recurrent infections nasal in spite of salivary secretions might have been determined and have induced higher mean values and an increased variation.

Patients treated with the bacterial lysate always showed a numerical difference resulting in higher mean values compared with the placebo-treated group regarding either total sIgA or specific antibodies. These results correlate with different animal models, which proved an induction of specific antibodies by LW 50020 (3,11). In another pilot study in healthy volunteers, an increase of sIgA was induced by LW 50020. An open clinical trial in infection-prone patients showed an increase of salivary IgA in comparison with the baseline values (12).

Finally, patients with IgG subclass deficiency and passive smoking patients showed decreased levels of sIgA in both groups. This may reflect a lack of the first line of mucosal defense and can partly explain the high risk of recurrent respiratory tract infections.

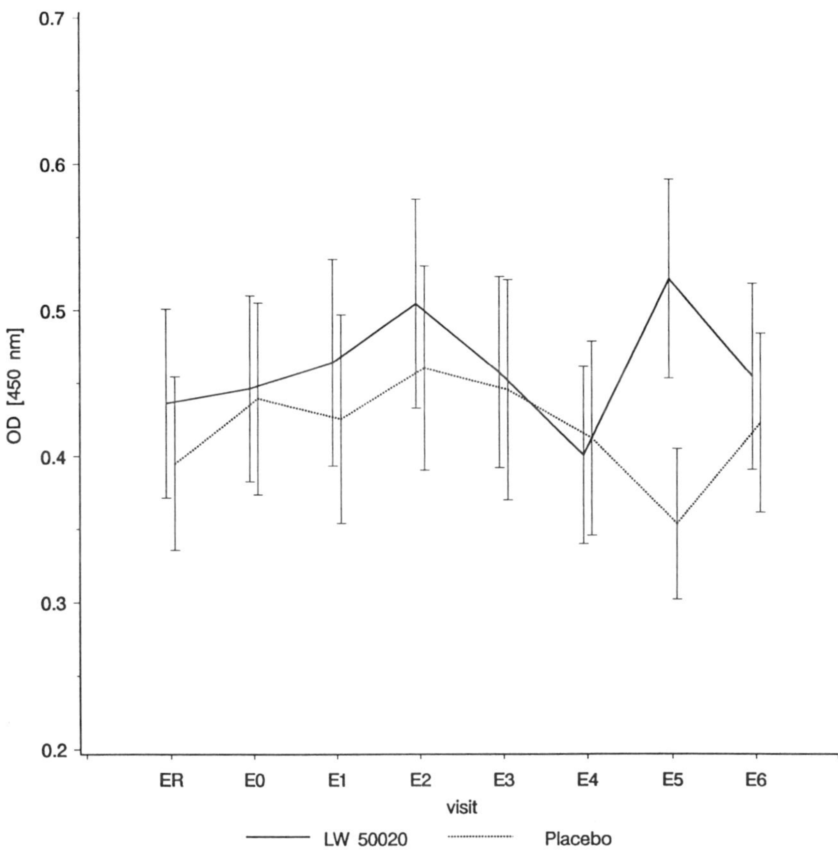

Figure 5 Specific antibodies against LW 50020 in the saliva (mean ± standard error of mean). Treatment groups: 34 patients LW 50020, 31 patients placebo.

V. CONCLUSIONS

A placebo-controlled double-blind pilot study was initiated in 65 children with recurrent infections to investigate the influence of LW 50020, a bacterial immunomodulator, on immunoglobulins in serum and saliva. The study revealed a numerical difference in favor of the LW 50020–treated group regarding the induction of secretory antibodies in the saliva. Children with IgG subclass deficiency and passive smoking children exhibited low IgA in the saliva.

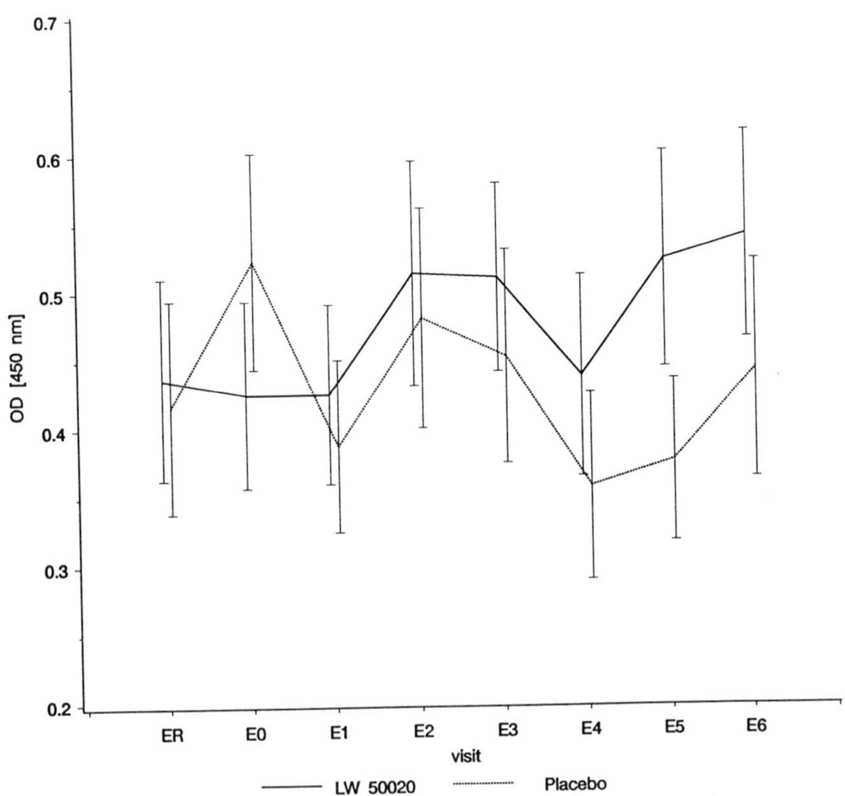

Figure 6 Specific antibodies against *Streptococcus pneumoniae* in the saliva (mean ± standard error of mean). Treatment groups: 34 patients LW 50020, 31 patients placebo.

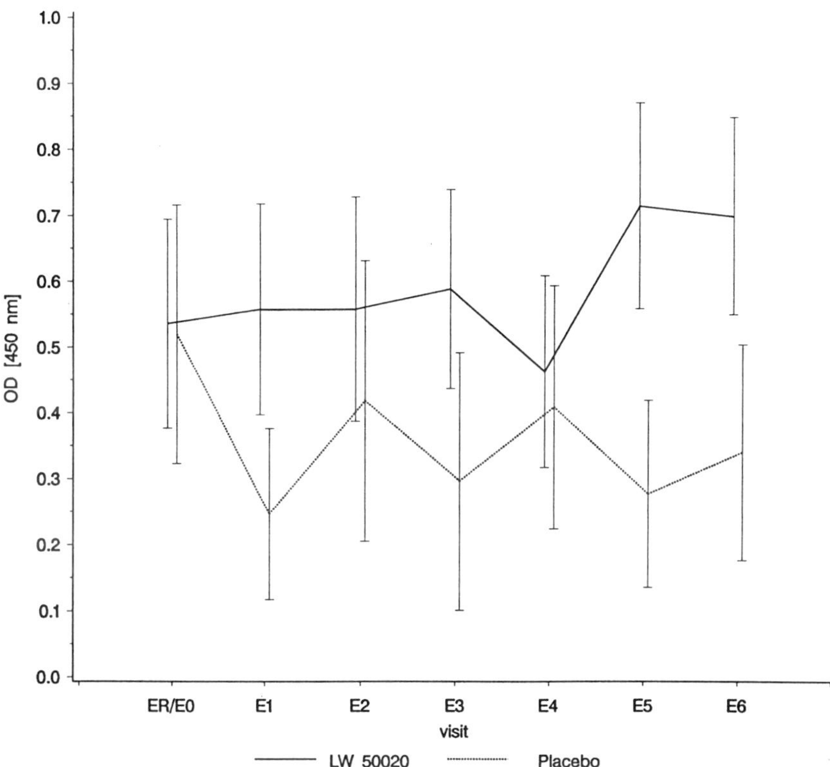

Figure 7 Specific antibodies against *Streptococcus pneumoniae* in the saliva, pooling of patients with baseline value below 0.2 or above 1.0 (mean ± standard error of mean). Treatment groups: LW 50020 11 patients, placebo 8 patients.

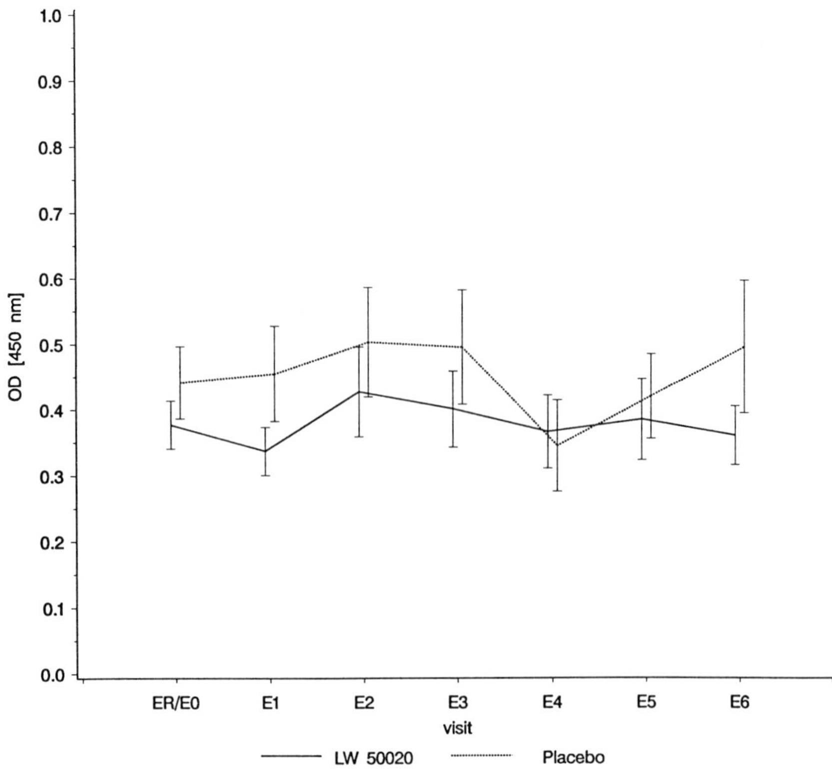

Figure 8 Specific antibodies against *Streptococcus pneumoniae* in the saliva, pooling of patients with baseline value between 0.2 and 1.0 (mean ± standard error of mean). Treatment groups: LW 50020 15 patients, placebo 16 patients.

REFERENCES

1. Mestecky J, Russel MW, Jackson S, Brown TA. Clin Immunol Immunopathol 1986; 105-14.
2. Van Daal GJ, de Jong PT, Mouton JW, Tenbrinck R, Petzoldt K, Bergmann KC, Lachmann B. Oral immunization with bacterial lysate against infection with S. pneumoniae in mice. Respiration 1990; 57:229-32.
3. Van Dijk H et al. (in press).
4. Van Daal GJ, Beusenberg FD, So KL, Fièvez RBAM, Sprenger MJW, Mouton JW, van't Veen A, Lachmann B. Protection against influenza virus infection in mice by oral immunization with a polyvalent bacterial lysate. Int J Immunopharmacol 1991; 13(7):831-40.
5. Van Daal GJ. Oral immunization with polyvalent bacterial lysate and infection with streptococcus pneumoniae: Influence on interferon-gamma and PMN elastase concentrations in murine bronchoalveolar lavage fluid. Int Arch Allergy Immunol 1992; 97:173-77.
6. Van Daal GJ, Management of pulmonary infection. Dissertation, Erasmus University Rotterdam, 1991.
7. Helmberg A, Böck G, Wolf H, Wick G. An orally administered bacterial immunomodulator primes rabbit neutrophils for increased oxidative burst in response to opsonized zymosan. Infect Immunity 1989; 57(11):3576-80.
8. Fischer H, Eckenberger H-P, van Aubel A, Kämmereit A, Elsasser U. Prävention von Infektrezidiven der oberen und unferen Luftwege. Atemw Lungenkrkh 1992; 18(4):146-55.
9. Debelic M, Eckenberger HP. Prävention von Infektrezidiven der oberen und unteren Luftwege. Fortschritte der Medizin 1992; 110(30):49/565-56/570.
10. Riedl-Seifert RJ, van Aubel A, Kämmereit A, Elsasser U. Recurrent infections of the respiratory tract—still a "crux medicorum"? Results of a placebo-controlled double-blind study in infection prone children receiving an oral bacterial lysate. J Pharmakol Ther 1993; 2(3):108-17.
11. Ruedl C et al. (in preparation).
12. Montez J et al. (in press).

34

Oral Immunotherapy in Women with Chronic Genital Infections by Means of a New Microbial Immunomodulator

Catalina-Elena Lupusoru, Sava Dumitrescu, Marie Jeanne Aldea, Elena Albu, Ovidiu Bredetean, and Florin Tuluc
Grigore T. Popa University of Medicine and Pharmacy, Iasi, Romania

Sylvia Hoisie
Cantacuzino Institute, Iasi, Romania

I. INTRODUCTION

The concept of the oral immunization is very ancient. This notion is found in the works of Alexandru Besredka (1921), as noted by Bernard in 1979 (2).

The lymphoid organs and the lymphoid areas of the gut (Peyer's patches) and the airways have a central role in the immune response via the immune net of macrophages and T and B lymphocytes (3,17–20).

The concept of a common defense system of mucous membranes had been promoted by Scrobogna and Bienenstock in 1980 (3,20), Nosubo in 1981 (12), and Husband in 1984 (11). This concept is a major step for the cognition of the action mechanism of oral immunomodulators and also of the connections between the lymphatic tissues of mucous membranes.

The Peyer's patches—gut-associated lymphoid tissue (GALT)—determine the production of immunoglobulins by means of their superficial M cells and by contact with the immune net of the macrophages and T and B lymphocytes. The immunological cells move between GALT and the bronchial-associated lymphoid tissue (BALT) or the lymphoid tissues associated with the urogenital tract (1,3,6, 9,11,12,22). This is the mechanism of action of oral immunomodulators (13).

The active immunomodulating principles in OROSTIM are standardized fractions extracted from 14 selected bacterial strains (*Escherichia coli,* two strains; *Klebsiella pneumoniae,* two strains; *Serratia marcescens,* one strain; *Neisseria catarrhalis,* one strain; *Staphylococcus aureus,* two strains; *Streptococcus pneumoniae,* two strains; *S. pyogenes,* one strain; *S. viridans,* one strain; *S. faecalis,* two strains). These bacterial strains are commonly responsible for respiratory and urogenital tract chronic infections in Romania. These fractions are lysed, ultrasound treated, diluted to contain 3×10^9 germs/ml saline solution and finally dried with an atomizer. One capsule of extract contains 10 mg immunostimulatory fractions and adjuvants for a final weight of 0.2 g.

The action of this product is likely to be based on the stimulation of the different humoral and cellular immune defense mechanisms. The drug acts by amplifying connections of the lymphatic tissues of the mucous membranes: GALT, BALT, and lymphatic tissues of the urogenital tract (1).

OROSTIM was patented in 1990 by one of us (C-EL) and coworkers from Cantacuzino Institute and marketed in Romania at the beginning of 1993. Very good results have been obtained in chronic sinusitis, rhinopharyngitis, or bronchitis and in intrinsic bronchial asthma when adding OROSTIM to the usual antibiotic treatment (own unpublished results).

Former immunopharmacological investigations in a small number of patients have shown that OROSTIM increases the efficiency of classic treatment with antibiotics also in women with chronic genital infections.

We resumed these clinical investigations with the aim of obtaining a statistically significant confirmation or refutation of OROSTIM efficiency in these types of gynecological diseases.

II. PATIENTS AND METHODS

A. The Immunomodulator

The batches of OROSTIM were prepared by Catalina-Elena Lupusoru and coworkers in the facilities of the Iasi Division of Cantacuzino Institute while working on a project she directed and was supported by the Academy of Medical Sciences (Bucharest, Romania).

B. Selection of Patients

The trial included 24 women (24–44 years old), each with a chronic genital infection with more than a 4-year history. These women had been hospitalized for acute episodes two to four times each year.

Each woman entered the trial in 1989, during an acute exacerbation of her chronic genital infection, as an in-patient of the Clinical Hospital of Obstetrics and Gynaecology No. 1, Iasi, Romania.

The separation of the gynecological patients in different groups of ages and gynecological diseases is shown in Table 1.

Table 1 Diseases of Gynecological Patients in the Trial

Disease	Age (years)				Total
	24–29	30–35	36–40	41–44	
Chronic pelvic cellulitis	1	3	2	2	8
Bilateral adnexitis	2	1	1	0	4
Cervicitis	1	0	0	1	2
Bilateral metroadnexitis	0	1	0	1	2
Cervicitis and metritis	1	1	0	0	2
Pelvic cellulitis and vaginitis	0	0	1	1	2
Chronic pelviperitonitis and both legs phlebitis	0	1	3	0	4

C. Schedule of Treatment and Follow-up

The patients were divided into two similar groups of 12 women each (Table 2). The duration of the acute exacerbation existing when the patients joined the trial (moment 0) was registered for every participant (Fig. 1). Patients of group I had been treated concomitantly with antibiotics (cephalosporins, ampicillin, or tetracyclines) and OROSTIM. OROSTIM was administered orally one capsule twice daily for 10 days. The antibiotics were administered for 10–14 days. After leaving the clinic, the patients continued to receive OROSTIM for 2 months (one capsule twice daily for 10 days every month) and after other 10 months they were subjected to a clinical reexamination.

Table 2 Two Groups of Gynecological Patients Treated with Antibiotics and OROSTIM (Group I) and with Antibiotics Only (Group II)

Disease	No. of patients	
	Group I	Group II
Chronic pelvic cellulitis	4	4
Bilateral adnexitis	2	2
Cervicitis	1	1
Bilateral metroadnexitis	1	1
Cervicitis and metritis	1	1
Pelvic cellulitis and vaginitis	1	1
Chronic pelviperitonitis and both legs phlebitis	2	2

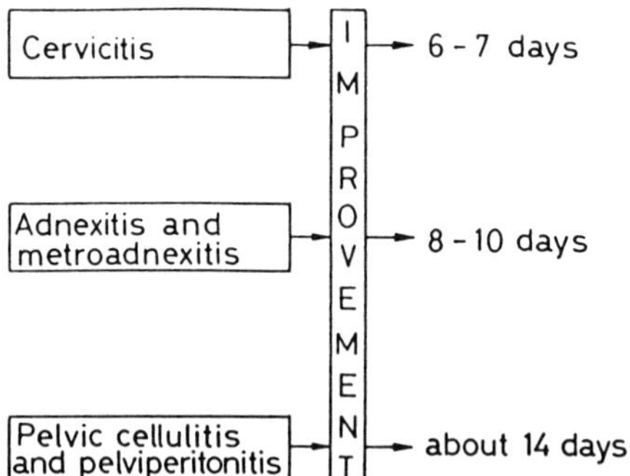

Figure 1 Mean duration of acute episodes of chronic gynecological infections at the beginning of the trial. The acute episodes in both groups had a similar duration. The acute episodes of different gynecological diseases had different durations.

Patients of group II (the controls) were treated only with antibiotics (cephalosporins, ampicillin, or tetracyclines) according to the same schedule. After leaving the clinic, they also were followed-up for 1 year. The number of relapses during the trial year was registered in all patients.

In both groups, the following immunological parameters were investigated when the patients joined the trial (moment 0) and 10 days later (moment 1): the T lymphocytes, the phagocytosis capacity of peripheral blood polymorphonuclear (PMN) cells, and the serum complement activity. Moment 1 coincides with the end of the first OROSTIM administration.

D. Methods for Immunological Assay

1. Phagocyte Analysis

Nitro-blue-tetrazolium (NBT) Test. This was performed according to the method described by Peacock and Tomar (24).

Phagocytosis Capacity of PMNs in the Presence of Staphylococcus aureus *(Oxford).* *Staphylococcus aureus* (Oxford) was diluted 1/100,000 in saline solution. The samples of blood from gynecological patients were collected in tubes containing 0.05 ml heparin (50 U/ml), and in the next step the following mixture was made in test tubes: 0.2 ml blood + 0.2 ml of diluted culture of *S. aureus* 1/100,000 in saline solution. The test tubes were incubated for 45 min at 37°C and during this time stirred for a few seconds every 5 min. Afterwards the samples were Giemsa stained. PMNs with phagocytosis capacity were microscopically counted as the percentage of the total. The opsonic index also was determined.

2. Rosetting Technique for T-Lymphocyte Quantitation

T lymphocytes obtained from human peripheral blood as described previously (24) were mixed in culture medium until attaining a cell density of 2×10^6/ml.

Simultaneously, defibrinated sheep blood was washed with saline solution by centrifuging at 2000 r/min for 5 min. The sheep red blood cells (SRBCs) sediment was diluted 1/100 in phosphate-buffered saline (PBS) and preserved for 24 h at 4°C. The following mixture was made in the test tubes: 0.25 ml of the T-cell mixture in culture medium (2×10^6 cells/ml) + 0.25 ml SRBCs 1/100 in PBS. The mixture was centrifuged for 2–3 min at 2000 r/min and the sediment preserved for 15 min at room temperature. The samples were Giemsa stained. The "rosettes" (five or more erythrocytes attached to a lymphocyte) were counted for 100 lymphocytes.

3. *Serum Complement Activity*

The determination was performed according to the method of Brécy and Hartmann (4).

E. Statistical Analysis

The statistical comparisons between both groups and those within each group were performed with Student's *t*-test (23).

III. RESULTS

A. Clinical Evolution

No difference was observed when the two groups were compared in the initial period of the trial. Well-intended, similar internal differences existed in both groups because of internal the various conditions of the included patients (Table 2; Fig. 1).

Important differences surged when the number of relapses registered in the two groups were compared at the end of the follow-up period (Fig. 2). The OROSTIM-treated group registered 8 relapses as compared with 21 relapses in the control group. The difference has statistical significance ($0.01 < P < 0.02$). The real difference is still greater, because the relapses in the OROSTIM-treated group were visibly milder and needed no hospitalization, whereas in the control group, all the patients with relapses had to be hospitalized.

None of the patients presented untoward effects caused by OROSTIM.

B. Immunological Parameters

The obtained values of the immunological parameters in both groups before and after treatment are shown in Tables 3 and 4.

In the first group, OROSTIM produced a statistically significant enhancement of the number of T lymphocytes and of opsonic index ($P < .01$). OROSTIM induces also an increase of phagocytosis capacity of PMNs in peripheral blood and activates the serum complement but without statistical significance (Fig. 3).

In the second group, the immunity was not stimulated. Some immunological parameters of the second group decreased during acute episodes (Fig. 4).

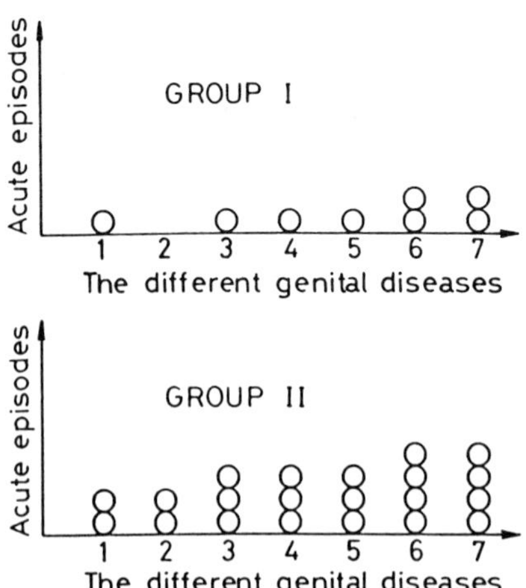

Figure 2 Mean number of acute episodes in groups I and II during the follow-up year. Group I includes women treated concomitantly with antibiotics and OROSTIM and group II the women treated with antibiotics only. Numerical symbols: 1, chronic pelvic cellulitis; 2, bilateral adnexitis; 3, cervicitis; 4, bilateral metroadnexitis; 5, cervicitis and metritis; 6, pelvic cellulitis and vaginitis; 7, chronic pelviperitonitis and both legs phlebitis. A circle represents an acute episode. A significant reduction in the number of acute exacerbations was observed in women with chronic genital infection when OROSTIM was added to the classic antibiotic treatment. Women treated only with antibiotics manifested a high number of acute episodes.

IV. DISCUSSION

The results of this trial showed the efficacy of OROSTIM in 12 women with chronic genital infections. This study showed a statistically significant reduction in the number, severity, and duration of acute exacerbation episodes of chronic gynecological diseases.

These clinical improvements were paralleled by increases of some immunological parameters. Statistically significant increases were observed for the opsonic index and the rosetting capacity of T lymphocytes.

The low number of treated cases could be an explanation for the lack of statistical significance concerning the determined increases in the other immunological parameters.

Our results confirm other published data relating beneficial effect when immunomodulators were combined with the antibiotic treatment in chronic urinary tract infections (1,8,21) or in chronic respiratory tract infections (2,5,7, 10,14,15,17,20,25).

OROSTIM is an orally administered immunostimulator that induces a statistically significant reduction in the number of acute exacerbations of chronic gynecological infections.

Table 3 Immunological Parameters of the Second Group of Gynecological Patients Before (0) and After (1) Treatment

No.	Patient	Age	Mom	NBT (%)	Phage (%)	OI (%)	LT (%)	Compl uCH_{50}
1	B.E.	34	0	16	24	1.6	20	72
			1	17	24	1.2	20	66.6
2	N.F.	40	0	18	28	1.4	36	67
			1	16	24	1.2	34	66.6
3	V.A.	40	0	17	50	3.6	40	100
			1	16	51	3.2	38	72
4	A.V.	34	0	12	28	1.6	38	100
			1	9	21	1.6	33	100
5	A.A.	24	0	19	21	1.0	31	72
			1	18	20	0.8	29	66.6
6	I.E.	24	0	22	45	3.5	37	167
			1	20	42	2.5	33	167
7	C.E.	37	0	28	40	3.0	14	50
			1	21	40	2.5	12	50
8	B.A.	33	0	20	24	1.4	40	67
			1	20	20	1.2	38	50
9	G.T.	28	0	20	32	1.4	40	67
			1	20	28	1.0	46	50
10	M.E.	38	0	12	14	1.1	44	120
			1	10	12	1.2	38	100
11	M.M.	36	0	20	40	2.8	32	100
			1	17	36	2.2	31	72
12	A.C.	35	0	20	22	1.8	30	72
			1	18	21	1.5	29	66.6

Abbreviations: Mom, moment of assay (0, initial assay; 1, assay after 10 days of treatment); NBT, nitro-blue-tetrazolium; Phag, phagocytosis; OI, Opsonic index; LT, T lymphocytes; Compl, serum complement.

The product increases in an impressive manner the efficiency of the antibiotic treatment in chronic genital infections in women.

V. CONCLUSIONS

OROSTIM is an orally administered bacterial product. The immunomodulating active principles are lysed and ultrasound-treated fractions extracted from 14 bacterial strains that are most frequently responsible for respiratory and urogenital tract chronic infections in Romania. It acts by amplifying connections of the lymphatic tissues of the mucous membranes (GALT and BALT).

Table 4 Immunological Parameters of the First Group of Gynecological Patients Before (0) and After (1) Treatment

No.	Patient	Age	Mom	NBT (%)	Phag (%)	OI (%)	LT (%)	Compl uCH_{50}
1	N.M.	40	0	16	38	2.0	30	111
			1	20	45	2.5	36	83
2	M.E.	30	0	16	32	1.2	26	111
			1	26	48	2.5	50	40
3	M.M.	30	0	16	40	1.8	28	55.5
			1	29	40	2.8	45	45
4	P.T.	32	0	16	28	1.0	29	59
			1	25	38	2.5	38	45
5	V.M.	34	0	15	26	1.2	32	100
			1	18	30	1.8	38	85
6	L.L.	44	0	28	30	1.5	74	67
			1	28	38	2.0	85	100
			0	8	17	0.5	21	80
			1	28	25	1.0	40	67
8	L.A.	44	0	28	40	1.2	48	100
			1	31	45	2.5	56	58
9	H.M.	40	0	32	50	5.0	39	45
			1	28	48	2.4	47	67
10	H.C.	42	0	32	50	4.0	36	67
			1	24	48	2.4	52	111
11	S.M.	28	0	10	20	1.0	35	55
			1	18	24	1.2	54	48
12	A.E.	28	0	18	20	1.0	16	76
			1	24	22	1.2	22	45

Abbreviations: Mom, moment of assay (0, initial assay; 1, assay after 10 days of treatment); NBT, nitro-blue-tetrazolium; Phag, phagocytosis; OI, Opsonic index; LT, T lymphocytes; Compl, serum complement.

A clinical assay was made in 24 women (24–44 years old) with chronic genital infections: 12 were treated with antibiotics and OROSTIM and 12 were treated with antibiotics only. We followed the clinical evolution and immunological parameter levels (tests of T lymphocytes, phagocyte analysis, complement determinations).

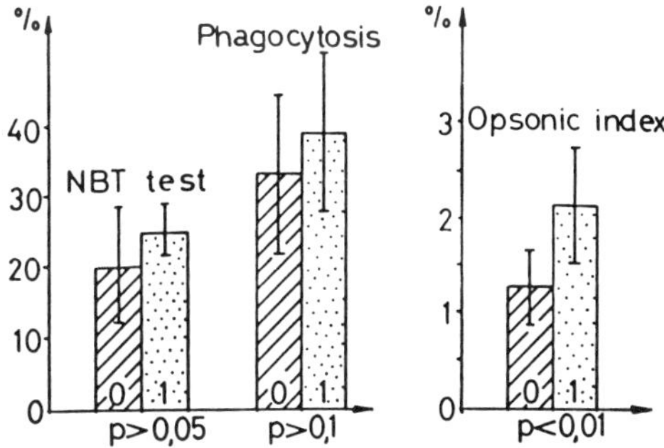

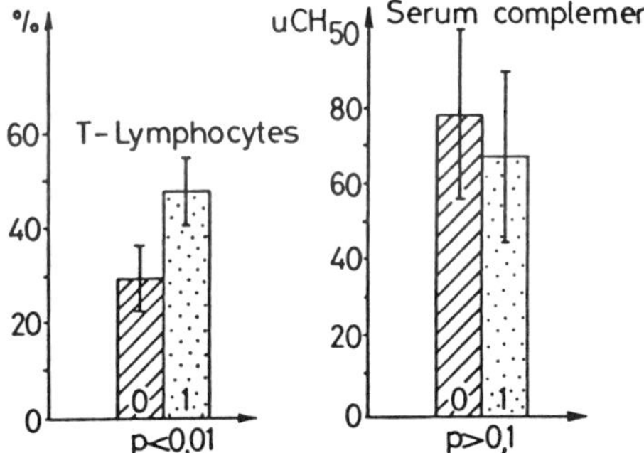

Figure 3 Immunological parameters of group I patients. NBT means nitro-blue-tetrazolium. The hatched columns represent the values of immunological parameters before the treatment. The dotted columns represent the values of immunological parameters after treatment. OROSTIM determines a statistically significant enhancement of T lymphocyte number and of opsonic index ($P < .01$). This product also determines an enhancement of PMN phagocytosis capacity and activates the serum complement but without statistical significance ($P > .05$ for NBT test, $P > .1$ for phagocytosis, $P > .1$ for serum complement).

The results demonstrate that the treatment with antibiotics and OROSTIM is more efficient than antibiotics only.

ACKNOWLEDGMENTS

We thank the Senate Committee of the University of Medicine and Pharmacy and the Brooke Foundation for sponsoring the participation of C.-E.L at the International Symposium on Immunotherapy of Infections in Berlin (May 4–7, 1993). We also thank Danut Lupusoru for drawing the figures.

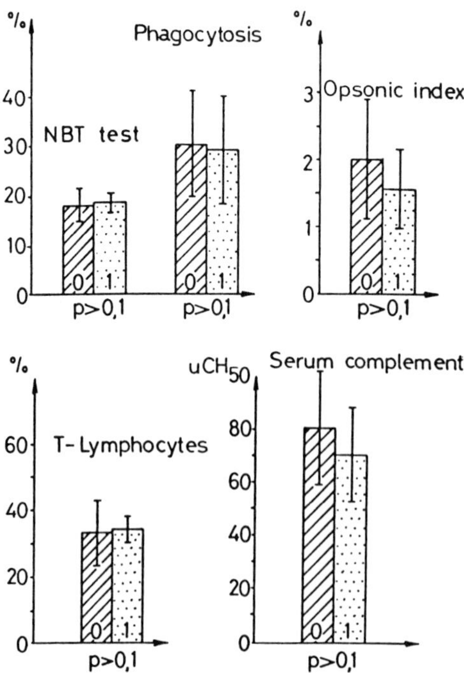

Figure 4 Immunological parameters of group II patients. The hatched columns represent the values of immunological parameters before treatment. The dotted columns represent the values of the immunological parameters after treatment. There are no statistically significant differences between the values obtained before and after treatment. Some of the determined values diminished slightly during the acute episodes (phagocytosis, opsonic index).

REFERENCES

1. Bessler WG, Beck P, Konetznick U, Loleit M, Sedelmeier E, Hoffmann P, Strecker M, Stocklin S. Biological activity of bacterial surface components immunogenicity and immunomodulatory properties of a bacterial extract from Escherichia coli. Arzneimittel Forschung Drug Res 1991; 41(I)3:274–79.
2. Bernard JG, Schafer J, Bernard JJ. l'immunotherapie microbienne en pathologie bronchique. Rev Fr Allergol 1979; 4–5.
3. Bienenstock J, Befus AD. Mucosal immunology. Immunology 1980; 41:249.
4. Brécy H, Hartmann L. Ann Biol Clin 1965; 27:559–70.
5. Djuric O, Mihailovic-Vucinic V, Stojcic V. Effect of Broncho-Vaxom on clinical and immunological parameters in patients with chronic obstructive bronchitis. A double-blind, placebo-oriented study. Int J Immunother 1989; V(3):139–43.
6. Finger H, Hofh H. Development of immunity following oral immunization with bacterial and viral vaccines. Aktuelle Probleme der Immunologie 1977; 71.

7. Gsell HO. Immunobiotherapie mit Broncho-Vaxom bei 20 Kindern. Schweiz Rundschau Med (Praxis) 1980; 69:1106–1108.
8. Hachen HJ. Oral immunotherapy in paraplegic patients with chronic urinary tract infections: a double blind, placebo-controlled trial. J Urol 1990; 143:759–763.
9. Hajicek V, Hajicek V. Utilisation du Broncho-Vaxom, dans le traitement de la bronchite astmatiforme. Acta Therapeutica 1980; 6:167–76.
10. Hoisie S, Pencea I, Potorac E, Ofelia T, Popescu C, Grigoras E, Anton D, Dumitrescu-Leibovici A. Strengthening the antiinfectious resistance following the oral administration of "Zoodin" in laboratory animals. Arch Roum Pathol Exp Microbiol 1982; 41(1):27–33.
11. Husband AJ. Kinetics of extravasation and redistribution of IgA-specific antibody-containing cells in the intestine. J Immunol 1982; 128:1355.
12. Juzaburo N, Yoshio H, Kozna T, Tamioki M. Effect of oral administration of lysosyme or digested bacterial cell walls on immunostimulation in guinea-pigs. Infect Immun 1981; 2(31):580–3.
13. Maestroni GJM, Losa GA. Clinical and immunobiological effects of an orally administered bacterial extract. Int J Immunopharmacol 1984; 6:111.
14. Marialigeti T, Szekely E, Kereki E. Broncho-Vaxom therapy for children with chronic bronchitis. Therapia Hungarica 1989; 33(1):28–31.
15. Martin du Pan R-R, Martin du Pan R-C. Etude clinique de prevention des infections des voies respiratoires superieures de l'enfant de l'age prescolaire. Schweiz Rundschau Med (Praxis) 1992; 71:36.
16. Martin du Pan R, Kochli B. Interferon induction by the bacterial lysate Broncho-Vaxom: a double blind, clinical study in children. Kinderarzt 1984; 15:646.
17. Munich D, Dalmi L. Attempts with oral immunobiotherapy (Broncho-Vaxom + Neomicin) for the elimination of bacteria in symptom-free "other salmonella" carriers. Therapia Hungarica 1989; 37(4):209–15.
18. Pitzura L, Franceschini S, Tricarico L, Cifarelli F. IgA secretoire intestinali et antigeni batterici somministrati per via orale. Boll Soc St Biol Sper 1979; 55:2111.
19. Rosenthal M. Effect of a bacterial extract on cellular and humoral immune responses in humans. J Immunopharmacol 1986; 8:315.
20. Scrobogna A. Aspetti generali dell'immunita locale nele bronchite croniche. Relazioni ira immunita locale e immunita umorale. Vantagi della vaccinazione parenterale con vaccino batterici polivalenti. Giorn di Malattie Infective Parasitarie 1980; 8:32.
21. Tammen H, Frey C. Behandlung rezidivierender Harnwegsinfekte mit Uro-Vaxom. Offene Multizenterstudie sint 521 Patienten. Urologe 1988; 28:294.
22. Vulcarone G, Brugo MA, Tassi GC. Immunomodulatori. Bolletino dell Istituto Sieroterapico Milanese 1980; 3:173–97.
23. CIBA-GEIGY SA, Bâle, Switzerland. Tables scientifiques, 7th ed. Redaction: Konrad Diem et Cornelius Lentner, 1972.
24. Peacock E, Thomar J, Rusell H. Manual of laboratory immunology. Philadelphia: Lea & Febiger, 1980.
25. Fontanges R, Bottex C, Muckart MF, Gerasimo P, Mathieu J. Broncho-Vaxom's mechanism of action, Asbtr. 7th Congress of the European Society of Pneumology (SEP), Budapest, Hungary, September 5–7, 1988.

35

Therapeutic Immunostimulatory Effects of Polyvalent Bacterial Vaccines: Laboratory Correlates in Humans and Mice

Jiří Franěk
Institute of Experimental Medicine, Academy of Sciences, Prague, Czech Republic

I. INTRODUCTION

In spite of the remarkable progress in the research of immunomodulators, only a few groups of preparations have been firmly established in the practice of immunotherapy of infectious diseases. Among them, the polyvalent bacterial vaccines (PBV) have been used most frequently.

In the therapy of chronic and recurrent respiratory infections, this group comprises commercial products such as IRS19, Broncho-Vaxom, Ribomunyl, Olimunostim, and a number of others.

The reported results of clinical studies are impressive: For example, excellent and good therapeutic effects in 65–85% of cases, substantial reduction in the number and duration of infectious episodes and of their severity, significant decrease in the need for antibiotic treatment, and compensation of ATB-induced immunosuppression.

At the same time, immunostimulatory therapy with PBV meets many criticisms, since many of the studies included only a very limited number of patients and lacked well-defined endpoints and there were problems with control groups. Also, the fact that 42–52% of "good therapeutic effects" were observed in patients treated with placebo only throws some shadow on the reported therapeutic efficacy. Many doubts stem from the fact that the principles of PBV immunostimulatory activities have never been fully elucidated.

According to the manufacturers, PBV combines three types of activities: (1) induction of a specific immune response against the most important bacterial pathogens of the respiratory tract selected for their preparation; (2) stimulation of mucosal immunity; and (3) activation of nonspecific effector elements such as neutrophils, macrophages, and natural killer (NK) cells and other mechanisms of innate immunity.

All three points provoke some comments:

None of the PBV covers the full spectrum of etiological agents.

All bacteria selected for the production of PBV exist in numerous antigenic types. Where identified, the protective antibodies are type specific. The protective value of cross-reactive antibodies is limited, if any.

Very little attention was paid to the bacteriological examination of patients treated with PBV. Causal links between an etiological agent, the antibody produced in response to PBV, and the therapeutic effect have never been documented. This is true for the systemic as well as for the mucosal immune responses.

As regards the most common components of PBV (e.g., pneumococci, streptococci, staphylococci), very little is known about the protective role of the secretory immunoglobulin A (IgA)–mediated defensive mechanisms.

Immediate short-lasting activation of nonspecific effector cells by bacterial compounds and products is an event commonly observed under experimental conditions. However, its biological relevance remains questionable.

No reliable laboratory test suitable for monitoring of the therapeutic effects of PBV in patients has been proposed by any of the manufacturers.

II. RESULTS AND DISCUSSION

Our own study focused on finding laboratory data that would allow a realistic assessment of the above-defined PBV concept.

In mice, a model of an inhalatory *Klebsiella pneumoniae* infection was developed (1,2) in which both specific and nonspecific immunostimulatory effects of bacterial lysates were examined. Experiments were designed to elucidate the following questions:

Induction of specific humoral (IgM, IgG, IgA) responses by soluble bacterial antigens (ELISA)

Kinetics of the specific IgA response in bronchoalveolar lavage (BAL) (ELISA)

Changes in the total IgA level in BAL (immunodiffusion)

Acceleration of a 4-h clearance of bacteria in BAL (quantitative bacteriology)

Dynamics of klebsiella proliferation within the respiratory tract in later intervals (quantitative bacteriology and immunofluorescence in BAL and in lung tissue homogenates)

Types and activity of phagocytic cells (Giemsa and immunofluorescence staining of cells in BAL and tissue homogenates)

Survival of infected mice (effectivity of protection expressed as enhancement of LD_{50})

We found the following specific effects:

Bacterial lysates repeatedly administered into the airways (5–10 consecutive daily doses) induce specific humoral responses both in BAL and in serum.

Level of protection equals 80–100 inhalatory LD_{50}.

Immunity is type specific.

Total IgA level in BAL is enhanced in two intervals: 1–2 days after the first administration (a nonspecific peak) and on day 7 (the specific peak); no long-lasting enhanced production was observed.

Four-hour clearance of bacteria in BAL is accelerated and results in a complete elimination of infection within 24 h.

The nonspecific stimulatory activities of heterologous lysates were characterized by

Their short duration (1–2 days after administration)

Slight immediate protective effect (1–2 LD_{50})

Irregular acceleration of 4-h clearance

Limited and irregular prevention of the secondary regrowth and proliferation of klebsiellae within the respiratory tract

A nonspecific increase in total IgA level in BAL 1–2 days after administration

Marked adjuvanticity when administered shortly before the *K. pneumoniae* antigen.

In contrast to in vitro assays, in the infection model, we failed to demonstrate any augmentation of the phagocytic activity of alveolar macrophages, not only with bacterial lysates but even in experiments with glucan, which is considered as a typical reticuloendothelial system (RES)–activating compound (3,4).

In humans, we focused on functional effects that could be reasonably considered as secretory IgA (sIgA)–mediated. The main group of over 400 patients consisted of children with recurrent respiratory infections. A limited study was arranged in adults with chronic bronchitis. This group of 11 in-ward patients was included to demonstrate the presumed effect of massive antigenic stimulation on the appearance of IgA plasma cells in the bronchial mucosa. A mixture of lysates was administered for 10 days for 30 min daily using a clinical apparatus for aerosol application. The presence of IgA plasma cells was investigated in frozen sections of mucosal specimens obtained by biopsy before and 14 days after the treatment by application of anti-IgA fluorescent antibody. Although no plasma cells were detected in the first interval, clusters of plasma cells could be demonstrated after the therapy. Also, IgA bound to the surface of bronchial epithelium was visualized.

Four parameters were examined in children before and after a 14-day treatment with a PBV:

Bacteria in the sediment of saliva
Adherence of bacteria to the buccal epithelial cells in vitro
Total and specific (relevant to the composition of PBV) sIgA in saliva
sIgA binding in vivo to bacterial surface structures

Findings that could be regarded as characteristic:

Bacterial monoculture (most frequently staphylococcal) is replaced by a bacterial mixture that is typical for the oral cavity after the treatment. Besides, the agglomeration of bacteria in saliva, facilitating their removal, is greatly enhanced.

The activity of sIgA, demonstrated by its binding to the agglomerated bacteria, is significantly improved. The treated organism also acquires a remarkable ability to respond to invading pathogenic bacteria promptly, as demonstrated in several cases of sore throat by the finding of extremely long streptococcal chains with characteristic fluorescence of cell wall segments covered by IgA. When the patients were examined after the treatment with PBV (i.e., before the outbreak of streptococcal infection), antibodies against *Streptococcus pyogenes* had not been detected.

No changes were found in the adherence of bacteria to buccal epithelial cells in vitro, as well as in the total IgA content in saliva (a vast majority of children suffering from recurrent infections produce normal amounts of sIgA).

III. CONCLUSIONS

We confirmed that at least some bacterial lysates when administered locally induce local as well as systemic antibody responses. Other data to support the hypothetical trias of PBV active principles have not been found.

However, PBV do exert some positive effects on the immune system, which are reflected by its renewed ability to seed the bronchial mucosa with IgA plasma cells, to control the microbial colonization of the mucous membranes, and to mount an effective sIgA response against invading pathogenic bacteria. The bacterial compounds responsible for the normalization of immune responsiveness are not known. In general terms, the concept of the host defense potentiators, as proposed by Chihara (5), seems to be the best explanation that can be offered at this time.

REFERENCES

1. Franěk J, Malina J, Libich J. Local antibacterial immune response: limits of its effectiveness. In: Rýc M, Franěk J, eds. Bacteria and the host. Prague: Avicenum, 1986:333.
2. Malina J, Hofmann J, Franěk J. Informative value of a mouse model of *Klebsiella pneumoniae* infection used as a host-resistance assay. Folia Microbiol 1991; 36:183.

3. Franěk J, Malina J, Krátká H. Bacterial infection modulated by glucan: a search for the host defense potentiation mechanisms. Folia Microbiol 1992; 37:146.
4. Chihara G. Immunopharmacology of lentinan, a polysaccharide isolated from *Lentinus edodes*: its application as a host defense potentiator. Int J Oriental Med 1992; 17:57.
5. Chihara G, Maeda YY, Suga T, Hamuro J. Lentinan as a host defense potentiator (HDP). Int J Immunother 1989; 5:145.

36

Immunotherapy from Principles of Traditional Chinese Medicine

Xiao-yu Li
Shanghai Institute of Materia Medica, Chinese Academy of Sciences, Shanghai, People's Republic of China

I. INTRODUCTION

About four-fifths of the population in the world lives in the developing countries, where the majority of people benefit from traditional medicine and only 15% of them benefit from Western medicines. Asia is the most important place for traditional medicine. In China, there were 6499 approved traditional Chinese medicine recipes; among them about 1000 set recipes are commonly used. Until now chemical ingredients have been analyzed in nearly 300 of the plant species used in these medicines. As traditional Chinese medicine always pays particular attention to strengthening the patient's nonspecific resistance against infections and illness, many Chinese herbs used for thousands of years are considered as tonics for the improvement of general health. Animal experiments and clinical trials have shown that quite a few herbs are immunologically active and most of the tonics are excellent immunomodulating agents. Some examples from a recent investigation by my group will be reviewed here briefly.

II. *TRIPTERYGIUM WILFORDII*

Tripterygium wilfordii Hook F. is a perennial twining vine belonging to the family Celastraceae. The plant has a long history in traditional medicine for the treatment of fever, chills, and edema, as well as inflammations. The crude extracts of its xylem

Table 1 Effects of Celastrol on Response of Mouse Splenocytes to Three Mitogens In Vitro (cpm $\times 10^{-2}$, $\bar{x} \pm$ SD of quadruplicate cultures)

		Celastrol (μg/ml)			
Mitogen	Dose (μg/ml)	0	0.01	0.1	1.0
ConA	2.5	353 ± 155	230 ± 107[a]	31 ± 17[b]	19 ± 5[b]
PHA	50	40 ± 7	44 ± 9[a]	13 ± 2[b]	9 ± 3[b]
PWM	50	240 ± 15	91 ± 8[b]	10 ± 3[b]	7 ± 2[b]
LPS	10	81 ± 10	62 ± 7[b]	19 ± 5[b]	1 ± 1[b]

Celastrol inhibited the proliferations of mouse splenocytes induced by mitogens, including PHA, Con A, PWM, and B-cell mitogen LPS. It indicated that celastrol suppressed both T and B cells.
[a] $P > .05$.
[b] $P < .01$.

have been used to treat rheumatoid arthritis, chronic nephritis, and various skin disorders, including psoriasis, systemic lupus erythematosus, allergic angitis, and lepra reactions, as well as a number of other immunological disorders. Clinical observations suggest promising results in a large number of cases. Celastrol, a triterpene compound isolated from this plant, was proved to be an identical immunosuppressor. Celastrol (0.01–1.0 μg/ml) inhibited the proliferations of mice splenic cells induced by mitogens, including concanavalin A (ConA), phytohemagglutinin (PHA), pokeweed mitogen (PWM) and lipopolysaccharide (LPS). It also inhibited the proliferation of lymphonodus cells, but no significant effects were seen on thymus cells (1,2).

In ICR inbred mice, intraperitoneal celastrol, 1–3 mg/kg daily, reduced antibody formation very significantly when assayed by spleen plaque-former counting (PFC) and serum hemolysin analysis after mice were challenged by sheep red blood cells (SRBCs). At the same time, celastrol increased the serum complement level and decreased the circulating immune complex level very markedly. Intraperitoneal celastrol, 1 and 3 mg/kg for 5 days, decreased LPS-induced interleukin-1 (IL-1) production of peritoneal macrophages, and intraperitoneal celastrol, 1, 3, and 5 mg/kg for 2 days, markedly decreased ConA-induced IL-2 production of splenic lymphocytes in mice and showed dose dependency. The above inhibitions are reversible after medication is stopped; thus celastrol is considered to be a strong immunosuppressor with mild cytotoxic actions (3).

III. *POLYPORUS UMBELLATUS*

Polyporus umbellatus Pers. is a basidiomycete that grows on the roots of trees. A polysaccharide composed of glucan with a molecular weight of 600,000 has been

Table 2 Immunosuppressive Activities of Celastrol ($\bar{x} \pm SD$)

Group[a]	Dose (mg/kg)	Thymus wt. (mg/10 g)	Hemolysin HC50	PFC/10^6 cell	[^{3}H]TdR (cpm × 10^{-2})	C3 (μg/10 μl)
Control	–	38 ± 6	843 ± 47	533 ± 85	208 ± 65	7 ± 1
Celastrol	1	17 ± 7[d]	744 ± 91[b]	465 ± 59[b]	129 ± 95[b]	14 ± 2[d]
	3	7 ± 3[d]	564 ± 40[c]	330 ± 39[c]	98 ± 17[d]	17 ± 3[d]
	5	9 ± 4[d]	261 ± 25[d]	305 ± 73[c]	13 ± 4[d]	20 ± 5[d]

[a] Each group of eight mice was injected with celastrol intraperitoneally for 5 days.
[b] $P > .05$.
[c] $P < .05$.
[d] $P < .01$ vs control.

isolated from it and identified. This polysaccharide (PUP) has been proved to increase macrophage phagocytosis and elevate antibody production and graft-versus-host (GVH) responses in mice, especially in immunodeficient animals such as in tumor-bearing mice. Clinical trials showed good results in patients with chronic hepatitis B. PUP was given to patients by subcutaneous injection, 40 mg/day for 20 days, then stopped for 10 days, and repeated for 3 months. During this period, serum hepatitis B antigen (HBsAg) vaccine, 30 μg, was given once every 2 weeks. A significant reduction in serum HBsAg and HBeAg titer was detected and about one-third of the patients with these viral antigens turned negative in randomized studies. The efficiency was even better in well-controlled double-blind studies when using other routine drugs as placebo. Liver histological examination by liver puncture in patients also showed improvement after 3 months of treatment. One hundred cases were followed up for 2–6 years; 14% of the patients with chronic active hepatitis relapsed within 1 year. Sixty-six percent of the HBeAg-negative patients remained negative during this period. No toxic side effects were found during immunotherapy except a few cases had low fever at the beginning of PUP injection.

There are only a few kinds of animals sensitive to hepatitis B virus (HBV) infection. Such animals include the chimpanzee, marmoset, and woodchuck, which are used as experimental animals. In east China near Shanghai, infectious duck hepatitis is found frequently. It is caused by a hepadna virus that is similar to HBV in many biological characteristics, including DNA polymerase and replication ways, and has been designated DHBV. Ducks in which DHBV was vertically transmitted were used as models in our laboratories. PUP and HBsAg vaccine were given to these ducks the same way as in clinical patients. Nine weeks after treatment, serum DHBV-DNA decreased in PUP groups from 86.0 to 18.5 pg/10 μl serum and remained at the low level after medication was stopped. Pathological exami-

Table 3 Effect of *Polyporus umbellatus* Polysaccharide on Serum HBV Antigens in Patients with Chronic Hepatitis B

		No. of cases	HBsAg negative (%)	HBsAg Cases	HBeAg negative (%)	HBeAg Cases	ALT Effective (%)
Randomized	PUP + Vaccine	296	36.8[b]	120	34.2[a]	279	77.4[b]
	Control	120	12.5	36	2.8	133	58.6
Double blind	PUP + Vaccine	32	62.4[a]	12	83.3[a]	36	69.4
	Control	32	31.3	11	18.2	29	44.8

[a] $P < .05$.
[b] $P < .01$ vs control.

nation in control ducks showed liver cell swelling and degeneration as well in Kupfer cells. Electronmicroscopy showed mitochondria swelling, rupture and lysis of inner cristae, and dilation and degranulation of the rough endoplasmic reticulum (RER). No significant changes in the mitochondria, RER, or nucleus were seen in PUP and HBsAg vaccine therapeutic ducks (4).

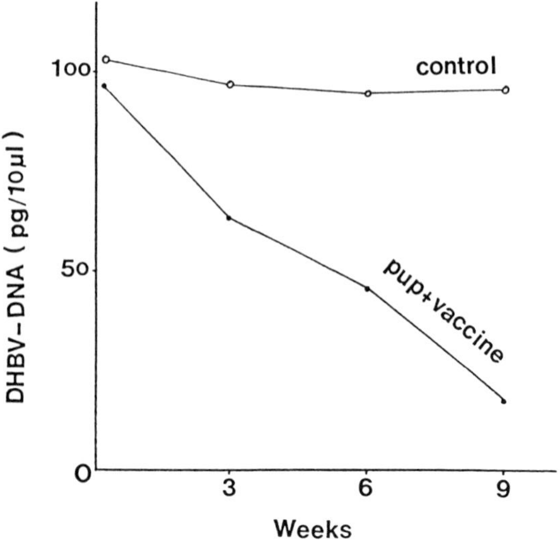

Figure 1 Effect of *Polyporus umbellatus* polysaccharide (PUP) and HBsAg vaccine treatment on duck hepatitis B virus DNA content. PUP was given 40 mg/kg im qod for 6 weeks, HBsAg vaccine 4 μg im every 3 weeks. Control ducks im normal saline.

IV. *CORIOLUS VERSICOLOR*

PSP, the protein-bound polysaccharide extracted from a strain of *Coriolus versicolor,* has been proved to be effective against tumors both experimentally and clinically. When PSP was combined with chemotherapy, radiotherapy, or surgery in cancer patients, it lessened the side effects of radiation, relieved pain, and improved appetite and thus improved the whole condition of patients. The antitumor effects of PSP were related to the potentiation of immunological responses, especially T-cell–mediated immune responses of tumor-bearing hosts. PSP, 100–800 μg/ml, caused a promoting action of lymphocyte proliferation by increasing DNA synthesis. When T-cell mitogen ConA was added, the proliferation of T cells appeared to be more evident. This effect was dose dependent both in the presence and in the absence of ConA, which indicated that PSP per se is a T-cell stimulator and also has a synergetic action with ConA. Intraperitoneally injected PSP, 25 mg/kg in mice caused a tendency to increase IL-2 production from splenocytes. Cyclophosphamide inhibited IL-2 production and when combined with PSP treatment, the inhibition is totally abolished.

PSP, 10–1000 μg/ml, increased the interferon alpha (IFN-α) level four to eight times higher than that of the classic group or two to four times higher than that of the priming group. The effect of PSP on IFN-α induction revealed a dose-dependent relationship. IFN-γ titer in PSP (10–1000 μg/ml)–treated cultured supernatants was four times higher than that of priming groups. It is therefore concluded that PSP has the ability to induce IFN-α and IFN-γ production. Since natural killer cells play an important role in host defense, the antitumor effect of PSP may be mediated through increasing production of IL-2 and IFN that then augments cytotoxic activity of natural killer cells synergetically. The immune-enhancing actions of PSP are more evident when given in vivo than in vitro and most significant on immun-

Table 4 Effect of PSP on IL-2 Production from Mouse Splenocytes In Vivo

Treatment	Dose (mg/kg)	IL-2 activity (dpm)
Control	—	4228 ± 110
PSP	25 × 5d	5048 ± 258
Cyclophosphamide	25q2d	2595 ± 717[a]
PSP + cyclophosphamide		3957 ± 200[b]

[a] $P < .01$ vs control.
[b] $P < .01$ vs cyclophosphamide group.

osuppressed animals suggesting that PSP is a host-mediated biological response modifier (5).

V. *ARTEMISIAE ANNUA*

The herb *Artemisiae annua* Linn. has been used as an antimalarial drug traditionally. Recently, artemisinine (Qinghaosu) has been isolated and shown to possess potent inhibitory effects on malarial infections both in animals and clinical patients. It is higher in efficacy than chloroquine, especially in treating those patients with cerebral malaria, with a quicker effect and lower toxicity. Immunological studies showed that artemisinine inhibited mice spleen cell proliferation and hemolysin formation with no cytotoxicity. Since the intensity of the immune response and the associated complement activation may be important factors in the pathogenesis of cerebral and fulminating malaria, we studied a semisynthesized derivative, artemether, on mice red blood cell immunity. The results showed that artemether, 2 mg/kg, lowered serum circulating immune complexes, maintained normal RBC membrane C3b receptors and serum complement levels, and protected the infected mice from death. It is suggested that artemether exerts some modulating effects on RBC immunity other than its schizont-killing activity (6).

VI. CONCLUSIONS

Traditional Chinese medicine, as well as the folk medicine in other countries, is a great treasure of the people. It offers significant potential for new drug discovery

Table 5 Effect of im Artemether-2 mg/kg daily for 3 days on Circulating Immune Complex (CIC, mg protein/ml serum) and Total Complement (C, complement hemolysis U/ml serum)

		Days after infection		
	Mice	0	7	9
CIC	Malarial	0.53 ± 0.12(8)	0.82 ± 0.12(7)[a]	1.14 ± 0.35(6)[a]
	Treated		0.68 ± 0.18(7)[b]	0.60 ± 0.20(6)[c]
C	Malarial	30 ± 2(5)	22 ± 5(3)[a]	8 ± 3(3)[a]
	Treated		49 ± 21(3)[b]	40 ± 14(3)[c]

Number of Balb/c mice in the parentheses ($\bar{x}$ ± SD).
[a] $P < .01$ compared with normal mice.
[b] $P < .05$.
[c] $P < .01$ compared with malarial mice.

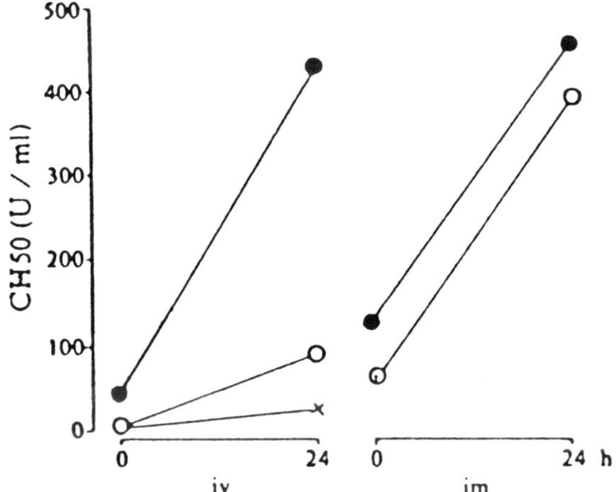

Figure 2 Serum total complement level increased after artemether emulsion 32 mg/kg iv or oil solution 8 mg/kg im on 3 monkeys inoculated with *P. cyonomolgi.*

and development. It has been only in the last two or three decades that serious efforts have been undertaken to carry out such programs in broader scope and depth. Thus, traditional Chinese medicine has gradually been integrated with Western medicine and has become a part of modern medicine. Apparently there is a pressing demand for an authoritative work on the evaluation of herbal drugs and methods of treatment according to the standards of modern scientific research in pharmacology, pathophysiology, and clinical medicine. However, we believe that new drugs that cure infections and cancer and increase human longevity related to immunomodulators will be developed in the near future and will make great contributions to the health of mankind (7).

REFERENCES

1. Lei W, Li XY. Immune suppressive actions of celastrol, a triterpene compound from *Tripterygium wilfordii*. Pharmacol Clin Chin Mat Med 1991; 7:18.
2. Zheng JR, et al. Studies on pharmacological actions of total glycosides in *Tripterygium wilfordii*. Acta Med Sinica 1983; 5:1.
3. Lei W, Li XY. Effect of celastrol on IL-1 and IL-2 production. Pharmacol Clin Chin Mat Med 1991; 7:15.
4. Nian SZ, et al. Experimental and clinical studies of *Polyporus umbellatus* polysaccharides on chronic viral hepatitis. J Chin West Med 1988; 8:141.
5. Li XY, et al. Immune enhancement of a polysaccharide peptide isolated from *Coriolus versicolor*. Acta Pharmacol Sinica 1990; 11:542.

6. Li XY, Liang HZ. Effects of artemether on red blood cell immunity in malaria. Acta Pharmacol Sinica 1986; 7:471.
7. Li XY. Immunomodulating Chinese herbal medicines. Mem Inst Oswaldo Cruz 1991; 86(Suppl. II):159.

37

Neem (*Azadirachta indica*): Immunomodulatory and Antimicrobial Properties

Shakti Upadhyay, Suman Dhawan, and Nalini Wali
National Institute of Immunology, New Delhi, India

I. INTRODUCTION

Neem (*Azadirachta indica* A Juss., synonym *Melia azadirachta*) is one of the most widely used plant materials in Indian traditional medicine. For centuries, the derivatives of this tree have been utilized for treatment of a variety of infections (1,2) and it is now universally accepted as a "wonder tree" (3). Scientific investigations carried out during the last two decades have, in fact, substantiated many of the traditional claims for neem (4). Neem extracts have been found to have potent antifungal (5) and antibacterial activities against a wide range of gram-positive and gram-negative microorganisms (6,7). Similarly, it has been reported that neem leaf extract inhibits the multiplication of vaccinia and fowlpox viruses (8,9). However, the exact mechanism(s) of biological actions of neem is not known. One of the therapeutic strategies in the Indian traditional medicine, the Ayurveda, has been to increase the body's natural resistance to infections using plant products rather than directly neutralizing the disease-causing agent. This can now be interpreted as a method to enhance immune responsiveness of the organism against a pathogen by nonspecifically activating the immune system using immunomodulatory agents of plant origin. It is possible that neem extracts may exert their wide-ranging antimicrobial effects by stimulating the immune system. In fact, neem has been reported to possess immunostimulatory properties (10,11).

This chapter presents the recent investigations carried out in our laboratory on the immunomodulatory and antimicrobial properties of the extracts prepared from the leaves, bark, and seeds of the neem tree. The plant materials were first subjected to methanol extraction and the residual material was further extracted with water. The aqueous extracts were freeze dried and used in all subsequent experiments. These extracts contained mainly polysaccharides and glycosides.

II. MODULATION OF MACROPHAGE FUNCTIONS

The neem extracts were evaluated for their effect on the phagocytic activity of mouse peritoneal macrophages following both in vivo and in vitro treatment. Peritoneal macrophages were either collected 72 h after intraperitoneal administration of neem extracts (100 mg/kg body weight) or treated in vitro overnight with the neem extracts (100 μg/ml). Treated macrophages were then incubated with sheep red blood cells (SRBCs) for 2 h. The phagocytic activity was measured by lysing the cells and recording the optical density. The results showed that leaf extract was most potent in enhancing the phagocytic activity of the macrophages (Fig. 1) following both in vivo and in vitro treatment. We had earlier reported that intraperitoneal treatment with neem induces expression of major histocompatibility class II (MHC-II) antigens, prerequisite for antigen presentation, on the macrophages (11). It appears that neem components primarily act on the macrophages and enhance their phagocytic and antigen-presenting ability.

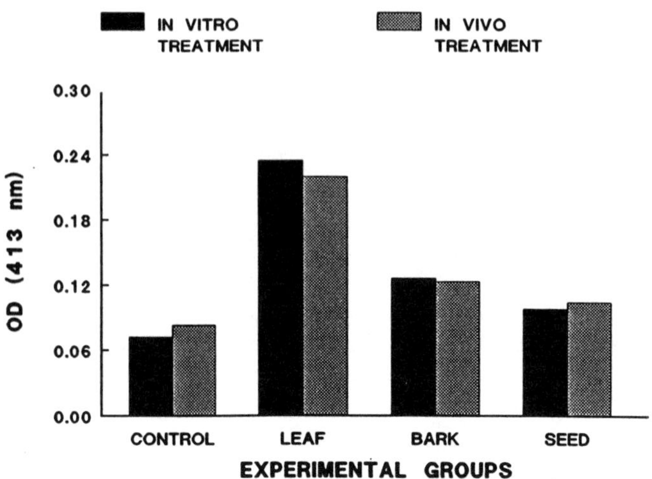

Figure 1 Effect of neem extracts on the phagocytic activity of mouse peritoneal macrophages, in vivo and in vitro, as measured by SRBC uptake. Results are expressed as mean OD values of assay run in triplicates.

III. MODULATION OF MITOGENIC RESPONSE OF SPLENOCYTES

Splenocytes from mice treated intraperitoneally with neem extracts (100 mg/kg body weight) were challenged after a week with the T-cell mitogen concanavalin A (5 μg/ml) in vitro; the lymphocyte proliferative response was measured by [^{3}H]thymidine incorporation assay. Results showed that pretreatment with neem leaf extract had the most significant T-cell proliferative response (Fig. 2). The enhancement of the proliferative response following neem treatment is not limited to mitogens; a similar stimulatory response has also been noted with tetanus toxoid (11). These results indicate that treatment with neem extracts nonspecifically activates the immune system, which responds more dynamically to subsequent mitogenic or antigenic challenge.

IV. INDUCTION OF CYTOKINE PRODUCTION

The immunomodulatory effect of neem leaf extract was further evaluated by measuring the cytokine production by human peripheral blood leukocytes (HPBLs) in vitro. HPBLs (1×10^7/ml) were incubated with neem leaf extract (100 μg/ml) for 72 h. The culture supernatants were tested for the presence of interleukins

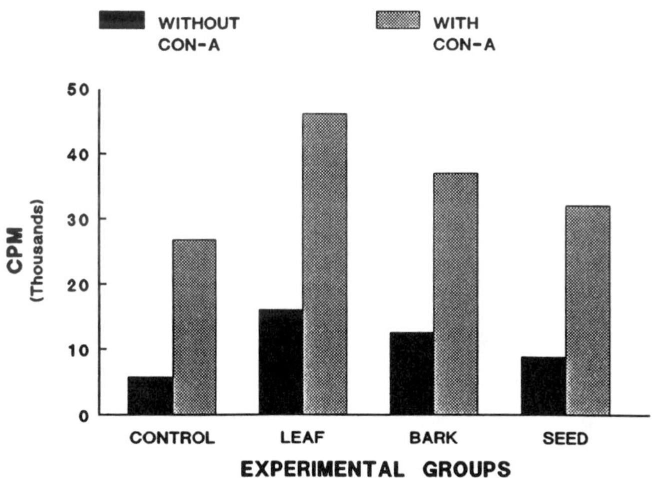

Figure 2 Effect of neem extracts on the lymphocyte proliferative response of mouse splenocytes to ConA, as measured by [^{3}H]thymidine incorporation. Results are expressed as mean CPM values of assay run in triplicates.

(IL)-1α and β, IL-2, interferon gamma (IFN-γ), tumor necrosis factor alpha (TNF-α), and granulocyte-macrophage colony-stimulating factor (GM-CSF) (Fig. 3), using ELISA kits (Genzyme). Induction of IL-1α and IL-1β and TNF-α following neem treatment confirms functional activation of macrophages by neem extract. The production of IL-2, IFN-γ, and GM-CSF indicates that neem component(s) specifically activate TH_1 cells, which are considered to be involved in cell-mediated immune responses (12). Since IFNs and TNF are known to have antimicrobial and antitumor activity, stimulation of these cytokines by neem may explain some of the biological effects of this plant.

V. ANTIMICROBIAL EFFECT

Antibacterial and antiviral effects of neem extracts were tested in vivo. Monolayer cultures of L-929 and VERO cells were infected with *Chlamydia trachomatis* and herpes simplex virus-1, respectively. The infected cells were incubated either directly with the neem extracts (100 μg/ml) or with the culture supernatants of mouse spleen cells treated in vitro with neem extracts (100 μg/ml). The results showed that the neem extracts had no direct effect on *Chlamydia* or herpes simplex virus. However, supernatants of mouse spleen cells treated with neem leaf and bark extracts significantly inhibited the intracellular multiplication of chlamydiae (Figs. 4 and 5) and the cytopathic effect of herpes simplex virus (Fig. 6). Neem leaf extract appeared to have the most significant effect. These results show that the antimicrobial effects of neem are mediated by the immune cells.

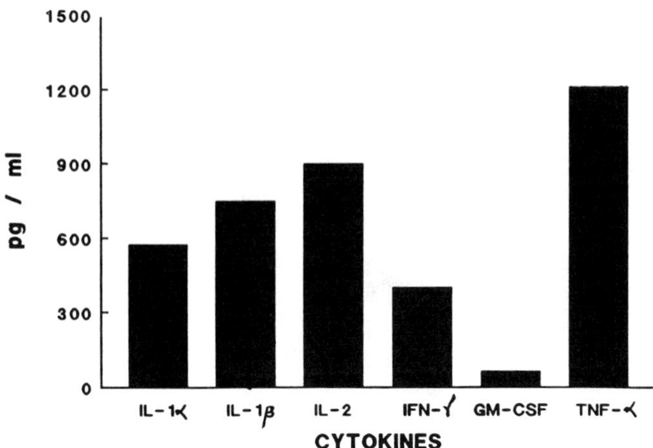

Figure 3 Effect of neem leaf extract on the cytokine production by human peripheral blood leukocytes in vitro. Results indicate amount of cytokines (pg/ml) present in the culture supernatant following 72 h of neem treatment as measured by ELISA.

Figure 4 Fluorescence micrographs showing inhibitory effects of neem treated spleen cell culture supernatants on the multiplication of *Chlamydia trachomatis* in vitro.
Figure 5 Control.
Figure 6 Plaque assay showing inhibitory effect of neem treated spleen cell culture supernatants on the cytopathic effect of herpes simplex virus 1, in vitro. Left column: control; middle column: neem bark–treated culture supernatant; right column: neem leaf–treated culture supernatant.

VI. CONCLUSIONS

This study demonstrates that neem possesses immunostimulatory properties and it activates the immune system nonspecifically, which responds more dynamically to subsequent antigenic challenge. The neem extracts primarily stimulate the macrophages, which become highly phagocytic and better antigen-presenting cells. Neem also appears to stimulate production of cytokines that are typical of TH_1 type of response implicated in cell-mediated immunity (12). It is therefore reasonable to believe that neem extracts nonspecifically enhance the cellular immune responses to pathogens. In fact, this study has shown that neem extracts do not have any direct microbicidal effects and that the antimicrobial action of neem is mediated by the immune system. It can be concluded that neem components have immunostimulatory properties that help the host to boost its immune defense mechanisms, especially cell-mediated immunity, to provide protection against a variety of infections. Neem extracts have been used in Indian traditional medicine for diseases that are now known to be caused by a lack of cell-mediated immune response; for example, leprosy and tuberculosis (1). The principal components of neem are now being isolated, characterized, and tested in our laboratory for their immunomodulatory properties and therapeutic applications.

REFERENCES

1. Pandey VN. Ancient human medicine. In Randhawa NS, Parmar BS, eds. Neem research and development. Society of Pesticide Science, India, 1993:199–207.
2. Zillur Rahman S, Shamim Jairajpuri M. Unani medicine. In: Randhawa NS, Parmar BS, eds. Neem research and development. Society of Pesticide Science, India, 1993; 208–219.
3. National Research Council (NRC). Neem: A tree for solving global problems. Washington, D.C.: National Academy Press 1992:141
4. Dhawan BN, Patnaik GK. Pharmacological studies for therapeutic potential. In: Randhawa NS, Parmar BS, eds. Neem research and development. Society of Pesticide Science, India, 1993:242–9.
5. Khan M, Wassilew SW. The effect of raw material from neem tree, neem oil and neem extract. In: Schumutterer H, Ascher KRS, eds. Natural pesticides from Neem tree (*Azadirachta indica*) and other tropical plants. GTZ, Eschborn, Germany, 1987: 645–50.
6. Chopra IC, Gupta KC, Nazir BN. Preliminary study of anti-bacterial substances from Melia azardirachta. Indian J Med Res 1952; 40:511–5.
7. Rao DVK, Singh K, Chopra P, Chambra PC, Ramanujalu G. In vitro anti-bacterial activity of neem oil. Indian J Med Res 1986; 84:314–16.
8. Rao AR, Sukumar S, Paramasivam TV, Kamalkshi, Parashuraman AR, Shanta M. Study of anti-viral activities of tender leaves of margosa tree on vaccinia and variola virus, a preliminary report. Indian J Med Res 1969; 57:495–502.

9. Rai A, Sethi MS. Screening of some plants for their activity against vaccinia and fowl pox viruses. Indian J Anim Sci 1972; 42:1066–70.
10. Labadie RP, Van Der Nat JM, Simons JM, Kroes BH, Kosasi S, Van Der Berg AJJ, Hart LA, Van Der Sluis WG, Abesekera A, Bamunuarachchr A, De Silva TD. An ethanopharmacognostic approach to the search for immunomodulators of plant origin. Planta Medica 1989; 55:339–48.
11. Upadhyay SN, Dhawan S, Garg S, Talwar GP. Immuno-modulatory effects of neem (*Azadirachta indica*) oil. Int J Immunopharmacol 1992; 14:1187–93.
12. Mossman TR, Cherwinski H, Bond MW, Giedlin MA, Coffman RL. Two types of murine helper T-cell clone. I. Definition according to profiles of lymphokine activities and secreted proteins. J Immunol 1986; 136–2348.

38

Treatment of Respiratory Syncytial Virus Infections Using Intravenous Gamma Globulins

Zhihong Wang and Xiqiang Yang
Children's Hospital, Chongqing University of Medical Sciences, Chongqing, People's Republic of China

I. INTRODUCTION

The respiratory syncytial virus (RSV) has proved to be a common cause of human respiratory infections, especially in infants and young children. Wherever sought, annual epidemics or endemic RSV infections have been documented throughout the world. RSV infections of the very young or elderly often involve the lower respiratory tract and may be sufficiently severe to require hospitalization. Respiratory RSV infections of immunocompromised patients are commonly severe and may disseminate resulting in the death of the patients. Children with chronic lung disease or congenital cyanotic heart disease also experience substantial morbidity and mortality when infected with RSV. There is no specific therapy available for treating these infections. Hemming et al. (1) have experimented in cotton rats using intravenous gamma globulins (IVGGs) in the treatment of RSV infections. The result showed that immunotherapy of RSV infections in laboratory animals was well tolerated, safe, and induced a highly significant reduction in RSV shedding from the lower respiratory tract (1). We have further investigated the immunotherapy of RSV infections by IVGG in children; the results have shown that children tolerated this therapy well and that it alleviated patients' suffering. The RSV-specific immunoglobulin (IgG) level was increased as compared with the control group.

II. MATERIALS AND METHODS

A. Patients

Conforming to the national standard in the diagnosis of RSV infections (2), 40 cases were selected. They included 22 cases younger than 6 months of age, 12 cases age 6 months to 1 year, and 6 cases age 1–2 years. There were 24 male cases and 16 female cases. All cases were admitted and treated within 5 days after infections. Except for conventional treatment, a double-blind method using IVGGs was employed in 20 cases and plasma was given in the other 20 cases. Patients' sera, before first and second infusions of IVGGs and at day 14 of hospitalization, were collected and stored at –20°C for further examination. Sera from 20 normal children 2–3 years of age also were collected. Mixed serum from 1000 adult donors was purchased from Shanghai Biological Products Institute.

B. Immunoglobulin Preparations

Chengdu IVGGs (Lot No. 8904020), pH 4.2, IgG single and double 100%, anticomplement activity 20 CH/50 mg IgG. The dosage schedule was 250–500 mg/kg/time for 3 consecutive days. Plasma, 10 ml/kg/time, for 3 consecutive days, was used in the control group.

C. Immunoglobulin Levels and RSV-Specific IgG (RSV-secretory IgG [sIgA]) Level of Sera

Total IgG, IgA, and IgM levels were determined by the standard radial diffusion method. The RSV-specific IgG levels were assessed by ELISA, and the results expressed by ΔOD $\times 1000$.

D. Clinical Observations

The main parameters consisted of wheezing, respiratory distress, and fever.

1. Efficacy

1. Markedly improved: main parameters of infection disappeared in 3–5 days.
2. Improved: main parameters of infection alleviated in 3–5 days.
3. Nonefficacious: main parameters were not changed or worsened.

E. Side Effects

During infusion, observation of the respiratory and pulse rates, blood pressure and rash, cyanosis, nausea, vomiting, and so forth was conducted every 5–10 min. Blood, urine, and sputum samples were examined daily.

Table 1 IgA, IgG, and IgM Levels in Serum Samples (× ±SD g/l)

Ig		1st Day[a]	3rd Day[b]	14th Day
A	Control group	0.36 ± 0.13	0.34 ± 0.13	0.32 ± 0.10
	Treatment group	0.34 ± 0.10	0.39 ± 0.20	0.34 ± 0.10
	P	>0.5	>0.5	>0.5
G	Control group	4.77 ± 1.24	5.37 ± 1.22	5.92 ± 1.36
	Treatment group	4.88 ± 1.23	6.99 ± 1.28	8.46 ± 1.55
	P	> 0.5	<0.01	<0.01
M	Control group	0.75 ± 0.13	0.78 ± 0.20	0.77 ± 0.17
	Treatment group	0.81 ± 0.28	0.76 ± 0.27	0.67 ± 0.16
	P	> 0.1	>0.5	>0.5

[a] Before infusion.
[b] Before 2nd infusion.

III. RESULTS

A. Immunoglobulins and RSV-sIgG Levels

Before infusion, IgA, IgG, IgM, and RSV-sIgG levels showed no difference between the two groups ($P > 0.5$). After infusion, IgA and IgM levels were not changed ($P > 0.1$), whereas IgG and RSV-sIgG levels of the treatment group were increased as compared with the levels of normal children 2–3 years of age and mixed serum (Tables 1 and 2).

Table 2 RSV-sIgG Levels in Serum Sample (ΔOD × 1000)

	No.	1st Day	3rd Day	14th Day
Control group	20	218.4 ± 104.0	314.9 ± 105.0	576.8 ± 156.8
Treatment group	20	254.0 ± 134.6	847.4 ± 300.9	1397.2 ± 563.7
2–3 years of age	20	646.9 ± 26.3		
Mixed sera from adult donors				
*P*1	>0.1	<0.01	<0.01	
*P*2	<0.01	>0.1	<0.01	
*P*3	<0.01	>0.5	<0.01	

Table 3 Improvement of Clinical Manifestations

	Marked improvement	Improved	Nonefficacious
Control group	11/20 (55%)	8/20 (40%)	1/20 (5%)
Treatment group	16/20 (80%)	4/20 (20%)	0/20
P	<0.05	>0.05	>0.1

B. Clinical Observations

See Tables 3 and 4.

IV. DISCUSSION

RSV is a common cause of serious bronchiolitis and pneumonia in infancy, especially in infants with underlying diseases such as bronchopulmonary dysplasia or congenital heart disease. The role of humoral immunity in RSV disease has been controversial (3). Early observations suggested that circulating RSV antibodies might exacerbate RSV lower respiratory tract disease. Poor outcome of a trial utilizing formalin-inactivated RSV vaccine provided further support for this observation. However, recent epidemiological and serological studies of RSV immunity showed that infants with high levels of maternally derived RSV antibodies were protected from RSV infections (4,5). Until the recent introduction of ribavirin, only supportive therapy has been available for treating RSV infections. Hemming has investigated the IVGG treatment of the cotton rat infected with RSV. The data showed that IVGGs containing high titers of RSV-neutralizing antibodies was efficacious (6). In this study, immunotherapy of RSV-infected infants were studied by infusing IVGGs, and the efficacy of this therapy was evaluated by the improvement of clinical manifestations. Furthermore, the IgG and RSV-sIgG levels were measured before and after IVGG infusions. The main parameters of RSV infection were alleviated, and the marked improvement rate was 80% (control 55%, $P < 0.05$). IVGG treatment was well tolerated and no case had side effects. A significant rise in the IgG levels of the patients ($P < 0.01$) and RSV-sIgG levels ($P < 0.01$) also were observed. Little is known of the mechanism of action of immunoglobulin in

Table 4 Days of Symptoms from Beginning of RSV Infections

	Fever	Wheezing	Respiratory distress
Control group	4.85	8.23	6.15
Treatment group	2.95	6.12	3.17
P	<0.01	<0.01	<0.01

this setting; possibly, immunoglobulins containing high titers of RSV-neutralizing antibodies sharply reduce or prevent the replication of RSV in the lower respiratory tracts of infants (7). Since the anticomplement activity and polymer of IgG can cause side reactions, the result of this study suggests that Chengdu IVGG consists of natural and nondisintegrated IgG and is safe to use clinically.

V. CONCLUSIONS

RSV is a common cause of bronchiolitis and pneumonia in infants and young children. Throughout the world, epidemics caused by RSV resulted in numerous hospitalizations and substantial morbidity and mortality. No specific therapy is available. Some animal models for observing the use of IVGGs are available, and the role of humoral immunity in RSV disease is controversial. By means of a double-blind method, treatment of respiratory syncytial virus infections in 40 children was conducted using IVGGs (250–500 mg/kg/time) or plasma (10 ml/kg/time) for 3 days. It showed that the main parameters of this infection (wheezing, respiratory distress, fever) disappeared within 3–5 days in 16 of 20 ill infants, and all infants with RSV bronchiolitis tolerated the immunotherapy with IVGGs well. Nine of 11 ill infants without IVGG infusion continued the clinical manifestations beyond 5 days after admission. The RSV-specific IgG level was increased in the IVGG group as compared with the control group.

REFERENCES

1. Hemming VG, et al. Respiratory syncytial virus infection and intravenous gamma-globulins. Pediatr Infect Dis J 1988; 7:s103.
2. Chen H, et al. A follow-up study of 104 cases of bronchiolitis. Chin J Pediatr 1984; 22:209.
3. Zhu Z, et al. Clinical and immunological study of infections with respiratory syncytial virus in children. Chin J Pediatr 1984; 22:198.
4. Wang W, et al. Research on special serum IgG level for respiratory syncytial virus. Chin J Pediatr 1984; 22:202.
5. Qian Y, et al. Longitudinal study on specific IgG level in serum for respiratory syncytial virus in healthy infants. Chin J Pediatr 1988; 26:348.
6. Prince GA, et al. Mechanism of antibody-mediated viral clearance in immunotherapy of respiratory syncytial virus infection of cotton rats. J Virol 1990; 64:3091.
7. Harsten, et al. Serum antibodies against respiratory tract virus. J Laryngol Otol 1989; 103:904.

V

HOST DEFENSE MECHANISMS

39

Immunity to Intracellular Bacteria: Lessons Learned from Listeriosis

Stefan H.E. Kaufmann
University of Ulm, Ulm, Germany

I. INTRODUCTION

Intracellular bacteria are a diverse group of microbial pathogens (1,2). This includes the etiological agents of tuberculosis, leprosy, typhus, legionnaire's disease, and listeriosis: *Mycobacterium tuberculosis, Mycobacterium leprae, Salmonella typhi/ S. paratyphi, Legionella pneumophila,* and *Listeria monocytogenes,* respectively. These pathogens are capable of surviving and replicating inside host cells. The mononuclear phagocyte, in its resting state, is their preferred habitat. Despite their predilection for an intracellular life style, these bacteria are capable of multiplying in the extracellular environment. In fact, *M. tuberculosis* flourishes in the cellular detritus of liquefied lesions. Although extracellular growth has not been formally proven for *M. leprae,* this bacterium probably possesses all the necessary essentials for this type of growth. Therefore, these intracellular bacteria are sometimes called facultative intracellular bacteria to distinguish them from the obligate intracellular pathogens that, among others, comprise *Rickettsia prowazekii, R. rickettsii, R. tsutsugamushi, Coxiella burnettii,* and *Chlamydia trachomatis,* which are the etiological agents of louse-borne typhus, Rocky Mountain spotted fever, scrubb typhus, Q-fever, and trachoma/lymphogranuloma venerum, respectively (3). These microbes are only viable inside cells and, furthermore, they are not at all focused on the mononuclear phagocyte but rather infect a variety of host cells. This chapter will focus on facultative intracellular bacteria with a major emphasis on *L. monocytogenes.*

L. monocytogenes is a gram-positive, non–spore-forming, facultatively anaerobic rod that was first isolated by E.G.D. Murray in 1924 from rabbits that suffered from septicemia (4). Human listeriosis is a rather rare disease, the clinical manifestation of which appears to depend on a compromised immune system. Of particular risk are patients suffering from Hodgkin's lymphoma and patients receiving immunosuppressive therapy, particularly renal transplant patients. In these immunocompromised patients, meningoencephalitis is the most common form of listeriosis. Most importantly, *L. monocytogenes* is responsible for prenatal and neonatal infections. Prenatal listeriosis leads to granulomatosis infantiseptica often resulting in stillbirth, and neonatal infection frequently results in sepsis and ultimately in meningoencephalitis. More recently, food-borne outbreaks of listeriosis were noted in numerous countries of the world that were typically traced back to contaminated animal products such as milk, cheese, or meat. Presumably these outbreaks of listeriosis also afflicted apparently healthy individuals. Yet, in most individuals, *L. monocytogenes* infection will remain asymptomatic and, in fact, many healthy individuals carry *L. monocytogenes* in the gastrointestinal tract.

The incidence of listeriosis has been estimated as low as three cases per million. Although improved detection methods will show higher incidences in the near future, listeriosis is a rare infectious disease. Particular interest in *L. monocytogenes* comes from a more scientific standpoint, since this bacterium has proven its great usefulness for studies concerned with the host response to intracellular bacteria. More recently, *L. monocytogenes* has also been used successfully for the identification of virulence factors responsible for host cell invasion and intracellular persistence (5). These molecular genetic studies have led to the identification of listeriolysin as a sulfhydril-activated pore-forming cytolysin that is active at the low pH existing in the phagosome. This molecule is crucial for in vivo virulence and persistence in macrophages. It allows regression of *L. monocytogenes* from the endosomal compartment into the cytoplasm, which is less hostile for the pathogen. As will be discussed later, this evasion mechanism has a decisive influence on the type of immune response that will be evoked: Antigens from bacteria existing in the cytoplasm resemble "endogenous" antigens such as viral antigens that associate with the class I pathway dictated by the major histocompatibility complex (MHC). Yet, recent evidence suggests that cytoplasmic invasion is not a unique function of listeriolysin and other cytolysins, including phospholipase C and lecithinase, may contribute to this step in addition. Intracellular persistence may be further promoted by a metalloprotease. Expression of numerous virulence factors involved in intracellular survival, including those mentioned above, is controlled and coordinated by a positive regulon termed prfA. Once in the intracellular milieu, *L. monocytogenes* spreads to other cells without contacting the extracellular milieu. *L. monocytogenes* organizes its own transport system that pushes the pathogen to the outside regions of the infected cell and subsequently induces pseudopod formation. Ultimately, this pseudopod penetrates the adjacent cell and the bacterium is released into a phagosomal vacuole. Listeriae then trespass the surrounding membrane thus reentering the cytoplasm. In this way, *L. monocytogenes* is well shielded from humoral and other extracellular defense mechanisms. Internalin and a secreted 60-kD product of the iap gene contribute to the invasion of nonphagocytic host cells.

II. GENERAL FEATURES OF INTRACELLULAR BACTERIAL INFECTIONS

The major features of intracellular bacterial infections are:

Intracellular habitat of the pathogen
T-cell dependence and antibody independence of protection
Low toxicity of the pathogen
Chronic course of disease
Granulomatous tissue reaction
Delayed-type hypersensitivity

Experimental listeriosis of mice fulfills most, although not all, of these features. In particular, experimental listeriosis is an acute, rather than a chronic, disease that reaches its peak approximately 5 days after infection and rapidly declines thereafter. In immunocompetent mice, the disease has generally vanished by day 8. It is, therefore, not surprising that the tissue reaction, although granulomatous in nature, is arrested at a rather early stage; that is, histiocytes and multinucleated giant cells do not develop. Otherwise, however, murine listeriosis paradigmatically satisfies the characteristics of intracellular bacterial infections.

III. IMMUNITY AGAINST *L. MONOCYTOGENES*

T cells are central to immunity against intracellular bacteria. Recent studies have provided compelling evidence that all three major T-cell populations participate in the acquisition of optimum protective immunity. These are CD4 α/β T cells that recognize their antigens in the context of MHC class II; CD8 α/β T cells that respond to antigens presented by the MHC class I, and γ/δ T cells that lack both the CD4 and the CD8 molecule and recognize foreign antigens through incompletely understood mechanisms that probably also involve MHC-encoded molecules (6,7).

CD4 T cells are biased toward so-called extracellular antigens; that is, proteins that are present in the endosomal compartment. CD8 T cells are focused to endogenous antigens; that is proteins localized in the cytoplasmic compartment (8). *L. monocytogenes* organisms are engulfed by macrophages via classic phagocytosis and, therefore, end up in the phagosome. At this stage, listerial proteins are introduced into the MHC class II pathway thus leading to CD4 α/β T-cell stimulation. Subsequently, *L. monocytogenes,* by using cytolysins, in particular listeriolysin, egresses into the cytoplasm (see above). There, its proteins have ready access to the MHC class I pathway and thus cause stimulation on CD8 T cells. It is, therefore, not surprising that both CD4 T cells and CD8 T cells are equally activated during experimental listeriosis of mice. Although cloned α/β T cells of either phenotype have been shown to confer protection against listeriosis in adoptive transfer experiments, evidence available thus far suggests that both T-cell populations are required for optimum protection. Furthermore, it has been shown that both CD4 and CD8 T cells produce cytokines of the TH_1 cell type and express specific cytolytic activities against *L. monocytogenes*–infected target cells (9,10). Among these, TH_1-like cytokines, interferon-gamma (IFN-γ) is of utmost importance. IFN-γ is a potent activator of various macrophage functions (11). In vitro, it induces production of reactive nitrogen intermediates that are major effector molecules of antibacterial defense (12). The decisive role of IFN-γ in acquired resistance against listeriosis has been underlined by three types of experiments: (1) administration of rIFN-γ confers increased resistance against listeriosis (13); (2) treatment of mice with anti–IFN-γ antibodies markedly impairs antilisterial resistance (14); and (3) mice in which the IFN-γ receptor had been deleted by homologous recombination suffer severely from listeriosis (15). Nevertheless, arguments against a central role of T-cell derived IFN-γ in protection have also been published (16,17). IFN-γ is produced very rapidly after listeriosis and its most likely source at this stage is the natural killer cell (18).

IV. CYTOLYTIC T-CELL FUNCTIONS

Based on in vitro studies, it has been proposed that cytolytic T-cell functions contribute to antilisterial protection (19). When tested directly ex vivo, T cells from *L. monocytogenes*–infected mice fail to express any cytolytic activity. However, soon after prestimulation with listerial antigens, both CD4 and CD8 T lymphocytes become potent killer cells that are able specifically to destroy *L. monocytogenes*–infected macrophages. *L. monocytogenes* is not only encaptured in macrophages but also resides in nonprofessional phagocytes with the notable example of hepatocytes. Although these cells possess a certain degree of antibacterial potential, they are less well equipped for microbial destruction. It may, therefore, be necessary to release bacteria from such cellular niches and make them accessible to more potent effector cells; that is, adequately activated macrophages. Recent evidence against IFN-γ as central mediator of antilisterial immunity may be taken as further support for a role of cytolytic T-cell functions (20).

V. γ/δ T CELLS

After infection with *L. monocytogenes,* γ/δ T cells rapidly accumulate at the site of listerial growth (21). Treatment of mice with anti-γ/δ monoclonal antibody (MAb) transiently exacerbates listeriosis in mice suggesting a partial contribution of γ/δ T cells to the acquisition of antilisterial resistance (22). More recently, the role of γ/δ T cells has been extensively studied using so-called knock-out (KO) mice in which the β-chain of the T-cell receptor (TCR) or the TCR δ-chain had been disrupted by homologous recombination. Mice with a deleted TCR β-chain are devoid of any α/β T cells, and conversely mice with a deleted TCR δ-chain are totally devoid of γ/δ T cells (23,24). During the first days, experimental listeriosis is well controlled in mice devoid of α/β T cells (25). In contrast, double mutants devoid of α/β T cells and γ/β T cells suffer enormously from experimental listeriosis. These findings suggest that γ/δ T cells partially compensate for the α/β T-cell defect. In the long run, however, TCR β–KO mice are unable completely to eliminate *L. monocytogenes* organisms, at least when higher inocula were used for infection. This experiment clearly shows the importance of α/β T cells, which becomes even more obvious in secondary infection studies. In this type of experiment, mice were first vaccinated with *L. monocytogenes* and subsequently infected with a high and normally lethal dose of listeriae. Although TCR β–deleted mice suffered substantially from listeriosis, the disease was worse in TCR β × TCR δ double mutants. These findings are consistent with a central role for α/β T cells and an auxiliary function of γ/δ T cells in antilisterial protection. Granulomatous lesions in TCR β mutant mice were smaller and less frequent than in normal control littermates. These findings substantiate the role of T cells, in particular of α/β T cells, in the development of *L. monocytogenes*-induced granulomatous lesions. Interestingly, lesions in TCR δ mutants were qualitatively different from those in control littermates. They were markedly larger and resembled abscesses. These findings suggest a regulatory role for γ/δ T cells in the development of granulomatous lesions in experimental listeriosis in mice. Thus, it appears that in experimental listeriosis, γ/δ T cells (1) compensate for α/β T cells but only partially, and (2) are functionally distinct from α/β T cells in that they perform a unique function during lesion development.

VI. LOCAL IMMUNITY

Although naturally acquired listeriosis is food borne, most experimental animal studies utilizing *L. monocytogenes* have employed intravenous or intraperitoneal infections. This is done with the aim of bypassing the first step of infection, namely, invasion through the gut epithelia, and to focus on the immune mechanisms going on in the central organs, that is, the spleen and liver. To study the early phase of immunity to *L. monocytogenes,* we have infected mice orally with this pathogen and subsequently analyzed responses of intraepithelial lymphocytes (IELs) during the course of infection (26).

Mice were infected orally with *L. monocytogenes* and IELs characterized phenotypically and functionally at various days after infection. Already 1 day after infection, *L. monocytogenes* had colonized the gut and invaded through the epithelia into the mesenteric lymph nodes. By days 2–3, the central organs were stably infected after oral inoculation of 2×10^8 *L. monocytogenes* organisms. Distribution of the CD3, CD4, and CD8 molecules as well as of the α/β and γ/δ TCRs remained virtually unaltered by infection. Thus, listeriosis had no significant effect on the distribution of IELs. Furthermore, gastrointestinal listeriosis did not induce listeria-specific killer activities as assessed on target cells pulsed with killed or viable *L. monocytogenes*. However, restimulation of IELs with listeria-pulsed target cells caused a significant IFN-γ production that was not seen when IELs from naive mice were employed. These findings suggest that gastrointestinal listeriosis primed IELs for IFN-γ secretion. Treatment of mice with anti-γ/δ TCR MAbs markedly reduced the percentage of γ/δ IELs. Furthermore, this treatment diminished IFN-γ production by IELs from *L. monocytogenes*-infected mice suggesting that *L. monocytogenes*-induced IFN-γ production is a function of γ/δ T cells. In further support of this notion, it was found that the number of IFN-γ–producing cells as determined by the ELISPOT technique was increased after intestinal listeriosis and reduced after concomitant administration of anti-TCR γ/δ MAb. In summary, these findings provide first evidence for a protective role of γ/δ IELs in the local immune response against *L. monocytogenes*. Secretion of IFN-γ could contribute to macrophage activation in the gut. Alternatively, IFN-γ could induce MHC class II expression and thus promote local antigen presentation to CD4 α/β T cells. Finally, IFN-γ could contribute to increased immunoglobulin A (IgA) secretion.

VII. CONCLUSIONS

In this chapter, evidence has been presented that a coordinated crosstalk between different T-cell types is required for optimum resistance against *L. monocytogenes*. Without doubt, this crosstalk is not restricted to the lymphocyte lineage but also involves macrophages as important effector cells. Finally, evidence has been presented that both cytokines and direct cell-cell interactions contribute to this crosstalk.

ACKNOWLEDGMENTS

This work received financial support from SFB 322 and Landesschwerpunkt Chronische Infektionskrankheiten. Thanks to R. Mahmoudi for secretarial help.

REFERENCES

1. Kaufmann SHE. Immunity to intracellular bacteria. In: Paul WE, ed. Fundamental immunology, 3rd ed. New York: Raven Press, 1993:1251.
2. Kaufmann SHE. Immunity to intracellular bacteria. Ann Rev Immunol 1993; 11:129.
3. Moulder JW. Interaction of chlamydiae and host cells in vitro. Microbiol Rev 1991; 55:143.

4. Kaufmann SHE. Listeriosis: new findings—current concern. Microbial Pathogen 1988; 5:225.
5. Portnoy DA, Chakraborty T, Goebel W, Cossart P. Molecular determinants of *Listeria monocytogenes* pathogenesis. Infect Immun 1992; 60:1263.
6. Raulet DH. The structure, function, and molecular genetics of the γ/δ T cell receptor. Ann Rev Immunol 1989; 7:175.
7. Janeway CA. The T cell receptor as a multicomponent signalling machine: CD4/CD8 coreceptors and CD45 in T cell activation. Ann Rev Immunol 1992; 10:645.
8. Kaufmann SHE, Reddehase MJ. Infection of phagocytic cells. Curr Opin Immunol 1989; 2:43.
9. Kaufmann SHE, Hahn H. Biological functions of T cell lines with specificity for the intracellular bacterium *Listeria monocytogenes* in vitro and in vivo. J Exp Med 1982; 155:1754.
10. Kaufmann SHE, Rodewald HR, Hug E, De Libero G. Cloned *Listeria monocytogenes* specific non-MHC-restricted $Lyt2^+$ T cells with cytolytic and protective activity. J Immunol 1988; 140:3173.
11. Nathan CF, Murray HW, Wiebe ME, Rubin BY. Identification of interferon-γ as the lymphokine that activates human macrophage oxidative metabolism and antimicrobial activity. J Exp Med 1983; 158:670.
12. Nathan CF, Hibbs, JB, Jr. Role of nitric oxide synthesis in macrophage antimicrobial activity. Curr Opin Immunol 1991; 3:65.
13. Kiderlen AF, Kaufmann SHE, Lohmann-Matthes M-L. Protection of mice against the intracellular bacterium *Listeria monocytogenes* by recombinant immune interferon. Eur J Immunol 1984; 14:964.
14. Buchmeier NA, Schreiber RD. Requirement of endogenous interferon-γ production for resolution of *Listeria monocytogenes* infection. Proc Natl Acad Sci USA 1985; 82:7404.
15. Huang S, Hendriks W, Althage A, Hemmi S, Bluethmann H, Kamijo R, Vilcek J, Zinkernagel RM, Aguet M. Immune response in mice that lack the interferon-γ receptor. Science 1993; 259:1742.
16. Dunn PL, North RJ. Limitations of the adoptive immunity assay for analyzing anti-Listeria immunity. J Infect Dis 1991; 164:878.
17. Lukas K, Kurlander RJ. $Lyt\text{-}2^+$ T cell–mediated protection against listeriosis. Protection correlates with phagocytic depletion but not with IFN-γ production. J Immunol 1989; 142:2879.
18. Bancroft GJ, Schreiber RD, Bosma GC, Bosma MJ, Unanue ER. A T cell independent mechanism of macrophage activation by interferon gamma. J Immunol 1987; 139:1104.
19. Kaufmann SHE. $CD8^+$ T lymphocytes in intracellular microbial infections. Immunol Today 1988; 9:168.
20. Harty JT, Schreiber RD, Bevan MJ. CD8 T cells can protect against an intracellular bacterium in an interferon γ-independent fashion. Proc Natl Acad Sci USA 1992; 89:11612.
21. Ohga S, Yoshikai Y, Takeda Y, Hiromatsu K, Nomoto K. Sequential appearance of γ/δ- and α/β-bearing T cells in the peritoneal cavity during an i.p. infection with *Listeria monocytogenes*. Eur J Immunol 1990; 20:533.

22. Hiromatsu K, Yoshikai Y, Matsuzaki G, Ohga S, Muramori K, Matsumoto K, Bluestone JA, Nomoto K. A protective role of γ/δ T cells in primary infection with *Listeria monocytogenes* in mice. J Exp Med 1992; 175:49.
23. Mombaerts P, Clarke AR, Rudnicki MA, Iacomini J, Itohara S, Lafaille JJ, Wang L, Ichikawa Y, Jaenisch R, Hooper ML, Tonegawa S. Mutations in T-cell antigen receptor genes α and β block thymocyte development at different stages. Nature 1992; 360:225.
24. Itohara S, Mombaerts P, Lafaille J, Iacomini J, Nelson A, Clarke AR, Hooper ML, Farr A, Tonegawa S. T cell receptor δ gene mutant mice: independent generation of α/β T cells and programmed rearrangements of γ/δ TCR genes. Cell 1993; 72:337.
25. Mombaerts P, Arnoldi J, Russ F, Tonegawa S, Kaufmann SHE. Differential roles of α/β and γ/δ T cells in immunity against an intracellular bacterial pathogen. Nature 1993; 365:53.
26. Yamamoto S, Ruß F, Conradt P, Kaufmann SHE. *Listeria monocytogenes* induced interferon-γ secretion by intestinal intraepithelial γ/δ T lymphocytes. Infect Immun 1993; 61:2154.

40

Mitogen-Activated Protein (MAP) Kinase Is a Common Element of Multiple Macrophage Activation Pathways: Activation of MAP Kinase by Lipopolysaccharide, Granulocyte-Macrophage Colony-Stimulating Factor, and Other Macrophage-Activating Factors

Gyorgy Frendl and Rong J. Guan
Immunology Research Unit, Boston University Medical Center, Boston, Massachusetts

I. INTRODUCTION

A. Macrophage Activation and Role of Macrophages in Protection Against Infections

1. Major Pathways of Macrophage Activation

Macrophages (MΦ) participate in the regulation of antigen-specific immune response via their antigen processing and presentation function (10) and the production of numerous immunoregulatory cytokines (e.g., interleukin-1 [IL-1], IL-6, tumor necrosis factor alpha [TNF-α], IL-8, macrophage inflammatory proteins (MIPs), IL-10) (11). MΦ, however, also display a variety of immunologically relevant nonantigen specific functions, like the removal of damaged cells and the production of complement proteins, lysozymes, and other important regulatory factors (12). The antimicrobial and tumoricidal functions of MΦ can be mediated by active oxygen radicals, NO, proteases, and/or TNF-α. The state of activation of MΦ is strictly controlled by biochemical and molecular regulatory mechanisms collectively termed MΦ activation (13,14). Although MΦ can be activated during T-cell–MΦ interaction via either (1) a lymphokine-independent pathway through direct-cell contact (involving cellular interaction molecules) (15), or (2) a variety of lymphokine-mediated pathways (4,16,17). Our studies have focused on the latter of these two MΦ-activation pathways; namely, the role of protein kinase–mediated signaling in the activation of MΦ by soluble mediators.

2. *Role of Macrophages in Protection Against Infections*

MΦ are crucial for the protection of the host against a variety of viral (e.g., influenza, respiratory syncytial virus), bacterial (i.e., *Listeria monocytogenes, Legionella pneumophila*), and parasitic (*Schistosoma mansoni, Leishmania major*) infections. Although cytokine-activated MΦ were shown to be more potent effector cells against these infections (18,19), interferon-gamma (IFN-γ) seems to be the most potent cytokine activator of MΦ against infectious agents (18). Granulocyte-macrophage colony-stimulating factor (GM-CSF) also was shown to activate macrophages to inhibit the growth of *Trypanosoma cruzi* and to enhance the production of hydrogen peroxide (20). Furthermore, IL-3 and GM-CSF also were demonstrated to enhance the protection against herpes simplex virus (HSV) infection by enhancing the IFN-α production of human blood leukocytes to HSV (21), whereas IL-3 also is known to enhance the candidacidal effects of human monocytes (22). Although we (23) and others (24) have shown that IL-3 and GM-CSF can enhance the production of TNF-α from both murine and human monocytes/macrophages, it was recently demonstrated that TNF-α is a key mediator of the macrophage defense against intracellular bacteria (1,2) and other intracellular parasites (e.g., *Toxoplasma gondii* [25]).

B. Macrophage-Activating Factors/Activation Pathways

1. *LPS-Induced Activation of Macrophages*

The gram-negative bacterial cell wall lipopolysaccharide (LPS) has a variety of effects on MΦ, including the induction of cytokines, cell interaction molecule expression, and the induction of active radical production (26). LPS also can contribute to killing of tumor cells by MΦ (27) while it has also a dose-dependent biphasic effect on class II MHC antigen expression (4,28).

2. *GM-CSF– or IL-3–Induced Activation of Macrophages*

Our earlier studies have demonstrated that IL-3 can function as a MΦ-activating cytokine (23) by inducing the expression of Ia antigens and β_2 integrins (4) and by contributing to the production of immunoregulatory cytokines (IL-1, IL-6, and TNF-α) of MΦ origin (3). GM-CSF was also shown to possess similar activities (4,17,29). Other investigators have also shown that GM-CSF and IL-3 have a priming effect on MΦ for the induction of both TNF-α and prostaglandin E_2 (PGE_2) (30), whereas they will also enhance the candidacidal activity of MΦ (22). Moreover, transgenic mice that overexpress GM-CSF have an overt MΦ infiltration and activation in various tissues resulting in blindness and the destruction of multiple organs (31).

3. *IFN-γ–Induced Activation of Macrophages*

Although a variety of T-cell cytokines can induce MΦ activation, IFN-γ is the most potent and best characterized MΦ-activating factor (16,32,33). It was demonstrated that IFN-γ can activate many MΦ functions (33), including tumoricidal activity, and also can enhance surface class I and class II major histocompatibility class (MHC) antigen expression and intracellular microbicidal activity (28,34). IFN-γ also has antiproliferative and antiviral effects on MΦ (35). The MΦ-activating properties of IFN-γ were reviewed in detail by Schreiber et al. (33).

4. *Interactions Between the Various Macrophage-Activation Pathways*

Our earlier studies have further demonstrated that IL-3 has a significant synergy with IFN-γ in the induction of Ia and CD11a (3,4). However, the synergistic interactions between IFN-γ and IL-3 were not manifest in all MΦ functions, since we found that IL-3 neither induced MΦ cytotoxicity nor enhanced the cytotoxic capacity of MΦ elicited by IFN-γ (4). Furthermore, negative interactions between IL-3 and IFN-γ also exist. To this end, it was shown that IL-3 can enhance the sensitivity of MΦ to cytotoxic T-cells, whereas IFN-γ seems to render them refractory to such effects (36,37). Additionally, IFN-γ can inhibit the growth-promoting effects of CSF-1, IL-3, or GM-CSF in murine bone marrow–derived MΦ (38). The existence of both opposing and synergistic interactions between LPS, GM-CSF, IFN-γ, or IL-3 on MΦ further suggest that these MΦ activation pathways represent distinct, but interacting, intracellular signaling processes. Such (positive and negative) interactions between these MΦ-activation pathways can provide the mechanisms for fine tuning of the MΦ functions in order to optimize the ability of MΦ to cope with environmental stimuli.

C. Structure of Cell Surface Receptors Involved in Macrophage Activation

1. *Structure of the Functional LPS Receptor CD14*

Although the profound effects of LPS on myeloid cells have been well-documented, early attempts to identify the cell surface receptor for LPS have repeatedly failed. However, recent binding-inhibition studies using antibodies to cell surface antigens and interaction molecules to block the effect of LPS have demonstrated that LPS can bind to more than one cell surface molecule, including acetylated low-density lipoprotein (LDL) (scavenger) receptors (39) and CD18 (the β-chain of the β_2 integrin family) (40), as well as CD14, a 55-kd glycoprotein previously described as a myeloid-differentiation antigen on MΦ (41). Subsequent studies also demonstrated that the prevention of LPS binding to CD14 is sufficient to eliminate its ability to induce MΦ cytokines indicating that CD14 is a (or perhaps the) primary functional, signal-transducing receptor for LPS in MΦ (42). In addition, it also was recognized that binding of LPS to CD14 requires an additional component, either the LPS binding serum protein (43) or septin (an abundant component of serum) (44), and these LPS + binding protein/septin complexes will bind to CD14 on the surface of MΦ. CD14 can be shed from the surface of MΦ (45) as well as its expression can be regulated by various cytokines (46,47). CD14 also was suggested to contribute to the adherence of human monocytes to cytokine-activated endothelial cells (48).

2. Structure of GM-CSF and IL-3 Receptors

The presence of both high- and low-affinity GM-CSF and IL-3R on monocytes was demonstrated earlier (49,50). Both of these receptors are known to bind their respective ligands in a strictly species-specific manner. Recent studies have demonstrated that both the GM-CSFR and IL-3R are composed of two subunits. The IL-3R has a 140-kd α-chain protein (AIC2A) that is able to bind IL-3 with low affinity (51), and the 120- to 140-kd β-chain (52) that confers high-affinity binding to the IL-3 receptor (53). A second IL-3R α-chain variant (AIC2B), highly homologous to AIC2A, also was cloned from murine cells (54). Although the IL-3R β-chain does not itself bind IL-3, it will confer high-affinity binding to the low-affinity binding IL-3Rα subunit when coexpressed (54). Furthermore, GM-CSFR, IL-5R, and IL-3R were shown to share a common β-chain (55,56), which is likely to be responsible for the overlapping physiological activities of GM-CSF, IL-3, and IL-5 (52). Thus, the GM-CSFR, IL-3R, and IL-5R receptor families have a common β-chain that confers high-affinity binding to these receptors, although they do not participate in the direct binding of ligands, whereas the unique α-chains determine the ligand (e.g., GM-CSF, IL-3, or IL-5) specificity of these receptors.

3. Structure of the IFN-γ Receptor

A high-affinity, species-specific IFN-γ receptor (IFN-γR) is expressed constitutively on virtually all cells of the body, including MΦ (57). Recent cloning and purification of the IFN-γR (58) has identified a 80- to 95-kd ligand binding transmembrane protein rich in Ser and Thr residues (58,59). It also was demonstrated that the 80- to 95-kd IFN-γR, responsible for high-affinity ligand binding, is not sufficient to provide effective signaling when transfected into CHO cells (60) or into L cells (59,61). Further experiments utilizing interspecies hybrid cell lines have revealed that an additional protein (encoded by human chromosome 21 and murine chromosome 16), interacting with two essential intracellular domains of the IFN-γR (61), is necessary for the effective functioning of the IFN-γR (62). The biochemical nature of this IFN-γR–associated chain (responsible for intracellular signaling) is yet unknown.

D. Signaling Pathways Implicated in Mediation of Effects of the Following Macrophage-Activating Factors

1. LPSR/CD14–Associated Signaling Events

Although a series of studies have identified activation events in LPS-activated MΦ as well as demonstrated the role of protein phosphorylation in this process (63–65), the protein kinase involved in this process has yet to be identified. CD14 is anchored to the membrane via a glycosylphosphatidyl inositol (GPI) linkage (66); thus it has no intracellular portion capable of transducing biochemical signals across the membrane. Furthermore, it was recently demonstrated that soluble CD14 can also bind LPS-LPS binding protein complexes and initiate a soluble CD14-dependent response to LPS from cells not expressing CD14 at all (67). These data strongly suggest that the association of such CD14 + LPS + LPS binding protein complexes with as yet unidentified transmembrane signaling molecules is necessary for effective signaling by CD14 through the cellular membrane. Additional studies have demonstrated the activation of *lck* (a *src*-type intracytoplasmic Tyr kinase enzyme) in response to cross linking of cell surface CD14 molecules (68) or other GPI-linked membrane antigens (68).

2. *GM-CSFR- and IL-3R-Associated Signaling Events*

Protein tyrosine and serine phosphorylation seems to be the primary event of GM-CSF- or IL-3-induced cellular activation (69). Furthermore, the tyrosine phosphorylation events mediated by tyrosine protein kinases seem to be independent of protein kinase C (PKC) (70). It was demonstrated that the activation of myeloid cells by GM-CSF or IL-3 will result in the phosphorylation of both unique as well as similar cytoplasmic proteins (71).

However, the role of PKC in IL-3-induced cell activation has yet to be determined. Although earlier evidence suggested a rapid and transient redistribution of PKC activity from the cytosol to the membrane on IL-3-induced activation (72,73), there also is evidence for enhanced membrane-associated PKC activity in the absence of any changes in the level of cytosolic PKC (74). Although the role of PKC in the IL-3-induced activation process was questioned (75), preliminary data with the use of phospho-amino-acid analysis seems further to support the involvement of PKC in the IL-3R-mediated signaling process (71). The involvement of cyclic adenosine monophosphate (cAMP)-dependent protein kinase A (PKA) in GM-CSF- or IL-3-induced cellular activation is unlikely, because GM-CSF has been shown to downregulate adenylate cyclase activity (76). Additionally, guanine nucleotide binding proteins were shown to be involved in the activation of leukocytes by GM-CSF (77). Although it was demonstrated that GM-CSF activation of leukocytes does not activate the phospholipase A, it also was shown that it activated phospholipase D (78).

3. *IFN-γR-Associated Signaling Events*

It was shown earlier that ligand binding of the IFN-γR will activate the adenyl cyclase and thus augment the level of cAMP in Ly cells and in thyroid cells (79). Earlier studies on the IFN-γ-activated protein kinase signaling mechanisms have not been conclusive (80). Although the activation of a Ser/Thr protein kinase enzyme, distinct from PKC (81), was demonstrated earlier in MΦ by IFN-γ (82), the specific characteristics of this Ser/Thr kinase were not identified. Additional studies using inhibitors of PKC have suggested a potential role for PKC in mediating the effects of IFN-γ (83), but the involvement of PKC in this process has yet to be directly demonstrated. The activation of (yet unidentified) calmodulin-dependent kinases via the mobilization of intracellular Ca^{2+} has also been suggested to mediate the effects of IFN-γ in MH-60 cells (84). It was also shown that the IFN-γR itself undergoes a rapid phosphorylation following treatment with IFN-γ (81).

E. Lack of Typical Protein Kinase Domain or Kinase-Interaction (SH_2 or SH_3) Domains in CD14 and in GM-CSFR β-Chain

Although the activation of MΦ by LPS was shown to be mediated via the binding of LPS-LPS binding protein complex to cell surface CD14 antigens, the CD14 antigen is phosphatidyl-inositol linked to the cell membrane, and thus possesses no intracellular domain. Furthermore, neither the common GM-CSFR and IL-3R β-chain (responsible for intracellular signaling) nor the IFN-γR possess any typical protein kinase interaction domain (SH_2 or SH_3) or intrinsic (tyrosine) kinase activity. Therefore, these structures provide no easily identifiable clues to the investigator with regard to their potential signaling mechanisms.

F. Goals of Current Studies

The aim of our studies presented here was to identify protein kinases involved in the intracellular signaling after LPS or GM-CSF activation of MΦ and, in particular, to identify common protein kinase elements of these MΦ activation pathways and of the IFN-γ–induced activation pathway of MΦ that we studied earlier (8). The studies presented here and elsewhere (8,9) have assessed the activation of protein kinases in both normal murine MΦ (isolated from the peritoneum of normal mice) and in the murine MΦ cell line P388.D1 and have revealed the activation of MAP kinases in these cells in response to either LPS, GM-CSF, or IFN-γ treatment.

II. MATERIALS AND METHODS

A. Reagents

1. Proteins, Cytokines, Peptides, and Other Chemicals

Escherichia coli (0111:B4) LPS was purchased from DIFCO (Detroit, MI). Murine recombinant GM-CSF was a kind gift of Dr. S. Clark (Genetics Institute, Cambridge, MA). Bovine myelin basic protein (MBP) was obtained from Sigma (St. Louis, MO). The nonapeptide (APRTPGGRR), containing amino acids 95–98 (PRTP) of the MBP, was obtained from Upstate Biotechnology, Inc. (Lake Placid, NY). Sepharose-bound Protein A (#P3391) was purchased from Sigma and was prepared by the following procedure: Lyophilized beads were resuspended in distilled water and swollen for 10 min and then equilibrated and blocked in a buffer containing 10 mM Tris (pH 7.5) and 1% (w/v) bovine serum albumin (BSA) for 30 min. The beads were then extensively washed in distilled water and stored in phosphate-buffered saline (PBS) containing 0.02% (w/v) sodium azide at 4°C. Sodium orthovanadate (Na_3VO_4; #S-6508), NP-40 (N-3516), protein inhibitors: pepstatin (#P-4265), antipain (#A-6271), and leupeptin (#L-2884) were all purchased from Sigma. Sodium orthovanadate solution was prepared according to the procedure described by Gordon (85). The source of all other chemicals are identified in the appropriate section of Materials and Methods.

2. *Radioisotopes*

$^{32}P\gamma ATP$ (3000 Ci/mmol; NEG-002H) and ^{32}P-orthophosphate (8500-9120 Ci/mmol, NEX-053H) were both purchased from New England Nuclear (North Billerica, MA).

3. *Antibodies*

The antirat MAP kinase R1 antibody (IgG, reactive to the N-terminal kinase subdomain III region of the MAP kinase, from Upstate Biotechnology, Inc., Lake Placid, NY) and the anti-PTyr monoclonal antibody (4G10, mouse IgG2b, from Upstate Biotechnology, Inc.) was used for either Western blotting and/or immunoprecipitation. Alternately, a purified rabbit polyclonal anti-MAP kinase (ERK) antibody (from Santa Cruz Biotechnology, Inc., Santa Cruz, CA) was used for immunoprecipitation of the MAP kinases and/or for Western blotting as indicated in the figure legends. Both of these antibodies were shown to react with the appropriate murine MAP kinase proteins by the manufacturers. HRPO-coupled antimouse IgG (H+L) antibody was used to develop the anti-PTyr Western blots, whereas horseradish peroxidase (HRPO)-coupled antirabbit IgG (H+L) (HyClone Laboratories, Logan, UT) was used to develop the MAP kinase Western blots.

B. Preparation of Cells

1. *Macrophages and Culture Conditions*

Thioglycollate-elicited murine peritoneal exudate cells (PEC) or the murine P388.D1 macrophage cell line (from American Type Culture Collection, Rockville, MD) were cultured in RPMI 1640 supplemented with 5% fetal bovine serum (FBS), 2 mM L-glutamine, 100 U/ml penicillin, 100 μg/ml streptomycin, and 5 mM HEPES buffer. Complete Hank's balanced salt solution (HBSS) was used for the washing steps. Tissue culture media (RPMI 1640, HBSS), as well as the PBS, were purchased from Cellgro/Mediatech (Washington, D.C.). FBS was purchased from HyClone. Only reagents that were proved to have LPS levels below 15 pg/ml (0.15 EU/ml; by the *Limulus* polyphamus lysate assay (E-Toxate; Sigma) were used in our experiments.

2. *Activation of Macrophages and Preparation of Cellular Lysates*

MΦ were plated into 3.5-cm tissue culture dishes at 5 × 10^6 MΦ cell/dish density or into 10-cm tissue culture dishes at 15 × 10^6 MΦ cell/dish density unless otherwise indicated (all tissue culture dishes from Costar, Cambridge, MA). Cells were placed into tissue culture the night before the experiment in order to eliminate the contribution of adherence-induced short-term signals in our system. Nonadherent cells were washed away before the stimulation of MΦ and cells were then incubated in 3 or 5 ml of fresh serum-free RPMI for the remaining time of the experiment in the presence or absence of MΦ-activating stimuli. These stimuli were provided to the cells at doses indicated in the figures and their legends for the indicated time periods. Culture medium was carefully removed from the cells at the end of incubation period and cell were lysed on ice in lysis buffer containing 20 mM Tris-HCl (pH 7.5), 137 mM NaCl, 1 mM $MgCl_2$, 10% glycerol, 1% NP-40, 2 mM Na_3VO_4, 50 mM NaF, 1 mM phenylmethyl sulfonyl fluoride (PMSF), 5 mM iodoacetamide, 50 μg/ml leupeptin and pepstatin, and 0.15 U/ml aprotinin unless otherwise indicated in the figure legends.

C. Biochemical Analysis

1. *SDS-PAGE and Electrophoretic Transfer*

MΦ lysates were supplemented with one-sixth volume of 6× SDS sample buffer (Laemmli buffer) containing a final concentration of 62.5 mM Tris-HCl (pH 6.8), 20% glycerol, 2% SDS, and 5% β-mercaptoethanol. Samples were then boiled for 5 min and subjected to 10% SDS-PAGE in a Bio-Rad (Richmond, CA) Protean IIxi electrophoresis system for the renaturation kinase assays and the Western blots. The loading of equivalent amount of proteins into the various lanes was controlled by running replicate gels with identical samples for Coomassie blue staining. The electrophoretically separated proteins were transferred to polyvinylidine difluoride (PVDF) membrane (Immobilon P, 0.45 μM pore size; Millipore, Milford, MA) membrane as detailed below. Prestained molecular weight standards (rainbow MW markers from Amersham, Arlington Heights, IL) containing: myosin, 200 kd; phosphorylase (b), 97.4 kd; BSA, 69 kd, ovalbumin, 46 kd; carbonic anhydrase, 30 kd; trypsin inhibitor, 21.5 kd; and lysozyme, 14.3 kd were used as molecular weight markers routinely to identify the location of the protein kinase bands on the membranes. Unstained Bio-Rad molecular weight markers (#1610317) were used to confirm the molecular weight of the protein kinases initially estimated using prestained markers.

2. *Western Blotting*

Following SDS-PAGE electrophoresis, the samples were electrophoretically transferred to PVDF membranes (Immobilon P from Millipore), in a transfer buffer containing 20 mM Tris and 150 mM glycine (pH 8.0). Proteins were transferred either by overnight blotting in a Bio-Rad transblot apparatus with 55 volts or by a 4- to 6-h transfer using a Bio-Rad transblot apparatus with an electroplate (supplemented with a cooling system) at 70–80 V and at 4°C as indicated in the Results section (see figure legends). Following transfer, the membranes were preincubated with blocking buffer (5% dry milk, 0.05% Tween-20, 1× TBS [containing 50 mM Tris-Cl in 0.9% saline, pH 7.6]) for three times for 20 min, then incubated with the appropriate dilutions of either the anti-MAP kinase antibody or monoclonal anti-PTyr IgG in binding buffer (1% milk, 0.05% Tween-20, 1× TBS) overnight at room temperature. The blots were then washed with wash buffer containing 1× TBS and 0.05% Tween-20 for three times for 20 min at room temperature with gentle agitation. 1:5000 dilution of HRPO-coupled secondary (antirabbit or antirat immunoglobulin [Ig]) antibody was used for the detection of the binding of the specific (first) antibodies, and subsequently the enhanced chemiluminescent reagent kit of Amersham International (UK, RPN2106) was used to develop the blots.

3. *Immunoprecipitation*

Immunoprecipitates of MΦ lysates were prepared on ice using 500 μl of lysis buffer (containing 20 mM Tris-HCl [pH 7.5], 137 mM NaCl, 1 mM $MgCl_2$, 10% glycerol, 1% NP-40, 2 mM Na_3VO_4, 50 mM NaF, 1 mM PMSF, 5 mM iodoacetamide, 50 μg/ml leupeptin and pepstatin, as well as 0.15 U/ml aprotinin) per sample followed by an additional 30-min lysis at 4°C. Nuclei were then removed by centrifugation with 2000*g* for 10 min at 4°C. Cell lysates were incubated with 2.5 μl of normal rabbit serum at 4°C for 1 h and then for an additional hour in the presence of 40 μl of sepharose-bound protein-A. Samples were microcentrifuged at 200*g* for 10 min at 4°C for preclearing. Subsequently, the MAP kinase was immunoprecipitated by incubating the cell lysates with anti-MAP kinase antibody (Santa Cruz Biotechnology, Inc.) at 4°C for 5 h on a rocking platform. Forty μl of sepharose-bound protein A was used then to immunoprecipitate the antibody-bound MAP kinase at 4°C for 1 h. Immunoprecipitates were pelleted by centrifugation (200*g* for 5 min) and were subsequently washed twice with each of the following wash buffers: (1) PBS + 1% NP-40; (2) Tris-LiCl (50 mM Tris, 0.5 M LiCl, pH 7.5); (3) TNE (10 mM Tris (pH 7.5) containing 100 mM NaCl, 1 mM EDTA). The protein A sepharose-bound MAP kinase antibody pellet was then aliquoted and used for further analysis.

4. *Two-Dimensional (2-D) Electrophoretic Analysis of Macrophage Proteins Phosphorylated In Vivo*

SDS-PAGE was combined with isoelectric focusing technique to perform 2-D analysis (as originally was described by O'Farrell [86]) of the cellular proteins phosphorylated in activated MΦ using a Bio-Rad Protean IIxi 2-D electrophoresis system. The lysates of 1.5 × 10^6 MΦ, following the in vivo labeling and activation of the cells (described above), were first subjected to isoelectric focusing on a gel containing 9.2 M urea in a 4:1 mixture of pH 5–7 and pH 3–10 ampholytes (BioLyte 3/10 and 5/7 Ampholyte 40%; Bio-Rad) at 700 V for 10 h (7000 Vh). The pI calibration kit (2-D–SDS–PAGE standard, Bio-Rad) was used to determine the pH profile of the isoelectric focusing step. Marker proteins were detected in the tubular electrofocusing gels, used for the separation in the first dimension, by Coomassie blue staining of a gel run in parallel to the samples. Nine percent SDS-PAGE gel was used for the further characterization of the sample proteins in the second dimension. Molecular weight markers (2-D–SDS–PAGE standard; Bio-Rad) were used for the estimation of the molecular weight of the proteins.

D. Kinase Assays

1. Renaturation and Assay of Protein Kinases on PVDF Membrane

For these studies, a technique based on the denaturation and subsequent renaturation of the cellular proteins isolated by SDS-PAGE and electrophoretically transferred to PVDF membrane (originally described by Ferrell and Martin [87]) was used. Specifically, for denaturation the PVDF membrane was incubated for 1 h at room temperature in 7 M guanidine HCl (Fisher, Pittsburgh, PA), 50 mM dithiothreitol, 2 mM EDTA, and 50 mM Tris (pH 8.3). The membrane was subsequently rinsed briefly in Tris-buffered saline (pH 7.4) and then incubated overnight at 4°C in a buffer containing 140 mM NaCl, 10 mM Tris-HCl (pH 7.4), 2 mM dithiothreitol, 2 mM EDTA, 1% (w/v) BSA (Sigma), and 0.1% (u/v) NP-40 (Sigma) to provide optimal conditions for the renaturation. Following overnight renaturation, the PVDF membrane was blocked with 2.5% (u/v) BSA in 30 mM Tris-HCl (pH 7.4) at room temperature for 1 h. The kinase reaction was performed subsequently by incubating the membranes in a reaction buffer containing 30 mM Tris-HCl (pH 7.4), 10 mM $MgCl_2$, 2 mM $MnCl_2$ plus 12.5 μCi/ml of $^{32}P\gamma$ATP (NEN Research Products, Boston, MA, 3000 Ci/mmol) at room temperature for 45 min. Subsequently, the membranes were washed for 3 × 10 min with 30 mM Tris-HCl (pH 7.4), then with 1 M potassium hydroxide (KOH), and finally with 30 mM Tris-HCl (pH 7.4) again to reduce the background radioactivity. Membranes were exposed to Kodak X-Omat AR film (Eastman Kodak, Rochester, NY) at –70°C for the period indicated in the figure legends. A potential technical limitation of this system is that enzymes that require the presence of multiple subunits (which would be separated on PAGE) for their active state will not display protein kinase activity.

2. Renaturation Kinase Assay Using MBP Substrate Incorporated into SDS-PAGE Gels for Measuring Phosphotransferase Activity

In some experiments, 50 μg/ml of myelin basic protein (MBP) was mixed with the gel solution and polymerized together with the polyacrylamide gels. MΦ lysates were subjected to electrophoresis in this MBP-containing 9% SDS-PAGE and the renaturation assay was performed in the gel. The basic buffer used in the following steps was 50 ml Tris-HCl, pH 7.4, and 5 mM mercaptoethanol. In certain steps there were additional additives, as noted. Gels were first washed twice with 100 ml basic buffer containing 20% isopropanol; each wash 1 h. Gels were then washed with 100 ml basic buffer for an additional hour. Proteins in the gels were denaturated by soaking the gels in 50 ml of buffer containing 6 M guanidine hydrochloride for 2 h in room temperature and renatured in 100 ml basic buffer containing 0.04% Tween-20 overnight at 4°C. Gels were then incubated with $^{32}P\gamma ATP$ in the same phosphorylation buffer as described above for 1 h at room temperature. Gels were rinsed with water, stained with Coomassie blue, destained, dried, and exposed to Kodak X-Omat AR film to detect ^{32}P-labeled bands. Although this assay can be used to assay the phosphotransferase activity of MAP kinases and some additional kinases as well, certain limitations for this assay exist: for example, many Tyr kinases, including p65 *lck* and p60 c-*src,* are not able to regain activity after denaturation, and this assay is not suitable for the detection of a PK that requires two different subunits for activity (87,88).

3. *In Vitro Substrate-Specific Kinase Assay*

A nine–amino acid synthetic peptide (APRTPGGRR), which includes amino acid 95–98 (PRTP) of the bovine MBP, was used as MAP kinase specific substrate (89). The PXTP/PXSP sequence was demonstrated earlier to be the consensus sequence for recognition and phosphorylation by the MAP kinases (89). The in vitro kinase assay containing 10 μg of the nonapeptide in 30 mM Tris-HCl (pH 7.4) buffer supplemented with 10 mM $MgCl_2$, 2 mM $MnCl_2$, and 20 μM $^{32}\gamma P$-ATP ($1.5–3 \times 10^3$ cpm/pmol) in a final volume of 40 μl, and the MAP kinase immunoprecipitated from 5×10^6 MΦ with the anti-MAP kinase antibody. The reaction was initiated with the addition of $^{32}P\gamma ATP$ and incubated at 30°C for 15 min. The samples were centrifuged briefly (with 200*g* for 1 min) and 30 μl supernatant of each sample was transferred onto a 2-cm diameter circle of P-81 (Whatman, Maidstone, England) phosphocellulose paper. The papers were washed six times in 200 ml of 1% cold phosphoric acid for 10 min per each wash, rinsed with 95% ethanol, dried, and the bound radioactivity was determined using a Beckman β-counter and Ultima Gold scintillation fluid (Packard, Meriden, CT)

4. *In Vivo ^{32}P-Labeling of Macrophage Cellular Proteins*

^{32}P-orthophosphate (NEN Research Products) was used to load the cells with ^{32}P for 4 h (1 mCi/ml/sample) before the experiment in phosphate-free Dulbecco's modified Eagle's medium (DMEM) (GIBCO, Grand Island, NY). Following this loading period, the MΦ were activated with 100 ng/ml of LPS for 20 min and cells were lysed with the urea solubilization buffer described by O'Farell (86). Cell lysates from 1.5×10^6 MΦ of each sample were subjected to 2-D electrophoresis in order to analyze the phosphorylation pattern of the intracellular proteins.

III. RESULTS

A. LPS Stimulates Activation of a Number of Protein Kinase Enzymes in P388.D1 Macrophages in a Time-Dependent Manner

In order to analyze the involvement of protein kinase enzymes in signal transduction in murine MΦ, we utilized an assay system based on the renaturation and subsequent evaluation of protein kinase activity of cellular proteins that had been separated by SDS-PAGE and transferred to PVDF membrane (87). This in vitro membrane-bound protein kinase system allows for the simultaneous evaluation of the activity of multiple protein kinases that differ in molecular mass.

When P388.D1 MΦ were activated by 1 μg/ml of LPS, we have identified three PKs (PK44, PK50, and PK56) that were activated in a time-dependent manner (Fig. 1). Based on our earlier studies with normal murine MΦ (9), we have suspected that the LPS-activated PK44 may be identical with 44-kd MAP kinase, a hypothesis that was further supported by Western blot colocalization experiments and by phospho-amino-acid analysis of the PK42 and PK44 bands, which indicated that both PK42 and PK44 are Ser/Thr kinase enzymes that are able to incorporate ^{32}P from $^{32}P\gamma ATP$ to Ser and Thr residues (data not presented here). Therefore, a modified renaturation kinase assay, utilizing the incorporation of a typical MAP kinase substrate protein (e.g., MBP) into SDS-PAGE gels, was used to identify PK42 and PK44.

B. LPS-Activated PK42 and PK44 Phosphorylates MBP, the Typical MAP Kinase Substrate, Incorporated into the SDS-PAGE Gel Matrix

In order to identify definitively PK42 and PK44 as MAP kinases, based on their molecular weight and their ability to phosphorylate the MBP, further experiments using the technique of renaturation of cellular proteins in the gel, following their separation by SDS-PAGE, were performed with SDS-polyacrylamide gels containing MBP incorporated in the gel (Fig. 2). By renaturing the PK and assessing their activity in an SDS gel containing MBP, we were able to detect a strong radiolabeled band with a molecular weight at 44 kd in lysates of LPS-activated (as well as PMA activated) but not of control (MED) MΦ (see Fig. 2, "+ MBP"). Furthermore, a substantial enhancement in the intensity of the 42-kd band was observed in lysates of LPS- and phorbol myristate acetate (PMA)–treated cells in the same assay (Fig. 2, "+MBP"), indicating the activation of both 42- and 44-kd PK in MΦ by both LPS and PMA treatment. Additionally, we observed no radiolabeled band in this region when a control assay was performed in the absence of MBP (see Fig. 2, "–MBP"). We have to emphasize that although MBP is optimal for this assay, it is not the most specific available substrate for MAP

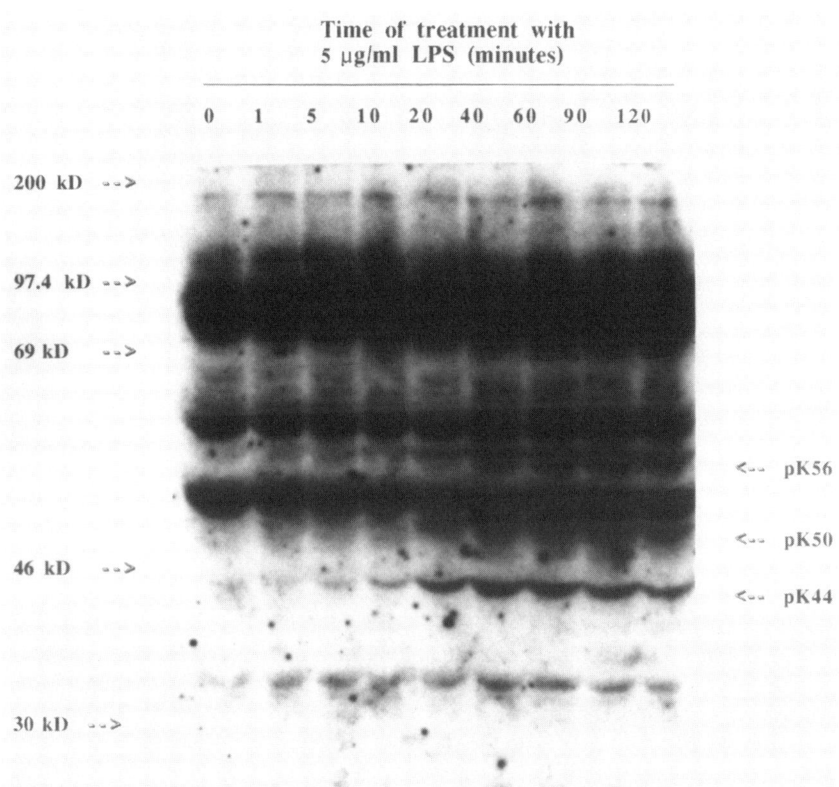

Figure 1 The kinetics of the activation of PK44 and other protein kinase enzymes in MΦ by LPS. 5×10^6 P388.D1 MΦ were activated with 5 μg/ml of LPS for the indicated periods of time and then cellular lysates were prepared as detailed in Materials and Methods. Cellular proteins were separated by 9% PAGE and transferred to PVDF membrane. The membrane-bound proteins were denatured for 1 h as detailed in Materials and Methods. Following overnight renaturation the PVDF membrane was blocked with BSA for 1 h. In order to assess the phosphotransferase activity of the renatured proteins on the membrane, a 40-min kinase reaction was performed in the presence of 10 μCi/ml of ^{32}PγATP followed by repeated washing steps. The autoradiogram was exposed for 48 h in the presence of an intensifying screen.

kinases, because it can be phosphorylated by PKC and calmodulin-dependent protein kinase II (Ca^{2+}/CaM KII) as well (90). However, Ca^{2+}/CaM KII has a molecular weight of 49–55 kd and PKC has a molecular weight of about 80 kd (or about 51 kd if cleaved by Ca^{2+}-dependent proteases) (91). Therefore, our experiments using MBP incorporated into SDS-PAGE gels had the capacity to differentiate between these PK based on their molecular mass. Therefore, based on the criteria of

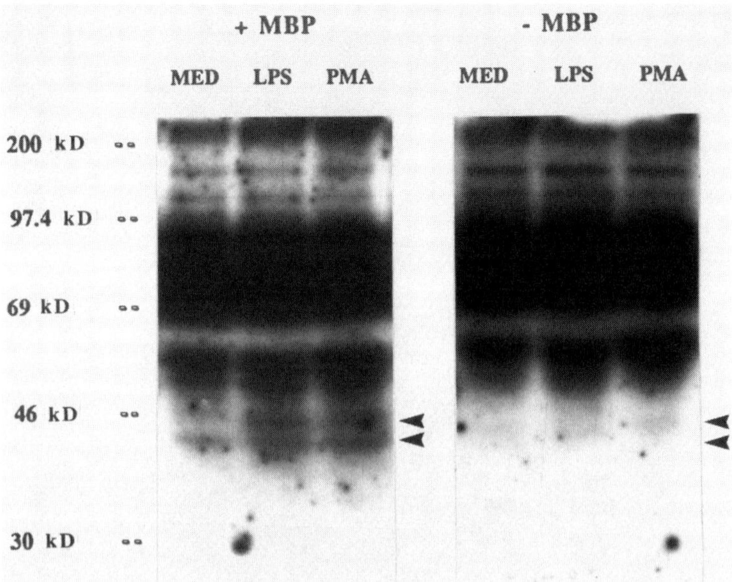

Figure 2 LPS-activated MΦ PK 42 and PK 44 can phosphorylate MBP, a typical MAP kinase substrate protein. A modified renaturation kinase assay, using MBP incorporated into SDS-PAGE gels as substrate for the phosphotransferase activity, was performed with cellular lysates from MΦ stimulated for 20 min with medium alone (Med), 100 ng/ml of LPS (LPS), or 100 nM of PMA (PMA). These cell lysates were subjected to 9% SDS-PAGE in the presence (+MBP) or absence (–MBP) of 50 μg/ml of bovine MBP and the gels were then processed through a denaturation and renaturation step as described in Materials and Methods. Subsequently, a protein kinase assay was performed with the proteins in these gels as described in Materials and Methods. The gels were extensively washed, dried, and visualized by autoradiography with Kodak X-Omat AR film (Eastman Kodak, Rochester, NY) for 24 h. The positions of PK42 and PK44 are denoted by arrows, indicating that LPS-activated PK42 and PK44 possess a key feature of 42- and 44-kd MAP kinases to be able to phosphorylate MBP.

molecular mass and the ability to phosphorylate MBP, we have identified the 42- and 44-kd MAP kinases as PK activated in MΦ by LPS (as well as PMA). Furthermore, that MΦ PK42 and PK44 (isolated either from normal murine MΦ or P388.D1 cells) in fact function as a MAP kinase was confirmed by additional immunoprecipitation and substrate-specific kinase experiments (8,9). These additional studies have not only further proved this notion but have also shown that the activation of MAP kinase by LPS in MΦ is a dose-dependent phenomenon and is accompanied by the enhanced phosphorylation of the MAP kinase upon activation (Fig. 3) (9).

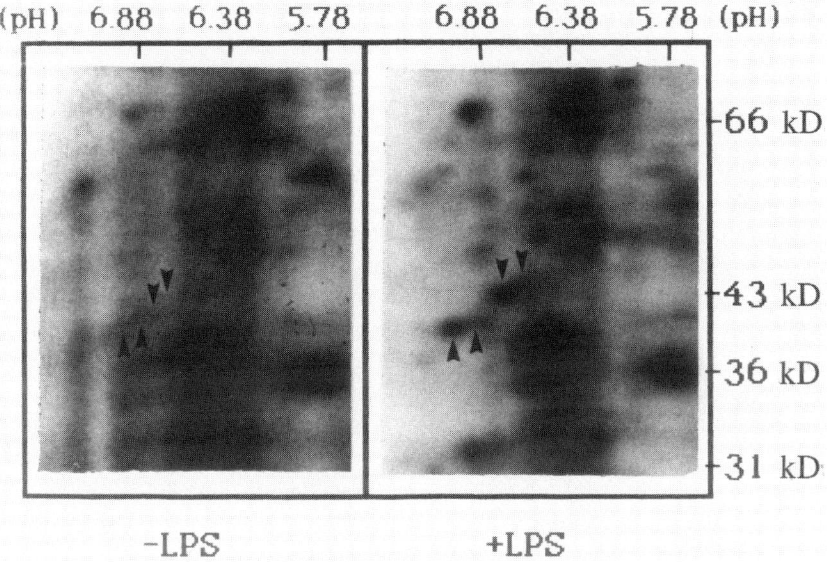

Figure 3 LPS-activated PK42 and PK44 show enhanced in vivo phosphorylation on activation of MΦ by LPS and have biochemical characteristics identical with the 42- and 44-kd MAP kinases in 2-D electrophoretic analysis. 1.5×10^6 MΦ (PEC) were preincubated with 1 mCi/ml of ^{32}P-orthophosphate for 4 h in phosphate-free DMEM for sufficient ^{32}P labeling of their adenosine triphosphate (ATP) pool. Following this period, the cultures were washed and left untreated (–LPS) or activated by 100 ng/ml of LPS (+LPS) for 20 min. Cells were lysed and the cellular lysates were subjected to 2-D electrophoresis as described in Materials and Methods. The gels were dried and were exposed to Kodak X-Omat AR film (Eastman Kodak, Rochester, NY) for 48 h. The position of pH marker proteins and molecular weight markers is indicated in the figure. The position of 42-kd (pI 7.0 and pI 6.9) proteins is labeled with arrows facing upward, whereas the position of 44-kd (pI 6.7 and pI 6.6) is indicated by arrows facing downward. These proteins have identical molecular weight and pI characteristics with the known 42- and 44-kd isoforms of MAP kinases. The autoradiograms indicate an enhanced in vivo phosphorylation of both forms of 42-kd MAP kinase (pI 7.0 and pI 6.9) as well as of the 44-kd MAP kinase (pI 6.7 and pI 6.6) on activation of the MΦ by LPS.

C. LPS-Activated PK42 and PK44 Show Biochemical Characteristics in 2-D Electrophoresis Similar to 42- and 44-kd MAP Kinases

Because the experiments presented above have demonstrated (under in vitro conditions) the presence of functionally active 42- and 44-kd MAP kinases in lysates of LPS-activated MΦ, it was of interest to further examine whether the activation of MAP kinases can be demonstrated in intact cells.

For these studies, in vivo ^{32}P labeling was combined with 2-D electrophoresis to characterize the LPS-activated phosphoproteins in MΦ. These studies have determined that both PK42 and PK44 proteins show a substantially enhanced phosphorylation on activation by 100 ng/ml of LPS for 20 min (see Fig. 3) under conditions when adherent peritoneal MΦ were prelabeled with ^{32}P-ortho-phosphate (1 mCi/ml/sample) for 4 h before the treatment of MΦ by LPS. We have further demonstrated that (similar to earlier observations [92]) PK42 separated into two proteins with an isoelectric point (pI) of 7.0 and 6.9, whereas PK44 separated into two proteins with a pI of 6.7 and 6.6. Such characteristics of the 42- and 44-kd isoforms of the MAP kinases were described earlier using in vivo labeling, immunoprecipitation, and two-dimensional electrophoresis (92). The pI 7.0 and 6.9 doublets of 42 kd and the pI 6.7 and 6.6 doublets of 44-kd MAP kinases were demonstrated to differ only in their phosphorylation on the Thr residue (92,93). The more acidic forms (pI 6.6 and 6.9) were found to be phosphorylated on both Tyr and The residues (92,93), whereas the more basic forms (pI 6.7 and 7.0) were found to be phosphorylated on their Tyr residues only (92). Our data also indicate that in our MΦ the more basic (pI 6.7 and 7.0) isoforms were dominant (see Fig. 3) in contrast to Swiss 3T3 fibroblast cells, where the more acidic (pI 6.6 and 6.9; containing both Tyr and Thr phosphorylated residues) forms seemed to be the dominant type of MAP kinases (92).

D. GM-CSF Stimulation of Macrophages Will Activate MAP Kinases Suggesting that PK42-PK44 MAP Kinases May Be Common Elements of Multiple Macrophage-Activation Pathways

In our earlier studies, we have demonstrated that GM-CSF and IL-3 can regulate a series of macrophage functions (3,4). We have also shown that some effects of GM-CSF and IL-3 on MΦ, such as the induction of class II MHC expression, can be synergistically enhanced by IFN-γ (4). Because our follow-up studies have demonstrated the time- and dose-dependent activation of MAP kinases in IFN-γ–stimulated MΦ, in the studies presented here we have also addressed the issue whether GM-CSF can activate protein kinase enzymes similar to those activated in MΦ as a potential biochemical mechanism for the interaction between these cytokines.

In our earlier studies (8,9) and in the studies presented above, we have used a series of approaches to demonstrate that the PK42–PK44 kinases, which can be detected in the renaturation kinase assays, are identical with 42- and 44-kd MAP kinases. These approaches have included (1) the partial purification of MAP kinases in diethyl aminoethyl (DEAE) sepharose and phenyl sepharose (8,9); (2) substrate kinase assay using MBP, a typical but less specific MAP kinase substrate; (3) the demonstration of enhanced kinase activity of PK42–PK44 in gels that had MBP incorporated in the SDS-PAGE gels; and finally (4) the immunoprecipitation of MAP kinase proteins with anti-MAP kinase antibody and the subsequent substrate kinase assay using a highly specific MAP kinase substrate nonapeptide (8) (Fig. 4). Because a series of renaturation kinase experiments have suggested that PK44 can be activated by GM-CSF (data not presented), the immunoprecipitation of MAP kinases was combined with the MAP kinase–specific peptide substrate kinase assay in order to determine whether GM-CSF activation of MΦ can result in the activation of MAP kinase.

MAP kinases were immunoprecipitated (with the rabbit anti-MAP kinase antisera, Santa Cruz Biotechnology, Inc.) from MΦ activated by GM-CSF and the immunoprecipitates were subjected to an in vitro substrate kinase assay (see Fig. 4A) as well as Western blotting with anti-MAP kinase antibody (see Fig. 4b). Lysates of 15×10^6 GM-CSF–activated P388.D1 MΦ were prepared in the presence of protease inhibitors and Na orthovanadate. The substrate utilized for these experiments is a synthetic nonapeptide (APRTPGGRR) of the MBP, which contains a typical recognition sequence for the MAP kinase (PXTP). This substrate was demonstrated to be specifically phosphorylated by 42- and 44-kd MAP kinases, whereas other protein kinases, including the catalytic subunit of the cAMP-dependent PKA, PKC, or Ca^{2+}/calmodulin-dependent protein kinase, were all unable to phosphorylate the ^{97}Thr residue of the MBP (94). Our results presented in Figure 4a demonstrate that GM-CSF treatment will enhance the phosphotransferase activity of MAP kinases in P388.D1 MΦ following their treatment with 1000 U/ml of GM-CSF for 40 min. We also demonstrated (see Fig. 4b) that the immunoprecipitation procedure yielded equivalent amounts of MAP kinase proteins from untreated MΦ or MΦ activated by GM-CSF, whereas the phosphotransferase activity of MAP kinases was significantly enhanced in GM-CSF–activated MΦ over the activity in the untreated controls. Additionally, our Western blot experiments of the MAP kinases, immunoprecipitated from MΦ, further suggested that the enhanced MAP kinase activity in GM-CSF–activated MΦ is likely to be due to the activation of 42- and/or 44-kd isoforms of MAP kinases, which were the predominant isoforms of MAP kinases in both intact MΦ (8,9) and in the immunoprecipitates (see Fig. 4b).

IV. DISCUSSION

A. Activation of MAP Kinases in Macrophages by LPS, GM-CSF, and IFN-γ

LPS, GM-CSF, and IFN-γ have all been shown to serve as potent activators of various (often overlapping) MΦ functions. They can all contribute to the production

(a)

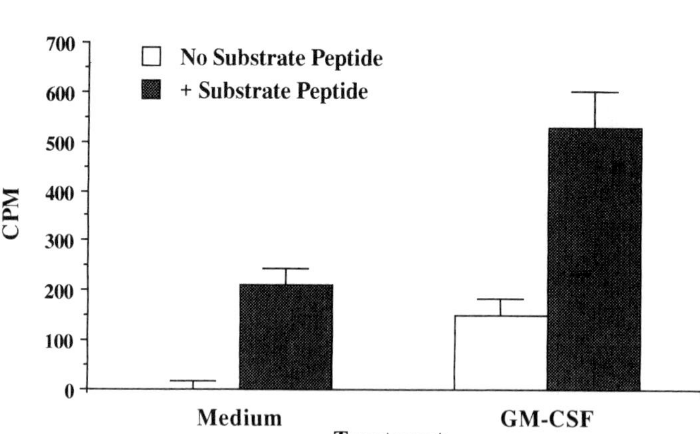

Figure 4 The stimulation of MΦ by GM-CSF will result in the functional activation of MAP kinases. (a) 2×10^7 P388.D1 MΦ/sample were cultured in the absence (Medium) or presence of 1000 U/ml murine recombinant GM-CSF for 40 min, lysed, and immunoprecipitated with anti-MAP kinase antibody and sepharose-bound protein A. The immunoprecipitates were subsequently aliquoted and tested for their ability to phosphorylate the substrate nonapeptide ("+ Substrate Peptide") containing the typical target sequence (PXTP) for phosphorylation by MAP kinases. Control samples were set up to test the level of phosphorylation in the immunoprecipitates in the absence of typical MAP kinase substrate nonapeptide ("No Substrate Peptide"). Additional details of the assay were described in Materials and Methods and the Results sections. Data presented here represent the results of triplicate samples of one of three independent experiments showing a similar increase in peptide phosphorylation by the MAP kinase after GM-CSF activation of MΦ. The level of background phosphorylation (represented by the counts per minute value [average of 210 cpm] of the in vitro kinase reaction obtained in the absence of substrate peptide with immunoprecipitates from nonactivated MΦ) was substrated from the data presented here. The data presented here demonstrate that MAP kinases immunoprecipitated from GM-CSF–activated MΦ are able to phosphorylate the MAP kinase specific substrate nonapeptide (APRTPGGRR); hence MAP kinases are functionally activated in MΦ by GM-CSF treatment. (b) Western blot determination of the level of MAP kinase proteins immunoprecipitated from untreated and GM-CSF–activated MΦ. One-quarter of the immunoprecipitates (representing 5×10^6 P388.D1 MΦ) were aliquoted from the immunoprecipitation reaction and were subjected to Western blots after 9% SDS-PAGE isolation and transfer to PVDF membrane. The membrane was preincubated with blocking buffer then incubated with 1:1000 dilution of the polyclonal anti-MAP kinase antibody in binding buffer (1% milk, 0.05% Tween-20, 1× TBS) overnight at room temperature. The blots were then washed with wash buffer for three times for 20 min at room temperature with gentle agitation. HRPO-coupled antirabbit IgG was used at 1:5000 dilution to detect the binding of anti-MAP kinase ab, and chemiluminescent technique was used for detection. Luminogram was prepared by exposing blots to Kodak X-Omat AR film (Eastman Kodak, Rochester, NY) for 1 min.

(b)

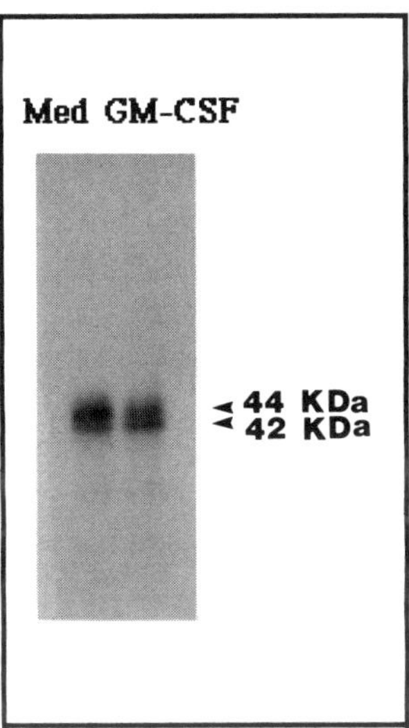

and secretion of newly synthesized MΦ products (e.g., cytokines, enzymes) or alter the morphology of MΦ. Furthermore, the binding of any of these MΦ-activating factors to their respective receptors on MΦ can result in both short-term and long-term functional changes, some of which are completed within a few hours (e.g., active radical production), whereas others may last for 24–96 h (e.g., the induction of intracellular enzymes or MHC class I and class II expression [4,95]). It has been suggested that a divergent signaling mechanism involving multiple secondary pathways (80) as well as the induction of distinct secondary mediators (e.g., cytokines) may be responsible for the pleiotropic effects of these MΦ-activating factors. An understanding of the regulation of these MΦ functions and the biochemical processes involved requires a determination of the sequence of events between binding of ligand to the cell surface receptors and the ultimate activation events (activation of gene transcription, intracytoplasmic enzyme activation, or changes of structural proteins).

Our studies presented here demonstrate the time-dependent activation of three protein kinase enzymes (PK44, PK50, PK56) in MΦ in response to LPS treatment. In studies presented earlier (9), we showed that these enzymes have Ser/Thr phosphotransferase activity. Furthermore, we have specifically identified the 42–44 MAP kinases as one of the protein kinases activated in MΦ by both LPS (see Figs. 2 and 3) and GM-CSF (Fig. 4), as well as have shown earlier the activation of MAP kinases in MΦ by IFN-γ (8).

Our identification of PK42 and PK44 as the 42- and 44-kd MAP kinase was based on the use of a combination of approaches. Initially comigration experiments of MAP kinase Western blots and membrane-bound renaturation kinase assays were used to define protein kinases activated in MΦ by either LPS, IFN-γ, or GM-CSF (see Fig. 1) (8,9). The ultimate identification of the activated MAP kinases in our stimulated MΦ was accomplished by either (1) in vitro kinase assays utilizing MAP kinases immunoprecipitated from MΦ with MAP kinase–specific antibodies and MAP kinase–specific synthetic substrate peptide (APRTPGGRR), or (2) renaturation kinase assays where the MAP kinase substrate protein MBP was incorporated into SDS-PAGE gels in which the cellular kinases were separated by SDS-polyacrylamide electrophoresis. These techniques have allowed the efficient isolation of MAP kinases from other cytoplasmic kinase enzymes and the demonstration of their enhanced phosphotransferase activity toward their specific substrates. Furthermore, a series of phospho-amino-acid analysis were performed on the proteins phosphorylated in the membrane-bound renaturation kinase assays to (1) further verify that this assay measures phosphotransferase enzyme activity, and (2) to determine the substrate specificity (Ser, Thr, or Tyr) of these kinases. These studies (8,9) have determined that all of the 17 different PKs identified in our renaturation kinase assays in either GM-CSF–activated MΦ, IFN-γ–activated, or LPS-treated cells have Ser and Thr phosphotransferase activity. One possible interpretation of this finding is that the renaturation kinase assay preferentially detects Ser/Thr protein kinases. It is noteworthy that Ferrell et al. have also found that the renaturation kinase assay will preferentially detect Ser/Thr kinases (96) and have specifically shown the loss of kinase activity of purified p56 *lck* and p60 *src* Tyr kinase enzymes in this system (likely due to their inability to regain their active conformation during the renaturation process) (87). Furthermore, those investigators have also demonstrated that the renaturation kinase assay system detects protein kinases with the potential for both autophosphorylation and the phosphorylation of other proteins (such as the BSA used for blocking in our system or proteins comigrating with the protein kinase) (87).

Our studies, presented here and earlier (8,9) have also determined that the activation of MAP kinase in MΦ is regulated via posttranslational modification of the MAP kinase, because activated and nonactivated MΦ possess equivalent amounts of this protein, whereas the phosphotransferase activity of MAP kinase was found to be enhanced in LPS-, GM-CSF–, or IFN-γ–activated MΦ. Therefore, we propose that MAP kinases serve as a common element of multiple MΦ activation pathways and are likely to be involved in regulating those MΦ functions that can be activated by either LPS, GM-CSF, or IFN-γ.

B. MAP Kinases

MAP kinases (42-, 44-, 54-, and 80-kd MPA kinase; see Ref. 97 for MAP kinase nomenclature) are Ser/Thr phosphotransferase enzymes that appear to require both

Tyr and Thr phosphorylation for their optimal activation (98,99). To this end, we have also demonstrated (see Fig. 3) that LPS-activated PK42 and PK44 both displayed the typical biochemical characteristics of 42- and 44-kd MAP kinases (92) and showed an enhanced in vivo phosphorylation on activation of MΦ by LPS. It is noteworthy that in both intact MΦ and immunoprecipitates from MΦ we have found that the 42- and 44-kd forms of MAP kinases were the prevalent isoforms. Moreover, dephosphorylation of either one of the Thr or Tyr regulatory residues of the MAP kinases can inactivate these enzymes (99). These regulatory residues of MAP kinases are highly conserved in the $T^{183}EY^{185}$ sequence motif (100) and serve as targets for phosphorylation by upstream kinases and for dephosphorylation by phosphatases. The MAP kinases have also been designated as members of the extracellular signal-regulated protein kinase (ERK) family because of their activation by mitogens or cytokines (101).

1. A Variety of Stimuli (Growth and Differentiation Factors as Well as Cross Linking of Cell Surface Molecules) Can Activate MAP Kinases

MAP kinases were demonstrated to be activated by a variety of growth factors/mitogens (102) as well as by differentiation-inducing factors (103). The activation of MAP kinases was demonstrated after insulin stimulation of fibroblasts (104) and nerve growth factor treatment of PC12 cells (103) as well as after stimulation of cells by various mitogens (102). Additionally, surface IgE cross linking in B cells was shown to activate MAP kinases (105). Although the activation of MAP kinase could also be accomplished by either CD3 or CD2 cross linking in Jurkat cells, only the effects of the latter stimulus can be reversed by the CD45 Tyr phosphatase (106). Additionally, we have shown here and earlier (8,9) that MAP kinases can also be activated in MΦ by LPS, GM-CSF, or IFN-γ stimulation.

2. Two Distinct Pathways of Activation of MAP Kinases

Several lines of evidence suggest that the activation of MAP kinases is unlikely to be a membrane proximal event. These include the fact that (1) MAP kinases require a dual phosphorylation on both of their regulatory Thr and Tyr residues for their activation; (2) none of the Tyr kinase–type membrane receptors (e.g., FGFR, EGFR, or NGFR), through which MAP kinases can be activated, were demonstrated to be able to activate MAP kinases directly; (3) a MAP kinase kinase (MEK/MEPKK) enzyme was recently identified and cloned (107) that seems to have the unique Tyr and Ser/Thr phosphotransferase activity required for the full activation of MAP kinases (102); (4) a more upstream kinase, the MEK kinase (MEKK/MEPKKK), also was identified recently (108), which has Ser/Thr phosphotransferase activity and is able directly to activate the MEK when it itself is activated through a guanine-binding protein-mediated mechanism (Fig. 5) (102,108). Thus, it has become apparent that the activation of MAP kinases in vertebrate cells can be accomplished through at least two distinct pathways (Fig.

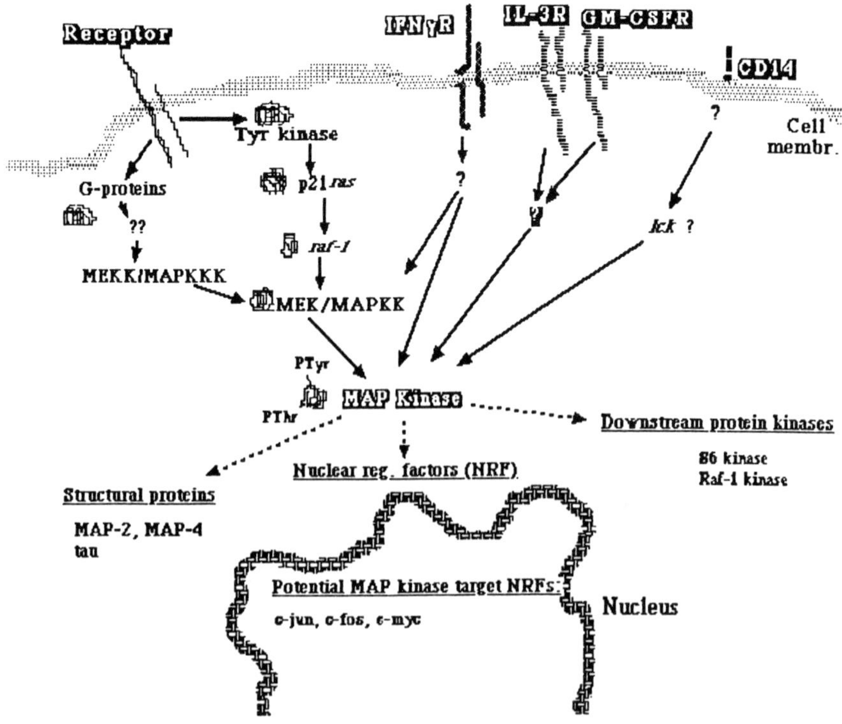

Figure 5 Signaling pathways and the potential role of MAP kinases in the activation of MΦ by LPS, GM-CSF, IL-3, or IFN-γ.

5), namely, via either (1) the activation of Tyr kinase–type membrane receptors (e.g., epidermal growth factor [EGF] or nerve growth factor [NGF] [103,109]) or membrane receptors activating intracytoplasmic Tyr kinases (110), or (2) guanine nucleotide–binding protein-coupled receptor (e.g., thrombin receptor [87]) systems (102,108). The activation of MAP kinase through tyrosine kinases was shown to be mediated by p21 *ras* (103), which in turn activates *raf*-1, resulting in the direct activation of MEK (111), and via MEK the ultimate activation of MAP kinases. In contrast, the activation of MAP kinases through guanine nucleotide–binding protein-coupled receptors is mediated via the sequential activation of the G-protein, MEKK, MEK, MAP kinase cascade.

3. Mechanisms of Deactivation of MAP Kinases

Multiple deactivation mechanism, involving both Tyr and Ser/Thr phosphatase enzymes, have been identified for the MAP kinase in vivo and in vitro. In vivo it has been shown that treatment of T cells with cross-linking antibodies against the transmembrane glycoprotein-CD45 phosphatase is able to reverse the activation of

MAP kinases induced by the stimulation (cross linking) of CD2 antigens (106), a deactivation that occurs via Tyr dephosphorylation of the MAP kinase. MAP kinases also can be inactivated by Ser/Thr specific phosphatases, like protein phosphatase 2A (104) and protein phosphatase 1 (112). These studies indicate that both Tyr and Ser/Thr specific phosphatases are involved in regulating the state of activation of MAP kinases.

C. Functional Role of MAP Kinases

Although the involvement of MAP kinases in the regulation of cell cycle has become evident, the role of MAP kinases in regulating various functions of fully differentiated cells has yet to be determined. However, there are multiple potential mechanisms by which MAP kinases could participate in these processes via the regulation of the activation of various intracellular proteins, like other protein kinases (e.g., rsk); cell surface receptors (e.g., EGF), and enzymes (e.g., phospholipase A_2), as well as nuclear regulatory factors (e.g., c-*fos*, c-*myc*, c-*jun*).

1. Potential Involvement of MAP Kinases in Regulation of Gene Expression via Activation or Deactivation of Nuclear Regulatory Factors

The direct phosphorylation of several nuclear regulatory factors, including c-*jun* (113), c-*fos* (114), c-*myc* (115), and p62TCF (116) by MAP kinases was demonstrated. Although the phosphorylation of c-*jun*, c-*myc*, and p62TCF by MAP kinases will result in enhanced functional activity of these nuclear regulatory factors, and hence in the enhanced transcription of their target genes, the phosphorylation of the C-terminus of *fos* protein was shown to result in the transcriptional transrepression of the c-*fos* promoter itself as a negative feedback mechanism (114). Because, for most if not all genes, the regulation of gene transcription is determined by the concerted interaction of a series of activated nuclear regulatory factors with the promoter region of the gene, it is likely that MAP kinase activated nuclear factors (e.g., c-*jun*, c-*fos*, c-*myc*, and p62TCF) will be but one (and perhaps by themselves insufficient) contributors for the activation of a particular gene.

2. Phosphorylation of Cytoplasmic Proteins Involved in Determining the Structure/Morphology of Cells by MAP Kinases

Although one of the best known characteristics of the MAP kinases is their ability to phosphorylate the microtubule-associated proteins (MAP-1 and MAP-2), which are likely to play a role in the destabilization of microtubules, the direct contribution of MAP kinases to the morphology of the cells, the role of MAP kinases in regulating the homo- and heterotypic interaction of cells, and their locomotion is yet unknown. To this end however, it was recently demonstrated that fMLP activation of peripheral blood neutrophils, which induces the chemotactic locomotion of these cells, will result in the activation of MAP kinases (117).

3. *Activation of Additional (Downstream) Kinases by MAP Kinase as Part of the Kinase-Activation Cascade*

Ribosomal protein S6 kinase II (Rsk 90) was shown to be activated by MAP kinases in insulin-stimulated fibroblasts (104). It has also been shown that the functional MAP kinase is able to activate the *raf*-1 kinase in insulin-activated cells, and thus MAP kinase has also been referred to as *raf*-1 kinase kinase (118). It is of our interest to determine the more distal elements of the activation pathway that are regulated by activated MAP kinases in MΦ. There is indirect evidence that p21 *ras* and *ras*-activated MAP kinases (or more downstream kinases of this cascade) may be involved in the expression of cell surface antigens, since class II MHC expression on the surface of cells that were partially deficient in class II MHC expression was found to be enhanced following transfection with *ras* (119).

4. *Activation of Cytoplasmic Enzymes Other Than Protein Kinases by MAP Kinases*

Although it was shown earlier that LPS activation of MΦ will result in the activation of PLA_2 (120), it was recently demonstrated that MAP kinases can directly activate this enzyme (the cytosolic PLA_2 (121)).

D. Potential Significance of MAP Kinase as a Common Element of Multiple Macrophage-Activation Pathways

Although our identification of three additional receptors (LPSR/CD14, GM-CSFR, and IFN-γR) has added to the multiplicity of the activation mechanisms of MAP kinases, we believe that these data, together with data demonstrating (1) the existence of highly conserved regulatory (TEY) sequences of MAP kinases (100), and (2) the substantial similarities between the prokaryotic and eukaryotic forms of these kinases and their activation pathways (102) point to their essential role in regulating cell physiology. Although MAP kinases seem undoubtedly to serve as integration points of signaling from a variety of receptor systems, we believe that they are part of a signaling pattern (of a spectrum of activated protein kinases) from each one of these receptors. In such a complex signaling system, the specificity of the response to the specific stimulus is provided by the pattern of activated protein kinases (integration of the signals) rather than by the individual kinases activated. We propose that the MAP kinases are likely to play a role in the expression of a variety of MΦ functions and are most likely to contribute to the activation of those functions common to LPS, GM-CSF, or IFN-γ stimulation of MΦ (e.g., class II MHC expression, cytokine production).

As detailed above, we and others have shown that MAP kinases can be activated by a large number of soluble mediators (growth factors, cytokines, differentiation factors, and LPS) or triggering of cell surface antigens (e.g., CD2, CD3, and sIgE). It also was shown that the activation of MAP kinases can be accomplished by at least two alternative pathways, a Tyr kinase–mediated and a G-protein–mediated pathway. Furthermore, MAP kinases can be inactivated by multiple (intracytoplasmic or membrane-associated) phosphatases. The abundance of these regulatory pathways provides an exceptional versatility for the regulation of the activation of MAP kinases in the various cell types. However, it is also likely that cell type–specific regulatory processes of the activation of MAP kinases will soon be revealed.

V. CONCLUSIONS

The state of activation of MΦ can be regulated by a variety of factors, including cytokines (e.g., IFN-γ, GM-CSF, IL-3) as well as bacterial products (e.g., LPS, MDP). Activated macrophages provide a more effective protection against both bacterial and viral infections and are more potent in eliminating intracellular pathogens. Although TNF-α has been identified as a key mediator of MΦ protection against intracellular pathogens (1,2) and by Nacy et al. (Chap. 4), other yet unidentified factors that are induced by MΦ by the MΦ-activating factors (e.g., IFN-γ) also seem to be crucial for this protective mechanism. Although we (3,4) and others (5–7) have demonstrated that many of these MΦ-activation pathways can synergize with each other, the precise intracellular-signaling mechanisms of these processes have yet to be elucidated. Our current studies were aimed at identifying PKs that can be mutually activated in MΦ by LPS, GM-CSF, or other MΦ-activating factors.

For our studies presented here and earlier (8,9), we have used an in vitro PK assay, based on the renaturation of electrophoretically isolated cellular proteins, to define a number of PK enzymes that were involved in mediating intracellular signaling in activated MΦ. We showed earlier (8) that the 42- to 44-kd MAP kinase was activated by IFN-γ in MΦ by demonstrating that 42- and 44-kd MAP kinases immunoprecipitated from IFN-γ activated MΦ with an anti-MAP kinase antibody possess an enhanced phosphotransferase/kinase activity toward the synthetic nonapeptide (APRTGGRR) substrate containing the typical MAP kinase target sequence (PRTP) (8,9). The studies presented here further demonstrated that 42- to 44-kd MAP kinases can also be activated by LPS or GM-CSF and thus function as a common element of multiple MΦ-activation pathways. These studies have also determined that the activation of MAP kinase in MΦ is regulated via posttranslational modification of the MAP kinase proteins, because the functional activation (enhanced phosphotransferase activity) of MAP kinases in activated MΦ were not paralleled by any changes in the amount of MAP kinase proteins.

The activation of common protein kinases by various MΦ-activation pathways may not only provide the mechanisms for synergy between these factors but may also be the mechanistic basis of overlapping activities of cytokines. Identification of such common signaling elements of multiple MΦ-activation pathways will further help us define key protein kinase enzyme targets for the effective and potentially more specific modulation of the immune response.

ACKNOWLEDGMENTS

G.F. was supported by a grant from the American Cancer Society (IN97-0), R.J.G. was a postdoctoral fellow supported by the Immunology Training Grant IA-07039.

REFERENCES

1. Green SJ, Crawford RM, Hockmeyer JT, Meltzer MS, Nacy CA. Leishmania major amastigotes initiate the L-arginine-dependent killing mechanism in IFN-gamma–stimulated macrophages by induction of tumor necrosis factor-alpha. J Immunol 1990, 145:4290.
2. Nacy C, Meierovich AJ, Belosevic M, Green SJ. TNF-alpha—central regulatory cytokine in the induction of macrophage anti-microbial activities. Pathobiology 1991, 59:182.
3. Frendl G, Fenton MJ, Beller DI. Regulation of macrophage activation by IL-3. II. IL-3 and LPS interact synergistically in the regulation of IL-1 expression. J Immunol 1990; 144:3400.
4. Frendl G, Beller DI. Regulation of macrophage activation by IL-3. I. IL-3 is a macrophage activating factor with unique properties inducing Ia and LFA-1 but not cytotoxicity. J Immunol 1990; 144:3392.
5. Ruco LP, Meltzer MS. Macrophage activation of tumor cytotoxicity: development of macrophage cytotoxic activity requires completion of a sequence of short-lived intermediary reactions. J Immunol 1978; 121:2035.
6. Ucla C, Roux-Lombard P, Fey S, Dayer S, Mach B. Interferon gamma drastically modifies the regulation of interleukin 1 genes by endotoxin in U937 cells. J Clin Invest 1990; 85:185.
7. Gerrard TL, Siegel JP, Dyer DR, Zoon KC. Differential effects of interferon alpha and interferon gamma on interleukin 1 secretion by monocytes. J Immunol 1987; 138:2535.
8. Frendl G, Guan R, Beller DI. Intracellular signaling in IFNγ-activated macrophages involves mitogen-activated protein kinase (MAP-kinase) and other serine/threonine kinases. J Immunol 1993 (submitted).
9. Guan RJ, Frendl G, Beller DI. Lipopolysaccharide activates multiple protein kinases including mitogen-activated protein kinase in macrophages. Int J Immunopharmacol 1994 (submitted).
10. Unanue ER. Antigen-presenting function of the macrophage. Ann Rev Immunol 1984; 2:395.
11. Unanue ER, Allen PM. The basis for the immunoregulatory role of macrophages and other accessory cells. Science 1987; 236:551.
12. Nathan CF. Secretory products of macrophages. J Clin Invest 1987; 79:319.
13. Mizel SB, Oppenheim JJ, Rosenstreich DI. Characterization of lymphocyte-activating factor (LAF) produced by a macrophage cell line, P388D1. I. Enhancement of LAF production by activated T lymphocytes. J Immunol 1978; 120:1497.

14. Meltzer M, Oppenheim JJ. Bidirectional amplification of macrophage-lymphocyte interactions: enhanced lymphocyte activation factor production by activated mouse peritoneal cells. J Immunol 1977; 118:77.
15. Weaver CT, Unanue ER. T cell induction of membrane IL-1 on macrophages. J Immunol 1986; 137:3868.
16. Schreiber RD, Altman A, Katz DH. Identification of a T cell hybridoma which produces large quantities of macrophage-activating factor. J Exp Med 1982; 156:677.
17. Morrissey P, Bressler L, Park LS, Alpert A, Gillis S. Granulocyte-macrophage colony-stimulating factor augments the primary antibody response by enhancing the function of antigen-presenting cells. J Immunol 1987; 139:1113.
18. Murray HW. IFN-gamma, the activated macrophage, and host defense against microbial challenge. Ann Int Med 1988; 108:595.
19. Nash TW, Libby DM, Horwitz MA. Interaction between legionnaires disease bacterium (*Legionella pneumophila*) and human alveolar macrophages. Influence of antibody, cytokines, and hydrocortisone. J Clin Invest 1984; 74:771.
20. Reed SG, Nathan CF, Pihl DI, Rodricks P, Shanebeck K, Conlon PJ, Grabstein KH. Recombinant macrophage/granulocyte colony-stimulating factor activates macrophages to inhibit T *Trypanosoma cruzi* and release hydrogen peroxide. J Exp Med 1987; 166:1734.
21. Cederblad B, Alm GV. Interferon and the colony-stimulating factors IL-3 and GM-CSF enhance the IFN-alpha response in human blood leukocytes induced by herpes simplex virus. Scand J Immunol 1991; 34:549.
22. Wang M, Friedman H, Djeu JY. Enhancement of human monocyte function against Candida albicans by the colony-stimulating factors (CSF): IL-3, granulocyte-macrophage-CSF, and macrophage-CSF. J Imunol 1989; 143:671.
23. Frendl G. Interleukin-3: From colony-stimulating factor to pluripotent immunoregulatory cytokine. Int J Immunopharmacol 1992; 14:421.
24. Cannistra SA, Vellenga E, Groshek P, Rambaldi A, Griffin JD. Human granulocyte-macrophage colony-stimulating factor and interleukin 3 stimulate monocyte cytotoxicity through a tumor necrosis factor-dependent mechanism. Blood 1988; 71:672.
25. Langermans JAM, v. Hulst MEB, Nibbering PH, Heimstra PS, Fransen L, vanFurth R. IFN-gamma-induced L-arginine-dependent toxoplasmastatic activity in murine peritoneal macrophages is mediated by endogenous tumor necrosis factor-alpha. J Immunol 1992; 148:568.
26. Raetz CRH, Ulevitch RJ, Wright SD, Sibley CS, Ding A, Nathan CF. Gram-negative endotoxin: an extraordinary lipid with profound effects on eucaryotic signal transduction. FASEB J 1991; 5:2652.
27. Hibbs JB, Jr, Taintor RR, Chapman HA, Jr, Weinberg JB. Macrophage tumor killing: influence of the local environment. Science 1977; 197:279.
28. Steeg PS, Johnson HM, Oppenheim JJ. Regulation of murine macrophage Ia antigen expression by an immune interferon-like lymphokine: inhibitory effect of endotoxin. J Immunol 1982; 129:2402.
29. Smith PD, Lamerson CL, Wong HL, Wahl LM, Sahl SM. GM-CSF stimulates human monocyte accessory cell function. J Immunol 1990; 145:3829.
30. Heidenreich S, Gong JH, Schmidt A, Nain M, Gemsa D. Macrophage activation by GM-CSF priming for enhanced release of TNF-alpha and PGE_2. J Immunol 1989; 143:1198.

31. Lang RA, Metcalf D, Cuthbertson RA, Lyons I, Stanley E, Kelso A, Kannourakis G, Williamson DJ, Klintworth GK, Gonda TJ, Dunn AR. Transgenic mice expressing growth factor gene (GM-CSF) develop accumulations of macrophages, blindness, and a fatal syndrome of tissue damage. Cell 1987; 51:675.
32. Nathan CF, Murray HW, Wiebe ML, Rubin BY. Identification of IFN-gamma as the lymphokine that activates human macrophage oxidative metabolism and antimicrobial activity. J Exp Med 1983; 158:670.
33. Schreiber RD, Celada A. Molecular characterization of IFN-gamma as a macrophage activating factor. Lymphokines 1985; 11:87.
34. Celada A, Gray PW, Rinderknecht E, Schreiber RD. Evidence for a gamma-interferon receptor that regulates macrophage tumoricidal activity. J Exp Med 1984; 160:55.
35. Vilcek J. Interferon-gamma: a lymphokine for all seasons. Lymphokines 1985; 11:1.
36. Djeu JY, Widen R, Blanchard DK. Susceptibility of monocytes to lymphokine-activated killer cell lysis: Effect of granulocyte-macrophage colony-stimulating factor and Interleukin-3. Blood 1989; 73:1264.
37. Djeu JY, Blanchard K. Interferon-gamma–induced alterations of monocyte susceptibility to lysis by autologous lymphokine-activated killer (LAK) cells. Int J Cancer 1988; 42:449.
38. Vairo G, Argyriou S, Knight KR, Hamilton JA. Inhibition of colony-stimulating factor–stimulated macrophage proliferation by TNFα, IFNγ, and LPS is not due to a general loss of responsiveness to growth factor. J Immunol 1991; 146:3469.
39. Hampton RY, Golenbock DT, Panman M, Krieger M, Raetz CRH. Recognition and plasma clearance of endotoxin by scavenger receptors. Nature 1991; 352:342.
40. Wright SD, Jong MTC. Adhesion-promoting receptors on human macrophages recognize *E. coli* by binding to LPS. J Exp Med 1986; 164:1876.
41. Goyert SM, Ferrero E, Seremetis SV, Winchester RJ, Silver J, Mattison AM. Biochemistry and expression of myelomonocytic antigens. J Immunol 1986; 137:3909.
42. Wright SD, Ramos RA, Tobis PS, Ulevitch RJ, Mathison JC. CD14, a receptor for complexes of Lipopolysaccharide (LPS) and LPS binding protein. Science 1990; 249:1431.
43. Schumann RR, Leong SR, Flaggs GW, Gray PW, Wright SD, Mathison JC, Tobias PS, Ulevitch RJ. Structure and function of lipopolysaccharide binding protein. Science 1990; 249:1429.
44. Wright SD, Ramos RA, Patel M, Miller DS. Septin: a factor in plasma that opsonizes LPS-bearing particles for recognition by CD14 on phagocytes. J Exp Med 1992; 176:179.
45. Bazil V, Strominger JL. Shedding as a mechanism of down-modulation of CD14 on stimulated human monocytes. J Immunol 1991; 147:1567.
46. Landmann R, Wesp M, Obrecht JP. Cytokine regulation of myeloid glycoprotein CD14. Pathobiology 1991; 59:131.
47. Ruppert J, Friedrichs D, Xu H, Peters JH. IL-4 decreases the expression of monocyte differentiation marker CD14, paralleled by an increasing accessory potency. Immunobiology 1991; 182:449.
48. Beckhuizen H, Blokland I, Tilburg JC, Koning F, vanFurth. CD14 contributes to the

adherence of human monocytes to cytokine-stimulated endothelial cell. J Immunol 1991; 147:3761.
49. Elliott MJ, Vadas MA, Eglinton JM, Park LS, Bik To L, Cleland LG, Clark SC, Lopez AF. Recombinant human Interleukin-3 and granulocyte-macrophage colony-stimulating factor show common biological effects and binding characteristics on human monocytes. Blood 1989; 74:2349.
50. Urdal DL, Price V, Sassenfeld HM, Cosman D, Gillis S, Park LS. Molecular characterization of colony stimulating factors and their receptors: human interleukin-3. Ann NY Acad Sci 1989; 554:167.
51. Itoh N, Yonehara S, Schreurs J, Gorman DM, Maruyama K, Ishii A, Yahara I, Arai K-I, Miyajima A. Cloning of an interleukin-3 receptor gene: a member of a distinct receptor gene family. Science 1990; 247:324.
52. Hayashida K, Kitamura T, Gorman DM, Arai K-I, Yokota T, Miyajima A. Molecular cloning of a second subunit of the receptor of human granulocyte-macrophage colony-stimulating factor (GM-CSF): reconstitution of a high-affinity GM-CSF receptor. Proc Natl Acad Sci USA 1990; 87:9655.
53. Miyajima A, Hara T, Kitamura T. Common subunits of cytokine receptors and the functional redundancy of cytokines. Trends Biochem Sci 1992; 17:374.
54. Gorman DM, Itoh N, Kitamura T, Schreurs J, Yonehara S, Yahara I, Arai KI, Miyajima A. Cloning and expression of a gene encoding an interleukin 3 receptor-like protein: identification of another member of the cytokine receptor gene family. Proc Natl Acad Sci USA 1990; 87:5459.
55. Takaki S, Mita S, Kitamura T, Yonehara S, Yamaguchi N, Tominaga A, Miyajima A, Takatsu K. Identification of the second subunit of the murine interleukin 5 receptor: interleukin 3 receptor-like protein, AIC2B is a component of the high affinity interleukin 5 receptor. EMBO J 1991; 10:2833.
56. Kitamura T, Sato N, Arai K, Miyajima A. Expression cloning of the human IL-3 receptor cDNA reveals a shared beta subunit for the human IL-3 and GM-CSF receptors. Cell 1991; 66:1165.
57. Pestka S, Langer JA, Zoon KC, Samuel CE. Interferons and their actions. Annu Rev Biochem 1987; 56:727.
58. Auget M, Dembic Z, Merlin G. Molecular cloning and expression of the human IFNγ receptor. Cell 1989; 55:273.
59. Schreiber RD, Farrar MA, Hershey GK, Fernandez-Luna J. The structure and function of IFN-gamma receptors. Int J Immunopharmacol 1992; 14:413.
60. Jung V, Jones C, Kumar CS, Stefanos S, O'Connell S, Pestka S. Expression and reconstitution of a biologically active human IFN-gamma receptor in hamster cells. J Biol Chem 1990; 265:1827.
61. Farrar MA, Fernandez-Luna J, Schreiber RD. Identification of two regions within the cytoplasmic domain of the human IFN gamma receptor required for function. J Biol Chem 1991; 266:19626.
62. Jung V, Rashidbaigi A, Jones C, Tishfield JA, Shows TB, Pestka, S. Human chromosomes 6 and 21 are required for sensitivity to human IFN-gamma. Proc Natl Acad Sci USA 1987; 84:4151.
63. Introna M, Hamilton TA, Kaufman RE, Adams DO, Bast RC Jr. Treatment of

murine peritoneal macrophages with bacterial lipopolysaccharide alters expression of c-FOS and c-MYC oncogenes. J Immunol 1986; 137:2711.

64. Hamilton TA, Adams DO. Molecular mechanisms of signal transduction in macrophages. Immunol Today 1987; 8:151.
65. Tannenbaum CS, Hamilton TA. Lipopolysaccharide-induced gene expression in murine peritoneal macrophages is selectively suppressed by agents that elevate intracellular cAMP. J Immunol 1989; 142:1274.
66. Goyert SM, Rerrero E, Rettig WJ, Yenamandra AK, Obata F, LeBeau MM. The CD14 monocyte differentiation antigen maps to a region encoding growth factors and receptors. Science 1988; 239:497.
67. Frey EA, Miller DS, Jahr TG, Sundan A, Bazil V, Espevik T, Finlay BB, Wright SD. Soluble CD14 participates in the response of cells to LPS. J Exp Med 1992; 1:1665.
68. Stefanova I, Horejsi V, Ansotegui IJ, Knapp W, Stockinger H. GPI-anchored cell-surface molecules complexes to protein tyrosine kinases. Science 1991; 254:1016.
69. Sorensen P, Mui ALF, Krystal G. Interleukin 3 stimulates the tyrosine phosphorylation of the 140 kD IL-3 receptor. J Biol Chem 1989; 264:19253.
70. Morla AO, Schreurs J, Miyajima A, Wang JY. Hemopoeitic growth factors activate the tyrosine phosphorylation of distinct sets of proteins in interleukin-3-dependent murine cell lines. Mol Cell Biol 1988; 8:2214.
71. Linnekin D, Evans G, Garcia G, Michiel D, Farrar W. IL-2, IL-3, and GM-CSF signal transduction pathways converge on similar protein kinases which have conserved and lineage specific substrates. Lymphokine Res 1990; 9:576.
72. Farrar WL, Thomas TP, Anderson WB. Altered cytosol/membrane enzyme redistribution on IL-3 activation of protein kinase C. Nature 1985; 315:235.
73. Evans SW, Rennick D, Farrar WL. Multilineage hemopoeitic growth factor interleukin-3 and direct activators of protein kinase C stimulate phosphorylation of common substrates. Blood 1986; 68:906.
74. Pelech SL, Paddon HB, Charest DL, Federsppiel BS. IL-3 induced activation of protein kinases in the mast cell/megakaryocyte R6-XE.4 line. J Immunol 1990; 144:1759.
75. Farrar WL, Linnekin D. Regulation of protein kinases and gene expression by immunocytokines. Ann NY Acad Sci 1990; 594:240.
76. Coffey RG, Davis JS, Djeu JY. Stimulation of guanylate cyclase activity and reduction of adenylate cyclase activity by granulocyte-macrophage colony-stimulating factor in human blood neutrophils. J Immunol 1988; 140:2695.
77. McColl SR, Kreis C, DiPersio JF, Borgeat P, Naccache PH. Involvement of guanine nucleotide binding proteins in neutrophil activation and priming by GM-CSF. Blood 1989; 73:588.
78. Bourgoin S, Plante E, Gaudry M, Nacchache PH, Borgeat P, Poubelle PE. Involvement of phospholipase D in the mechanism of action of granulocyte-macrophage colony-stimulating factor (GM-CSF): priming of human neutrophils in vitro with GM-CSF is associated with accumulation of phosphatidic acid and diradylglycerol. J Exp Med 1990; 172:767.
79. Meldolesi MF, Friedman RM, Kohn LD. An interferon-induced increase in cyclic

AMP levels precedes the establishment of the antiviral state. Biochem Biophys Res Commun 1977; 79:239.
80. Adams DO, Hamilton TA. Molecular transduction mechanisms by which gamma-interferon and other signals regulate macrophage development. Immunol Rev 1987; 97:5.
81. Mao C, Merlin G, Ballotti R, Metzler M, Aguet M. Rapid increase of the human IFN gamma receptor phosphorylation in response to human IFN gamma and phorbol myristate acetate. J Immunol 1990; 145:4257.
82. Celada A, Schreiber RD. Role of protein kinase C and intracellular Ca mobilization in the induction of macrophage tumoricidal activity by IFN gamma. J Immunol 1986; 137:2373.
83. Gumina RJ, Freire-Moar J, DeYoung L, Webb DR, Devens BH. Transduction of the IFN-gamma signal for HLA-DR expression in the promonocytic line THP-1 involves a late-acting PKC activity. Cell Immunol 1991; 138:265.
84. Koide Y, Ina Y, Nezu N, Yoshida TO. Calcium influx and the Ca^{2+}-calmodulin complex are involved in interferon-gamma-induced expression of HLA class II molecules on HL-60 cells. Proc Natl Acad Sci USA 1988; 85:3120.
85. Gordon JA. Use of vanadate as protein-phosphotyrosine phosphatase inhibitor. In: Hunter T, Sefton BM, eds. Methods in Enzymology. Protein phosphorylation. Boston: Academic Press, 1991:477.
86. O'Farrell PH. High-resolution two-dimensional electrophoresis of proteins. J Biol Chem 1975; 250:4007.
87. Ferrell JE, Martin GS. Thrombin stimulates the activities of multiple previously unidentified protein kinases in platelets. J Biol Chem 1989; 264:20723.
88. Celenza JL, Carlson M. Renaturation of protein kinase activity on protein blots. In: Hunter T, Sefton BM, eds. Methods in enzymology. Boston: Academic Press, 1991:423.
89. Clark-Lewis J, Sanghera JS, Pelech SI. Definition of a consensus sequence for peptide substrate recognition by p44 mpk, the meiosis-activated myelin basic protein kinase. J Biol Chem 1991; 266:15180.
90. Anderson NG, Kilgour E, Sturgill TW. Activation of mitogen-activated protein kinase in BC3H1 myocytes by fluoroaluminate. J Biol Chem 1991; 266:10131.
91. Edelman AM, Blumenthal DK, Krebs EG. Protein serine/theronine kinases. Ann Rev Biochem 1987; 56:567.
92. Chatani Y, Tanaka E, Tobe K, Hattori A, Sato M, Tamemoto H, Nishizawa N, Nomoto H, Takeya T, Kadowaki T, Kasuga M, Kohno M. Mitogen-induced tyrosine-phosphorylated 41- and 43-kDa roteins are family members of extracellular signal-regulated kinases/microtubule-associated protein 2 kinases. J Biol Chem 1991; 267:9911.
93. Kohno M. Diverse mitogenic agents induce rapid phosphorylation of cellular proteins at tyrosine in quiescent mammalian cells. J Biol Chem 1985; 260:1771.
94. Erickson AK, Payne DM, Martino PA, Rossomando AJ, Shabanowitz J, Waber MJ, Hunt DF, Sturgill TW. Identification by mass spectrometry of threonine 97 in bovine myelin basic protein as a specific phosphorylation site for mitogen-activated protein kinase. J Biol Chem 1990; 265:19728.

95. Lew DJ, Decker T, Darnell JE, Jr. Alpha IFN and gamma IFN stimulate transcription of a single gene through different signal transduction pathways. Mol Cell Biol 1989; 9:5404.
96. Farrell JE, Martin GS. Identification of a 42 kD phosphotyrosyl protein as a serie (threonine) protein kinase by renaturation. Mol Cell Biol 1990; 10:3020.
97. Blenis J. Growth regulated signal transduction by MAP kinases and RSKs. Cancer Cells 1991; 3:445.
98. Ray LB, Sturgill TW. Insulin-stimulated microtubule-associated protein kinase is phosphorylated on tyrosine and threonine in vivo. Proc Natl Acad Sci USA 1988; 85:3753.
99. Anderson NC, Maller JL, Tonks NK, Sturgill TW. Requirement for integration of signals from two distinct phosphorylation pathways for activation of MAP kinase. Nature 1990; 343:651.
100. Posada J, Cooper JA. Requirements for phosphorylation of MAP kinases during meiosis in Xenopus oocytes. Science 1992; 255:212.
101. Boulton TG, Yankopoulos GD, Gregory JS, Slaughter C, Moomaw C, Hsu J, Cobb MH. An insulin-stimulated protein kinase similar to yeast kinase involved in cell cycle control. Science 1990; 249:64.
102. Nishida T, Gotoh Y. The MAP kinase cascade is essential for diverse signal transduction pathways. Trends Biochem Sci 1993; 18:128.
103. Wood K, Sarnecki C, Roberts TM, Blenis J. Ras mediates nerve growth factor receptor modulation of three signal-transducing protein kinases: MAP Kinase, Raf-1, and RSK. Cell 1992; 68:1041.
104. Sturgill TW, Ray LB, Erikson E, Maller JL. Insulin-stimulated MAP-2 kinase phosphorylates and activates ribosomal protein S6 kinase II. Nature 1988; 334:715.
105. Casillas A, Hanekom C, Williams K, Katz R, Nel AE. Stimulation of B cells via the membrane immunoglobulin receptor or with PMA induces tyrosine phosphorylation of a 42kDa microtubule-associated protein-2 kinase. J Biol Chem 1991; 266:19088.
106. Nel EA, Ledbetter JA, Williams K, Ho P, Akerley B, Franklin K, Katz R. Activation of MAP-2 kinase activity by CD2 receptor in Jurkat cells can be reversed by CD45 phosphatase. Immunology 1991; 73:129.
107. Matsuda S, Kosako H, Takenaka K, Moriyama K, Sakai A, Akiyama and Gotoh Y. Xenopus MAP kinase activator: identification and function as a key intermediate in the phosphorylation cascade. EMBO J 1992; 11:973.
108. Lange-Carter CA, Pleiman CM, Gardner AM, Blumer KJ, Johnson GL. A divergence in the MAP kinase regulatory network defined by MEK kinase and raf. Science 1993; 260:315.
109. Sturgill TW, Ray LB, Anderson NG, Erickson AK. Purification of MAP kinase from Epidermal Growth factor-treated 3T3-L1 fibroblasts. In: Hunter T, Sefton BM, eds. Protein kinases: assays, purification, antibodies, functional analysis, cloning and expression. Boston: Academic Press, 1991:342.
110. Ettehadieh E, Sanghera JS, Pelech SL, Hess-Bienz D, Watts J, Shastri N, Aebersold R. Tyrosine phosphorylation and activation of MAP kinase by p56lck. Science 1992; 255:853.
111. Kyriakis JM, App H, Zhang X, Benerjee P, Brautigan DL, Rapp UR, Avruch J. Raf-1 activates MAP kinase kinase. Nature 1992; 358:417.

112. Kyriakis JM, Brautigan DL, Ingebritsen TS, Avruch J. pp54 MAP-2 kinase requires both tyrosine and serine/threonine phosphorylation for activity. J Biol Chem 1991; 266:10043.
113. Pulverer BJ, Kyriakis JM, Avruch J, Nikolakaki E, Woodgett JR. Phosphorylation of c-jun mediated by MAP kinases. Nature 1991; 353:670.
114. Ofir R, Dwarki VJ, Rashid D, Verma IM. Phosphorylation of the C terminus of fos protein is required for transcriptional transrepression of the c-fos promoter. Nature 1990; 348:80.
115. Seth A, Alvarez E, Gupta S, Davis RJ. Phosphorylation site located in the NH_2 terminal domain of c-Myc increases transactivation of gene expression. J Biol Chem 1991; 266:23521.
116. Gille H, Sharrocks AD, Shaw PE. Phosphorylation of transcription factor p62TCF by MAP kinase stimulates ternary complex formation at c-fos promoter. Nature 1992; 358:414.
117. Torres M, Hall FL, O'Neill K. Stimulation of human neutrophils with fMLP induces Tyr phosphorylation and activation of two distinct MAP kinases. J Immunol 1993; 150:1563.
118. Lee RM, Cobb MH, Blackshear PJ. Evidence that extracellular signal-regulated kinases are the insulin-activated Raf-1 kinase kinases. J Biol Chem 1992; 267:1088.
119. Hume CR, Accolla RS, Lee JS. Defective HLA class II expression in a regulatory mutant is partially complemented by activated ras oncogens. Proc Natl Acad Sci USA 1987; 84:8603.
120. Mohri M, Spriggs DR. Kufe D. Effects of lipopolysaccharide on phospholipase A_2 activity and tumor necrosis factor expression in HL-60 cells. J Immunol 1990; 144:2678.
121. Lin LL, Wartmann M, Lin AY, Knopf JL, Seth A, Davis RG. Cytosolic PLA_2 is activated by MAP kinase. Cell 1993; 72:269.

41

Legionella pneumophila Interferes with Signaling in Human Monocytes

T. Jacob, J. C. Escallier, M. V. Sanguedolce, P. Bongrand, C. Capo, and J. L. Mege
Hôpital de Sainte-Marguerite, Marseille, France

I. INTRODUCTION

Legionella pneumophila is the causative agent of legionnaires' disease (1). The bacteria of the genus *Legionella* are facultative intracellular pathogens that multiply within monocytes and alveolar macrophages (2,3).

The intracellular pathogens must employ mechanisms that allow them to evade to host defenses and multiply inside phagocytes (4). It is now well established that only *L. pneumophila* serogroup 1 (Philadelphia 1) is contained in phagosomes that do not become highly acidified (5) and do not fuse with lysosomes (6). In contrast to *L. pneumophila* Philadelphia, no defect in phagosome-lysosome fusion has been found in monocytes that have phagocytozed other species of *Legionella* such as *L. micdadei* or *L. pneumophila* Knoxville 1 (7). Indeed, the survival of *Legionella* within monocytes and macrophages must require other strategies to subvert phagocyte defenses. There is growing evidence that several microorganisms and parasites may interfere with the oxidative metabolism of phagocytes and thus impair their antimicrobial function (4). In the case of *Legionella,* this property may permit the intracellular proliferation, since the bacteria are sensitive to reactive oxygen intermediates (8). It has been reported that different species of *Legionella* inhibit the oxidative metabolism in neutrophils (9–11). The effect of *Legionella* on the oxidative function of monocytes and macrophages has rarely been reported, although these cells constitute their privileged hosts (7).

The study was undertaken to investigate the oxidative response of human monocytes infected with *Legionella pneumophila* Knoxville 1, a virulent strain that does not inhibit phagosome-lysosome function.

II. MATERIALS AND METHODS

A. Infection of Human Monocytes

Mononuclear cells were isolated by MSL gradient centrifugation as previously described (12) and suspended in supplemented RPMI 1640. Mononuclear cells at 2×10^6/ml were mixed with *L. pneumophila* Knoxville 1 (harvested at the log phase of growth after 36 h cultivation on charcoal-yeast extract agar) for a colony-forming unit (CFU)/cell ratio of 10:1 and incubated on a giratory shaker for 1 h at 37°C as described elsewhere (2). Then monocyte adhesion was obtained by incubating 2×10^7 cells in 100-mm Petri dishes (phosphorylation experiments) and 2×10^6 cells in flat-bottom 24-well culture plates (superoxide experiments) for 45 min at 37°C. Nonadherent cells and free bacteria were removed by washing and remaining adherent cells were designated as monocytes.

B. Superoxide Generation

Superoxide production was assayed as previously described (13). Briefly, control and infected monocytes were incubated in Hank's balanced salt solution (HBSS) containing 120 μM cytochrome *c* and 2 mM sodium azide. The reaction was carried out for 60 min in the presence of 50 ng/ml phorbol myristate acetate (PMA) or 1.2 mg/ml zymosan. The reaction was stopped by the addition of 1 mM N-ethylmaleimide. The results were expressed in nanomoles of superoxide per milligrams of protein and per hour by using an extinction coefficient of 21,000 M^{-1}/cm (14).

C. Measurement of Protein Phosphorylation by ^{32}P Radiolabeling

Protein phosphorylation was assessed by ^{32}P radiolabeling of monocytes as previously described with some modifications (15). Briefly, control and *L. pneumophila*–infected monocytes were incubated with 250 μCi orthophosphoric acid in HBSS containing 0.025% bovine serum albumin (BSA) and then stimulated with 100 ng/ml PMA. The reaction was stopped by adding HBSS containing protease and phosphatase inhibitors. The cell extract was boiled and electrophoresed through SDS-polyacrylamide (9%) gel. After electrophoresis, the gels were dried and exposed to Kodak XAR-5 x-ray films (Eastman Kodak, Rochester, NY).

D. Measurement of Cellular PKC Isoforms by Western Blotting

Control and *L. pneumophila*–infected monocytes were incubated in HBSS containing protease and phosphatase inhibitors. After SDS-PAGE electrophoresis and transfer of cell extract to nitrocellulose sheets, blots were washed with phosphate-buffered saline (PBS) containing 0.05% Tween-20 (16). They were then incubated with 1:500 dilution of each monoclonal antibody MC3a, MC2a, and MC1a (Seikagaku, Rockville, MA), respectively, directed to protein kinase C (PKC) α, β, and γ isotypes overnight at room temperature and under shaking. After washing, horseradish peroxidase–conjugated antibodies to mouse immunoglobulins (1:20000; Amersham, Les Ulis, France) were incubated with nitrocellulose for 60 min at room temperature in PBS containing 0.05% Tween-20. PKC isoforms were revealed with a detection kit (ECL, Amersham).

III. RESULTS

A. Effect of *L. pneumophila* on Stimulated Superoxide Generation

Monocytes were infected with *L. pneumophila* and cultivated at 37°C for 48 h, with time corresponding to the maximal intracellular proliferation of *L. pneumophila*. Infected monocytes did not release superoxide in the absence of agonists. When they were stimulated with 50 ng/ml PMA, superoxide generation was reduced to 40% of control values. In contrast, zymosan-induced superoxide was similar in infected monocytes and control cells. It is noteworthy that monocytes having ingested heat-inactivated *L. pneumophila* exhibited the same oxidative response as did control monocytes (Table 1).

Table 1 Effect of *L. pneumophila* on Superoxide Generation

Stimulation	None	Zymosan	PMA
Control cells	15 ± 3	153 ± 17	296 ± 38
Cells infected with *L. pneumophila*	17 ± 3	139 ± 15	118 ± 22
Cells infected with heat-inactivated *L. pneumophila*	16 ± 5	145 ± 11	277 ± 35

Mononuclear cells were incubated with *L. pneumophila* (LP) or heat-inactivated LP (CFU/cell ratio 10:1) for 1 h at 37°C. Once adherent, monocytes having ingested LP and control monocytes were cultivated for 48 h in supplemented RPMI 1640. After washing, cells were stimulated with PMA (50 ng/ml), zymosan (1.2 mg/ml), or 0.01% dimethylsulfoxide for 60 min at 37°C in HBSS containing 120 μM cytochrome *c* and 2 mM sodium azide. Superoxide generation was measured by superoxide dismutase (SOD)-inhibitable ferricytochrome *c* reduction and the results are expressed in nanomoles of superoxide per milligram protein and per hour. The values are the means ±SE of 10 experiments.

The inhibition of PMA-stimulated oxidative response in *L. pneumophila*-infected monocytes was evident after 8 h of infection and reached a maximal value between 24 and 48 h (data not shown). During all this period of time, infected monocytes did not spontaneously release superoxide.

B. Effect of *L. pneumophila* on the Phosphorylation of Monocyte Endogenous Substrates

PMA at 100 ng/ml stimulated the increase in the phosphorylation levels of several substrates in control monocytes. They mainly included 34-, 47-, 62-, 68-, 80-, and 90-kd phosphoproteins (Fig. 1, lanes 1 and 2). When infected monocytes were stimulated by 100 ng/ml PMA for 15 min, a significant decrease (60%) in the phosphorylation levels of 34-, 47-, 62-, 68-, and 80-kd proteins was found (Fig. 1, lane 4). When monocytes were incubated with heat-inactivated *L. pneumophila,* we found the same pattern of protein phosphorylations as in control cells (Fig. 1,

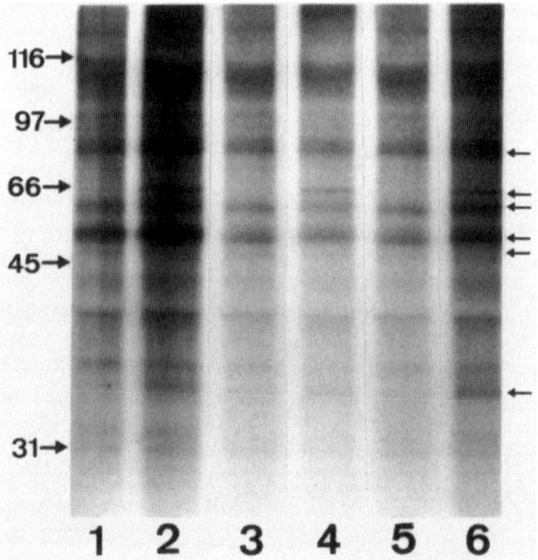

Figure 1 Effect of *L. pneumophila* on the phosphorylation of endogenous substrates. Control monocytes (lanes 1 and 2) and monocytes having ingested *L. pneumophila* (lanes 3 and 4) or heat-inactivated *L. pneumophila* (lanes 5 and 6) were cultivated for 48 h in supplemented RPMI 1640. After washing, they were incubated with 250 μCi orthophosphoric acid, washed, and stimulated with PMA (lanes 2, 4, and 6) or 0.01% dimethylsulfoxide as vehicle (lanes 1, 3, and 5) for 15 min at 37°C. Cell extracts were electrophoresed through SDS-polyacrylamide (9%) gels. The gels were exposed to Kodak XAR-5 x-ray films (Eastman Kodak, Rochester, NY). Molecular masses were determined by using Bio-Rad (Richmond, CA) standards (in left margin) and stimulated phosphoproteins were indicated with arrows in the right margin. A representative autoradiogram of five distinct experiments is shown.

lanes 5 and 6). Moreover, the inhibition of superoxide generation and protein phosphorylations in response to PMA exhibit a similar time course (data not shown).

C. Downmodulation of PKC Isotypes by *L. pneumophila*

MC2a monoclonal antibody, which recognizes PKC β, labeled a band migrating as 81–82 kd in control monocytes cultivated for 48 h (Fig. 2, lane 2). MC3a monoclonal antibody directed against PKC α labeled a faint band migrating as 82 kd (Fig. 2, lane 1). In contrast, MC1a monoclonal antibody–recognizing PKC γ did not detect any substrate in control monocytes (Fig. 2, lane 3). In *L. pneumophila*–infected monocytes cultivated for 48 h, there was a partial decrease (30 ± 10%) of PKC α expression (Fig. 2, lane 4) and a more pronounced downmodulation (55 ± 6% inhibition) of PKC β expression (Fig. 2, lane 5) as compared with control monocytes. The decreased expression of both PKC isotypes started after 8 h of monocyte culture with *L. pneumophila* and reached a maximal inhibition between 24 and 48 h (data not shown). It is noteworthy that the monocytes having ingested heat-inactivated *L. pneumophila* did not exhibit any decrease in PKC isoform expression (data not shown).

IV. DISCUSSION

We showed in this chapter that the infection of human monocytes with *L. pneumophila* Knoxville 1 resulted in the inhibition of superoxide generation stimulated by PMA but not by a particulate agonist, zymosan.

Different mechanisms may account for the selective limitation of the superoxide generation. They include a direct effect of *L. pneumophila* on reactive oxygen

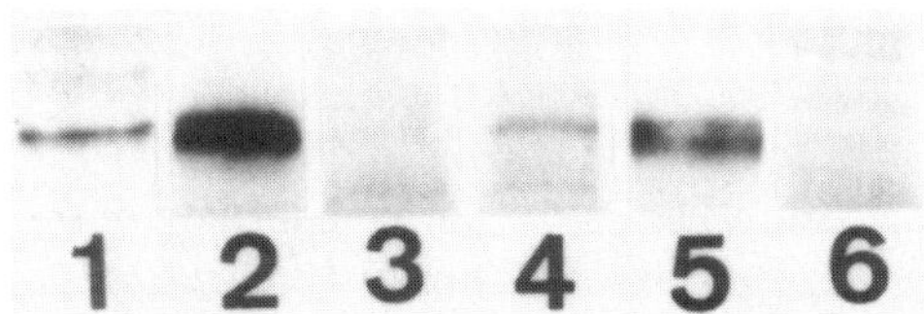

Figure 2 Effect of *L. pneumophila* on the expression of PKC isotypes. Control monocytes (lanes 1–3) and monocytes having ingested *L. pneumophila* (lanes 4–6) were cultivated for 48 h at 37°C and then in HBSS containing vanadate (1 mg/ml) and protease inhibitors before being scraped. Cell extracts were electrophoresed through SDS-polyacrylamide (9%) gel and transferred onto nitrocellulose sheet. PKC isotypes were assessed by using mouse monoclonal antibodies directed to PKC α (lanes 1 and 4), PKC β (lanes 2 and 5), and PKC γ (lanes 3 and 6), and horseradish peroxidase–conjugated antibodies to mouse immunoglobulins. Western blots were revealed with the ECL detection kit. A representative luminogram of five experiments is shown.

intermediates, a change in NADPH oxidase activity, and a deficiency in the signaling of monocytes leading to superoxide. Although the different strains of *Legionella* possess the enzymes that scavenge reduced oxygen metabolites (17), the maintain of zymosan-stimulated superoxide in infected monocytes as well as the inability of supernatants from these monocytes to inhibit superoxide (data not shown) provided a clearcut evidence against the scavenging hypothesis. We also provided evidence that *L. pneumophila* did not affect the NADPH oxidase assembly but interfered with some transductional pathways leading to the activation of the enzyme complex. As PMA binds directly to the regulatory domain of PKC, we hypothesized that the defect in PMA-stimulated superoxide generation would be located at the level of PKC or some downstream steps. The study of cellular phosphorylations confirmed the first hypothesis. The inhibition of cellular phosphorylations have been already described in macrophages infected with *Leishmania donovani* or treated with its virulence factor, lipophosphoglycan (LPG) (18).

The alteration of PMA-stimulated protein phosphorylations might result from either a deficiency in PKC levels or a defect in its translocation. Infectious agents have been reported to affect PKC in different manners. In *L. pneumophila*-infected monocytes, we noted a decrease in the expression of both PKC isoforms. The time course of the downmodulation of PKC was superimposable to the decrease of protein phosphorylations and superoxide generation; thus suggesting these events are tightly related. This mechanism was clearly distinct from the inhibition of protein phosphorylations in *Leishmania donovani*-infected monocytes where the expression of PKC α and β isoforms was not affected (18).

Several putative mechanisms may account for the downmodulation of PKC expression. They include either an impairment of PKC synthesis or a degradation of PKC. We favored the second hypothesis. It is well established that the long-term treatment of several cell types with PKC-activating agonists such as phorbol esters leads to a PKC depletion. It is likely that some virulence factors at the surface of *L. pneumophila* such as lipopolysaccharide (LPS), which is known to activate PKC (19), may induce the downmodulation of PKC.

In conclusion, we showed here that the infection of monocytes with *L. pneumophila* Knoxville 1 resulted in the selective inhibition of PMA-stimulated oxidative metabolism. This deficiency was related to the downmodulation of the expression of α and β PKC isotypes. The responsible factor in *L. pneumophila* remains still speculative and its knowledge will be useful for the development of specific therapeutic strategies.

V. CONCLUSIONS

L. pneumophila may subvert monocyte defenses via several mechanisms, including the inhibition of phagosome-lysosome fusion or the impairment of the oxidative metabolism. We have investigated the effect of *L. pneumophila* Knoxville 1, a virulent strain that does not inhibit phagosome-lysosome fusion, on the oxidative responsiveness of human monocytes. We showed that the infection of monocytes with *L. pneumophila* for 48 h resulted in the marked inhibition of superoxide generation stimulated by PMA but not by a particulate agonist, zymosan. The oxidative deficiency was evident after 8 h of infection, reached a maximal inhibition between 24 and 48 h, and involved all infected cells. In contrast, heat-inactivated *L. pneumophila* were ingested by monocytes but did not downmodulate PMA-stimulated superoxide generation. We provided evidence that *L. pneumophila* interfered with the transductional pathway (i.e., protein kinase C) leading to the activation of the NADPH oxidase in monocytes. The phosphorylation of 34-, 47-, 62-, and 80-kd proteins stimulated by PMA was markedly inhibited in infected monocytes. The phosphorylations on serine/threonine residues were decreased but not those on tyrosine residues. In addition, the expression of both α and β PKC isotypes was partially inhibited in infected monocytes. Taken together, our data suggest that the downmodulation of PKC isotypes provides the means for understanding the inhibition of PMA-stimulated superoxide generation and may contribute to the virulence of *L. pneumophila* Knoxville 1.

REFERENCES

1. Fraser DW, Tsai TR, Orenstein W, Pardin WE, Beecham HJ, Sharrar RG, Harris J, Mallison GF, Martin SM, McDade E, Shepard CC, Brachman PS. The Field Investigation Team Legionnaires' disease: description of an epidemic of pneumonia. N Engl J Med 1977; 297:1189.
2. Horwitz MA, Silverstein SC. Legionnaires' disease bacterium (*Legionella pneumophila*) multiplies intracellularly in human monocytes. J Clin Invest 1980; 66:441.
3. Nash TW, Libby DW, Horwitz MA. Interaction between the legionnaires' disease bacterium (*Legionella pneumophila*) and human alveolar macrophages. Influence of antibody, lymphokines and hydrocortisone. J Clin Invest 1984; 74:771.
4. Speert DP. Macrophages in bacterial infection. In: Lewis CE, McGee JOD, eds. The macrophage. Oxford, England: University Press, 1992:216.
5. Horwitz MA, Maxfield FR. *Legionella pneumophila* inhibits acidification of its phagosomes in human monocytes. J Cell Biol 1984; 99:1936.
6. Horwitz MA. The legionnaires' disease bacterium (*Legionella pneumophila*) inhibits phagosome-lysosome fusion in human monocytes. J Exp Med 1983; 158:2108.
7. Dowling JN, Saha AK, Glew RH. Virulence factors of the family *Legionellaceae*. Microbiol Rev 1992; 56:32.
8. Locksley RM, Jacobs RF, Wilson CB, Weaver WM, Klebanoff SM. Susceptibility of *Legionella pneumophila* to oxygen microbicidal systems. J Immunol 1982; 129:2192.
9. Friedman RL, Lochner JE, Bigley RH, Iglewski BH. The effects of *Legionella pneumophila* toxin on oxidative processes and bacterial killing on human polymorphonuclear leukocytes. J Infect Dis 1982; 146:328.
10. Saha AK, Dowling JN, Lamarco KL, Das S, Remaley, AT, Olomu N, Pope MT,

Glew RH. Properties of an acid phosphatase from *Legionella micdadei* which blocks superoxide anion production by human neutrophils. Arch Biochem Biophys 1985; 243:150.

11. Donowitz GR, Reardon I, Dowling JN, Rubin L, Fochr D. Ingestion of *Legionella micdadei* inhibits human neutrophil function. Infect Immun 1990; 58:3307.
12. Sanguedolce MV, Capo C, Bongrand P, Mege JL. Zymosan-stimulated tumor necrosis factor-α production by human monocytes. Down-modulation by phorbol ester. J Immunol 1992; 148:2229.
13. Andre P, Capo C, Fossat C, Bongrand P, Mege JL. Effect of botulinum D toxin on human neutrophilic leukocytes and localization of its substrate. Membrane Biochem 1991; 9:203.
14. Andre P, Capo C, Mege JL, Benoliel AM, Bongrand P. Zymosan but not phorbol myristate acetate induces an oxidative burst in rat bone marrow-derived macrophages. Biochem Biophys Res Commun 1988; 151:641.
15. Gomez-Cambronero J, Mege JL, Molski TFP, Naccache PH, Becker EL, Sha'afi RI. Actions of the protease inhibitor phenylmethylsulfonyl fluoride on granule enzyme secretion and superoxide production induced by fMet-Leu-Phe and phorbol-12-myristate-13-acetate. Int Archiv Allergy Appl Immunol 1989; 89:362.
16. Sanguedolce MV, Capo C, Bouhamdan M, Bongrand P, Huang CK, Mege JL. Zymosan-induced tyrosine phosphorylations in human monocytes: role of protein kinase C. J Immunol 1993; 151:405.
17. Pine L, Hoffman PS, Malcolm GB, Benson RF, Keen MG. Determination of catalase, peroxidase and peroxide dismutase within the genus *Legionella*. J Clin Microbiol 1984; 20:421.
18. Olivier M, Brownsey RW, Reiner NE. Defective stimulus-response coupling in human monocytes infected with *Leishmania donovani* is associated with altered activation and translocation of protein kinase C. Proc Natl Acad Sci USA 1992; 89:7481.
19. Mege JL. Role of protein kinases in macrophage functions. Huang CK, Sha'afi RI, eds. Protein kinases in blood cell functions. Boca Raton, FL: CRC press, 1993:117.

42

Regulatory Areas of the Central Nervous System for the Immune Response

K. Mašek
Academy of Sciences of Czech Republic, Prague, Czech Republic

P. Petrovický
Charles University, Prague, Czech Republic

I. INTRODUCTION

Within the past years, anatomical, physiological, and pharmacological evidence has accumulated to support the concept of a close interaction and bidirectional communication between the central nervous system (CNS) and the immune systems (1–3).

The best understood mechanism by which the central nervous system can influence the immune response involves a number of hormones that are under the control of the hypothalamopituitary axis (4). Another mechanism by which the brain can influence the immune system is via the autonomic nervous system. This system is known to secrete not only classic neurotransmitters such as norepinephrine (noradrenaline) and acetylcholine but also a variety of peptides (2,5). On the other hand, much less is known about the specific areas of brain that might be involved in the immune response. Many studies that investigated the involvement of different brain areas in the immune response have utilized the method of the stereotaxic lesion of different areas of brain followed by measurement of a variety of immunological reactions. The area studied most frequently was the anterior hypothalamus/preoptic area. The lesions in this area in most cases resulted in a decrease of a variety of immune parameters (6,7). The other areas of the CNS that have been studied with the aid of stereotaxic lesions were subcortical structures such as the hippocampus or amygdala, or structures located in the brain stem in different areas. Depending on the place of the lesion, the immune response was either enhanced or inhibited (8–10). In some of the literature, however, the exact site of the lesion was not made clear and also the anatomical structure that was lesioned was not mentioned. In addition, different methods for the evaluation of the immune response were used. Thus, it is not easy to make a clear conclusion as to which areas of the brain or anatomical structures might be important in relation to the followed immune response. It also is important to keep in mind that in some of these experiments with extensive lesions a number of brain structures and nerve pathways might be destroyed.

The possible involvement of different brain regions in the immune response, particularly after the administration of immunostimulators, was the subject of our research for more than a decade. With the aid of small electrolytic lesions placed in different specific brain regions, we have localized a number of structures that might be involved in the immune response. Two methods that evaluate the effect on immune response were employed: a method of delayed skin hypersensitivity (DSH) and a method that investigates the effect on proliferative processes in immunocompetent organs with labeled precursor [^{14}C]thymidine (11,12). To detect the neuronal pathways and connections between lesioned structures, we have used the horseradish peroxidase or dextra-fer (D-F) technique (13).

Our results clearly indicate that the lesions placed in the more rostral parts of the reticular formation (RF) always increased the delayed type of hypersensitivity to the administered immunostimulator muramyl dipeptide (MDP). The caudal lesions, on the other hand, had the opposite effect and suppressed this type of reaction. The effective rostral lesions of the reticular formation were located in its rapheal part, that is, in the nucleus raphe dorsalis, nucleus linearis caudalis, and area ventralis tegmenti, Tsai (B_{6-8}). It is important to stress that in our experiments the lesion placed in the raphe system caudally did not influence the delayed skin hypersensitivity reaction. On the contrary, the caudal lesions with the opposite effect were located more laterally in the lateral part of the reticular formation, that is, in the nucleus parvocellularis, nucleus parabrachialis, and locus coeruleus, and more caudally coincides with the catecholaminergic cell groups A_{1-5} (11,13). It is important to point out that the number of lesions placed in the other areas of brain stem had no influence on the effect.

Besides the structures in the brain stem, we have also localized different structures in the amygdaloid complex belonging to the limbic system and also structures in the septum. The septum, which is a medial part of telencephalon, is connected with limbic and olfactory systems.

The lesions that diminished the immune response (DHS) were located in the dorsal and posterior parts of lateral septum (nucleus septalis lateralis pars intermedia and dorsalis and nucleus septohippocampal). The lesions that were placed in the ventral and anterior parts of the medial septum (nucleus septalis medialis, nucleus of diagonal band of Broca) had the effect opposite. The number of other lesions in the septum had no effect on DHS (14).

In the amygdala, we have observed that the lesions of the anteriorly located nucleus centralis amygdala led to the suppression of DHS and the lesions of the posteriorly located basomedial nucleus of the amygdala enhanced the DHS effect. The lesions of other nuclei of the amygdala or in surrounding areas of the amygdaloid complex were much less effective or had no effect (10).

On the basis of our results as well as the results reported by the other groups (15,16), we have also investigated the possibility of the participation of structures located in the medial frontal interhemispheric cortex. We have investigated these regions mainly because of described existing neuronal connections between the nucleus, amygdala, structures of the septum, and medial frontal cortex (14,17). The lesions were placed in different areas, but the only effective lesions were those placed in the cingulate cortex areas 1 and 2 (3,14).

The effect of lesions in the areas of the brain described above on the stimulatory effect of MDP on DNA synthesis in immunocompetent organs measured through the effect on the utilization of [^{3}H]thymidine principally led to the similar results as with the method of DHS.

Using the method of small electrolytic lesions placed in the different areas of the CNS, we could thus localize the areas that are evidently somehow involved in the immune response. The structures that we have found to be so far involved in the immune response are:

1. Catecholaminergic cell groups A_{1-7} and lateral parts of reticular formation
2. Serotonergic cell groups B_{6-8} and the rapheal part of reticular formation
3. Nucleus parabrachialis, a structure known to participate in the transmission of gustatory and visceral afferents as well as in emotional behavior and sleep regulation
4. Amygdaloid complex, belonging to the limbic system, which regulates vegetative functions and emotions
5. The septum, a medial part of the telencephalon, which is connected with the limbic and olfactory systems
6. Medial frontal cingulate cortex, particularly areas 1 and 2.

On the basis of our studies, we can suggest the existence of three circuits that seem to be involved in the immune response. One circuit represents the reticular formation and its catecholaminergic cell groups A_{1-7}, nucleus parabrachialis, and central nucleus amygdala. These structures are connected mainly with descending neuronal pathways. A lesion in this circuit decreases the immune response. The second circuit represents the rapheal reticular formation with its serotonergic cell groups B_{6-8}, hypothalamus, and basomedial amygdala nucleus. These structures are connected mainly with ascending neuronal pathways. Lesions in this circuit increase the immune response. The third circuit, which is the highest circuit, is represented by limbic telencephalic structures such as the medial frontal cortex and the septum and by the amygdala nucleus. Lesions in these mutually interconnected structures modulate the immune response. Figure 1 presents a schematic diagram showing the connections among the structures with lesions that are effective on the immune response in our experiments.

Of course, the lesion experiments do not allow us to make a clear conclusion about the exact role of these areas in the intact brain during the immune response. Nevertheless, with this method, one can locate the important structures or "effec-

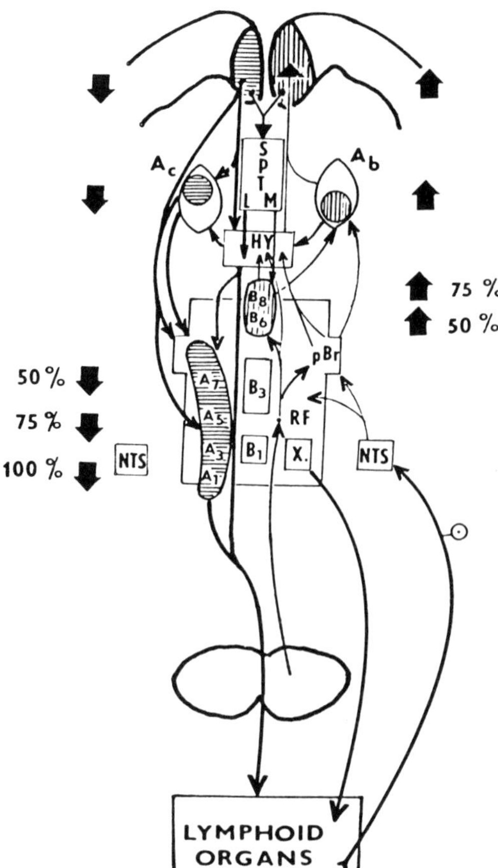

Figure 1 A schematic diagram demonstrating the connection among the structures lesioned in our experiments. The lesions effective on the immune response are depicted by shaded areas and are expressed as a percentage of controls.

tive areas" and circuits that might be important in the CNS for immune response. These can be further investigated by special neuroimmunopharmacological methods.

II. CONCLUSIONS

With the aid of small electrolytic lesions placed in different areas of the brain, the effect on the immune response can be measured by a method of delayed skin hypersensitivity (DHS) and the utilization of [^{14}C]thymidine in the immunocompetent organs. On the basis of the obtained results with the lesion and neuronal interconnection detected by the horseradish peroxidase and dextra-fer methods, the hypothesis was formulated suggesting the possible existence of three brain circuits. One circuit represents the reticular formation and its catecholaminergic groups A_{1-7}, the nucleus parabrachialis, and the central amygdala nucleus. These structures are interconnected mainly with descending pathways. Lesions in this circuit decrease the immune response. The second circuit represents the rapheal RF with its serotonergic cell groups B_{6-8}, the hypothalamus, and the basomedial amygdala nucleus. These structures are interconnected mainly with ascending pathways. Lesions in this circuit increase the immune response. The third—highest—circuit is represented by limbic telencephalic structures such as the medial frontal cortex and the septum and by the amygdala nucleus. Lesions in these mutually interconnected structures modulate the immune response.

REFERENCES

1. Weihe E, Nohr D, Michel S, Mueller S, Zentel HJ, Fink T, Krekel J. Molecular anatomy of the neuroimmune connections. Int J Neurosci 1991; 59:1–24.
2. Bellinger DL, Lorton D, Felten SY, Felten DL. Innervation of lymphoid organs and implications in development, aging and autoimmunity. Int J Immunopharmacol 1992; 14:329–44.
3. Mašek K, Petrovický P, Seifert J. An introduction to the possible role of central nervous system structures in neuroendocrine immune system interaction. Int J Immunopharmacol 1992; 14:317–22.
4. Besedovsky HO, Del Rey A, Sorkin E. Immune-neuroendocrine interactions. J Immunol 1985; 135:750–4.
5. Bellinger DL, Lorton D, Romano T, Olschowka JA, Felten SY, Felten DL. Neuropeptide innervation of lymphoid organs. Ann NY Acad Sci 1990; 594:17–33.
6. Stein M, Schiavi RC, Camerino M. Influence of brain and behaviour on the immune system. Science 1976; 191:435–40.
7. Roszman TL, Brooks WH. Neural modulation of immune function. J Neuroimmunol 1985; 10:59–69.
8. Isakovič K, Jankovič BD. Neuro-endocrine correlates of immune response. II. Changes in the lymphatic organs of brain-lesioned rats. Int Arch Allergy 1973; 45:373–84.

9. Cross RJ, Brooks WH, Roszman TL, Markesberry WR. Hypothalamic immune interactions. Effect of hypophysectomy on neuroimmunomodulation. J Neurol Sci 1982; 53:557–66.
10. Kadlecová O, Mašek K, Petrovický P, Seifert J. Mostecká H. The central anatomical structures involved in the interaction between neuroendocrime and immune systems. In: Hadden JW, Mašek K, Nistico G, eds. Interactions among CNS, neuroendocrine and immune system. Rome, Milan: Pythagora Press, 1989:67–75.
11. Mašek K, Kadlecová O, Petrovický P. The involvement of brain structures in the adjuvant effect of muramyl dipeptide. Brain Res Bull 1985; 15:443–6.
12. Seifert J, Šebestová L, Mašek K. The stimulatory effect of muramyl dipeptide (MDP) on the proliferative processes in parenchymal organs of rats. Int J Immunopharmacol 1988; 10:67–72.
13. Petrovický P. Thalamic afferents fromt he brain stem. An experimental study using retrograde single and double labelling with HRP and irondextran in the rat. I. Medial and lateral reticular formation. J Hirnforschung 1990; 31:359–74.
14. Petrovický P, Mašek K. Structures and connections in CNS which influence the immune response. Brain Res Bull 1993;
15. Renoux G, Biziere K, Renoux M, Gullamin JM, Degenne D. A balanced brain asymmetry modulates T cell mediated events. J Neuroimmunol 1983; 5:227–38.
16. Renoux G, Biziere K, Renoux M, Bardos P, Degenne D. Consequences of bilateral brain neocortical ablation on immuthiol-induced immunostimulation in mice. Ann NY Acad Sci 1987; 496:346–53.
17. Llamas A, Avendano C, Reinoso-Suares F. Amygdaloid projections to the motor, premotor, and prefrontal areas of the cat's cerebral raddish peroxidase. Neuroscience 1985; 15:651–7.

43

The Role of Chorionic Gonadotropin in Regulation of Mitogen- and Lymphokine-Activated Splenocytes: Immunotherapeutic Aspect in Infection

Sergei V. Shirshev and Nikolaj N. Kevorkov
Perm State Medical Institute, Institute of Ecology and Genetics of Microorganisms, Russian Academy of Sciences, Perm, Russia

I. INTRODUCTION

At present, there is a significant amount of experimental and clinical data that make it possible to discuss the neuroendocrine control of lymphoid tissue function as well as to discuss the reverse bond in this regulation via the immunocompetent cell mediators. The presence of specific hormonal receptors on lymphoid and macrophageal cells (1,2) as well as their ability to secrete protein-peptide hormones (3,4) has been proven. It has been established that interleukins (IL) can regulate the functional activity of the endocrine glands (5,6). From this point of view, chorionic gonadotropin (CG) is of a special interest, because its synthesis is associated not only with pregnancy and carcinogenesis (7) but also with the processes of antigen-dependent lymphocyte differentiation (4). Such microbes of *Streptococcus faecalis* and *Staphylococcus simulans* are known to produce CG (8) and this allows it to be regarded as a regulating factor of immune processes in the presence of infections. Consequently, we may have a better understanding of both the reproduction and oncology as well as the immune response regulation accompanying infections aggravating pregnancy and carcinogenesis by the investigation of the effects of CG on the immune system.

This study was designed to investigate the influence of CG on the ability of intact and activated splenocytes to form the humoral immune response and the possibility of its correction by recombined IL-2 and sodium diclofenac inhibiting cyclooxygenase.

II. MATERIALS AND METHODS

Female mice of the CBA strain and (CBA·C57BL/6)F_1 hybrids weighing 18–22 g were used. Chorionic gonadotropic (Profasi, Italy) in doses of 40 or 200 IU corresponding to its level, respectively, of the second to third and first trimesters of pregnancy (9) was added to mice splenocyte culture. The splenocytes (2×10^7) were preincubated in 4 ml of medium 199 for 60 min at 37°C and then concentrated (2×10^7/0.5 ml) by centrifugation. Then the cells with antigen (sheep erythrocytes—2×10^8) were intravenously transferred to syngeneic recipients that had been lethally irradiated by a 850 R (219.3 mCi/kg) dose. The effect was defined by the number of recipient spleen plaque-forming cells (PFCs) determined by the local hemolysis method in agarose gel using Jerne methods (10) on days 4 and 5.

The following T lymphocyte–activating agents have been added to splenocytes in vitro with or without CG: concanavalin A (ConA) (Calbiohem) in a dose of 20 μg/4 ml and recombinant human IL-2 (Diagnosticum) in a dose of 150 IU/4 ml, as well as cyclooxygenase inhibitor of the key enzyme of eicosanoid synthesis—diclofenac sodium ($C_{14}H_{11}Cl_2NO_2$; 2-[(2.6-dichlorophenyl)amino] benzeneacetic acid, PLIVA, Zagreb, Yugoslavia; or voltaren; diclofenac's mode of action is similar to that of indomethacin) in a dose of 0.06 mg/4 ml (11).

The cell viability after cultivation estimated by means of vital stain (trypan blue) averaged 95%.

In control experiments, the donor splenocytes were cultivated in similar conditions and then they were transferred to the recipients.

The significance of differences between two values was assessed by t (Student) and F (Fisher) criteria. The results are presented in all tables and text as means ±SE.

III. RESULTS

In the investigation, we established that the CG in the doses corresponding to the mean values of hormone concentration during the physiological pregnancy produces statistically significant suppression of the ability of the spleen lymphocytes to form an adoptive immune response. The hormone is likely to destroy the mechanisms of antigen-independent differentiation of immunocompetent cells into the PFC precursors.

It has been previously demonstrated (12,13) that the immunomodulating CG effect on the cooperative cell processes in vivo is realized by endogenous PGs (prostaglandins), with their main producers being macrophage (14). Cyclooxygenase is known to transform arachidonic acid into PG and its enzyme blockade by diclofenac results in a considerable decrease in PFCs (Table 1). Endogenous PG may be of great importance in the processes of antigen-independent differentiation of lymphocytes in PFC precursors. The combined addition of hormone and cyclooxygenase inhibitor into the spleen cell culture has revealed different mech-

Table 1 CG Influence on Ability of Intact Splenocytes to Form PFCs and Its Dependence on Cyclooxygenase Activity

		PFC/2 × 10^7 cells	
Group	Experimental influence	No. of recipients	Mean $\log_{10}$ ± SE
1	Control	40	3.237 ± 0.033
2	CG (40 IU)	8	2.893 ± 0.103[a]
3	CG (200 IU)	12	2.949 ± 0.047[b]
4	Diclofenac (0.06 mg)	11	2.859 ± 0.079[c]
5	Diclofenac + CG (40 IU)	10	3.232 ± 0.036[d]
6	Diclofenac + CG (200 IU)	10	2.684 ± 0.109[e]

[a] $F(12.4) = 17.06, P < .001; t(46) = 3.18, P < .002$ in relation to group 1.
[b] $F(12.2) = 21.02, P < .001; t(50) = 5.01, P < .001$ in relation to group 1.
[c] $F(12.3) = 27.34, P < .001; t(49) = 4.41, P < .001$ in relation to group 1.
[d] $F(8.5) = 11.77, P < .001; t(16) = 3.10, P < .01$ in relation to group 2; $F(15.1) = 18.77, P < .001; t(19) = 4.29, P < .01$ in relation to group 4.
[e] $F(12.3) = 46.88, P < .001; t(48) = 4.85, P < .001$ in relation to group 1; $F(4.3) = 5.86, P < .05; t(20) = 2.23, P < .05$ in relation to group 3.

anisms of the induction of an immunodepressive effect for low (40 IU) and high (200 IU) CG doses; for example, in the presence of inhibitor, a dose of 200 IU causes a more expressed immunosuppressive activity, whereas a dose of 40 IU does not have the ability to suppress the adoptive immune response. It can be assumed that the low-dose CG either initiates an alternative transmitter mechanism of a signal transfer or prevents diclofenac action. The high-dose CG probably interacts with a variety of target cells, some of which have CG-dependent suppression not connected with cyclooxygenase enzyme activity.

In spite of the fact that ConA used in concentration of 5 μg/ml induces the synthesis of IL-2 by T helper lymphocytes (15), its addition to the splenocyte suspension does not lead to a statistically reliable PFC increase in recipients (Table 2). Preincubation of splenocytes in the presence of ConA and CG (40 IU) abolishes the hormone immunosuppressive effect. However, this does not apply to the immunosuppressive effect of a high CG dose.

Cyclooxygenase inhibitor addition to the culture containing ConA or ConA and CG (40 and 200 IU) suppresses the adaptive immune response. Thus, both an independent immunosuppressive effect of the CG (40 IU) and the leveling effect of ConA are equally connected with the cyclooxygenase enzyme activity. In the presence of ConA or ConA and diclofenac CG (200 IU) is shown to suppress the formation of antibody producers that may be connected to IL-2 synthesis inhibition by splenocytes induced by ConA as it was described previously (12).

Table 2 CG Effect on Splenocyte Ability to Form PFCs in Conditions of Simultaneous Influence on ConA and Its Dependence on Cyclooxygenase Activity

Group	Experimental influence	PFC/2×10^7 cells	
		No. of recipients	Mean $\log_{10}$ ± SE
1	Control	18	3.261 ± 0.062
2	ConA (20 μg)	10	3.380 ± 0.101
3	ConA + CG (40 IU)	10	3.205 ± 0.048
4	ConA + CG (200 IU)	10	3.077 ± 0.081[a]
5	Control	10	3.416 ± 0.076
6	Diclofenac (0.06 mg)	11	2.859 ± 0.079[b]
7	Diclofenac + ConA + CG (40 IU)	12	2.986 ± 0.100[c]
8	Diclofenac + ConA	10	3.078 ± 0.089[d]
9	Diclofenac + ConA + CG (200 IU)	10	2.936 ± 0.106[e]

[a] $F(4.4) = 5.61, P < .05$; $t(18) = 2.34, P < .05$ in relation to group 2.
[b] $F(15.1) = 26.15, P < .001$; $t(19) = 5.08, P < .001$ in relation to group 5.
[c] $F(8.1) = 11.45, P < .01$; $t(20) = 3.42, P < .01$ in relation to group 5.
[d] $F(8.3) = 10.74, P < .01$; $t(18) = 2.88, P < .01$ in relation to group 5.
[e] $F(8.3) = 14.03, P < .01$; $t(18) = 3.68, P < .002$ in relation to group 5.

Although ConA is known to be a mitogen capable of imitating the process of T-lymphocyte antigenic activation, IL-2 contributes to the differentiation and proliferation of activated T lymphocytes (second signal) (16). Splenocyte preincubation in the presence of IL-2 does not change significantly the cellular ability to form the adoptive immune response. As with ConA, in the presence of CG (40 IU) and IL-2, splenocyte preincubation is found to decrease immunodepression produced by a low hormone dose. A marked immunostimulating action is produced by CG (200 IU) and IL-2 combination. There is no dependence of this high hormone dose effect on cyclooxygenase activity. Despite the immunostimulating action of IL-2 in the presence of cyclooxygenase inhibitor, no additive effect is observed with the combined use of CG (200 IU), IL-2, and diclofenac (Table 3).

IV. DISCUSSION

Thus, the influence of ConA or IL-2 on T cells abolishes the immunodepressive effect of a low CG dose. At the same time, the immunomodulating activity of a high CG dose depends on the type of signal-activating T cells. In the case of an afferent signal (ConA), CG (200 IU) has been found to suppress antigen-independent differentiation of cells into the PFC precursors. On efferent signal (IL-2), CG

Table 3 CG Effect on the Splenocyte Ability to Form PFC in Conditions of Simultaneous Influence of IL-2 and Its Dependence on Cyclooxygenase Activity

Group	Experimental influence	PFC/2 × 10^7 cells No. of recipients	Mean $\log_{10}$ ± SE
1	Control	40	3.237 ± 0.033
2	IL-2 (150 IU)	9	3.282 ± 0.095
3	IL-2 + CG (40 IU)	9	3.299 ± 0.110
4	IL-2 + CG (200 IU)	10	3.587 ± 0.071[a]
5	Diclofenac (0.06 mg)	11	2.859 ± 0.079[b]
6	Diclofenac + IL-2 + CG (40 IU)	7	3.321 ± 0.081[d]
7	Diclofenac + IL-2	9	3.485 ± 0.064[c]
8	Diclofenac + IL-2 + CG (200 IU)	9	3.606 ± 0.083[e]

[a] $F(12.3) = 23.9$, $P < .001$; $t(48) = 4.47$, $P < .001$ in relation to group 1.
[b] $F(12.3) = 27.34$, $P < .001$; $t(49) = 4.41$, $P < .001$ in relation to group 1.
[c] $F(7.2) = 11.57$, $P < .01$; $t(47) = 3.44$, $P < .001$ in relation to group 1; $F(15.4) = 36.51$, $P < .001$; $t(18) = 6.15$, $P < .001$ in relation to group 5.
[d] $F(8.5) = 15.69$, $P < .01$; $t(17) = 4.08$, $P < .001$ in relation to group 5.
[e] $F(12.4) = 25.33$, $P < .001$; $t(47) = 4.13$, $P < .001$ in relation to group 1; $F(4.5) = 6.68$, $P < .05$; $t(16) = 2.56$, $P < .05$ in relation to group 2; $F(15.4) = 42.49$, $P < .001$; $t(18) = 6.51$, $P < .001$ in relation to group 5.

activates the immune response. In both cases, CG (200 IU) effects do not demonstrate any dependence on the cyclooxygenase enzyme activity.

The results obtained from the present study provide a new interpretation of the previously documented data of CG activity. In particular, hormonal synthesis in antigen lymphocyte activation (4) may be considered as the mechanism increasing immunocompetent cell sensitivity to IL-2, which evidently activates expansion of antigen-activated T-cell helpers. From this point of view, CG can be regarded as a lymphokine contributing to the immune protection and particularly during formation of infection immunity.

In discussing the significance of CG as an immune reaction modulator during pregnancy, it should be pointed out that this hormone suppresses the functional activity of immunocompetent cells that have not come into contact with the antigen. Those lymphocytes that manage to recognize the antigen and go into further differentiation are selected by CG for IL-2–dependent activation. It should be noted that it is in the first trimester of pregnancy that these processes can take place, as the CG dose possessing the costimulating effect is extrapolated from this particular period. The biological significance of this phenomenon needs further investigation. A costimulating CG effect may probably provide the protective function of the organism in the perinatal infection. In addition to these theoretical conclusions, one should pay special attention to the practical value of the hormone, IL-2, and the nonsteroidal anti-inflammatory preparation diclofenac sodium for treatment of patients with urogenital infections or other infections complicating pregnancy and delivery. It is known that a low IL-2 dose does not produce an immunostimulating effect. However, in our studies, a low IL-2 dose in combination with CG (200 IU/4 ml) has produced marked immunostimulating effects. The combined use of diclofenac with IL-2 has a similar immunostimulating effect. It should also be stressed that the tested therapeutic diclofenac dose of 3 mg/kg of body mass has an independent immunodepressive action, being absent in combinations with CG, IL-2, or CG and IL 2.

Therefore, the combined use of pharmaceutic agents of differentiated action with regard to the mechanisms of their endogenous synthesis may be helpful in solving new immunotherapeutic problems and avoid certain complications.

V. CONCLUSIONS

A 1-h incubation of mouse splenic cells with CG in doses of 40 and 200 IU significantly decreases the number of PFCs, which is determined in the syngeneic transfer system. Preincubation of splenocytes with ConA or human recombinant IL-2 in vitro does not change the PFC response. The simultaneous use of ConA and CG in a 1-h preincubation abolishes immunosuppression, which is induced by a lower CG dose (40 IU), whereas a higher dose of CG (200 IU) in the presence of ConA has an expressed immunodepressive effect. The combination of IL-2 with CG leads either to abolition of the immunosuppressive effect (40 IU) or to stimulation of a more than twofold adoptive immune response (200 IU). It has been established that the immunosuppressive effect of only a lower CG dose depends on cyclooxygenase activity, but a costimulating effect of a high hormone dose in the presence of IL-2 is not connected with endogenic eicosanoid synthesis. The connection between the observations in this study of the immunomodulatory effects of CG and a resistance to infection have been discussed.

REFERENCES

1. Hadden JW, Hadden EM, Middeleton E. Lymphocyte blasttransformation. I. Demonstration of adrenergic receptors in human peripheral lymphocytes. Cell Immunol 1970; 1:583.
2. Arrenbrecht S. Specific binding of growth hormone to thymocytes. Nature 1974; 252:255.
3. Blalock JE. The immune system as a sensory organ. J Immunol 1984; 132:1067.
4. Harbour-McMenamin D, Smith EM, Blalock JE. Production of immunoreactive chorionic gonadotropin during mixed lymphocyte reactions: a possible selective mechanism for genetic diversity. Proc Natl Acad Sci USA 1986; 83:6834.

5. Uehara A, Gottschall PE, Dahl RR, Arimura A. Stimulation of ACTH release by human interleukin-1, but not by interleukin-1, in conscious, freely moving rats. Biochem Biophys Res Commun 1987; 146:1286.
6. Schettini G, Florio T, Meucci O, Landolfi E, Grimaldi M, Lombardi G, Scala G, Leong D. Interleukin-1-b modulation of prolactin secretion from rat anterior pituitary cells. Involvement of adenylate cyclase activity and calcium mobilization. Endocrinology 1990; 126:1435.
7. Surgova TM, Sidorenko MV, Vinnisky VB. Protein markers of pregnancy and cancer. Successes Mod Biol (USSR) 1989; 107:418.
8. Acevedo HF, Koide SS, Slifkin M. Coriogonadotropinlike antigen in a strain of *Streptococcus faecalis* and a strain of *Staphylococcus simulans* detection identification, and characterization. Infect Immun 1981; 31:487.
9. Wide Z. Immunological method for the assay of human chorionic gonadotropin. Acta Endocrinol (Copenh.) 1962; 70(Suppl.):1.
10. Jerne NK, Nordin AA. Plaque formation in agar by single antibody-production cells. Science 1963; 140:405.
11. Ku EC, Wasvary JM, Cash WD. Diclofenac sodium. A potent inhibitor of prostaglandin synthetase. Biochem Pharmacol 1974; 23:641.
12. Shirshev SV, Kevorkov NN, Shary NI. Influence of chorionic gonadotropin on the cooperation of spleen cells forming the primary immune response. Bull Expl Biol Med (USSR) 1987; 104:337.
13. Kevorkov NN, Shilov JI, Shirshev SV. Major reproduction hormones as regulators of cell-to-cell interactions in humoral immune responses. Brain Behav Immun 1991; 5:149.
14. Gromykhina NY, Kozlov VA. Prostaglandin as a factor regulating humoral immune response. Immunology (USSR) 1982; 5:11.
15. Klaus GGR. Lymphocytes. A practical approach. IRL Press, 1990:393.
16. Cheredeev AP, Kovalchuk LV. Cellular and molecular aspects of immune processes. Results of science and technology. Series Immunology (USSR) 1989; 19:240.

44

Complement C3 and the Immune Response

Reinhard Burger and Vera Köhler
Robert Koch-Institute, Berlin, Germany

I. INTRODUCTION

Complement proteins and their split products influence various elements and reactions within the immune system. These complement-mediated functions represent a major element of host defense and are involved in the specific and the nonspecific immune reactions (1–5). Complement fragments affect both the humoral and the cell-mediated immune reactivity. Among the various complement proteins, the component C3 has a central role because it represents an extremely versatile molecule (6–8). The various C3 fragments generated in the cause of complement activation affect cellular function via receptors present on lymphoid cells and on phagocytic cells.

II. MEMBRANE-BOUND C3 RECEPTORS

The various C3 receptors bind the C3 fragments generated through proteolytic cleavage from the circulating native molecule with a selective reactivity pattern (5,7). The cellular expression of the C3 receptors also occurs in a selective manner. CR1 (CD35): recognizing C3b; additional ligands iC3b, C3c, C4b; present on erythrocytes, B lymphocytes, T lymphocytes, monocytes, macrophages, and follicular dendritic cells, polymorphonuclear neutrophils (PMNs), eosinophils. CR2 (CD21): recognizing C3d and C3dg; additional ligand iC3b; present on B lymphocytes, thymocytes, and on cell lines of B- and T-cell origin, follicular dendritic cells. CR3 (CD11b/CD18): recognizing iC3b; additional ligand C3d; present on monocytes, macrophages, PMNs, eosinophils, follicular dendritic cells, and K cells. CR4 (CD11c/CD18): recognizing iC3b; present on monocytes, macrophages, PMNs, and K cells. CR5: recognizing C3dg, C3d; present on PMN and platelets. C3a receptor: recognizing C3a; additional ligand C4a; present on mast cells, monocytes, macrophages, T lymphocytes, PMNs, basophils, and chronic myelogenous leukemia (CML)-derived basophilic granulocytes.

III. C3 FRAGMENTS AS IMMUNOREGULATORY MEDIATORS

The biologically active C3 fragments represent a kind of endogenous immunomodulator. A heterogeneous spectrum of experimental systems were used to define the biological functions of the various C3 fragments as listed in Table 1 (1,5,7).

IV. IN VIVO C3 DEPLETION AND GENETIC DEFICIENCIES OF C3

Cobra venom factor (CVF) induces a temporary depletion of C3 in vivo. In the initial studies in the murine system from Pepys et al. (9), a markedly impaired humoral immune response to T-cell–dependent antigens was observed after previous injection of CVF. However, the contribution of C3 to the physiological immune response is difficult to analyze in this artificial system. The depletion of C3 in the circulation is associated at the same time with massive generation of the biologically active C3 fragments that might affect the functional status of a number of immunocompetent cell populations. It is difficult therefore to ascertain whether an

Table 1 Experimental Systems for Analysis of C3 Functions in the Immune Response

C3 depletion in vivo with cobra venom factor
Spontaneously occurring genetic complement deficiencies of C3 or C3-related proteins (regulatory proteins or receptors)
In vitro culture systems in the presence of antibodies to C3 or its fragments
Inhibition or activation of cellular functions in vitro in the presence of purified C3 fragments
Influence of antibodies to C3 receptors on cellular functions
Application of soluble receptors or of hybrid proteins containing the binding domains of C3 receptors

observed effect is due to the absence of C3 per se or is a consequence of cellular-activation processes. Therefore, inherited C3 deficiencies occurring as a rare event spontaneously in humans and in animals offer an important model to analyze the contribution of C3 to immune reactivity without prior artificial treatment (6,10,11).

Experimental immunization of C3-deficient patients resulted in somewhat heterogeneous findings (11–13). In some patients, a more or less normal anamnestic response to bacterial antigens was found (e.g., diphtheria toxoid). In some patients, a normal primary response was found, but the switch from immunoglobulin IgM to IgG was impaired. The formation of IgG4 and in some experiments also IgG2 was particularly affected, whereas the formation of antibodies of a different subclass was not impaired (13). The molecular basis of the genetic C3 deficiencies differs between various patients (11,48). The biological consequences might depend on the nature of the particular C3 deficiency. In general, however, there is an impairment in the antibody response in C3-deficient humans. The outcome of immunization experiments might depend strongly on the previous history of exposure to the relevant antigen and to the concentration of the antigen used for immunization. A secondary response might be less dependent on C3 and an increase in antigen concentration might be able to compensate an impairment.

In contrast to the heterogeneous C3-deficient patients, animal models with a given, genetically controlled C3 deficiency provide a defined model to analyze in more detail the role of C3 in antibody formation (10,14). A genetic C3 deficiency was identified in inbred guinea pigs and dogs, and the two systems represent important experimental models (15–18). Both the C3-deficient guinea pigs and the C3-deficient dogs were immunized with a model antigen; namely, the bacteriophage ΦX174. In both species, there was an impaired formation of antibodies observed against this well-known T-cell–dependent antigen. The impairment was most prominent with a limited amount of antigen. A reduced formation of IgM occurred and on booster injection, there was no switch from IgM to IgG and no amplification of the antibody titer occurred. The impaired immune reactivity could be partially compensated by immunization with a higher concentration of the antigen. Even immunization with complete Freund's adjuvant only normalized the antibody formation to a limited extent. An impairment also remained after a partial reconstitution with C3 in vivo (16).

V. C3 IN B-CELL ACTIVATION AND DIFFERENTIATION

A large variety of in vitro systems were applied for analysis of C3 effects in the afferent limb of the immune response. Soluble or immobilized C3 fragments were used as ligands, or alternatively antibodies to the corresponding C3 receptor were used (19–23). Cross linking of CR1 led to the production of IgM, IgG, and IgA on costimulation of B cells with limited concentration of mitogen (24). Antibodies to CR2 induced proliferation and an increase of intracellular free calcium (25,26). The biological effect of the CR2 ligand depends on the physical state of the ligand. Immobilized C3d/C3dg led to an increased proliferation of anti-IgM–preactivated B cells. In contrast, soluble C3b ligand had an inhibitory effect (7,20,27,28).

CR2 is associated with other proteins and forms a multimolecular complex (CR2/CD19/TAPA-1) that is involved in signal transduction (29). CR2 effects might be of particular importance in situations where the available antigen concentration is limited (21,30,31). A series of elegant experiments used soluble recombinant CR2 protein. In this system, a hybrid protein containing IgG1 constant region in association with the receptor-binding domains of CR2 was applied. This hybrid protein inhibited the primary antibody formation to T-cell–dependent antigens and the isotype switch to IgG (22) presumably by preventing the binding of C3d to the corresponding cellular receptor.

VI. T-CELL REACTIONS AND C3

T cells express CR2 (CD21) and CR2-related, antigenetically similar proteins (32,33). These proteins are not only expressed on peripheral T cells but also on immature thymocytes and on T-cell lines (25,34–37). The reactivity of these molecules against their physiological ligands and to antibodies differ between T and B cells. Therefore, there are obviously structural differences between the CR2 molecules on T and B lymphocytes.

A number of studies suggest a functional relevance of C3 receptors on T cells. Incubation of T helper cells with aggregated C3 and interleukin-2 (IL-2) reveals a costimulatory effect (5). C3 fragments inhibit the antigen- and mitogen-induced proliferation of T cells in vitro. The production of cytotoxic T cells is similarly inhibited as the IL-2–dependent growth (5,38). On a human T-cell line, a complex between CR1 and CR2 was found (39). Similar complexes were found on follicular dendritic cells (33). The physiological importance of CR2 expression on T cells remains to be elucidated and requires further investigation. It might be involved in regulatory processes in T-cell differentiation.

An interaction of CR2 with CD23 (low-affinity receptor for IgE) and a regulatory effect on IgE production was recently shown (40). CD23 is also present on thymocytes and on thymic epithelial cells. The interaction between CD23 and CR2 might be an additional element controlling T-cell differentiation.

VII. C3 AND MACROPHAGE FUNCTIONS

The macrophage has a central function in complement-mediated effects within nonspecific host defense. The macrophage is able to synthesize and secrete most of the complement proteins. It bears on its surface most of the complement receptors. Complement fragments are able to bind to the corresponding receptors and trigger macrophage functions that contribute to the inflammatory reaction and defense against microorganisms (41–43). C3 fragments were shown to induce the production of prostaglandins and to modulate the expression of major histocompatibility complex (MHC) II proteins. Therefore, the contribution of the macrophage to the induction of the specific immune response, namely, its antigen presenting capacity, is also affected by complement fragments.

The macrophage might be considered a kind of autocrine system: Cleavage of the secreted complement proteins in the vicinity of the cell and generation of the biologically active fragments by proteolytic enzymes (e.g., lysosomal proteases) followed by binding to the C3 receptors might lead to stimulation of the macrophage (44). Stimulatory effects were not only observed with particles bearing C3 fragments but also with soluble C3 fragments. These activation products also influence the expression of CR1 and CR2 on the cell surface and the production of potent mediators like IL-1 (45).

C3-deficient guinea pigs provide an experimental system for analysis of the role of autocrine phenomena or of endogenous C3 effects. If endogenous, secreted C3 contributes to the control of macrophage functions, an impairment of macrophage function should be found with cells from C3-deficient animals. Macrophages from normal and C3-deficient guinea pigs were tested for chemiluminescence activity after stimulation with opsonized zymosan in a luminol-dependent assay (Fig. 1). The chemiluminescence response in the cells from the C3-deficient animals was reduced about 30–40% compared with the controls. This impairment was found both with peritoneal macrophages and with spleen cells (46). In addition, the phagocytic activity of cells from the C3-deficient animals was reduced. Also, the capacity to present soluble protein antigen to T cells obtained from immunized congenic animals was impaired (47).

This impairment might be the consequence of the formation of an abnormal, functionally inactive C3 protein produced by the C3-deficient macrophages. The formation of the biologically active fragments probably requires the presence of a functional protein. A reduced formation of cleavage products of C3 might interfere with regulatory processes that control the activation state of macrophages, including the expression of C3 receptors.

C3 does obviously affect the processing and presentation of antigen. These phenomena might be of particular importance when a limited amount of antigen is available. Covalent binding of C3b to antigen reduces the antigen amount required for the induction of proliferation with human T-cell clones.

VIII. CONCLUSIONS

C3 acts apparently at different and independent sites within the immune response: C3 fragments function as stimulatory elements and growth factors, bound C3

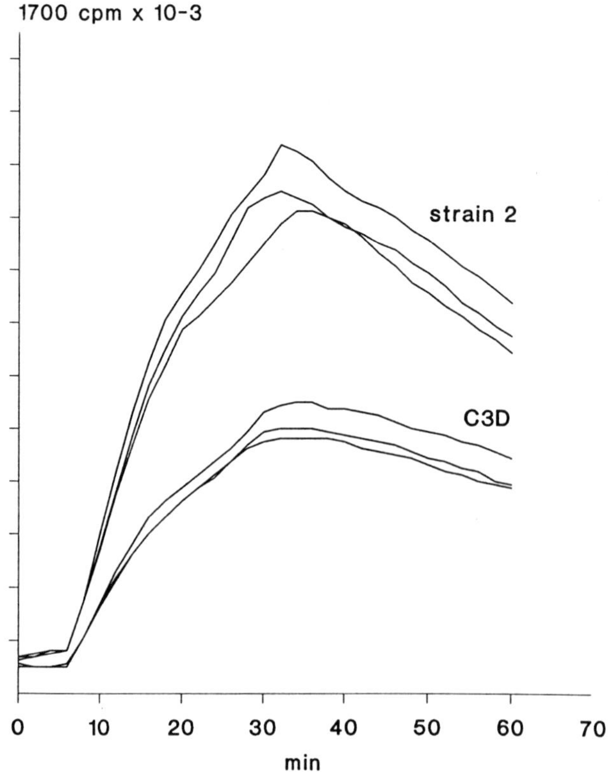

Figure 1 The capacity of macrophages for "oxidative burst" is impaired in cells from C3-deficient guinea pigs (C3D) compared with normal controls (strain 2).

increases adhesion between different lymphoid cell populations, and, finally, C3 fragments induce the release of additional immunoregulatory mediators.

REFERENCES

1. Böttger EC, Bitter-Suermann D. Complement and the regulation of humoral immune responses. Immunol Today 1987; 8:261.
2. Burger R. The complement component C3 as a mediator of the inflammatory reaction. Prog Appl Microcirc 1987; 12:108.
3. Dalmasso AP. Complement in the pathophysiology and diagnosis of human diseases. Crit Rev Clin Lab Sci 1986; 24:123.
4. Hugli TE. Structure and function of C3a anaphylatoxin. Curr Topics Microbiol Immunol 1990; 153:181.

5. Erdei A, Füst G, Gergely J. The role of C3 in immune response. Immunol Today 1991; 12:332.
6. Lambris JD (ed.). The third component of complement-chemistry and biology. Berlin: Springer, 1990.
7. Lambris JD. The multifunctional role of C3, the third component of complement. Immunol Today 1988; 9:387.
8. Müller-Eberhard HJ. Molecular organization of the complement system. Ann Rev Biochem 1988; 57:321.
9. Pepys MB. Role of complement in induction of antibody production in vivo. J Exp Med 1974; 140:127.
10. Bitter-Suermann D, Burger R. C3 deficiencies. Curr Topics Mikrobiol Immunol 1990; 153:233.
11. Botto M, Walport M. Hereditary deficiency of C3 in animals and humans. Intern Rev Immunol 1993; 10:37.
12. Alper CA, Colten HR, Gear JS, Rabson A, Rosen FS. Homozygous human C3 deficiency. The role of C3 in antibody production, C1s-induced vasopermeability and cobra venom-induced passive hemolysis. J Clin Invest 1976; 57:222.
13. Bird P, Lachman PJ. The reduction of IgG subclass production in man: low serum IgG4 in inherited deficiencies of the classical pathway of C3 activation. Eur J Immunol 1988; 18:1217.
14. Bitter-Suermann D, Burger R. Guinea pigs deficient in C2, C4 or C3 or C3a-receptor. Prog Allergy 1986; 39:134.
15. Burger R, Gordon J, Stevenson G, Ramadori G, Zanker B, Hadding U, Bitter-Suermann D. An inherited deficiency of the third component of complement C3 in guinea pigs. Eur J Immunol 1986; 16:7.
16. Böttger EC, Metzger S, Bitter-Suermann D, Stevenson G, Kleindienst S, Burger R. Impaired humoral immune response in complement C3-deficient guinea pigs: absence of secondary antibody response. Eur J Immunol 1986; 16:1231.
17. Auerbach HS, Burger R, Dodds A, Colten HR. Molecular basis of complement C3 deficiency in guinea pigs. J Clin Invest 1990; 86:96.
18. O'Neil KM, Ochs HD, Heller SR, Cork LC, Morris JM, Winkelstein JA. Role of C3 in humoral immunity; defective antibody production in C3 deficient dogs. J Immunol 1988; 140:1939.
19. Weigle WO, Goodman MG, Morgan EL, Hugli TE. Regulation of immune response by components of the complement cascade and their activated fragments. Springer Semin Immunopathol 1983; 6:173.
20. Melchers F, Erdei A, Schulz T, Dierich MP. Growth control of activated, synchronized murine B cells by the C3d fragment of human complement. Nature 1985; 317:264.
21. Heyman B, Wiersma EJ, Kinoshita T. In vivo inhibition of the antibody response by a complement receptor-specific monoclonal antibody. J Exp Med 1990; 172:665.
22. Hebell T, Ahearn JM, Fearon DT. Suppression of the immune response by a soluble complement receptor of B lymphocytes. Science 1991; 254:102.
23. Wiersma EJ, Kinoshita T, Heyman B. Inhibition of immunological memory and T-independent humoral responses by monoclonal antibodies specific for murine complement receptors. Eur J Immunol 1991; 21:2501.

24. Daha MR, Bloem AC, Ballieux RE. Immunoglobulin production by human peripheral lymphocytes induced by anti-C3 receptor antibodies. J Immunol 1984; 132:1197.
25. Cooper NR, Moore MD, Nemerow GR. Immunobiology of CR2, the B lymphocyte receptor for Epstein-Barr virus and the C3d complement fragment. Ann Rev Immunol 1988; 6:85.
26. Tsokos GC, Lambris JD, Finkelman FD, Anastassiou ED, June CH. Monovalent ligands of complement receptor 2 inhibit whereas polyvalent ligands enhance anti-Ig-induced human B cell intracytoplasmic free calcium concentration. J Immunol 1990; 144:640.
27. Erdei A, Melchers F, Schulz T, Dierich MP. The action of human C3 in soluble or cross-linked form with resting and activated murine B lymphocytes. Eur J Immunol 1985; 15:184.
28. Esparza I, Becherer JD, Alsenz J, De La Hera A, Lao Z, Tsoukas CD, Lambris JD. Evidence for multiple sites of interaction in C3 for complement receptor type 2 (C3d/EBV receptor, CD21). Eur J Immunol 1991; 21:2829.
29. Matsumoto AK, Kopicky-Burd J, Carter RH, Tuveson DA, Tedder TF, Fearon DT. Intersection of the complement and immune system: a signal transduction complex of the B lymphocyte–containing complement receptor type 2 and CD 19. J Exp Med 1991; 173:55.
30. Carter RH, Fearon DT. CD19: lowering the threshold for antigen receptor stimulation of B lymphocytes. Science 1992; 256:105.
31. Thyphronitis G, Kinoshita T, Inoue K, Schweinle JE, Tsokos GC, Metcalf ES, Finkelmann FD, Balow JE. Modulation of mouse complement receptors 1 and 2 suppresses antibody responses in vivo. J Immunol 1991; 147:224.
32. Tsoukas CD, Lambris JD. Expression of CR2/EBV receptors on human thymocytes detected by monoclonal antibodies. Eur J Immunol 1988; 18:1299.
33. Tsoukas CD, Lambris JD. Expression of EBV/C3d recpetors on T cells: biological significance. Immunol Today 1993; 14:56.
34. Toben HR, Smith RG. T-lymphocytes bearing complement receptors in a patient with chronic lymphocytic leukaemia. Clin Exp Immunol 1977; 27:292.
35. Sauvageau G, Stocco R, Kasparian S, Menezes J. Epstein-Barr virus receptor expression on human CD8+ (cytotoxic/suppressor) T lymphocytes. J Gen Virol 1990; 71:379.
36. Fischer E, Delibrias CC, Kazatchkine MD. Expression of CR2 (the C3dg/EBV receptor, CD21) on normal human peripheral blood T-lymphocytes. J Immunol 1991; 146:865.
37. Delibrias CC, Fischer E, Bismuth G, Kazatchkine MD. Expression, molecular association, and functions of C3 complement receptors CR1 (CD35) and CR2 (CD21) on the human T cell line HPB-ALL. J Immunol 1992; 149:768.
38. Meuth JL, Morgan EL, DiScipio RG, Hugli TE. Suppression of T lymphocyte functions by human C3 fragments. I. Inhibition of human T cell proliferative responses by Kallikrein cleavage fragment of human iC3b. J Immunol 1983; 130:2605.
39. Tuveson DA, Ahearn JM, Matsumoto AK, Fearon DT. Molecular interactions of complement receptors on B lymphocytes: a CR1/CR2 complex distinct from the CR2/CD19 complex. J Exp Med 1991; 173:1083.

40. Aubry JP, Pochon S, Graber P, Jansen KU, Bonnefoy JY. CD21 is a ligand for CD23 and regulates IgE production. Nature 1992; 358:505.
41. Anderson DC, Springer TA. Leucocyte adhesion deficiency: an inherited defect in the Mac-1, LFA-1, and p150,95 glycoproteins. Annu Rev Med 1987; 38:1975.
42. Graham IL, Gresham HD, Brown EJ. An immobile subset of plasma membrane CD11b/CD18 (Mac-1) is involved in phagocytosis of targets recognized by multiple receptors. J Immunol 1989; 142:2352.
43. Hartung HP, Hadding U. Synthesis of complement by macrophages and modulation of their functions through complement activation. Springer Semin Immunopathol 1983; 6:283.
44. Maison CM, Villiers CL, Colomb MG. Secretion, cleavage and binding of complement C3 by the human monocytic cell line U937. Biochem J 1989; 262:407.
45. Haeffner-Cavaillon N, Cavaillon JM, Laude M, Kazatchkine MD. C3a (C3a-desArg) induces production and release of interleukin 1 by cultured human monocytes. J Immunol 1987; 139:794.
46. Köhler V, Schäfer H, Burger R. Impaired macrophage function in C3-deficient guinea pigs (abstr). Compl Inflamm 1991; 8:175.
47. Erdei A, Köhler V, Schäfer H, Burger R. Macrophage-bound C3 fragments as adhesion molecules modulate presentation of exogenous antigens. Immunobiology 1992; 185:314.
48. Figueroa JE, Densen P. Infectious diseases associated with complement deficiencies. J Immunol 1991; 4:359.

45

Influence of High Doses of C1 Esterase Inhibitor Concentrate on Hemolytic Function of Complement and on Complement Activation Products in Capillary Leakage Syndrome

W. Nürnberger, K. Petrik, I. Michelmann, St. Burdach, and U. Göbel
Heinrich Heine University Medical Center, Düsseldorf, Germany

I. INTRODUCTION

A. Sepsis, Capillary Leakage Syndrome, and Complement Activation

The hemodynamic changes in the sepsis syndrome in humans have been studied in detail (1,2). However, the pathogenic mechanisms leading from uncomplicated bacteremia to the circulatory disturbance of the sepsis syndrome are less well understood. Both experimental and clinical findings support the concept that the balance between activation and inhibition of the Hageman factor (FXII)–dependent kallikrein-kinin pathway and of the complement system influences the course of systemic sepsis and thus the final outcome in patient with severe infections (2–4). Side effects of high-dose interleukin-2 (IL-2) therapy, such as low blood pressure and generalized edema, are paralleled by activation of FXII and prekallikrein (5) and of the complement system (6). Intravenous IL-2 therapy has been understood as a model for the sepsis syndrome (6). The typical side effects of intravenous IL-2 infusion were described as the capillary (or vascular) leakage syndrome (CLS) (7), with the following symptoms:

1. Weight gain and generalized edema by fluid retention; pleural and/or pericardial fluids, ascites

2. Tachycardia and/or low blood pressure, (pre-)renal failure, nonresponsiveness to furosemide
3. Hypoproteinemia, low sodium blood concentration

Plasma levels of C1 esterase inhibitor (C1 INH), roughly within normal range in uncomplicated sepsis, were often decreased in the sepsis syndrome—the most pronounced in patients who do not survive (4). Additionally, high levels of the anaphylatoxin C3a (as a parameter of complement activation) were found in severe sepsis. Patients in shock had significantly higher levels than patients not in shock, and patients who subsequently died had significantly higher levels than survivors (8). These findings indicate that complement activation is probably involved in the development of the sepsis syndrome and may contribute to fatal outcome.

B. Complement Activation, C1 INH, and CLS After Bone Marrow Transplantation

A number of oncological and hematological diseases may be cured or at least positively influenced by bone marrow transplantation (BMT). Following conditioning therapy (high-dose chemotherapy with or without total body irradiation) autologous or allogenic bone marrow (stem cells) is transfused. During the first 2–3 weeks after BMT, the patients are at risk for several problems:

1. Aplasia can cause bleeding and infections
2. Nonmyeloic toxicity, such as mucositis, endotheliitis, renal failure, and liver failure, in its most severe form as veno-occlusive disease,
3. Early onset of acute graft-versus-host disease

The clinical syndrome is usually characterized by CLS; the typical period for development of CLS is day +7 to day +21 after BMT. In a first approach, the complement system of 12 patients was studied in this critical period (9). Four of 12 patients developed severe CLS (more than a 3% increase in body weight within 24 h), whereas 8 of 12 patients did not. In patients presenting with CLS, the complement parameters showed activation of the classic pathway of complement: decrease of C3, C4 and hemolytic function of complement system (CH 50), increase of the activation products C4d (9) and of the terminal complement complex (TCC). Increase of the TCC is a result of the activation of the whole complement sequence. Thus, increase of the C5 activation product C5a is expected but not yet proved in these patients. The generation of C5a might be of value, because (1) C5a is able to induce tumor necrosis factor-alpha from monocytes (10) and (2) infusion of purified C5a induces a sepsi s-like syndrome in animal experiments (11). In addition to complement activation, C1 INH was found to have reduced activity in patients with CLS when compared with patients without CLS (9). An example of the clinical findings and complement parameters in one patient who developed CLS at day +5 after BMT is shown in Figure 1.

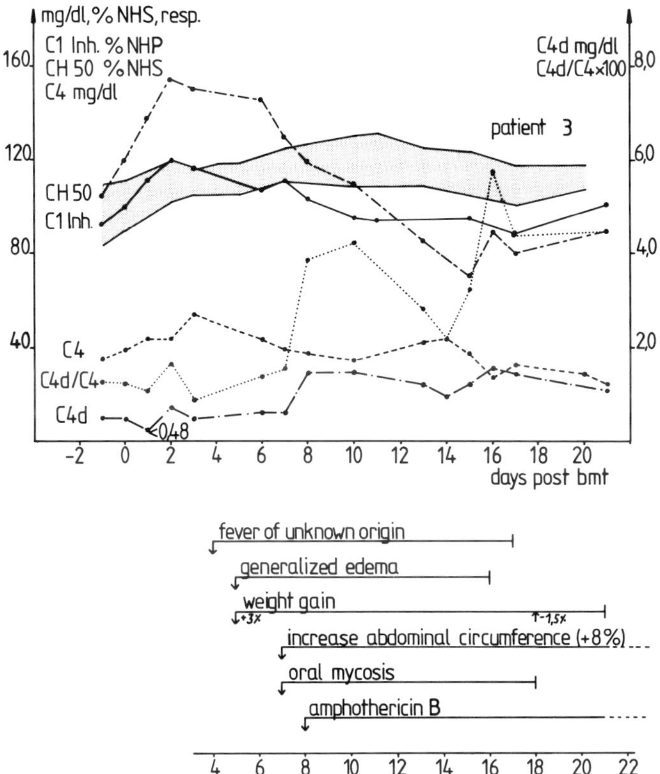

Figure 1 Demonstration of values for CH 50 (●— — — ●), C1 INH (●—●), C4 (●–•–•●), C4d (○– – – ○), and C4d/C4 quotient (●••••●) in a patient after BMT. The lower part shows relevant clinical findings: the patient developed fever from day +4 and CLS beginning with day +5 and ending at day +21. This period was paralleled by an oral candidiasis. The gray area shows the field of C1 INH values for patients after BMT without CLS.

C. C1 INH: Physiological Actions and Clinical Relevance

Human C1 INH is a member of the serine protease inhibitor (serpin) family (12) with a molecular weight of the functional active form of about 104,000 daltons (13). Its concentration in normal plasma ranges from 15 to 35 mg/dl (14). C1 INH is the main inhibitor of both the Hageman factor (FXII)–dependent coagulation-kinin pathway and the complement system (15–17). The inhibition occurs through stoichiometric complex formation between C1 INH and a given enzyme. Clinically, two kinds of C1 INH deficiencies have been described: first the congenital condition leading to hereditary angioedema (HAE-syndrome), and second, an acquired condition in clinical entities like the sepsis syndrome and disseminated intravascular coagulation (3,4,18).

D. Substitution of C1 INH Concentrate in the Sepsis Syndrome and CLS

An uncontrolled, preliminary study during the course of which C1 INH was substituted in five patients with the sepsis syndrome (without CLS), revealed the following results (19): Those patients who succeeded in increasing their functional C1 INH plasma levels during substitution with C1 INH concentrate revealed a steady decrease in anaphylatoxins C3a and C5a, whereas plasma concentrations of both anaphylatoxins almost exactly antiparalleled the fluctuations in C1 INH plasma levels. The administration of C1 INH concentrate was accompanied by an improvement in the clinical situation of the patients without any signs of toxicity. Functional measurements of FXII revealed an increase of FXII activity of 100% during C1 INH concentrate substitution. In addition, the cardiorespiratory performance was markedly improved in those patients (19).

In a newborn with sepsis syndrome–related CLS, C1 INH concentrate was effective (1) to stabilize cardiovascular function, as indicated by discontinuation of catecholamine infusion; and (2) to improve edema and normalize body weight (20).

These preliminary clinical data were the rationale to assess the effects of high doses of C1 INH concentrate on complement function and complement-activation products in patients with severe CLS after BMT.

IV. PATIENTS, OBJECTIVES, AND METHODS

A. Definition of "Severe CLS"

C1 INH concentrate was given to all patients undergoing bone marrow transplantation who developed *severe CLS* prior to day +21 after BMT. The definition of severe CLS was as follows: weight gain (>3% of body weight, but at least 500 g in 24 h) despite regular fluid administration *and* generalized edema. Five consecutive patients who fulfilled these criteria were included in this study.

B. Objectives and Parameters

The aim of the present study was to assess the influence of high doses of C1 INH concentrate on the hemolytic function of complement and on complement-activation parameters in patients with CLS after BMT. The following parameters were assessed:

C1 INH activity (functional assay) was assessed to prove the recovery of C1 INH concentrate

CH 50 (= classic pathway activity) was assessed to prove to which extent defined doses of C1 INH concentrate (given to the patients) might inhibit subsequent in vitro activation of the classic pathway.

Complement activation parameters (C4d, C5a, TCC) were assessed to prove the inhibition of early (C4d) and terminal (C5a and TCC) activation of the classic pathway in vivo.

C. Materials and Methods

CH 50 (9), C1 INH activity (9), and C4d (21) were determined as described. TCC (range of individual serum samples of 30 blood donors: 130–360 ng/l [median: 200 ng/l]) and C5a (range of individual serum samples of 30 blood donors: 0.1 to 0.8 ng/l [median: 0.34 ng/l]) were determined by ELISA according to the manufacturer's instructions (Behringwerke AG, Marburg, Germany). The TCC ELISA (sandwich technique) was based on a monoclonal antibody against a neoantigen of the TCC as a capture antibody (22). All data were corrected for serum protein concentration. Blood samples (serum and EDTA-blood) were taken immediately before and 1 h, 4 h, and 12 h after each administration of C1 INH concentrate (dosage scheme given in Table 1). Processing of the samples was described (9).

V. TREATMENT SCHEDULE FOR C1 INH CONCENTRATE

A. Preparation of C1 INH Concentrate

C1 INH concentrate (Berinert P; Behringwerke AG) is a purified, pasteurized, and lyophilized C1 INH concentrate for intravenous use. The concentrate is prepared from human plasma proven to be negative for anti–HIV-1 (human immunodeficiency virus 1) and also negative for hepatitis B surface antigen (HbsAg). Donors with more than two times the upper normal threshold ALT levels

Table 1 Dosage Scheme for C1 INH Concentrate

Initial Dose:	Day 0 60 U/kg bw, after 12 h:
Maintenance dose:	Day 1 60 U/kg bw (30 U/kg bw every 12 h)
	Day 2 30 U/kg bw (15 U/kg bw every 12 h)
	Day 3 30 U/kg bw (15 U/kg bw every 12 h)
Total dose:	80 U/kg bw

bw, body weight.

were included. C1 INH concentrate is purified from plasma by several chromatography and salt precipitation steps (23). The material is pasteurized by heat treatment in aqueous solution for 10 h at 60°C. The drug has been frequently used in several countries for treatment of hereditary angioedema, a disorder in which the inhibitor protein is either lacking or reduced in its activity. In more than 7 years of clinical experiment, no hepatitis B or hepatitis non-A/non-B could be causally related to the use of the drug.

B. Treatment Schedule

Reconstituted C1 INH concentrate was given by intravenous infusion (30 min) every 12 h. The administration of the study drug started within 2 h after the patient had been found to fulfill the criteria for "severe CLS" (see Sec. IV.A) and informed consent had been obtained.

VI. COMPLEMENT PARAMETERS UNDER INFUSION OF HIGH DOSES OF C1 INH CONCENTRATE

The data in this section focus on the initial dose of C1 INH concentrate (60 U/kg body weight) and on the two following doses (30 U/kg). These doses are two- to eight-fold higher than the dose of C1 INH used for hereditary angioedema. It was therefore of interest to assess the influence of high doses of C1 INH concentrate on the complement system (C1 INH serum levels, CH 50, complement activation parameters). The maximum alteration of complement parameters was seen at 4 h after infusion of C1 INH concentrate.

A. Influence of C1 INH Concentrate on C1 INH Activity

The 60 U/kg dose resulted in a significant increase of functional C1 INH plasma level of 16% compared with the starting values (126 ± 8% of normal human plasma [NHP] to 149 ± 6% of NHP; Table 2). The 30 U/kg dose (given at 12 h after the 60 U/kg dose) resulted in a mean increase of functional C1 INH plasma level of 5% compared with the starting values (144 ± 9% of NHP to 151 ± 5% of NHP; Table 2). This indicates a slight decrease of C1 INH activity (149 ± 6% of NHP to 144 ± 9% of NHS, not significant by paired *t*-test) in the interval in between the time points of C1 INH infusion. The C1 INH treatment (60 and 30 U/kg body weight, respectively) resulted in a high plasma C1 INH activity of about 140–150% of normal human plasma. The 15 U/kg doses maintained the C1 INH plasma activity in the range of 140–150% of NHP.

Table 2 Complement Parameters Before and After Administration of C1 INH Concentration (60 U/kg body weight)

	Prior to C1 INH infusion range; median	4 h after C1 INH infusion range; median	Paired *t*-test
C1 INH [% of NHS]	114–134; 126	138–158; 147	$P = 0.0004$
CH 50 [% of NHS]	111–146; 129	105–147; 124	NS
C4d [mg/dl]	4.0–8.0; 6.0	3.6–6.6; 4.7	$P = 0.03$
C5a [ng/ml]	0.77–1.70; 1.1	0.49–1.17; 0.87	$P = 0.034$
TCC [ng/ml]	291–673; 454	207–596; 384	NS

NS, not significant.
NHS, normal human serum.

B. Influence of C1 INH Concentrate on Hemolytic Activity

CH 50 was assessed in order to determine the degree of inhibition of in vitro–activated classic pathway of complement. The 60 U/kg dose resulted in a slight decrease of CH 50 level from 129 ± 14% of NHS to 124 ± 17% of NHS (see Table 2). No influence of the 30 U/kg dose on CH 50 test was demonstrable (see Table 3).

C. Influence of C1 INH Concentrate on Parameters of Complement Activation

C4d: The 60 U/kg dose resulted in a significant decrease of mean C4d plasma levels from 6 ± 2 to 4.7 ± 1.5 mg/l) $P = 0.03$; see Table 2). The 30

Table 3 Complement Parameters Before and After Administration of C1 INH Concentration (30 U/kg body weight)

	Prior to C1 INH infusion range; median	4 h after C1 INH infusion range; median	Paired *t*-test
C1 INH [% of NHS]	130–156; 144	143–158; 151	$P = .018$
CH 50 [% of NHS]	100–158; 130	91–157; 130	NS
C4d [mg/dl]	4.0–6.3; 5.2	3.6–5.8; 4.5	$P = .002$
C5a [ng/ml]	0.56–1.05; 0.88	0.51–1.02; 0.82	NS
TCC [ng/ml]	239–697; 394	223–592; 363	NS

NS, not significant.
NHS, normal human serum.

U/kg dose (given at 12 h after the 60 U/kg dose) resulted in a significant decrease of C4d levels from 5.2 ± 1.5 to 4.5 ± 1.2 mg/L(see Table 3).

C5a: The 60 U/kg dose resulted in a significant decrease of mean C5a plasma levels from 1.10 ± 0.4 to 0.87 ± 0.3 ng/ml (P = 0.034; see Table 2). The 30 U/kg dose (given at 12 h after the 60 U/kg dose) resulted in a slight decrease of C5a levels from 0.88 ± 0.2 to 0.82 ± 0.2 ng/ml (see Table 3).

TCC: The 60 U/kg dose resulted in a decrease of TCC plasma levels from 454 ± 149 to 384 ± 153 ng/ml (not significant by paired *t*-test; see Table 2); the 30 U/kg dose (given at 12 h after the 60 U/kg dose) resulted in a lower decrease of TCC concentrations from 394 ± 154 to 363 ± 132 ng/ml (not significant by paired *t*-test; see Table 3).

V. CONCLUSIONS

A. Clinical Course of CLS

Four of 12 patients developed CLS at days +8, +10, +11, and +13, respectively. In these patients, CLS reached its maximum at days +11 to +36 (range; median: day +23) with fluid retention in the range of 7–16% of body weight (median: 13%). These patients died of cardiovascular complications at days +12 to +78 (range; median: +36). The patients who received C1 INH concentrate in the present study showed improvement of body weight within 3–7 days after start of C1 INH infusion. C1 INH was well tolerated by all patients. Duration of CLS was shortened when compared with historical controls.

In conclusion, C1 INH appears to be a promising therapy for CLS. A randomized, double-blind trial is indicated to assess the effects of C1 INH on the overall outcome of patients with CLS after BMT.

B. High-Dose C1 INH Concentrate-Infusion and Complement Activation

The dose of C1 INH concentrate used in this study was sufficient to increase C1 INH plasma activity by about 20%. This is similar to the increase of C1 INH plasma levels that occurs spontaneously in patients without CLS after BMT (9). The activation parameter for the early classic pathway (C4d) was significantly reduced and/or normalized under infusion of C1 INH concentrate at doses of 60 and 30 U/kg body weight. C5a was only significantly reduced when the 60 U/kg dose was infused; TCC values tended to be lower. This discrepancy might be explained in part by alternative pathway activation in these patients, which might account for at least part of C5a and TCC generation.

There was no inhibition of the functional assay for the classic pathway, CH 50. The CH 50 assay is based on C1 activation via red blood cell–bound immunoglobulin G (IgG). This indicates that antibody-mediated activation of the classic pathway was not significantly inhibited by high doses of intravenous C1 INH concentrate. The fact that the early classic pathway is inhibited in vivo in patients with CLS could be explained by other mechanisms (than bound IgG antibody) of classic pathway activation (e.g., enzymatic activation of C1 by plasmin-induced complement activation).

ACKNOWLEDGMENTS

We appreciate the expert technical assistance of E. Oellers and K. Schirlau. We also are indebted to the medical and nursing staff for excellent primary care of the patients. This work was supported by the Elterninitiative Kinder krebsklinik e.V., Düsseldorf.

REFERENCES

1. Parker MM, Parillo JE. Septic shock. Hemodynamics and pathogenesis. JAMA 1983; 250:3324.
2. Aasen AO, Smith-Erichsen N, Gallimore MJ, Amundsen E. Studies on components of the plasma kallikrein-kinin system in plasma samples from normal individuals and patients with septic shock. Adv Shock Res 1980; 4:1.
3. Martinez-Brotòns F, Oncins JR, Mestres J, Amargòs V, Reynaldo C. Plasma kallikrein-kinin systems in patients with uncomplicated sepsis and septic shock—comparison with cardiogenic shock. Thromb Haemostas 1987; 58:709.
4. Kalter ES, Daha MR, ten Cate JW, Verhoef J, Bouma BN. Activation and inhibition of hageman factor–dependent-pathways and complement system in uncomplicated bacteremia and bacterial shock. J Infect Dis 1985; 151:1019.
5. Thijs LG, Hack CE, Strack von Schijndel RM, Nuijens NH, Wolbink GJ, Eerenberg-Belmer AJM, van der Vall H, Wagstaff J. Activation of the complement system during immunotherapy with recombinant interleukin-2: relation to the development of side effects. J Immunol 1990; 144:2419.
6. Hack CE, Wagstaff J, Strack van Schijndel RM, Eerenberg AJM, Pinedo HM, Thijs LG, Nuijens JH. Studies on the contact system of coagulation during therapy with high doses of recombinant IL-2: implications for septic shock. Thromb Haemostas 1991; 65(5):497.
7. Rosenberg SA, Lotze MT, Mule JJ. New approaches to immunotherapy of cancer using interleukin-2. Ann Intern Med 1988; 108:853.
8. Hack CE, Nuijens JH, Felt-Bersma RJF, Schreuder WO, Eerenberg-Belmer AJM, Paardekooper J, Bronsveld W, Thijs LG. Elevated plasma levels of the anaphylatoxins C3a and C5a are associated with a fatal outcome in sepsis. Am J Med 1989; 86:20.
9. Nürnberger W, Michelmann I, Petrik K, Holthausen S, Lauermann G, Eisele B, Delvos U, Burdach St, Göbel U. Activity of C1 esterase inhibitor in patients with vascular leak syndrome after bone marrow transplantation. Ann Hematol 1993; 67:17.
10. Okusawa S, Yancey KB, van der Meer JWM, et al. C5a stimulates secretion of tumor necrosis factor from human mononuclear cells in vitro. J Exp Med 1988; 168:443.

11. Lundberg C, Marceau F, Hugli TE. C5a-induced hemodynamic and hematological changes in the rabbit. Am J Pathol 1987; 128:471.
12. Bock SC, Skriver K, Nielsen E, Thorgesen H-C, Wilman B, Donaldson VH, Eddy RL, Marrinan J, Radziejewski E, Huber R, Shows TB, Magnusson S. Human C1 inhibitor: primary structure, cDNA cloning and chromosomal localization. Biochemistry 1986; 25:4292.
13. Sim RB, Reboul A. Preparation and properties of human C1 inhibitor. Methods Enzymol 1981; 80:43.
14. Dick W, Cullmann W. Gegenwärtige diagnostische Möglichkeiten beim hereditären Angioödem (HAE) und beim erworbenen Angioödem (AAE). Immun Infekt 1985; 13:113.
15. Kaplan AP, Silverberg M. The coagulation-kinin pathway of human plasma. Blood 1987; 70:1.
16. deAgostini A, Lijnen HR, Pixley RA, Colman RW, Schapira M. Inactivation of factor XII active fragment in normal plasma. J Clin Invest 1984; 73:1542.
17. Meijers JCM, Vlooswijk RAA, Bouma BN. Inhibition of human blood coagulation factor XIa by C1 inhibitor. Biochemistry 1988; 27:959.
18. Colten HR. Hereditary angioneurotic edema 1887–1987. N Engl J Med 1987; 317:43.
19. Hack CE, Voerman HJ, Eisele B, Keinecke HO, Nuijens JH, Eerenberg-Belmer AJM, Ogilvie A, Strack RM, van Schijndel, Delvos U, Thijs LG. C1-esterase inhibitor substitution in sepsis. Lancet 1992; 339:378.
20. Nürnberger W, Göbel U, Stannigel H, Eisele B, Janssen A, Delvos U. C1-inhibitor concentrate for sepsis related capillary leak syndrome. Lancet 1992; 339:990.
21. Nürnberger W, Stannigel H, Müntel V, Michelmann I, Wahn V, Göbel U. In vivo Aktivierung der vierten Komponente des Komplementsystems (C4) bei Früh- und Reifgeborenen mit generalisierten bakteriellen infektionen. Klin Pädiatr 1980; 202:141.
22. Mollnes TE, Lea T, Frøland SS, Harboe M. Quantification of the terminal complement complex in human plasma by an enzyme-linked immunosorbent assay based on monoclonal antibodies against a neoantigen of the complex. Scand J Immunol 1985; 22:197.
23. Haupf H, Heimburger N, Kranz T, Schwick HG. Ein Beitrag zur Isolierung und Charakterisierung des C1 inaktivators aus Humanplasma. Eur J Biochem 1970; 17:254.

46

Immunomodulating Effect of Quantum Chemotherapy in Patients with Chlamydial Infection

Daina Andersone
Latvian Medical Academy, Riga, Latvia

I. INTRODUCTION

Chlamydial infection is one of the most common sexually transmissible infection in Latvia. Examining 2360 patients with urogenital infections, *Chlamydia trachomatis* monoinfection was detected in 19.7% of cases, *C. trachomatis* with one infection, 37.3%; with two infections, 28.4%; and with three infections, 8.8% (1). Analysis of the acquired results shows that every second patient from the studied group suffers from infection with *C. trachomatis*. The most widespread extragenital manifestation of the infection is reactive arthritis. It closely connects these patients with specialists in rheumatology (2).

Like any infection, chlamydial infection is characterized by a dysfunction in the mechanism of immunoregulators. Taking as the basis the main immunologic axis relations between T helper (Th) and T suppressor (Ts) cells, the patients were divided into three groups, depending on the coefficient: 33% of patients with normal Th/Ts coefficient (1.6–2.0); 25% with increased Th/Ts coefficient (>2.0); and 42% with decreased Th/Ts coefficient (<1.6) (3). This, the most common type (42%) is with decreased Th/Ts coefficient, and in its framework, increased amounts of B lymphocytes are observed. With an increase of the activity of the process, the amounts of immunoglobulins (Ig) and circulating immunocomplex (CIC) increase.

Changes of the immune state made us look for new types of immunomodulating therapy, including quantum chemotherapy.

II. BIOLOGICAL EFFECT OF LASER RADIATION

It is known that wavelength, absorbable in the most important biological structures of the body, is 200–600 mm. It is this range of wavelength, where biological effectivity of radiation is the greatest, and the red light is the light which, due to its energetic character, is very close to energetic levels of living tissue. The laser beam has phototermic, photomechanical, photoablative, and photochemical effects on living tissue. It promotes changes in the immunological processes of the body, but data on its concrete effect on the immune system are controversial. The effect of laser therapy depends on many factors: first of all, on wavelength, regimen, and the kind of effect, radiation capacity, stage of the process, and patient's age. The most difficult task is to choose the adequate parameters of laser radiation. With a decrease of laser radiation wavelength, its anti-inflammatory effect increases.

III. GOAL OF THE WORK

In order to concretize the influence of intravenous laser radiation on the immune status in chlamydial infection, 100 patients were studied. Some immunological indices used in practical medicine were studied dynamically: quantitative indices of total number of lymphocytes and monocytes, quantitative indices of T and B lymphocytes and their subpopulations as well as the most Ig and CIC in blood.

IV. METHODS

The quantity of T and B lymphocytes and their subpopulation was determined with monoclonal antibodies using cytofluorograph SPECTRUM III (Ortho Diagnostics, Raritan, NJ). Ig in blood serum was determined by the method of radial immunodiffusion according to Mancini (4) using monospecific antiserum. CIC was determined by method of eapahobcknn using a solution of polyethyleneglycol. For quantum chemotherapy, we used laser equipment in the following regimen: radiation capacity 1.5 mV; generation regimen, persistent; length of session, 20 min; course of treatment, six procedures; wavelength, 620 nm. After vena cubitalis puncture, a metal furrule with laser radiation was inserted to a depth of 5–10 cm.

V. RESULTS

The results obtained may be divided into two groups—clinical and immunological. During the first days of radiation therapy, aggravation of the pathological process in patients was observed—arthralgia and malaise increased—but after the third procedure, signs of reactive arthritis caused by chlamydial infection disappeared and a prolonged remission was maintained.

In the results of quantum chemotherapy, we detected the following changes: the total number of T lymphocytes was increased 25% and monocytes 34%, but the total amount of B lymphocytes in blood had decreased 55%, with a simultaneous decrease in IgG of 30%, IgA of 27%, IgM of 28%, and CIC of 32%.

VI. DISCUSSION

When analyzing the results obtained, it is very important to evaluate them in relation to immunological changes of the chlamydial-infected patients. Our results suggest that in a chronic course, the patients have insufficient cellular immunity—decreased amounts of Th, Ts, and natural killer cells in the blood, as well as decreased myeloperoxydazse in leukocytes. When activity of the process increases, activation of humoral immunity is observed—the amount of CIC with a simultaneous increase in the number of B lymphocytes. The acquired immunological effect suggests that laser radiation has an immunomodulating effect, and correction of immune system, resulting from it, has the role of pathogenetic therapy in case of chlamydial infections.

The nonspecific immunomodulating effect of the neon laser most likely is due to intensification of cellular breathing, and a decrease of the lipid peroxidation on cellular membranes is the result of quantum chemotherapy (personal observations).

VII. CONCLUSIONS

Monitoring of quantum chemotherapy suggests that the immunomodulizing effect in patients with chlamydial infection is characterized by an increase in the number of phagocytic cells and a significant decrease in the indices of humoral immunity that leads to relatively prolonged remission of the disease.

REFERENCES

1. Miltinsh A, Andersone D, Ancshupane I, Ivdra P, Melka D, Zilevica A. Chlamydial infection in Latvia. Proceedings of the European Society for Chlamydial Research, Uppsala, Sweden, 1992:286.
2. Zeidler H. Chlamydial-induced arthritis, clinical features, diagnosis and therapy. In: Balint G, Gömör B, Hodinka L, eds. Rheumatology, state of the art. Budapest: Excepta Medica, 1991:49–62.
3. Andersone D. Diagnostical importance of changes of immune system in patients with reactive arthritis. Scand J Rheumatol 1992; 93(Suppl.):54.
4. Mancini G, Carbonara AO, Heremans JF. Immunochemical quantitation of antigens by single radial immunodiffusion. Immunochemistry 1965; 2:235.

Index

About the Editor

K. NOEL MASIHI is Head of the Laboratory of Immunomodulation at the Robert Koch-Institute, Berlin, Germany. A member of the International Society for Immunopharmacology, among other organizations, he is the editor of two previous books that reflect his research interests in immunomodulation and immunotherapy. Dr. Masihi received the Ph.D. degree (1972) in microbiology from the Czech Academy of Sciences, Prague, Czech Republic.